1
Genome Dynamics and Stability

Series Editor: Dirk-Henner Lankenau

Genome Integrity

Facets and Perspectives

Volume Editor: Dirk-Henner Lankenau

With 60 Figures

Springer

Series and Volume Editor:

Priv.-Doz. Dr. Dirk-Henner Lankenau
Hinterer Rindweg 21
68526 Ladenburg
Germany
e-mail: d.lankenau@t-online.de

Cover

The cover illustration depicts two key events of DNA repair: 1. The ribbon model shows the structure of the termini of two Rad50 coiled-coil domains, joined via two zinc hooks at a zentral zinc ion(sphere). The metal dependent joining of two Rad50 coiled-coils is a central step in the capture and repair of DNA double-strand breaks by the Rad50/Mre11/Nbs1 (MRN) damage sensor complex. 2. Immunolocalization of histone variant γ-H2Av in γ-irradiated nuclei of *Drosophila* germline cells. Fluorescent foci indicate one of the earliest known responses to DNA double-strand break formation and sites of DNA repair.
(provided by Karl-Peter Hopfner, Munich and Dirk-Henner Lankenau, Heidelberg)

Library of Congress Control Number: 2006935268

ISSN 1861-3373
ISBN-10 3-540-37528-7 Springer Berlin Heidelberg New York
ISBN-13 978-3-540-37528-9 Springer Berlin Heidelberg New York
DOI 10.1007/b104871

Springer is a part of Springer Science+Business Media

springer.com

Editor: Dr. Sabine Schreck
Desk Editor: Ursula Gramm, Heidelberg
Cover figures: Prof. Karl-Peter Hopfner and Dr. Dirk-Henner Lankenau
Cover design: WMXDesign GmbH, Heidelberg
Typesetting and Production: LE-TEX Jelonek, Schmidt & Vöckler GbR, Leipzig

Printed on acid-free paper 02/3100 YL – 5 4 3 2 1 0

Preface

Genome dynamics and stability are the *non plus ultra* requirements for cellular life. No matter whether life began with metabolism, with self-replicating genetic molecules, or as a cooperative chemical phenomenon, all cells and viruses maintain a genome capable of multiplication, variation and heredity. A population of living entities with these properties will evolve by natural selection, and while modern metabolism supplies the monomers from which genomes (i.e. replicators) are made, genomes alter the kinds of chemical reactions occurring in metabolism (Maynard Smith and Szathmary 1997). This book deals with DNA repair and replication. Together with two other planned volumes, one on transposable elements and genome dynamics and another on recombination and meiosis as a key issue of the metazoan germline development, this volume introduces the conceptual framework of the series.

An earlier review on the classic monograph *Mobile DNA* (Berg and Howe 1989) was entitled "On the Impossibility of Knowing More." It states: "This big book indeed tells us everything but says nothing. It provides no conceptual framework as to what the burgeoning bulk of molecular data means, not out of intent but because it is swept along by an attitude found increasingly in science of 'never mind the quality, feel the width' ... the book is essentially uninformative regarding the biological importance of transposable elements in ontogeny and phylogeny" (Dover 1990). The present book series tries to circumvent such criticism. Of course, there have been milder opinions of the monumental *Mobile DNA* book as well (Brookfield 1989; Fincham 1989). Actually, the 2002 publication of its successor *Mobile DNA II* (Craig et al. 2002) impressively demonstrates the swift progress in this significant research field, which now not only largely addresses questions of evolutionary relevance but pragmatically feeds additional knowledge applied in human gene therapy or helps to understand the somatic maturation of the immune system by V(D)J recombination. The latter actually demonstrates the closeness of transposable element transposition to DNA repair as the V(D)J recombination reaction is completed by the non-homologous end joining (NHEJ) DNA repair pathway in lymphocyte development where the DNA double-strand break (DSB) is generated through the transposase (i.e. endonuclease) activity of an ancient transposable element. This transposon inserted into an ancestral vertebrate genome some 450 million years ago (Yu et al. 1999). In line with this important interface between a vertebrate transposon and DSB repair, the second chapter of Part II of this book reports on a similar relationship of the *Drosophila* P elements triggering DSBs and facilitating the understanding of the mechanisms of replication-dependent DSB repair. Other molecularly fossilized but experimentally revitalized transposable elements which promise to be of biomedical relevance are planned for an upcoming book volume. As Carl Woese recently said, it seems to be about time that biology makes a choice between the comfortable path of continuing to follow molecular biology's lead or the more refreshing one seeking a new and inspiring vision of the living world (Woese 2004).

To accomplish this is my goal with the book series *Genome Dynamics and Stability*, where this first volume is dedicated to integrative aspects of replication and DNA repair providing an overview of some facets and perspectives of genome integrity. DNA integrity is relevant for all organisms, and therefore it opens avenues of curiosity ranging from viroids in applied plant research to grasping biodiversity. This vision however must include pragmatic aspects of biomedical relevance as well.

The book at hand is entitled *Genome Integrity: Facets and Perspectives*. It contains a rather broad spectrum of chapters representing key aspects of DNA repair with a slight bias towards DSB repair as justified by its importance. Actually, every chapter is self-sufficient and could serve as an independent entry point to the whole book. The sequence chosen starts with three chapters introducing replication as a fundamental aspect of life. Here, the first chapter gives a general introduction to replication worth to be read by undergraduate students as well as academics, while the second chapter attempts to present a concept towards an anatomy of the eukaryotic replication fork. The third chapter adds the aspect of human diseases to the two more fundamental aspects in Part I. Replication is then linked by two interface-chapters in Part II to the world of DSB repair. The second chapter of Part II first reviews the history of the discovery of the physical nature of the gene and gene mutations. Exploiting gene targeting as an experimental, technical pillar, it attempts to compose the different models of DSB repair into a unifying synthesis. This joins Part II with four key aspects of DSB repair representing Part III. These four key aspects review the structure and function of the Rad50/SMC protein complexes in chromosome biology, further focus on the simplest pathway for DSB repair, i.e. non-homologous endjoining (NHEJ), and focus on a central gatekeeper crucial to avoiding cancer development, i.e. p53, and the most complex role of chromatin in DSB repair. The chapter on DNA base damage recognition in Part IV introduces DNA repair pathways involving one-strand lesions and their pleiotropic interactions with cell physiological functions, such as cell cycle, apoptosis and examples of major human diseases. While DSBs can be triggered and their repair can be studied at precisely defined positions on nucleotide level within a given chromosome, DNA damage introduced through radiation and other genotoxic stress factors follows a slightly different research lead. This is the common theme of the four chapters in Part IV. Ion irradiation as a tool to reveal tracts of damage throughout the eukaryote nucleus reminds us of cloud or Wilson chamber experiments in atomic physics detecting elementary particles of ionizing radiation. Here, in the final chapter of Part V, the tract of damage in a cloud of chromatin is monitored using antibodies to proteins characteristic of specific DNA repair pathways, as discussed in the last chapter of Part III. The four final chapters are important for many reasons, ranging from a significance for irradiation treated cancer patients, or victims of the Chernobyl disaster to the exposure to cosmic radiation of astronauts on long-term space missions.

First of all I am very greatfull to all contributors for their dedication and trust to participate in the first volume of this new book series. The original idea for this book came from the 8th meeting of the DNA Repair Network in Ulm, Germany, and would not have been possible without the support of the Deutsche Gesellschaft für DNA-Reparaturforschung (DGDR). Here I would like to mention especially Jürgen Thomale, Alexander Bürkle, Lisa Wiesmüller, Bernd Kaina and Friederike Eckardt-Schupp, who supported the initial idea and acted in the background. Further I would like to thank the anonymous referees for doing a great job in peer reviewing and improving the manuscripts. I also thank the University of Heidelberg, which gave access to their electronic journal collection. Last but not least, I have to thank Sabine Schreck (Springer, Heidelberg) without whom I could never have engaged in this project. Ursula Gramm (Springer, Heidelberg) and Michael Reinfarth (LE-TeX GbR, Leipzig) did a fine job copy editing all manuscripts and the Springer team succeeded well in establishing the SpringerLink OnlineFirst version of this book series, which provides authors with more flexibility in the individual handling of their contributions.

References

Berg DE, Howe MM (eds) (1989) Mobile DNA. American Society for Microbiology, Washington

Brookfield JFY (1989) Sorting out the 'junk'. Trends Biochem Sci 14:428–429

Craig NL, Craigie R, Gellert M, Lambowitz AM (eds) (2002) Mobile DNA II. ASM Press, Washington

Dover G (1990) On the Impossibility of Knowing More. Cell 60:703–704

Fincham JRS (1989) DNA in flux. Nature 342:26–27

Maynard Smith J, Szathmary E (1997) The major transitions in evolution. Oxford University Press, Oxford, New York, Tokyo

Woese CR (2004) A new biology for a new century. Microbiol Mol Biol Rev 68:173–186

Yu W et al. (1999) Coordinate Regulation of RAG1 and RAG2 by Cell Type-Specific DNA Elements 5' of RAG2. Science 285:1080–1084

Contents

Part A
Replication

Genome Dyn Stab (1)
D.-H. Lankenau: Genome Integrity
DOI 10.1007/7050_012/Published online: 20 January 2006

Chromosomal DNA Replication: On Replicases, Replisomes, and Bidirectional Replication Factories

Richard Egel

Department of Genetics, Institute of Molecular Biology and Physiology,
University of Copenhagen, Øster Farimagsgade 2A, 1353 Copenhagen, Denmark
regel@my.molbio.ku.dk

Abstract This chapter deals with bulk DNA synthesis in the three "Domains" of cellular life – Bacteria, Archaea and Eukarya. Presenting less specific detail, in comparison with other chapters in this volume, it summarizes current views on the vital protein complexes that organize genomic replication: together with numerous accessory proteins, individual replicases build up each strand of newly formed DNA. At least two of these cooperate in a replisome to form both branches of a replication fork, and higher-order replication factories appear to gather the divergently operating forks that have been started at the same origin site. Moreover, the crucial control that limits replication to exactly once per cell cycle is manifested at the level of intricate origin recognition complexes, which in turn initiate the assembly of functional replisomes at appropriate intervals.

1 Introduction

While many aspects of DNA replication appear equivalent throughout the tree of life, there are profound domain-specific differences, as illustrated by representative examples. Moreover, certain virus systems bear striking resemblance to cellular replication mechanisms. The noticeable similarity between many components of eukaryal and archaeal replication systems, in particular, clearly indicates their evolutionary relatedness. The striking dissimilarity of many bacterial counterparts, on the other hand, is not yet completely understood in evolutionary terms. Accordingly, a comparison of bacterial and eukaryotic model systems will fill most of the specific sections in this chapter. The main emphasis here is put on a comparative overview of the replication process, where DNA synthesis proceeds along a flawless template. The accompanying chapter of Nasheuer presents in greater depth the buildup of a functional replication fork in S-phase of the eukaryotic cell cycle, as well as the regulation of origin firing during the cell cycle. The chapter of Rudolph et al., on the other hand, discusses what happens when a replication fork is obstructed by DNA damage or other obstacles, so as to rescue the stalled fork by partly overlapping repair mechanisms.

Chromosomal replication generally starts at internal replication origins, and the stretch of DNA responding to the same origin is often referred to as a replicon. At the origin, before replication can start, the double-stranded DNA must be unravelled locally by special unwinding enzymes. Characteristically, these replicative DNA helicases are organized as multisubunit complexes, forming hexameric rings around the DNA. These helicases cannot usually bind on their own at the potential origins of replication, but have to be properly installed by other proteins, such as origin recognition complexes and helicase loaders. Two oppositely oriented hexamers are loaded at each origin. Upon helicase activation, two single-strand loops of opposite polarity are extruded at the origin, and DNA primase is recruited to each of these loops to initiate leading-strand synthesis – pointing away from the origin in either direction. After having assembled a primer of several ribonucleotides, the primase is replaced by replicative DNA polymerase(s) for further extension of the growing chains.

The replicative DNA polymerases are highly processive enzymes, not leaving their substrate for considerable lengths of time: they are held in place by so-called sliding clamps, which consist of protein rings with internal sixfold symmetry – forming a closed ring around the nascent product strand of DNA together with the parental template strand. Accordingly, such enzyme complexes are often referred to as ring-type polymerases. The sliding clamps serve as mobile tethers, effectively tieing polymerases and other proteins to the substrate. To thread the DNA chains into the central cavity – or, in turn, release them when the job is done – the tethering rings have to be opened by force, as mediated by specific clamp loaders. This is another class of composite gadget, where typically pentameric subunits cooperate to convert chemical energy into physical work – initially locking the sliding clamps on target, and removing them later on, when adjacent chains of nascent DNA are being joined.

As the nascent leading strands start growing in either direction, corresponding single-stranded loops are generated by the displaced templates for the still missing lagging strands. These loops are rapidly and cooperatively covered by single-strand binding proteins (SSB), which actively participate in organizing the assembly line of discontinuous lagging-strand synthesis. In due course the first Okazaki fragment is initiated by a new primase working in the reverse direction, and another module of replicative DNA polymerase is added to complete this fragment. Now the so-called replisome is fully assembled. While the lagging polymerase is producing the first Okazaki fragment, the leading polymerase proceeds further on – leaving another single-stranded loop in its wake, where the second Okazaki fragment will be made shortly after. This cycle continues until the replication fork meets head-on with another fork or is stopping at a terminator sequence.

Replisomes are highly effective protein machines. Not only do they use the same polymerase molecules to synthesize the entire leading strands from start to finish, they also allow repetitive recycling of the second polymerase subcomplex – recurrently switching from an Okazaki fragment just being finished to the next one to follow suit. Thereby, at each replication fork, synthesis of leading and lagging strands is coordinated at a common site.

As measured along the length of newly synthesized DNA, the divergent forks of each replicon move away ever farther from the corresponding origins. Physically speaking, however, both replisomes stay in contact close together at a supportive structure, from where the duplicated strands of DNA are spooled as a pair of growing loops. The functional association of several replisomes is often referred to as a replication factory. In particular, the multiple replication origins found in eukaryal chromosomes tend to associate in clusters, and numerous adjacent replicons can be served by a common replication factory. In bacteria, too, such replication factories can accommodate multiple replisomes, engaged in the replication of the main chromosome and several plasmids simultaneously.

2 Replisomes Working

As to the effective recycling of lagging-strand polymerase upon the completion of each Okazaki fragment, most details have been worked out for bacterial replisomes. As the DNA polymerases for both strands work together in close proximity, the lagging strand has to form a temporary loop, containing the growing Okazaki fragment closest to the fork. This curled configuration has been termed the trombone model (Fig. 1). Two variations on this general scheme are presented in this section, both developed for bacteria. A comparison to eukaryal components will be given later on (Sect. 5). The various replication proteins in bacteria, eukarya and archaea are listed in Table 1.

2.1 Organization of a Single Replisome, the Basic Model

A lot about how different components are tied together is known for the *E. coli* replisome (Leu et al. 2003; Schaeffer et al. 2005). The important steps in the basic trombone model are illustrated in Fig. 1. The hexameric DNA helicase DnaB moves along the lagging-strand template, which is threaded through the central cavity, while the leading-strand template is displaced sideways – be it in front of the helicase, as it is generally assumed for DnaB of *E. coli*, or through one of the lateral voids between adjacent subunits, as suggested for other organisms. The functional replisome carries two ring-type DNA polymerases, comprised of the core enzyme Pol-III and the β sliding clamp. Both these complexes are held together and tethered to the helicase by mediation of the strategically placed clamp loader (Fig. 1A) – a pentameric

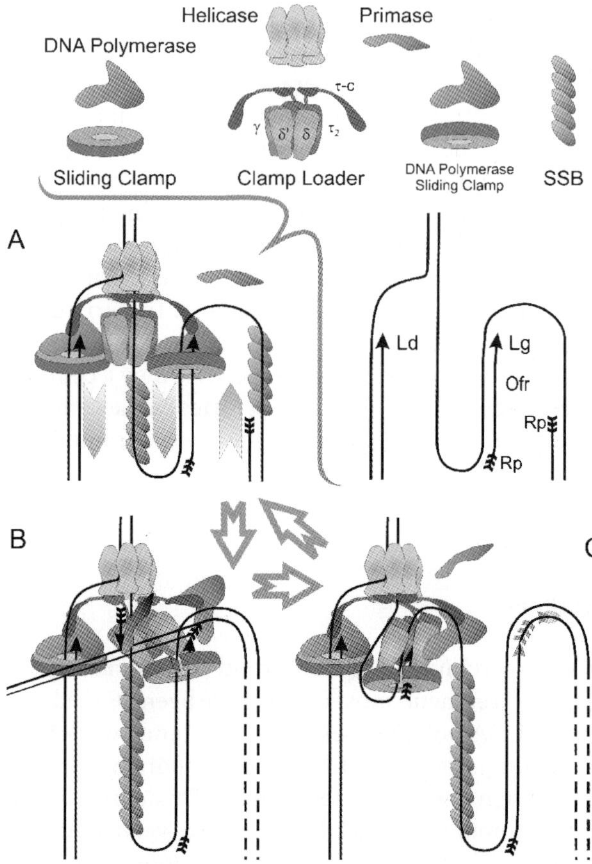

Fig. 1 The trombone model of a bacterial replisome. **A** This replisome is staged in the middle of an Okazaki fragment switching cycle. The *exploded view* represents the corresponding protein components (*top*) and participating strands of DNA (*right*). *Solid arrow heads* indicate the 3′-growing ends of leading (Ld) and lagging (Lg) strands – in the latter case by Okazaki fragments (Ofr), with RNA primers (Rp) at the 5′-ends. *Shaded arrows* indicate the movement of template DNA with respect to the replisome. **B** After having reached the preceding fragment, the lagging-strand polymerase has disengaged from both template and the sliding clamp, which just has been opened by the clamp loader. The primase has started the next primer at the helicase. **C** The open clamp loader senses the 3′-end of the primer–template duplex and loads the clamp behind the primer. Thereafter, the lagging-strand polymerase will reengage at the clamp (stage A). The joining of fragments (*right*) occurs outside the replisome. The repeated cycling through stages **A**, **B** and **C** is indicated by *unfilled arrows*

modul composed of four related but slightly different proteins ($\delta\delta'\gamma\tau_2$). Two additional subunits ($\chi\psi$) assist in primase loading.

Most importantly, the two τ-subunits carry particular C-terminal domains, which specifically bind to the replicative DNA helicase, as well as to

Table 1 Replisome components and associated proteins

Function	*E. coli*	*S. cerevisiae*	Archaea
Origin recognition	DnaA {> 10}[a]	ORC {6}[a]	ORC/Cdc6 {?}[a]
Helicase loader	DnaC[b] {iso 6}	ORC–Cdc6 {6+1}	
Replicative helicase	DnaB[b] {iso 6}	MCM {6}	MCM {iso 6}
Helicase support	?	Cdc45 {complex}	?
Single-strand DNA binding	SSB {1}	RPA {3}	RPA/SSB {1, 3}
Primer synthesis	DnaG {1}	Polα complex {4}	Primase {2}
Core polymerase, leading strand	Pol-III $\alpha\varepsilon\theta$ {3}	Polε {5}	PolD {2}; not all[c]
Core polymerase, lagging strand		Polδ {3}	PolB {1}
Clamp loader	$\delta\gamma\tau$ complex {5}	RFC {5}	RFC-L, -S {2+3}
Sliding clamp	β {iso 2}	PCNA {iso 3}	PCNA {3; or iso 3}[d]
Primer removal	RNase H	RNase H	RNase H
	Pol-I	Rad27 (~ Fen1)[e], Dna2	Fen1
Fragment joining	DNA ligase		
Torsional unswiveling	Type I and type II topoisomerases		

[a] Number of subunits {as indicated}
[b] Unrelated equivalents are used in gram-positive bacteria
[c] *Euryarchaeota* contain PolD as a second core polymerase; *Crenarchaeota* only have PolB
[d] PCNA is isotrimeric in *Euryarchaeota*, but heterotrimeric in *Crenarchaeota*
[e] In *S. cerevisiae*, the *FEN1* symbol happened to be assigned to a different gene beforehand

core DNA polymerases. Thereby, the advancing DNA helicase drags with it not only the directly attached clamp loader, but also two operational DNA polymerase complexes: one serving the leading strand all along, in pace with helicase movement, the other making Okazaki fragments pointed away from the helicase. The catalytic polymerase subunits, in turn, are riding piggyback on their sliding clamps – locked around the DNA – so that they stay on their template for the entire time it takes to produce a contiguous copy. Topologically, however, the helicase is immobilized in the cell, and the unwound strands of DNA are being pumped backwards. Likewise, the DNA polymerases are physically kept in place, so that the double-stranded products are moving backwards as well. Together, the relative movement of DNA is powered by energy consumption at the growing points by both the helicases and polymerases.

While movement of the leading-strand template is uniform through both helicase and DNA polymerase, the lagging-strand polymerase drives its template against the movement of DNA at the helicase. Temporarily, therefore, a loop is being spooled out at both these fixed points simultaneously. One half is double-stranded at the growing Okazaki fragment, whilst the other leg of similar length remains single-stranded for a while – spanning the distance between the helicase and the 5′-end of the Okazaki fragment. Correspondingly, the growing 3′-end fills up the single-stranded template towards the 5′-end of the preceding Okazaki fragment, the 3′-end of which in the meanwhile has been joined with the bulk of the lagging strand. Both areas of single-stranded template are temporarily covered by tiled arrays of single-strand binding protein (SSB), keeping the conformation of these stretches disentangled and appropriate for rapid utilization by the polymerases.

As long as the Okazaki fragment is being synthesized, the clamp loader modul is inactive, leaving the sliding clamp in place; it merely maintains structural connectivity to the replisome. This changes when the advancing 3′-end of the Okazaki fragment reaches the 5′-end of the previously assembled lagging strand (Fig. 1B). The core polymerase then disconnects from the sliding clamp, yet stays connected to the clamp loader. This activates the clamp loader to open the sliding clamp, removing it from that part of the DNA. It is now ready to reposition the clamp at a new primer in the near vicinity of the helicase. The primase protein itself, which is not permanently attached to the replisome, becomes the limiting factor at this point; it has to interact with both the helicase and SSB to make a new primer. Conceivably, the helicase becomes more readily accessible for the primase when the lagging-strand polymerase disengages from its template; this would ensure that the RNA primer for the next Okazaki fragment be made at the appropriate moment. The clamp loader is strategically placed to convey this signal from polymerase to helicase. An accessory modul of Pol-III subunits ($\chi\psi$) is held responsible for letting the primase in and throwing it off later on.

The clamp loader is detecting the presence of a new primer on the single-stranded template, when this is threaded into its central cavity (Fig. 1C). This causes the clamp to be locked behind the primer; the clamp loader disengages from the clamp, and the idling core polymerase can attach anew to the repositioned clamp. Significantly, clamp loader and core polymerase compete for the same binding site on the sliding clamp. When the polymerase resumes synthesis by extending the primer, this leads back to stage A of the cyclic process. The completed Okazaki fragment, of course, has to be connected to the bulk of the lagging strand. This happens outside the replisome, at least in bacteria. Together with RNaseH, the repair polymerase Pol-I replaces the RNA primer and a number of adjacent deoxyribonucleotides by newly synthesized DNA, before the remaining nick is closed by DNA ligase. Pol-I possesses $5' \rightarrow 3'$ exonuclease activity, which is absent in the replicative core polymerase Pol-III.

The helical twist of double-stranded DNA causes torsional stress piling up in all three branches of an advancing replication fork. In front of the helicase, at least, this tension has to be relieved by recurrent topoisomerase action, removing one helical turn per ten base pairs being unwound by the helicase. Behind the fork, however, relaxation can also occur at the replisome directly. As the processive DNA polymerases at both branches are safely tethered to the DNA by closed rings at the sliding clamps, they can periodically slip the 3′-end of the nascent strand – letting it swivel freely to relieve the torsional strain. This is yet another mechanical advantage of the ring-locked clamp system.

2.2
A Modified Model of a Bacterial Replication Factory

The schematic diagram of Fig. 1 emphasizes overall connectivity within the replisome – not necessarily representing a reliable view of its three-dimensional organization. In particular, the convoluted threading of the newly primed DNA into the clamp loader in reverse appears overly complicated. As demonstrated for *Bacillus subtilis*, the clamp loader can be found attached to the side of the cylindrical helicase hexamer, rather than covering one of its ends (Haroniti et al. 2004). This lateral configuration facilitates a more favorable arrangement of all the DNA branches being processed at a replication factory (Fig. 2). As the two helicases that have been installed at the same replication origin are thought to remain physically attached to the scaffold of a common replication factory, it is reasonable to assume that each helicase carries its clamp loader at the side facing away from the scaffold –

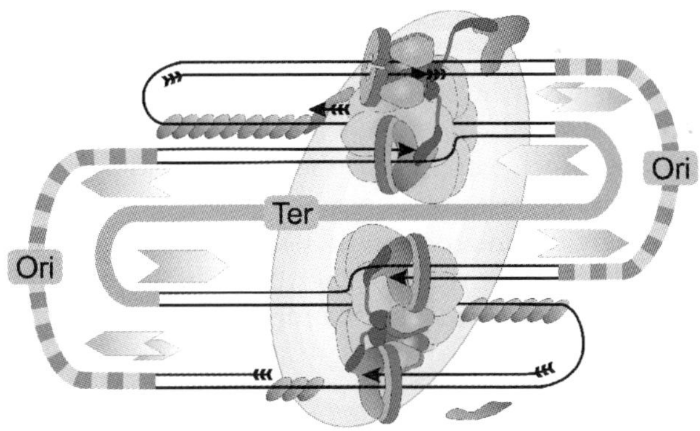

Fig. 2 The bacterial replication factory, supporting two replisomes. In this modified trombone model, the clamp loader, together with polymerases, is laterally attached to the helicase. The upper replisome is about to switch to the next primer, the lower one is running in the synthesizing mode. The movement of DNA strands at the replisomes, relative to origin (Ori) and terminator (Ter) regions, is indicated by *shaded arrows*. This movement is uniform at the leading strand, but shuttles back and forth for each Okazaki fragment of the lagging strand (*overlapping double arrows*) – the net movement is pointing away from the replisome

together with both the polymerases attached to this replisome. The clamp loader of *E. coli* and other bacteria may well attain a similar position at the helicase, but this remains to be ascertained by structural information.

This modified model suggests that leading and lagging strands leave the replisome in more or less opposite directions. This fits well with the general segregation mechanism of bacterial chromosomes, where the duplicated origins of replication are connected to other cellular structures, gradually moving away from each other (Graumann and Defeu-Soufo 2004). Significantly, upon completion of each Okazaki fragment, this bulk movement tends to thread the newly primed duplex into the open complex of clamp loader and sliding clamp – smoothly shifting in the right direction. Conversely, as the parental template strands are steadily being pulled towards both helicases in opposite directions behind the circular replicon, this movement ceases locally when a specialized terminator signal is driven into either replisome – due to inhibitory helicase–contrahelicase interaction (Mulugu et al. 2001).

Certain aspects of the models presented in Figs. 1 and 2 are more than others consolidated by experimental evidence. Above all, it has been inferred from crystallography how the clamp loader contacts the clamp and how it interacts with primer–template DNA (Bowman et al. 2005). This has first been achieved for the corresponding yeast proteins, but it appears applicable to bacteria as well. The activated clamp loader specifically recognizes the recessed 3′-end in partially duplex DNA, and several of the most conserved amino acid positions in the five subunits snugly follow the minor groove of the double helix. Also, it is generally believed that both clamp loader and core DNA polymerase contact the sliding clamp from the same side of the ring. For clarity, the flexible joints between clamp loader and polymerase are highly exaggerated in the drawing; their spatial configurations remain to be characterized.

Other aspects are less stringently supported by experimental evidence, but have been incorporated for consistency by circumstantial reasoning. One of these concerns the direct recycling of the sliding clamp from a completed Okazaki fragment to the next in line. Another one concerns the coordination between primase activity and processivity, presumed to be modulated by direct feedback through the helicase. As of yet, these hypothetical assumptions are not inconsistent with prevailing evidence, but further consolidation is desirable.

3
A Toolbox of Composite Appliances

3.1
DNA-Embracing ATPase Machines

Most of the accessory proteins assisting the DNA polymerases are organized as complexes of identical or similar subunits, where dynamic function is related to repetitive structure. In some of these, structural affinity alone is sufficient for self-assembly in characteristic patterns, such as filaments of SSB proteins on single-stranded DNA, or the regular, toroidal arrangement of subunits in sliding clamps. Many others belong to the versatile superfamily of AAA^+ proteins – the extended group of "ATPases associated with a variety of cellular activities" (Davey et al. 2002). Originally, classical AAA proteins were defined more narrowly, mainly comprising ring-shaped proteases related to proteasomes. Origin recognition complexes and helicase loaders are prominent examples of the AAA^+ category, where adjacent subunits cooperate to convert chemical energy into conformational change or physical work. The archaeal/eukaryal replicative helicases belong to this class as well, whereas bacterial counterparts belong to another branch of a yet larger superfamily (see below). Also, the clamp loaders described above consist of AAA^+ subunits, which are needed to position the ring-shaped sliding clamp around the newly formed DNA double helix.

These are the pumps and treadmills of the replication factories, where binding of ADP or ATP, and/or ATP hydrolysis are correlated with steric changes of protein folding, relevant for intra- or inter-subunit arrangement. Typically, AAA^+ proteins aggregate as hexameric complexes; pentamers and heptamers occur as well. Often the subunits surround an open channel in the middle of a highly symmetric ring, especially if all of them consist of identical protein chains. In other cases, slightly different protomers can form open, curved arrangements with some helical twist. Characteristically, adjacent subunits interact at a shared ATP-binding site; terminal subunits, however – in the more open arrangement of a clamp loader – may have lost the ability of binding nucleotides directly. Besides, in a large group of related ATPases – the RecA branch of the RecA/AAA^+ superfamily – identical subunits are able to assemble as long, open-ended spiral filaments, which can impose their own geometry on simplex, duplex and even triplex helices of DNA. These interactive filaments are instrumental in the initiation of homologous recombination. Notably, a RecA-related protein constitutes the hexameric, ring-shaped replicative DNA helicase in gram-negative bacteria, such as *E. coli.*

The functional association of sliding clamps and clamp loaders is an instructive example of how AAA^+ proteins work in general. The primary task of sliding clamps, of course, is to provide a lasting structural support to ring-type DNA polymerases for the entire period it takes to synthesize a growing chain. It is only before and after this productive phase that the AAA^+ clamp loaders come into play, installing the clamps on track at the beginning and removing them at last to reset the stage.

The full elaboration of structure–function relationships in hexameric rings of selected AAA^+ proteins is still under active investigation (Wang 2004; Hanson and Whiteheart 2005). These proteins share a characteristic nucleotide binding domain (NBD), together with a compact C-terminal domain (CTD), connected by a flexible hinge. The wedge-shaped NBDs tend to self-aggregate as hexameric rings around a central channel. A single ATP molecule usually makes contacts with the NBD of one subunit and a conserved argine residue in the CTD of the adjacent subunit in the ring. This so-called arginine finger is reminiscent of GTPase-activating partners of Ras-like G-proteins. It specifically contacts the γ-phosphate of bound ATP (or GTP in G-proteins), which is considered the active state of a subunit. The binding of ATP from the empty state is associated with relatively subtle movements within the NBD, but the adjacent CTD responds by considerable rotation at its hinge (20° and 30° angles have been observed). The splitting of ATP to ADP only causes smaller motions.

Different AAA^+ proteins can have very different substrate specificities, but their respective substrate binding surfaces are commonly exposed inside the central channel. The main substrates are repetitive biopolymers, such as unfolded peptide chains, single-stranded nucleic acids or double-helix DNA. The major components of intra- and inter-subunit motion are torsional twist around the ring axis and lateral slippage along this axis. While the subtle rearrangements caused by ATP hydrolysis appear to be appropriate to affect substrate binding, the angular movement of the lid-like CTD – upon ATP binding – likely represents the power stroke that translocates the substrate chain through the inner channel. The interlocked arrangement of subunits, together with successive ATPase activation from one subunit to the next, suggests that a cyclic wave of ATP binding and hydrolysis can freely run around the hexameric ring. Thereby, the substrate chain is being passed over from one subunit to the next and effectively pulled through the ring of AAA^+ proteins.

3.2 Self-Locking Rings of Sliding Clamps

Ring DNA polymerase sliding clamps are present in all three domains of living organisms, as well as in certain viruses. They are called PCNA (proliferating cell nuclear antigen) in eukaryotes and archaea, β-subunit of DNA polymerase III holoenzyme in bacteria, or gene 45 processivity factor in T4-like phages. Homology by protein similarity can still be recognized between archaeal and eukaryal PCNA sequences; bacterial and viral sequences, however, are highly divergent. Overall topology of sixfold symmetry is strikingly similar throughout the tree of life (Fig. 3), despite the divergence of protein sequence and number of subunits. The closed rings are formed by three identical subunits in archaea and eukaryotes (as well as in T4-like phages), but only two subunits in bacteria (Bruck and O'Donnell 2001).

The central tunnel is lined by 12 semi-parallel α-helices; the entire outer surface is formed by β-barrel sheets, undulating over six elevated crests and six lowered saddles in between. Accordingly, all the trimeric clamps are assembled from double-crested subunits, whilst the dimeric clamps in bacteria are assembled from larger and triple-crested subunits. In general, purified

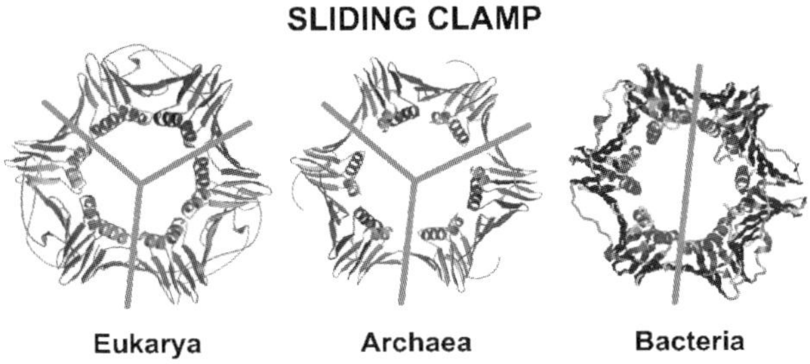

Fig. 3 Sliding clamps in the three domains of living organisms – trimeric in Archaea/Eukarya, dimeric in Bacteria. Three-dimensional structures were solved for human PCNA (Gulbis et al. 1996), archaeal PCNA in *Pyrococcus furiosus* (Matsumiya et al. 2001), and bacterial β clamp in *E. coli* (Kong et al. 1992)

subunits form very stable rings in vitro by self-assembly. On the other hand, purified δ-subunits of the bacterial clamp loader are able to open the ring spontaneously by hydrophobic interaction at the crested surface close to the interface between two subunits. So why doesn't the clamp loader keep the clamp open all the time in vivo?

In the inactive resting state, the assembled clamp loader does not allow any of its subunits to bind to the ring effectively. Rather, it requires binding of ATP to the inner subunits to change the conformation of the entire complex to this end (Johnson and O'Donnell 2003); only then can accessible rings be grabbed and opened – i.e., any ring that is not engaged by a working core polymerase. Thereafter, the inner surface of the clamp loader has to sense the 3′-end of a primer–template duplex to let the opened clamp snap back as a closed ring; and it takes ATP hydrolysis to release the clamp loader fully after that (Bowman et al. 2005). Thereby, the processes of unloading or reloading of sliding clamps are being directed according to demand – pulsating at the very heart of the replisome.

The hydrophilic inner lining of the clamp surrounds duplex DNA at a comfortable distance, not showing particular interactions. The more hydrophobic outer surface provides a docking platform for tight protein–protein contacts, not only with core polymerase or clamp loaders, but also with a host of ancillary proteins (Warbrick 2000). Archaeal or eukaryal clamps bind ligase and other processing enzymes. Moreover, certain eukaryal PCNA-related clamps have been diversified to serve DNA damage repair functions exclusively.

4 Origin Recognition and Firing

Eukaryotic cells in general, but also bacteria growing at sub-optimal rate, do not engage in DNA synthesis continuously. To cope with the prevailing rate of cellular growth and division, DNA replication is controlled by limiting initiation at special replication origins. Origin-binding proteins of the AAA^+ family play key roles throughout the living world, albeit in slightly varying tunes in different organisms. Their main role is to recognize the origin as such and to recruit the replicative helicases. As proposed for eukaryotes and archaea, certain hexameric helicases are assumed to be installed as double rings of opposite polarity (Sect. 4.2), which may work against each other – thereby putting torsional strain on the encircled DNA to be unwound (Mendez and Stillman 2003). This model has also been adapted to bacteria (Kaplan and O'Donnell 2004).

4.1 Origin Usage in Bacteria

Bacteria, such as *E. coli*, typically bear their entire genome on a circular chromosome, carrying a single origin of replication (*oriC*). By default, during the cell division cycle, this origin is functional during the θ-mode of bidirectional replication; the unidirectional σ-mode mobilizes other origins at special occasions, such as conjugational transfer to a different recipient cell. Notably, the bacterial cell cycle does not require that replication of a chromosome has to be completed first – before the next round can be started at the duplicated origins. Accordingly, at high rates of growth, rereplication is observed repeatedly at regular intervals; i.e., the lag time between reinitiation events at the same origin gets shorter than the processing period of replication forks to reach the terminal site at the other side of the circular replicon.

On average, of course, the rate of DNA replication should equal the rate of growth for other vital cell constituents. Therefore, if a chromosome is replicated faster than it takes to double cell mass as a whole, rereplication should be delayed accordingly. Likewise, if cells grow faster than the chromosomes can double from single replication origins, then rereplication should be advanced; alternatively, multiple origins could be activated simultaneously. *E. coli*, for that matter, has chosen to keep a single origin throughout – allowing rereplication before termination. The regulation of initiation depends on various proteins binding at *oriC*.

The critical sequence elements at *oriC* are clustered around a partly palindromic series of binding sites for the AAA^+ DnaA protein (A-box consensus TGTGAATAA), together with an AT-rich, low-melting region not far off (Fig. 4A). While some of the best matching A-boxes can bind DnaA with ATP or ADP, these are insufficient for origin activation. There are several less matching A-boxes scattered in between, which are highly specific for ATP-bound DnaA. When ~ 10 or more adjacent DnaA proteins, with ATP, have occupied these sites, the origin fires to initiate the next round of replication (Leonard and Grimwade 2005). Hence, DnaA acts as an origin recognition protein.

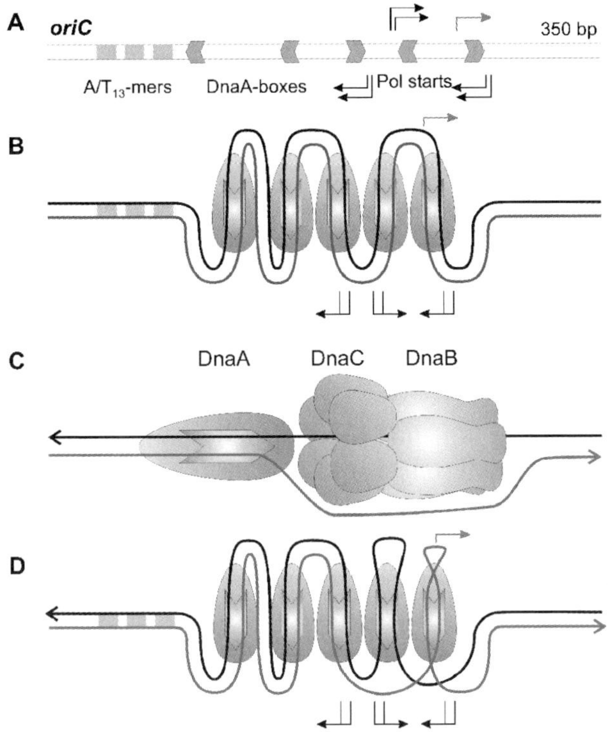

Fig. 4 Canonical binding sites for DnaA at *oriC* of *E. coli*. **A** The minimal fragment of *oriC* activity contains five DnaA boxes, as oriented, close to three repeats of 13-mers, rich in A/T. Polymerase start sites for leading-strand synthesis are indicated (*flagged arrows*; see text for details). **B** The orientation of DnaA boxes suggests a particular pattern of DNA bending through the DnaA cluster, or around it at its surface, although the relative positioning of interacting DnaA subunits remains to be determined. **C** The orientation of polymerase start sites relative to the adjacent DnaA box indicates a predominant order of protein–protein interactions towards helicase loader (DnaC) and replicative helicase (DnaB). Notably, three cases of closely spaced doublet sites are located at the same side of a neighboring DnaA box. Only the single unique site is oriented differently. **D** Helicase loading may involve preferential unwinding and hairpin looping around the rightmost DnaA boxes, but this is probably not yet the final answer to *orC* activation

It has long been assumed that binding of DnaA itself is sufficient to melt out the initial single-stranded loops to get the helicases under way, but the mechanism is somewhat enigmatic. It appears that the modular spacing and palindromic arrangement of DnaA sites is critical, perhaps by allowing extrusion of 1–2 pairs of hairpin loops, to which four of the five DnaA proteins can bind, so that a pair of helicase rings can be loaded at the base of these hairpins (Kaplan and O'Donnell 2004). As the major DNA polymerase start sites on either strand map close to three particular DnaA boxes of the palindromic cluster – the three rightmost ones of Fig. 4B (Messer and Weigel 1996), helicase loading or specific activation must be directed to those sites. How the flanking AT-rich region at the other side of cluster can affect helicase loading remains to be clarified. There are reasons to believe that most of the DNA at *oriC* is wound up at the surface of the DnaA complex, which may place the AT-rich opening site close to the sites of strand initiation (Leonard and Grimwade 2005).

At its time (Seufert and Messer 1987), the mapping of polymerase start sites at *oriC* was performed by cutting-edge procedures. Minichromosomes containing *oriC* were replicated in vitro, in the absence of ligase activity, and assayed for the presence of single-strand discontinuities, which could be assigned to leading-strand initiation. This mapping revealed four particular starting sites, two on each strand; the three of which consisted of closely spaced doublets (Fig. 4B). Somewhat surprisingly, it has never been tried to reevaluate the usage of initiation sites at *oriC* by modern primer technology – representative for in vivo conditions.

As for *E. coli*, the replicative DnaB helicase is a RecA-related ATPase, and the helicase loader DnaC belongs to the AAA^+ class, like DnaA itself. During the loading phase, monomeric DnaC chains bind 6 : 6 to DnaB chains in the hexameric rings; they are thought to function as "ring breakers" to thread DNA into the central channel (Davey and O'Donnell 2003). Subsequent dissociation of the loader proteins is necessary to activate the helicase. The peculiar distribution of polymerase start sites suggests that three of the five DnaA chains at *oriC* appear to be capable of loading a single helicase close by (Fig. 4B). This argues against the generality of two coupled rings being loaded simultaneously as a universal mechanism. Comparing the polarity of leading-strand initiation with the topology of the ternary complex of ssDNA–DnaC–DnaB (Galletto et al. 2003), the ring of DnaC subunits is placed between the corresponding DnaA site and the DnaB helicase. This indicates some critical interaction between DnaA and DnaC for helicase loading (Fig. 4C). Direct protein–protein interactions have been demonstrated for specific DnaA–DnaB and DnaB–DnaC contacts (Messer 2002; Schaeffer et al. 2005), but the ring of DnaC subunits is released upon successful loading of the ring of DnaB helicases.

It is still unclear how the initial unwinding of DNA in the crucial area is accomplished. May be the initial opening in AT-rich region has a catalytic role in the effective restructuring of the multi-subunit DnaA cluster, especially affecting the region around the rightmost pair of DnaA boxes. As suggested by Kaplan and O'Donnell (2004), DnaA binding could be reconfigured to two reannealed DnaA boxes surrounded by single-stranded hairpin loops (Fig. 4D) – appropriate for the loading of DnaC–DnaB; but this is not the only possibility as yet. Comparative experimental evidence indeed suggests that a multi-step process transforms the DnaA cluster from sequence-specific binding of individual DnaA proteins to co-operative protein–protein interactions (Felczak et al. 2005; Zawilak-Pawlik et al. 2005). Perhaps it is not entirely inconceivable that the activated core of the DnaA cluster attains a hexameric configuration of AAA^+ domains surrounding duplex DNA in the critical area. This may, in fact, be similar to the equivalent ORC–Cdc6 complex in eukaryotes (Sect. 4.2), when Cdc6 binding to the origin recognition complex completes the assembly of a hexameric AAA^+ configuration (Speck et al. 2005).

Once *oriC* has been fired, there is negative feedback upon DnaA to prevent immediate rereplication. Yet another AAA^+ protein, Hda, in a complex with the polymerase sliding clamp, inactivates DnaA by catalysing ATP hydrolysis. Also, the secondary binding sites for DnaA at duplicated *oriC* are temporarily sequestered by other proteins, which preferentially bind to hemi-methylated GATC sites of newly replicated DNA. This effect explains the local overabundance of GATC sites observed in the *oriC* region.

4.2
Origin Usage in Eukaryotes

The protein biochemistry of eukaryal replisomes, in general, has been studied most thoroughly in vertebrate systems, as pioneered by in vitro replication of SV40 DNA (Simian Virus 40). Conveniently, a single viral protein is sufficient to specifically direct host replisome components to the SV40 origin. This hexameric AAA^+ ATPase protein is termed large tumor antigen (LTag). It combines the properties of an SV40-specific origin recognition protein with an efficient replicative helicase. Its tight association as a double-hexameric complex has significantly contributed to developing the pumping model of double-ring helicase action (Gai et al. 2004; see below in this section). On the other hand, the characterization of chromosomal replication origins was pioneered in the genetically tractable yeast system.

The nuclear genome of eukaryotic cells is distributed over several linear chromosomes. Each chromosome carries multiple sites for potential initiation of biderectional DNA replication, often referred to as *ars* signals (autonomously replicating sequence). These signals come with various degrees of efficiency, and many of them are not actually used in every cell cycle. The stochastics of choosing between the many possibilities are yet poorly understood. Also, many *ars* signals can be deleted without deleterious effect on overall replication efficiency; i.e. other origins outside of the manipulated region can take over, and particular terminators to separate adjacent replicons are absent from most eukaryal chromosomes. A number of high-efficiency *ars* signals have been studied in detail in several model organisms, such as yeast. A notable exception comprises the nucleolar organizer region with its tandem arrays of multiple rDNA loci, all of which are separated by termination sites, as well as replication origins. These polarized terminator signals are thought to prevent potential conflicts of DNA synthesis running against the direction of rDNA transcription.

The budding yeast *Saccharomyces cerevisiae* has turned out very useful for studies of origin recognition, since most of its *ars* signals are organized more tightly around a narrowly defined core sequence, compared to other eukaryotes analysed in this respect. Accordingly, many of the key players were first identified and characterized in the yeast system – before their general importance was extended to other organisms (MacNeill and Burgers 2000; Bell and Dutta 2002). Genome sequencing has established the wide-spread conservation of the main components (Sect. 4.3), and many of them are referred to by yeast gene terminology, not only in other eukaryotes, but also for corresponding archaeal components (Table 1).

Roughly comparable to *oriC* recognition in bacteria, *ars* recognition in eukaryotes is also based on AT-rich specificity regions and DnaA-related binding proteins, but other factors have been added and developed to higher complexity. Efficient *ars* signals in *S. cerevisiae* can be as short as 150 bp, around a common 11-bp motif (core *ars* consensus sequence 5′-$^{A}/_{T}$ TTTA $^{T}/_{C}$ $^{A}/_{G}$ TTT $^{A}/_{T}$ -3′). The rather distantly related fission yeast *Schizosaccharomyces pombe*, as well as most other eukaryotes, use 5–10 times longer specificity regions, and strictly defined consensus signals cannot usually be identified. In *S. cerevisiae*, the *ars1* locus has been studied particularly well. Most remarkably, both leading strands originating from this site were found to initiate in either direction from adjacent base pairs, close to the core consensus at one side. This indicates that a pair of divergently oriented helicases has to be loaded at a uniquely defined position. In metazoans, however, effective initiation can be scattered over extended regions, so local mechanisms may be slightly different.

The proteins binding at DNA regions with *ars* activity have become known as origin recognition complex (ORC), containing six different subunits in yeast (Orc1–Orc6). Five of these belong to the family of AAA^+ proteins (Speck et al. 2005), although not all of these still retain ATPase activity. These complexes are loaded on to newly replicated origin sites during S-phase and remain bound there throughout the entire cell cycle. Being inactive on their own, they are pushed into action by additional proteins in the subsequent S-phase. At first, throughout G_2-phase and mitosis, the ORC alone is devoid of helicase loading activity; thus, no replicative helicase can be installed at a potential origin too early (Fig. 5A). After the Orc1-related AAA^+ ATPase Cdc6 has been added to the ORC as a complementary subunit, this supplemented complex acts as a helicase loader (Fig. 5B). Notably, among all the proteins mentioned in this section, only Cdc6 is synthesized periodically – controlled transcriptionally during G_1 and, in turn, degraded after ubiquitinylation during S-phase.

Various lines of circumstantial evidence suggest that the eukaryal replicative helicase consists of the hexameric MCM complex (Forsburg 2004), first identified as six different mini-chromosome maintenance factors (Mcm2–Mcm7). All six of these protein chains carry a mutually related AAA^+ ATPase domain, and intermediate subcomplexes are thought to aggregate as a closed ring in a fixed order of subunits. Prolific loading of MCM complexes is strictly

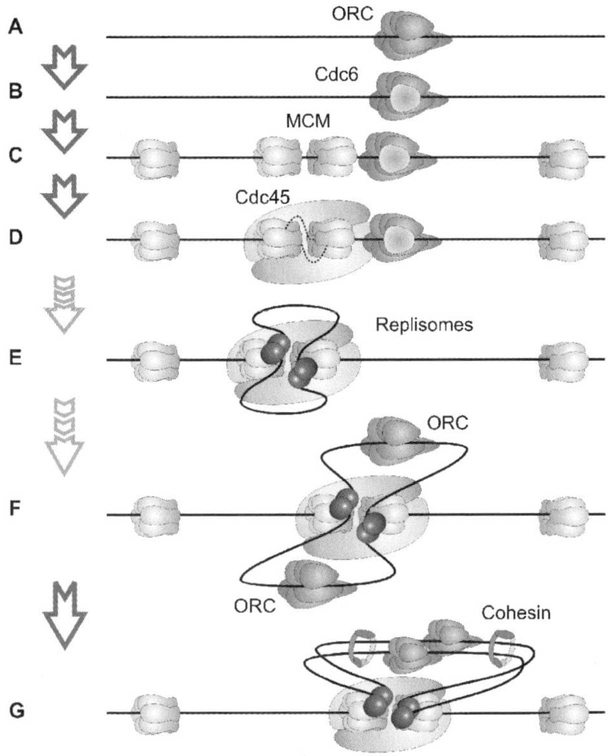

Fig. 5 Origin recognition and firing in eukaryotes – the yeast model. *Thick lines* are representing duplex DNA. **A** For most of the cell cycle, potential origins are bound by ORC proteins, yet to be activated. **B** Activation starts by binding of Cdc6 to the basal ORC subunits. **C** This allows repetitive loading of MCM hexamers, still inactive as helicases. **D** A divergently oriented pair of MCM hexamers is activated by binding to Cdc45 complexes. A pair of nascent single-stranded loops (*dotted lines*) is pumped out at this stage. **E** Replisomes start working at the emerging loops to make them double-stranded, and ORC proteins are displaced from the template. **F** New complexes of ORC proteins (again inactive) bind to the replicated origins. **G** More realistically, the sister chromatids remain physically connected by cohesin rings, which are loaded at newly activated origins (stage C/D)

dependent on Cdc6 having bound to ORC at an *ars* site (Fig. 5C). It is now assumed that the addition of Cdc6 to the basal subunits of ORC completes the formation of a hexameric assembly of AAA^+ domains (Speck et al. 2005), which is fitting to a hexameric ring of MCM proteins. Surprisingly enough, this loading quickly leads to a $\sim$ 40-fold excess of MCM around this site, as compared to the number of prospective replication forks – referred to as the "MCM paradox". Most of the chromatin-associated MCM complexes do not engage in unwinding of the DNA. These may participate in other important tasks, such as sweeping the DNA clear of chromatin proteins or monitoring for DNA integrity in front of replication forks (Shechter and Gautier

2004). Clearly, the total number of MCM is not limiting for the initiation of DNA synthesis as such. Eventually, a single pair of divergently oriented MCM complexes must be converted into an active pair of helicases, competent of priming the divergent leading strands.

The inability of MCM complexes to initiate unwinding of duplex DNA on their own has led to the torsional-strain model of helicase activation (Mendez and Stillman 2003; Kaplan and O'Donnell 2004). This model posits that duplex DNA is actively being pumped through the hexameric rings by the AAA^+ mechanism (Sect. 3.1), without breaking hydrogen bonds between the bases. If two such rings are immobilized face to face; however, the forceful torque and pressure applied at the interface might be sufficient locally to splay apart both strands, extruding nascent single-stranded loops of DNA from the immobilized double ring of MCM proteins. As mentioned earlier, this model was first suggested by the structural organization of the SV40 LTag helicase (Gai et al. 2004).

The most critical factor assisting in the activation of a double ring of MCM proteins is Cdc45, which is truly limiting by its low concentration. It interacts with both ORC and MCM initially, and remains bound to active replisomes later on. Conceivably, therefore, Cdc45 is part of the scaffold platform keeping replisomes together in a replication factory (Fig. 5D). A host of other Cdc45-interacting partners have been identified (Takayama et al. 2003), but details of their interactions remain to be clarified. Besides, the MCM complexes and associated proteins are subject to phosphorylation by Cdc7–Dbf4, a specialized protein kinase responding to cell cycle regulation (chapter of Nasheuer).

Cdc45-dependent activation of the replicative helicases then leads to local DNA unwinding, replisomes are organized at the pair of initially single-stranded loops, and replication to double-stranded DNA commences. As one of the helicases starts working against the ORC-binding region itself, the ORC proteins are thought to be displaced at this stage (Fig. 5E). When the ORC-binding region is replicated as well, ORC proteins can bind anew at both loops, albeit in the inactive state without Cdc6 (Fig. 5F). They will only be reactivated again in the next cell cycle.

In the diagram of Fig. 5, steps E and F are simplified to emphasize the emergence of new daughter molecules of DNA. In addition to that, a structurally important aspect of eukaryotic chromosome organization has to be accounted for, which happens to be linked to early events at replication origins. In the bacterial model of Fig. 2, the replicated origins are physically pulled apart – in early preparation for proper segregation at the upcoming cell division. In eukaryotic nuclei, however, the replicated DNA molecules remain intimately connected as undivided chromosomes throughout the remaining interphase of the cell cycle. This so-called sister chromatid cohesion is mediated by a set of specialized proteins, cohesin, which can form loosely fitting rings around two double-helix molecules of DNA (Nasmyth and Schleiffer 2004). The loading of these rings has been narrowed down to stages C or D of Fig. 5 – the establishment of the so-called prereplication complex of ORC–Cdc6–MCM (Takahashi et al. 2004). Hence, the newly installed cohesin rings are thought to encircle the template DNA before it gets unwound and replicated. At a slightly later stage, therefore, both branches of each replication fork will automatically be held together by the stable cohesin rings (Fig. 5G), which only are broken down during chromosome separation at mitosis. Notably, an alternative clamp loader related to Replication Factor C participates in the loading of cohesin rings (Majka and Burgers 2004).

4.3
Evolutionary Conservation of Origin Usage

While many molecular aspects of *oriC* utilization are considerably diverged in various branches of modern bacteria, the basic components can be traced back to the primordial stage of LUCA – the last universal common ancestor of cellular life on Earth. This hypothetical progenitor of bacteria, archaea and eukaryotes is presumed to have carried 6–8 different members of the important class of AAA^+ ATPases (Iyer et al. 2004). Three of those are relevant for origin recognition and replisome organization.

Above all, the clade of origin-binding proteins still exerts the same function of origin recognition and/or helicase loading throughout the three domains of living organisms: bacteria use DnaA to this effect, together with DnaA-related DnaC; archaea have one or more proteins similar to eukaryal ORC/Cdc6 proteins; in eukaryotes this clade has split up into several ORC proteins, as well as Cdc6, with mutually complementary properties. Likewise, the clamp loader clade has retained the same function in all three domains, and it has been diversified to two or more subunits: $\delta\delta'$ vs. $\gamma\tau$ of DNA polymerase III in bacteria, RfcL vs. RfcS in archaea, and Rfc1 vs. Rfc2–5 in eukaryotes (Majka and Burgers 2004). As to replicative helicases, only archaea and eukaryotes use directly related proteins for this role: typically a single MCM protein to form isohexamers in archaea, but heterohexamers in eukaryotes (Mcm2–7). Bacteria use a different ATPase helicase to this effect; another protein subfamily related to MCM is also widely present in bacteria as NtrC, but instead, its members serve as phosphorylation-activated transcriptional regulators for narrow groups of genes involved in nitrogen assimilation.

Very remarkably indeed, origin recognition has been conserved by evolution more stringently than the catalytic subunits of replisomes. DNA polymerase III of bacteria is, in fact, so different from archaeal/eukaryal replicative polymerases that only a few amino acids at the active center have been kept invariant. Sliding clamp subunits, too, appear entirely scrambled at the primary sequence level, comparing bacterial vs. archaeal/eukaryal lineages – even though structural similarity is strikingly conspicuous. As to triple-crested clamp proteins in bacteria, repetitive sequence similarity between N-terminal and central parts of a subunit is still relatively high. This indicates that dimeric bacterial clamps have evolved from trimeric ones by intragenic duplication, comprising the N-terminal domain of a double-crested precursor.

5 DNA Polymerases: Priming, Switching and Recycling

Inasmuch as replisome organization in bacteria has been introduced before (Sect. 2), this part mainly concerns eukaryal/archaeal components and mechanisms (MacNeill and Burgers 2000; Grabowski and Kelman 2003). By and large, the number of subunits in functional modules of replisomes is higher in eukaryotes than in bacteria (Table 1). To begin with, there are three different DNA polymerases: Polα, Polδ and Polε.

The Polα/primase complex cooperates with replication protein A (RPA), the eukaryal single-strand binding protein. This primase complex does not only deliver the initial RNA primer of each newly started strand, but also up to 40 nucleotides of DNA thereafter. Polα, however, does not share the proofreading ability of other replicative DNA polymerases – making its product less reliable. In consequence, this potentially ambiguous part of DNA is being removed before joining to the adjacent fragment (Sect. 6). Polδ is the working horse of the replisome, being absolutely required for synthesis of all the Okazaki fragments in the lagging strand. On the other hand, the role of Polε is somewhat enigmatic. While its catalytic domains for polymerase and exonuclease can be deleted in yeast, other parts of the main subunit are still essential for viability. In general, Polε is held responsible for leading-strand synthesis, although this activity alone can be supplemented by an additional module of Polδ in a functional replisome.

Evidently, besides making DNA, Polε is contributing another essential function: together with several associated proteins, it appears to give structural integrity to the entire replisome. This may resemble the supportive role of τ-subunits in the bacterial clamp loader, to which there is no archaeal/eukaryal equivalent by direct sequence homology. In bacteria, the extra C-terminal domains of τ-subunits serve as physical links between the replicative helicase at the stationary platform and two processive DNA polymerase modules per replisome (Fig. 1). Significantly, Polε subunits interact with the Cdc45-dependent scaffold, as well as MCM complexes (Takayama et al. 2003; Spiga and D'Urso 2004) – putting the Polε module into a strategic position for replisome organization. Whether Polε also interacts with Polδ modules directly, remains to be clarified.

When a newly primed strand of DNA is handed over from Polα to Polε or Polδ, this switching reaction is mediated by replication factor C (RFC), the heteropentameric clamp loader (MacNeill and Burgers 2000; Majka and Burgers 2004). Polα activity is interrupted by the open RFC–PCNA complex, the sliding clamp is locked around the RNA–DNA primer and the template, and RFC dissociates from the complex. The vacant binding site on the clamp is filled by recruitment of a core polymerase – typically Polδ in the assembly line of Okazaki fragments. This may be comparable to Fig. 1; however, it is not yet known for eukaryotes whether clamp loader and/or Polδ are actually reused within the same replisome or, in turn, recruited anew from unattached components.

6 Joining of Adjacent Fragments

The processing and joining of adjacent Okazaki fragments in eukaryotes and archaea requires the concerted action of three widely conserved proteins: the combined helicase/endonuclease Dna2, the flap endonuclease Fen1, and DNA ligase I (MacNeill and Burgers 2000; Kao and Bambara 2003). RNaseH is also

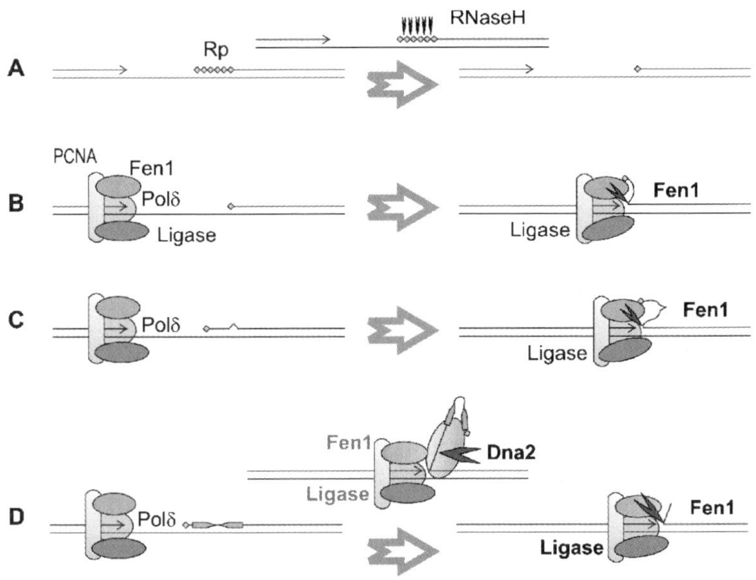

Fig. 6 Multiple pathways for the joining of adjacent Okazaki fragments. **A** Most of the RNA primer (Rp) can be removed by RNaseH (not always obligatory). This occurs in a RNA/DNA duplex, before Polδ has started to displace the 5′-end as a flap junction. **B** By default, a short flap is cleaved at its base by Fen1, directly forming a suitable substrate for ligation. **C** If mismatched bases are contained in the primer region, due to a faulty product of Polα, Polδ gets a chance to rectify this flaw; a somewhat longer flap is formed and processed as before. **D** If hairpin structures conceal the 5′-end from Fen1, the flap gets excessively long and Fen1 is unable to cut. Instead, Dna2 starts processing this flap, but the remaining stub is still not suitable for ligation. The final cleavage is again carried out by Fen1

capable of removing part of the RNA primer (Fig. 6A), but this activity is not always necessary. Both Fen1 and Dna2 can remove 5′-single-stranded overhangs, termed flaps, at DNA triple junctions. While Fen1 and ligase bind to and are stimulated by the PCNA sliding clamp, Dna2 closely interacts with RPA/SSB. Of special interest, certain archaea have diversified a heterotrimeric PCNA-like clamp, where different subunits bind DNA polymerase, flap endonuclease and ligase, respectively (Dionne et al. 2003). This may suggest that the Polδ-associated PCNA clamp in eukaryotes carries Fen1 and ligase simultaneously as well – thereby acting as a composite scanning and processing device (Fig. 6B,C).

It is still disputed whether Dna2, in fact, is involved in the processing of every Okazaki fragment; it rather appears to be required preferentially for the removal of excessively long flap structures, containing fold-back hairpins inhibitory to Fen1 (Kao and Bambara 2003). Such structured flaps can only be generated in partly palindromic regions (Fig. 6D). By default, therefore, Fen1 alone may be sufficient to provide a ligatable junction for most Okazaki fragments. It has been noted that Fen1 tends to cut at slightly longer flaps in the presence of mismatched bases close to the 5′-end at the potential junction, as compared to precisely base-paired regions (Fig. 6C). This suggests that Fen1 actively participates in a proof-reading mechanism to counteract the marginal infidelity of Polα at the priming stage of that fragment.

7
Concluding Remarks

This introductory chapter, preceding two more detailed and specific contributions, has tried to focus on the present level of understanding of chromosomal DNA replication – with lesser concerns about the countless experimental work that has paved the way. Very deliberately, therefore, most of the literature cited consists of a representative panel of authoritative reviews, whereby the devoted reader will be guided to relevant original articles on particular problems and mechanisms discussed above.

Genomic replication is not only a tantalizing biochemical process, it is also one of the cornerstones in the formidable edifice of life. The individual enzymes participating in these reactions have been integrated in sophisticated complexes, ensuring a high degree of processivity, coordination and perfection. While the basic mechanisms of replicases as such are firmly documented and well understood, the more subtle aspects of their integration into replisomes and higher-order replication factories are still subject to intensive investigation. As for eukaryotes, the scaffolding complexes tethering several replisomes to a supportive center are beginning to emerge and yield themselves to experimental approaches. In bacteria, however, the evidence for such factors is rather circumstantial yet. The near future will certainly show rapid progress on these lines.

Acknowledgements The critical comments to this manuscript by Olaf Nielsen and Michi Egel-Mitani are gratefully acknowledged.

References

Bell SP, Dutta A (2002) DNA replication in eukaryotic cells. Annu Rev Biochem 71:333–374

Bowman GD, Goedken ER, Kazmirski SL, O'Donnell M, Kuriyan J (2005) DNA polymerase clamp loaders and DNA recognition. FEBS Lett 579:863–867

Bruck I, O'Donnell M (2001) The ring-type polymerase sliding clamp family. Genome Biol 2(1):Reviews3001.1–3001.3

Davey MJ, Jeruzalmi D, Kuriyan J, O'Donnell M (2002) Motors and switches: AAA$^+$ machines within the replisome. Nat Rev Mol Cell Biol 3:826–835

Davey MJ, O'Donnell M (2003) Replicative helicase loaders: ring breakers and ring makers. Curr Biol 13:R594–596

Dionne I, Nookala RK, Jackson SP, Doherty AJ, Bell SD (2003) A heterotrimeric PCNA in the hyperthermophilic archaeon Sulfolobus solfataricus. Mol Cell 11:275–282

Felczak MM, Simmons LA, Kaguni JM (2005) An essential tryptophan of Escherichia coli DnaA protein functions in oligomerization at the E. coli replication origin. J Biol Chem 280:24627–24633

Forsburg SL (2004) Eukaryotic MCM proteins: beyond replication initiation. Microbiol Mol Biol Rev 68:109–131

Gai D, Zhao R, Li D, Finkielstein CV, Chen XS (2004) Mechanisms of conformational change for a replicative hexameric helicase of SV40 large tumor antigen. Cell 119:47–60

Galletto R, Jezewska MJ, Bujalowski W (2003) Interactions of the Escherichia coli DnaB helicase hexamer with the replication factor the DnaC protein. Effect of nucleotide cofactors and the ssDNA on protein-protein interactions and the topology of the complex. J Mol Biol 329:441–465

Grabowski B, Kelman Z (2003) Archaeal DNA replication: eukaryal proteins in a bacterial context. Annu Rev Microbiol 57:487–516

Graumann PL, Defeu-Soufo HJ (2004) An intracellular actin motor in bacteria? Bioessays 26:1209–1216

Gulbis JM, Kelman Z, Hurwitz J, O'Donnell M, Kuriyan J (1996) Structure of the C-terminal region of p21(WAF1/CIP1) complexed with human PCNA. Cell 87:297–306

Hanson PI, Whiteheart SW (2005) AAA$^+$ proteins: have engine, will work. Nat Rev Mol Cell Biol 6:519–529

Haroniti A, Anderson C, Doddridge Z, Gardiner L, Roberts CJ, Allen S, Soultanas P (2004) The clamp-loader-helicase interaction in Bacillus. Atomic force microscopy reveals the structural organisation of the DnaB-τ complex in Bacillus. J Mol Biol 336:381–393

Iyer LM, Leipe DD, Koonin EV, Aravind L (2004) Evolutionary history and higher order classification of AAA$^+$ ATPases. J Struct Biol 146:11–31

Johnson A, O'Donnell M (2003) Ordered ATP hydrolysis in the γ complex clamp loader AAA$^+$ machine. J Biol Chem 278:14406–14413

Kao HI, Bambara RA (2003) The protein components and mechanism of eukaryotic Okazaki fragment maturation. Crit Rev Biochem Mol Biol 38:433–452

Kaplan DL, O'Donnell M (2004) Twin DNA pumps of a hexameric helicase provide power to simultaneously melt two duplexes. Mol Cell 15:453–465

Kong XP, Onrust R, O'Donnell M, Kuriyan J (1992) Three-dimensional structure of the β subunit of E. coli DNA polymerase III holoenzyme: a sliding DNA clamp. Cell 69:425–437

Leonard AC, Grimwade JE (2005) Building a bacterial orisome: emergence of new regulatory features for replication origin unwinding. Mol Microbiol 55:978–985

Leu FP, Georgescu R, O'Donnell M (2003) Mechanism of the E. coli τ processivity switch during lagging-strand synthesis. Mol Cell 11:315–327

MacNeill SA, Burgers PMJ (2000) Chromosomal DNA replication in yeast: enzymes and mechanisms. In: Fantes P, Beggs J (eds) The yeast nucleus. Oxford University Press, p 19–57

Majka J, Burgers PM (2004) The PCNA-RFC families of DNA clamps and clamp loaders. Prog Nucleic Acid Res Mol Biol 78:227–260

Matsumiya S, Ishino Y, Morikawa K (2001) Crystal structure of an archaeal DNA sliding clamp: proliferating cell nuclear antigen from Pyrococcus furiosus. Protein Sci 10:17–23

Mendez J, Stillman B (2003) Perpetuating the double helix: molecular machines at eukaryotic DNA replication origins. Bioessays. 25:1158–1167

Messer W (2002) The bacterial replication initiator DnaA. DnaA and oriC, the bacterial mode to initiate DNA replication. FEMS Microbiol Rev 26:355–374

Messer W, Weigel C (1996) Inititation of chromosome replication. In: Niedhardt FC (ed) Escherichia coli and Salmonella typhimurium, Cellular and Molecular Biology. American Society for Microbiology, p 1579–1601

Mulugu S, Potnis A, Shamsuzzaman, Taylor J, Alexander K, Bastia D (2001) Mechanism of termination of DNA replication of Escherichia coli involves helicase-contrahelicase interaction. Proc Natl Acad Sci USA 98:9569–9574

Nasmyth K, Schleiffer A (2004) From a single double helix to paired double helices and back. Philos Trans R Soc Lond B Biol Sci 359:99–108

Schaeffer PM, Headlam MJ, Dixon NE (2005) Protein–protein interactions in the eubacterial replisome. IUBMB Life 57:5–12

Seufert W, Messer W (1987) Start sites for bidirectional in vitro DNA replication inside the replication origin, oriC, of Escherichia coli. EMBO J 6:2469–2472

Shechter D, Gautier J (2004) MCM proteins and checkpoint kinases get together at the fork. Proc Natl Acad Sci USA 101:10845–10846

Speck C, Chen Z, Li H, Stillman B (2005) ATPase-dependent cooperative binding of ORC and Cdc6 to origin DNA. Nat Struct Mol Biol 2005 Oct 16; [Epub ahead of print]

Spiga MG, D'Urso G (2004) Identification and cloning of two putative subunits of DNA polymerase ε in fission yeast. Nucleic Acids Res 32:4945–4953

Takahashi TS, Yiu P, Chou MF, Gygi S, Walter JC (2004) Recruitment of Xenopus Scc2 and cohesin to chromatin requires the pre-replication complex. Nat Cell Biol 6:991–996

Takayama Y, Kamimura Y, Okawa M, Muramatsu S, Sugino A, Araki H (2003) GINS, a novel multiprotein complex required for chromosomal DNA replication in budding yeast. Genes Dev 17:1153–1165

Wang J (2004) Nucleotide-dependent domain motions within rings of the RecA/AAA(+) superfamily. J Struct Biol 148:259–267

Warbrick E (2000) The puzzle of PCNA's many partners. Bioessays 22:997–1006

Zawilak-Pawlik A, Kois A, Majka J, Jakimowicz D, Smulczyk-Krawczyszyn A, Messer W, Zakrzewska-Czerwinska J (2005) Architecture of bacterial replication initiation complexes: orisomes from four unrelated bacteria. Biochem J 389:471–481

Genome Dyn Stab (1)
D.-H. Lankenau: Genome Integrity
DOI 10.1007/7050_016/Published online: 5 July 2006

Progress Towards the Anatomy of the Eukaryotic DNA Replication Fork

Heinz Peter Nasheuer[1] (✉) · Helmut Pospiech[2] · Juhani Syväoja[3]

[1]National University of Ireland, Galway, Dept. of Biochemistry, Cell Cycle Control Laboratory, University Road, Galway, Ireland
h.nasheuer@nuigalway.ie

[2]University of Oulu, Department of Biochemistry, P.O. Box 3000, 90014 Oulu, Finland

[3]University of Joensuu, Department of Biology, P.O. Box 111, 80101 Joensuu, Finland

Abstract During cell growth before each division, cells have to accurately duplicate their genome. The processes associated with DNA replication are tightly controlled; failure thereof can result in genome instability, which is a hallmark of cancer (Marte 2004). A large number of proteins are involved in this ambitious task to replicate the chromosomal DNA once and only once at a given time (Table 1).

Abbreviations

AAA	ATPases associated with a variety of cellular activities
ATR	Ataxia telangiectasia-mutated and Rad3-related
ATRIP	ATR interacting protein
BRCA	breast cancer associated
BRCT	BRCA1 C-terminal
CAF	chromatin assembly factor
Cdc	cell division cycle
Cdt	Cdc10-dependent
Cdk	cyclin-dependent kinase
CKI	Cdk inhibitor
Dbf	dumbbell former
DBD	DNA binding domain
DDK	Dbf4-dependent kinase
DPB	DNA polymerase epsilon subunit B
Drf	Dbf4-related factor
dsDNA	double-stranded DNA
FEN	flap endonuclease
GINS	Go Ichi, Nii, and San Japanese for five, one, two, and three; protein complex containing Sld5 and Psf1-3
Mcm	minichromosome maintenance
Mus	nitrogen mustard-sensitive
OB	oligonucleotide/oligosaccharide binding
ORC	origin recognition complex
PARP	poly[ADP-ribose]polymerase
PCNA	Proliferating cell nuclear antigen
pre-RC	pre-replicative complex
Pol	DNA polymerase

POL2	DNA polymerase epsilon catalytic subunit in S. cerevisiae
Psf	partner of Sld five
Rad	radiation-sensitive
rec	recombination defective
RF-C	replication factor C
RPA	replication protein A [eukaryotic ssDNA binding protein]
Srs	suppressor of radiation-sensitive mutations
ssDNA	single-stranded DNA
Sld	synthetically lethal with *DPB11*
SUMO	small ubiquitin-related modifier protein
SV	simian virus
TAg	T antigen
TopBP1	DNA topoisomerase II binding protein

1 DNA Replication and the Cell Cycle

The replication of the genome of eukaryotic cells occurs in a defined time separated from other cellular processes such as chromosome segregation in mitosis (Blow et al. 2005; Machida et al. 2005; Nasheuer et al. 2002). During their duplication eukaryotic cells follow a tightly controlled order of events summarized as the cell cycle. The eukaryotic cell cycle is divided into four phases. After cell division and before starting DNA replication, cells enter G1 phase where proteins are synthesized, and each cell depending on its state and environment will decide whether to continue cell division and replicate its DNA or leave the cell cycle. The latter is termed the resting phase, G0, if it is reversible or terminal differentiation if cells do not continue to divide. After all preparations for the duplication of the genome are finished, cells replicate their chromosomal DNA in S phase. Before cells enter mitosis and segregate their chromosomes, they synthesize the necessary proteins and extensively check their genome in the G2 phase to avoid the transfer of damaged DNA into the two daughter cells in mitosis. In the following we want to discuss the establishment of the replication forks and its components.

2 Give Cells the License to Replicate – The Assembly of the Pre-replicative Complex

2.1 Activities at Origins of DNA Replication

To ensure the timely duplication of genomic information eukaryotic, DNA replication starts from multiple sequences on each chromosome called ori-

Table 1 Eukaryotic DNA replication factors

Factor	Function during Pre-initiation	Initiation	Elongation Lagging strand	Leading strand
ORC	Origin binding, MCM loading	–	–	–
Cdc6	MCM loading	–	–	–
Cdt1	MCM loading	–	–	–
Geminin	Cdt1 inhibitor	–	–	–
Mcm2-7 (MCM complex & sub-complexes)	Licensing for DNA replication	Origin DNA unwinding	?	Potential replicative helicase [1]
Mcm8	–	?	(Putative replicative helicase?)	
Mcm10	–	Cdc45 & Pol α loading		Structural role?
Cdc45	–	Loading and activation of replication factors		Helicase cofactor
RPA	–	ssDNA and protein binding	ssDNA bdg., cooperation with Pol α & δ	ssDNA bdg.
Dpb11/Cut5/ Mus101/TopBP1	–	Establishing the replication fork (Pol ε loading)		Rescue of stalled replication forks
Sld2/RecQL4	–	See above	–	?
GINS	–	See above		Structural role?
Cdc7-Dbf4	–	Protein	Protein phosphorylation	
Cdk-Cyclins	–	phosphorylation	Protein phosphorylation	
Pol α	–	Primer synthesis	Primer synthesis	
Pol δ	–		DNA synthesis	Can replace Pol ε
Pol ε	–	Regulation?	?	DNA synthesis?
PCNA	–		Clamp for replication factors & DNA modifying enzymes	
RF-C	–		Loading of PCNA	
FEN-1	–		Maturation of Okazaki fragments	–
Dna2	–		see above	–
RNase H	–		see above	–
Ligase I	–		Ligating Okazaki fragments	Termination?
Topoisomerase I	–	Release of torsional stress in DNA		

[1] possibly a subset of the complex

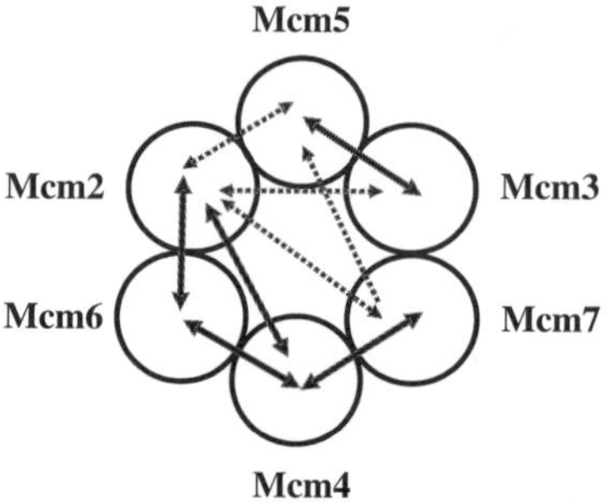

Fig. 1 Physical interactions within the MCM complex. The Mcm (minchromosome maintenance) proteins Mcm2, Mcm3, Mcm4, Mcm5, Mcm6 and Mcm7 form a complex (MCM complex), which is crucial for licensing (formation of the pre-replicative complex, pre-RC), initiation and elongation of eukaryotic DNA replication. It is thought that the MCM complex has a heterohexameric, ring-like structure (scheme adopted from Forsburg 2004). The *arrows* indicate protein-protein interactions found by yeast two-hybrid analysis of the mouse proteins with *solid lines* suggesting strong interactions and with *dashed lines* weaker ones (Kneissl et al. 2003). The *arrowheads* show interactions of bait towards prey, (two *arrowheads* represent interactions in both directions using the yeast two-hybrid technique). Additional studies suggest that the binding of Mcm4, Mcm6 and Mcm7 to Mcm2 is weaker than the interactions between the former and a complex of Mcm4, Mcm6 and Mcm7 without Mcm2 can be purified from eukaryotic cells (see text for more details)

gins of replication. These origins are particularly well studied in budding yeast – *Saccharomyces cerevisiae* (S. cerevisiae) – using genetic and biochemical techniques (for review see Stillman 2005 and the chapter by Egel in this book). This single cell organism has helped to understand various basic mechanisms surrounding the DNA replication in general. For higher eukaryotes the frog *Xenopus laevis* and mammalian cells in culture have served as model systems (Blow et al. 2005; Machida et al. 2005; Nasheuer et al. 2002)[1].

Investigations in yeast and higher eukaryotes showed that a specific conserved protein complex – the origin recognition complex (ORC) – binds to the DNA sequence of replication origins in an ATP-dependent manner. ORC consists of 6 subunits (Orc1–Orc6), which are all essential for the initiation of DNA replication, but only Orc1 to Orc5 are required for DNA binding in vitro (Machida et al. 2005; Stillman 2005). Several ORC subunits contain an AAA^+ domain (AAA, ATPases associated with a variety of cellular activities) and belong to the large family of AAA^+ ATPase similarly as Cdc6 (cell division cycle), the Mcm2 to Mcm7 (minichromosome maintenance) proteins, and replication factor C (RF-C) with the ATPase activity of ORC stimulated by single-stranded DNA (ssDNA; Machida et al. 2005; Stillman 2005). In *S. cerevisiae* ORC binds to the origin DNA during the entire cell cycle (Machida et al.

[1] A detailed discussion of eukaryotic origins can be found in the chapter by Egel.

2005; Stillman 2005). In activated egg extracts from *Xenopus*, ORC binds to the chromatin in the G1 phase. However, after the pre-replicative complex (pre-RC) assembly is completed, ORC can be removed from the chromatin without inhibition of DNA replication (Sun et al. 2002). In hamster and human cells ORC is unstable since Orc1 protein is not constantly present during the cell cycle and is degraded in S phase whereas in the *Xenopus* system Orc1 is not degraded (Maiorano et al. 2005b; Stillman 2005). In all cases a mechanism is established that prevents a re-replication at the level of the origin recognition by ORC.

ORC serves as a "landing platform" for other replication proteins (Table 1) that assemble prior to the beginning of DNA replication in late mitosis and early G1, to form a highly conserved complex called the pre-replicative complex (pre-RC). The pre-RC consists of proteins ORC, Cdc6, Cdt1 (Cdc10-dependent), and the MCM complex (consisting of Mcm2, Mcm3, Mcm4, Mcm5, Mcm6 and Mcm7; Figs. 1 and 2) (Blow et al. 2005; Machida et al. 2005;

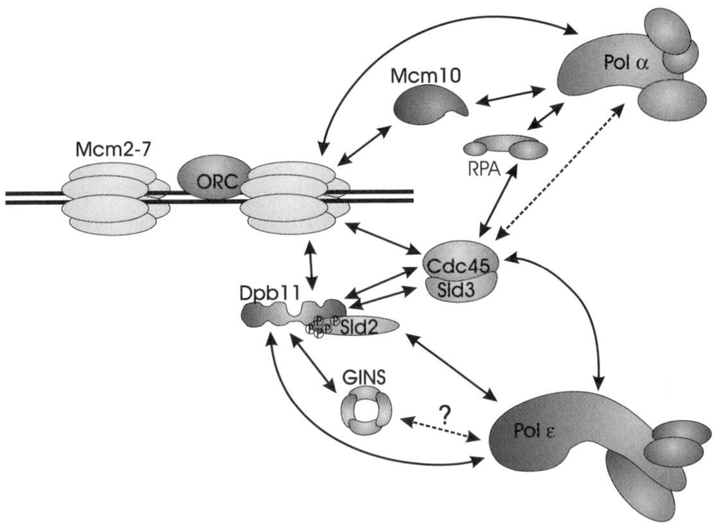

Fig. 2 Interactions of replication proteins during the initiation reaction. The initiation of eukaryotic DNA replication requires multiple protein-protein interactions, which are presented as *arrows* between proteins and protein complexes. *Solid lines* indicate interactions in the budding yeast, which are conserved between eukaryotic organisms. The interaction between Cdc45 and Pol α is weak or not conserved and therefore marked with a *dashed line*. An interaction between the GINS complex and Pol ε has been proposed but not proven and is represented by a *double arrow* with a *dashed line* and a *question mark*. Abbreviations: Cdc, cell division cycle; DPB11, suppressor of DNA polymerase epsilon subunit B mutant; GINS, Go, Ichi, Nii, and San [japanese for five, one, two and three; the protein complex consists of Sld5 and Psf1-3]; Mcm, minichromosome maintenance; ORC, origin recognition complex; Pol, DNA polymerase, RPA, replication protein A; Sld, synthetic lethal with DPB11

Madine et al. 2001). After pre-RC is assembled the chromatin is prepared for DNA replication, a process also called "licensing", the pre-RC associated with early in S phase active origins is converted to an initiation complex (IC) at the G1/S transition. This switch requires the activity of cyclin-dependent kinases (Cdks) and of Cdc7 kinase (additional discussions see below; for review Blow et al. 2005; Machida et al. 2005; Madine et al. 2001). In addition to its function in DNA replication, ORC is involved in the silencing of certain genes in yeast (Bose et al. 2004; Hsu et al. 2005). Orc2 interacts with the centrosome during the cell cycle and is required for its function whereas Orc6 binds to the kinetochore in mitosis, which is necessary for establishing cytokinesis (Prasanth et al. 2002, 2004).

2.1.1 Factors Cdc6, Cdt1 and Geminin Control Replication Early in the Cell Cycle

In the beginning of the eukaryotic cell cycle, ORC sequesters Cdc6 (in *Schizosaccharomyces pombe* {S. pombe} called Cdc18) and Cdt1 to chromatin (Table 1; reviewed in Bell et al. 2002). The functions of both proteins are conserved and tightly regulated in all eukaryotes although the control mechanisms might vary. At the onset of S phase in yeast, Cdc6 is phosphorylated and thereby targeted for proteolytic degradation. In higher eukaryotes Cdc6 is also phosphorylated, but phosphorylation induces its export from the nucleus rather than its degradation. Cdc6 belongs to the family of AAA^+ ATPases and binds ATP. The protein shares a high degree of sequence similarity with Orc1 and shows some similarity in sequence or structure to Mcm2 to Mcm7 as well as to Orc2 to Orc5. Cdc6 physically interacts with ORC and together they form a ring-like structure (Speck et al. 2005). In addition, Cdc6 modulates the binding of ORC with origin sequences (Speck et al. 2005). Moreover, Cdc6 binds to Cdt1 and both together load the Mcm2 to Mcm7 complex (MCM complex) onto chromatin, which represents a requirement for the licensing of cells for replication (Bell et al. 2002). In this way Cdc6, together with the ORC subunits, acts similar to RF-C, which loads the proliferating cell nuclear antigen (PCNA) onto DNA (for additional information see discussion below). Recent results in higher eukaryotes suggest that Cdc6 is also involved in the establishment of intra-S phase checkpoint response (Oehlmann et al. 2004). Interestingly this function is distinguishable from its loading activity of the MCM complex and does not require ORC and chromatin association of Cdc6.

Cdt1 is located in the cell nucleus and in presence of ORC-Cdc6 it is associated with DNA (Randell et al. 2006). It physically binds to Cdc6, and the MCM complex and Cdt1 and Cdc6 cooperate to load the MCM complex onto origins and adjacent sequences. The function of Cdt1 is tightly regulated. In budding yeast the Cdt1 concentration is constant during the entire cell cycle and Cdt1 is controlled via its export from the nucleus (Tanaka et al. 2002; Wohlschlegel et al. 2000). However, in *S. pombe*, *Drosophila*, and humans Cdt1 protein levels are high in G1 phase and become reduced in S and G2 phase. Cdt1 functions are exclusively regulated by proteolysis in fission yeast whereas in higher eukaryotes Cdt1 is also degraded in a cycle-dependent manner but, moreover, another protein called geminin binds Cdt1 to inactivate it (Bell et al. 2002; Saxena et al. 2005). In *Xenopus* the proteolysis of Cdt1 is regulated by an interaction of Cdt1 with PCNA (Bell et al. 2002; Saxena et al. 2005). Recently it was described that the addition of excess recombinant Cdt1 to G2 phase nuclei in the *Xenopus* DNA replication system causes the start of DNA replication without disruption of the nuclear membrane (Maiorano et al. 2005b).

Geminin is a nuclear protein that is degraded in M phase prior to pre-RC formation and is absent G1 when pre-RC is formed. It accumulates in S and G2 phase cells to inactivate Cdt1 (McGarry et al. 1998). In S phase geminin forms a stable complex with Cdt1, which prevents a renewed association of Cdt1 with the chromatin until the M phase. The structure of the mouse geminin-Cdt1 complex using truncated proteins revealed that the coiled-coil dimer of geminin mainly interacts with two helices in the central part of Cdt1 (Lee et al. 2004). In turn, by binding to Cdt1, geminin prevents the association of the MCM complex to Cdt1 through sterical hindrance (Lee et al. 2004). Moreover, the formation of the Cdt1-geminin complex stabilizes Cdt1 in G2/M phase, which is required for sufficient formation of the pre-RC in the G1 phase of the following cell cycle (Machida et al. 2005). In M phase geminin is degraded and Cdt1 can support pre-RC assembly in the following G1 phase (Tada et al. 2001). Neither in budding nor in fission yeast an orthologue of geminin has yet been identified (Blow et al. 2005; Machida et al. 2005).

Recent findings indicate a model describing the sequential steps during the loading of the MCM complex (Randell et al. 2006 summarized in Cvetic et al. 2006). The ATP-bound ORC, in the first step, is recognized by Cdc6, which also associates with ATP and enhances the sequence-specific binding of ORC as discussed above. The protein-ATP-DNA complex interacts with the Cdt1-bound MCM complex, which in turn activates the hydrolysis of ATP by Cdc6, followed by a release of Cdt1 from and the loading of the MCM complex to chromatin. At the same time, ATP hydrolysis destabilizes Cdc6 association with chromatin, and it dissociates from the protein-DNA complex. In the next step ORC also hydrolyses the associated ATP and additional round of Cdc6 binding and MCM loading might occur.

2.1.2 The MCM Complex

The proteins Mcm2 to Mcm7 were originally identified in genetic screens searching for proteins involved in plasmid stability, cell cycle progression and the distribution of chromosomes (Table 1; reviewed by Forsburg 2004; Tye 1999). The Mcm2 to Mcm7 proteins are conserved in all eukaryotes and form a heterohexameric complex (Mcm2-3-4-5-6-7) the so-called MCM complex (Forsburg 2004; Madine et al. 2001). However, multiple complexes of Mcm proteins, such as those containing the proteins Mcm2-4-6-7, Mcm3-5, or Mcm4-6-7, have been purified from various eukaryotic organisms[2] (Forsburg 2004).

These complex formations are in good agreement with yeast two hybrid studies using mouse Mcm proteins summarized in Fig. 1, which performed strong (Mcm2 as a bait to Mcm4 and Mcm6 as prey; Mcm4 to Mcm2 and Mcm6; Mcm6 to Mcm4; between Mcm3 and Mcm5 as well as between Mcm4 and Mcm7) and weak interactions in this analysis (these interactions are depicted from Kneissl et al. 2003).

Recent data showed that the archaeal bacterium *Methanobacterium thermoautotrophicum* produces only one Mcm-related protein and that the *N*-terminus of this protein forms stacked hexameric rings and has helicase activity (Forsburg 2004). These findings suggest that the heterohexameric MCM complex might also form a ring structure.

The coordinated activity of ORC, Cdc6 and Cdt1 is necessary for the association of the MCM complex with chromatin as summarized above. With the association of the MCM complex the assembly of the pre-RCs is established and the loading components ORC, Cdc6 and Cdt1 are dispensable for later DNA replication events (Forsburg 2004). The cellular DNA is fully licensed and prepared to start DNA replication. The yeast MCM complex shows an intrinsic ATPase activity, which is a requirement for helicase activity (Schwacha et al. 2001). The MCM complex exhibits various features of a DNA helicase, but only the complex containing Mcm4, Mcm6 and Mcm7 (Mcm4-6-7) has been shown to contain a weak DNA helicase activity in vitro (Ishimi 1997, summarized in Forsburg 2004). Together these three proteins also form a heterohexamer (two trimers) and show processive DNA helicase activity with a fork-like structure (Lee et al. 2001). The Mcm4-6-7 helicase translocates on the parental strand in 5′ to 3′ direction similarly to SV40 large T antigen (TAg) but moves in opposite direction compared to the *E. coli* replicative DnaB helicase (Fanning 1992; Ishimi 1997; Kornberg et al. 1992). Thus, the eukaryotic DNA helicases Mcm4-6-7 and SV40 TAg would move ahead of the leading strand replicase whereas the prokaryotic DnaB helicase translocates on the template for the lagging strand.

In *S. cerevisiae* and *S. pombe* the Mcm2 to Mcm7 proteins are nuclear in the G1 and S phase, whereas they are actively transported out of the cell nucleus during the G2 and M phase (Forsburg 2004). In higher eukaryotes Mcm2 to Mcm7 proteins are constitutive in the cell nucleus but their association with chromatin is regulated in S phase (Forsburg 2004). Genetic data in yeast and biochemical findings in the *Xenopus* DNA replication system indicate that this

[2] In some cases the term "sub-complexes" was used for the multiple complexes containing a subset of the Mcm2 to Mcm7 proteins (Diffley and Labib, 2002). The protein (sub-) complexes Mcm2-4-6-7, Mcm3-5, and Mcm4-6-7 consist of Mcm2-Mcm4-Mcm6-Mcm7, Mcm3-Mcm5, and Mcm4-Mcm6-Mcm7, respectively.

putative DNA helicase complex is involved in the initiation as well as in the elongation reaction of the DNA replication (Forsburg 2004). The phosphorylation of the MCM complex is apparently regulated by Cdks, Cdc7 kinase, Mcm10 and Cdc45 (Fig. 2), which might lead to the formation of the Mcm4, 6 and 7 complex followed by conformational and structural changes and activation of the intrinsic helicase activity (Forsburg 2004). An active helicase is the requirement for the double-stranded DNA (dsDNA) at an origin of replication to be melted and RPA (replication protein A) to be loaded onto chromatin (Forsburg 2004). To achieve its function the MCM complex interacts with a variety of proteins, which might be necessary to activate and to sustain helicase activity of the MCM complex in vivo. A detailed study of a mouse revealed that the MCM proteins interact with ORC, Cdc6, Dbf4-Cdc7, a cell cycle-regulating kinase (Dbf, dumbbell former; see below), Cdc45, and RPA (Kneissl et al. 2003; see also Fig. 2).

The functions of the MCM complex and its subunits are central for the initiation and elongation of eukaryotic DNA replication, but the molecular mechanism of their activity in DNA replication is still an enigma. Investigations of the function of the MCM complex are complicated by the abundance of its subunits in the nucleus. In budding yeast there are more than 10 complexes per origin (Forsburg 2004). Moreover, in the cell-free *Xenopus* replication system ORC, Cdc6 and Cdt1 load 20 to 40 MCM complexes per ORC onto chromatin whereas only two MCM complexes are necessary for full replication activity (Blow 2001). In addition, the amount of Cdc45, which is thought to activate the MCM complex to initiate DNA replication in S phase, is equivalent to two molecules per molecule of ORC in both model systems (Forsburg 2004). However, the DNA unwinding activities are not the only functions of the MCM complex and it participates in DNA damage signaling pathways during S phase. Depending on the genotoxic stress the Mcm2 to 7 proteins are involved in establishing the ATR (Ataxia telangiectasia-mutated and Rad3-related) signaling cascade or are also its target (Byun et al. 2005; Forsburg 2004; Luciani et al. 2004). Under replication stress conditions, the MCM complex, Cdc45 and GINS (Go, Ichi, Nii, and San Japanese for five, one, two and three; Sld5 and Psf1-3) seem to form a large complex independent of the additional components of the replication fork that unwinds DNA (Pacek et al. 2006).

As discussed above the Mcm2 to Mcm7 proteins show several features expected for a replicative eukaryotic DNA helicase. Genetic, cell biological and biochemical data in various model systems indicate that all six subunits of the MCM complex are equally necessary for the initiation and elongation reactions during DNA replication. Therefore, it is suggested that the MCM complex together with other factors such as Cdc45 and GINS forms the replicative DNA helicase (Pacek et al. 2006). In contrast, only the Mcm4-6-7 complex sharing the three subunits Mcm4, Mcm6, and Mcm7 with the MCM complex has been shown to carry DNA helicase activity in vitro. Therefore, an alternative model favours that Mcm4, Mcm6, and Mcm7 form a complex with helicase activity whereas Mcm2, Mcm3 and Mcm5 have regulatory functions (Ishimi 1997). To solve these apparent contradictions, various authors have presented specific models (Forsburg 2004; Laskey et al. 2003; Nasheuer et al. 2002; Stillman 2005): The hexameric MCM complex might "pump" dsDNA into the replication factories e.g., against rigid structures or proteins functioning as "ploughshares", which would support the unwinding of DNA. As an additional function the MCM complex might remove nucleosomes before the replication fork approaches these proteins and transfers them to the newly synthesized dsDNA strands. Alternatively, the Mcm4-6-7 complex might act as a replicative DNA helicase together with Mcm8 at the replication fork. Since the level of the replicative helicase is only a small fraction in comparison to the abundance of Mcm2 to Mcm7 and the replicative helicase might require additional activities for its activation/formation, discrimination between the full MCM complex and complexes sharing subunits with the MCM complex is hard to achieve in cell-based biochemical systems. Moreover, during purification auxiliary factors required for enzyme activity might easily be lost. This leaves the speculation open that the MCM complex might have additional essential functions to support eukaryotic DNA replication whereas Cdc45 and putatively other components cooperate to support the MCM complex or the Mcm4-6-7 complex to act as DNA helicase at the replication fork.

2.2 New Kids on the Block – Mcm8 and Mcm9 and Their Evolving New DNA Replication Functions

Mcm8 and Mcm9 recently emerged as new proteins of the Mcm family (Blanton et al. 2005; Gozuacik et al. 2003; Lutzmann et al. 2005; Maiorano et al. 2005a; Matsubayashi et al. 2003; Volkening et al. 2005; Yoshida 2005). Mcm8 and Mcm9 have a conserved Mcm domain and a zinc finger motif but are otherwise distinct from Mcm2-7. Despite their broad phylogenetic distribution, the Mcm8 and Mcm9 genes are missing in several eukaryotic lineages such as most fungi and the nematode *C. elegans* (Caenorhabditis elegans) whereas the genes coding for Mcm2-7 appear to be present in all eukaryotes analyzed so far (Blanton et al. 2005). In contrast to Mcm9, which has only been studied by sequence analyses, the function of Mcm8 has been investigated biochemically in humans and *Xenopus laevis* as well as genetically in *Drosophila melanogaster*. Human Mcm8 is a nuclear protein associated with chromatin in early S phase and depletion by RNA interference causes a delayed S phase (Gozuacik et al. 2003; Volkening et al. 2005). In *Xenopus* the chromatin association of Mcm8 is similar as that of Cdc45 and Mcm8 has DNA unwinding activity (Maiorano et al. 2005a). The depletion of Mcm8 allowed a normal initiation but a very slow elongation reaction (Maiorano et al. 2005a). These data suggest that the protein is possibly involved in the regulation of the replication fork movement. In contrast, the *rec* (recombination defective) gene, which is required for meiosis, codes for *Drosophila* Mcm8. Homozygous female *Drosophila* with null alleles of the *rec/Mcm8* gene produce offspring, which have high levels of chromosome disjunction, but which show no defects in proliferation or DNA replication (Blanton et al. 2005; Matsubayashi et al. 2003). Therefore, the nature of the Mcm8 function differs considerably depending on the eukaryotic organism, or Mcm8 has auxiliary functions in meiosis and DNA replication. The latter possibility appears more likely, considering the repeated losses of the MCM8 gene in several lineages during eukaryotic evolution (Blanton et al. 2005).

2.2.1
The Mcm10 Protein

The replication protein Mcm10 does not show structural similarity to the Mcm2-7 proteins (Forsburg 2004; Tye 1999). Mcm10 is needed for an efficient initiation of DNA replication, and it shows genetic and biochemical interactions with ORC (Fig. 2 and Table 1; Forsburg 2004). *Xenopus* and human Mcm10 are loaded together with the MCM complex onto chromatin. Moreover, Cdc45 and RPA can only be loaded onto chromatin after Mcm10 is already associated with chromatin (Forsburg 2004), and the protein is probably involved in the release of origin-bound factors (Forsburg 2004). In various eukaryotic organisms Mcm10 and the MCM complex interact with each other (Forsburg 2004). These interactions are necessary for Mcm10 phosphorylation, which in turn is apparently necessary for the activation of the MCM complex (Forsburg 2004). Recently it was determined that Mcm10 controls the stability and the chromatin association of DNA polymerase α (Pol α, Fig. 2; Ricke et al. 2004).

2.2.2
Transformation of the Pre-replication Complex to the Initiation Complex

The assembly of DNA replication proteins at replication origins and the activity of the initiation reaction are tightly regulated by the activity of at least two different kinases: Cdc7 and Cdks (Table 1; for review Bell et al. 2002; Diffley 2001; Diffley et al. 2002; Nasheuer et al. 2002). These kinases act at specific steps during DNA replication e.g., at the transition of pre-RC to the initiation complex (IC) and the elongation phase during S phase. One phosphorylation target of Cdks is Cdc6 as discussed. Phosphorylation of the MCM complex by Cdc7-Dbf4 kinase (DDK, Dbf4-dependent kinase) probably leads to structural changes within the complex, which might lead to the formation of the Mcm4,6,7 complex and the activation of its DNA helicase activity. Moreover, the phosphorylation of the MCM complex is required for the loading of Cdc45 onto chromatin. As a consequence of the IC formation and the replication start Cdt1 and Cdc6 diminish from the chromatin. The initiation complex consists at least of the replicative DNA helicase, comprising the MCM complex or the Mcm4,6,7 complex, Cdc45, the Sld (synthetically lethal with *DPB11* {DPB, DNA polymerase epsilon subunit B}) proteins, RPA, Dpb11/Cut5 (cut, cut phenotype, mutation causing cytokinesis in the absence of normal nuclear division in *S. pombe*), GINS, DNA polymerases α and ε (Fig. 2). The activity of topoisomerase I is required to release the torsional stress in front of the replication fork introduced during the unwinding reaction.[3]

2.2.3 Cyclin-dependent Kinases

Cyclin-dependent kinases (Cdks) are serine/threonine kinases, which are essential for the control of the individual cell cycle phases (Diffley 2001; Diffley et al. 2002; Sherr et al. 2004). They are activated and deactivated in a cell cycle-dependent manner (Sherr et al. 2004). Cdk represents the catalytic subunit, which is associated with an unstable positive regulatory subunit called cyclin. At their *N*-terminus cyclins possess a specific sequence (destruction box), which is the acceptor site for ubiquitin marking the cyclin for cell cycle-dependent proteolysis (Sherr et al. 2004). In budding and fission yeast, only one Cdk gene exists, and the expressed protein associates with specific cyclins. In higher eukaryotes a family of genes code for Cdks, which are expressed throughout the whole cell cycle. However, these Cdks associate in a cell cycle-dependent manner with cyclins e.g., Cdk2-cyclinA and Cdk2-cyclinE are necessary for progression into S phase. The Cdk activity is regulated by different mechanisms. Cdks require the association of a cyclin for activity. Moreover, conserved amino acids in the *N*-terminus of Cdks must be dephosphorylated, and a threonine in the active centre must be phosphorylated. On the other hand, the interaction of the Cdks with Cdk inhibitors (CKIs) inactivates the kinase complex (Sherr et al. 2004). ORC, Cdc6 and DDK regulate the recruitment of Cdks to origins (Sherr et al. 2004). Immunoprecipitations revealed interactions between Cdc6 and Cdks as well as Pol α and Cdks (Petersen et al. 1999; Schub et al. 2001).

2.2.4 Cdc7 Kinase

Cdc7-Dbf4 kinase is also called DDK. Cdc7 is a serine/threonine kinase, which is conserved from yeast to mammals (for review Masai et al. 2005). In vertebrates the protein concentration is constant over the cell cycle, but the kinase activity is highest at the transition from G1 to S phase (Jares et al. 2000). The activity of Cdc7 depends on the association of its regulatory sub-

[3] A more detailed discussion of the enzyme is presented by Søe et al. in this book.

unit Dbf4. In vertebrates an additional, Dbf4-related factor Drf1 was recently found to associate with and activate Cdc7 (Montagnoli et al. 2002; Takahashi et al. 2005). In *Xenopus* egg extracts, Cdc7-Drf1 levels exceed Cdc7-Dbf4 levels, and removal of Drf1 but not Dbf4 severely inhibits phosphorylation of Mcm4 and DNA replication (Takahashi et al. 2005). Dbf4 levels, and thus the cell cycle-dependent functions of DDK, have their maximum at the G1/S transition, which are controlled on the level of the gene expression and proteolysis (Masai et al. 2005). The Dbf4 chromatin association in *S. cerevisiae* is ORC-dependent but independent of Cdc6 and the MCM complex (Masai et al. 2005). In contrast, a dependence of Dbf4 chromatin loading on the MCM complex and an independence of ORC and Cdc6 was observed in *Xenopus laevis* (Jares et al. 2000; Walter 2000). The MCM complex is regarded as the primary substrate of DDK. Mcm2, 3, 4, 6 and 7 but not Mcm5 can be phosphorylated with Mcm2 being a preferential substrate of the DDK in vitro (Masai et al. 2005). In yeast and *Xenopus* the temporal kinase activity at specific origins strongly suggests that the DDK complex acts individually at particular origins after pre-RC formation and that it is involved in the activation of Cdc45 (Nasheuer et al. 2002). Moreover, a mutant of Mcm5, though not itself a substrate of DDK, in *S. cerevisiae* by-passes the requirements of DDK and emphasizes the central function to regulate the MCM complex by DDK (Lei et al. 2001).

2.3
Establishing the Replication Fork

2.3.1
Cdc45 and the Initiation Step

The replication factor Cdc45 is not a component of the pre-RC, but it is essential during the pre-RC to IC transition and in the elongation phase of DNA replication (Table 1, Figs. 2 and 3; reviewed in Bell et al. 2002; Diffley 2001, 2002; Nasheuer et al. 2002). Cdc45 assembles with origins in a temporal order, which corresponds with origin activation (Vogelauer et al. 2002). To fulfil its function Cdc45 participates in a complex multi-protein network e.g., in budding yeast the subunits of the MCM complex and Cdc45 cooperate. Expression of mutant forms of *cdc46* (*mcm5*) and *cdc47* (*mcm7*) rescues a *cdc45* mutant (Nasheuer et al. 2002). The Cdc45 protein interacts with the Mcm7 protein whereas the binding of Cdc45 to Mcm2 depends on the Cdc7-Dbf4 kinase and only exists in late G1 and S phase (Bell et al. 2002; Diffley 2001; Diffley et al. 2002; Nasheuer et al. 2002). The functional cooperation of Cdc45 and the MCM complex is highly conserved, but the interactions of Cdc45 with other replication factors seem to be not fully conserved in eukaryotes. For instance, formation of Cdc45-RPA and Cdc45-DNA polymerase ε (Pol ε) complexes can be detected from yeast to mammals. However, fis-

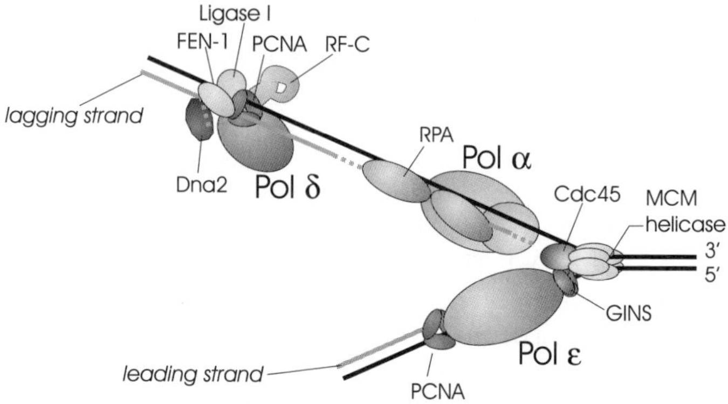

Fig. 3 Protein assembling to perform leading and lagging strand DNA synthesis. DNA synthesis of the leading strand and the lagging strand require different sets of proteins. Abbreviations: Cdc, cell division cycle; FEN-1, flap endonuclease; GINS, Go, Ichi, Nii, and San (japanese for five, one, two and three; the protein complex consists of Sld5 and Psf1-3); Mcm, minichromosome maintenance; ORC, origin recognition complex; PCNA, proliferating cell nuclear antigen; Pol, DNA polymerase, RPA, replication protein A; RF-C, replication factor C

sion yeast and *Xenopus* Cdc45 and Pol α physically interact with each other, whereas they do not associate in *Drosophila* and budding yeast (Nasheuer et al. 2002).

The finding that the same *mcm* mutants suppress mutant *cdc45* and *mcm10* suggests that both proteins target the same pathway and interact with the same regions of these MCM proteins. Since Mcm10 stabilizes the pre-RC in an early state, Cdc45 most likely activates the pre-RC complex. The sequential processes in *Xenopus* extracts are suggested as follows: MCM complex – Mcm10 – Cdc45 – RPA – Pol α (Walter et al. 2000; Wohlschlegel et al. 2000). This sequence of the protein associations requires the activity of the kinases Cdk and DDK (Chou et al. 2002; Walter 2000) and is followed by the activation of the replicative DNA helicase. Interestingly mouse Cdc45 exclusively interacts with Mcm3, 6 and 7 but not with Mcm2, 4 and 5 suggesting that these physical interaction might be involved in the activation of the helicase activity (Kneissl et al. 2003).

Cdc45 is also involved in the regulation of DNA replication after DNA is damaged. In *Xenopus laevis* DNA damage prevents the replication initiation by inhibiting Cdc45 loading (Luciani et al. 2004). This seems to be a conserved mechanism, since it was also observed in humans (Falck et al. 2002). Moreover, Cdc45 is essential for binding checkpoint proteins e.g., Claspin and Cut5, to origins and/or the replication forks (Costanzo et al. 2000). The presence of lesions blocking replication forks results in a Cdc45-dependent unwinding of DNA and an activation of Chk1 (checkpoint kinase 1), which starts processes to repair the lesion (Byun et al. 2005; Luciani et al. 2004).

The latter requires DNA polymerase and primase activity of Pol α (Byun et al. 2005; Michael et al. 2000).

2.3.2 Dpb11, GINS, and the Sld Family of Proteins

Araki et al. (Araki et al. 1995) identified the *DPB11* gene in budding yeast as a multicopy suppressor of mutations in the essential Pol ε subunits genes *POL2* (DNA polymerase epsilon catalytic subunit) and *DPB2* (DNA polymerase epsilon subunit B). *DPB11* was found to be homologous to the fission yeast Rad4/Cut5 protein (Rad, radiation-sensitive, various DNA repair proteins; cut, mutation causing cytokinesis in the absence of normal nuclear division in *S. pombe*; Table 1, Figs. 2 and 3). Both *DPB11* and *Cut5* are essential for cell viability and are required during DNA replication as well as cell cycle control (Araki et al. 1995; Saka et al. 1994). The proteins Dpb11 and Cut5 share a repetitive structure containing two pairs of BRCT (BRCA1 C-terminal) domains. Dpb11/cut5 orthologues, such as the human TopBP1 and Drosophila Mus101 (nitrogen mustard-sensitive), have also been identified and characterized in several metazoan eukaryotes (reviewed in Garcia et al. 2005). These homologues have acquired additional BRCT domains during evolution, which may be associated with additional functions. As their yeast counterparts, the metazoan proteins have been implicated in replication. Dpb11/Cut5 is required for the transition from the pre-replication to the pre-initiation complex. In particular, the loading of Cdc45, and subsequently of Pols α and ε depends on Dpb11/Cut5. Although the involvement of Dpb11/Cut5 in elongation reaction of DNA replication has been discussed in both yeasts, no experiment has confirmed that the protein moves with the replication fork in unperturbed cells. However, it has an important role in the stabilization or reinitiation of stalled DNA replication forks (Mäkiniemi et al. 2001; Parrilla-Castellar et al. 2003).

The *SLD* genes were discovered in a genetic screen for genes synthetically lethal with *DPB11* (SLD; Table 1, Figs. 2 and 3; Kamimura et al. 1998). *SLD1* encodes Dpb3, which is the third largest subunit of the Pol ε. *SLD6* is the same as the checkpoint factor *RAD53* whereas Sld2 is identical to Drc1 (DNA replication and checkpoint 1). Phosphorylation of the Cld2 protein facilitates interaction with Dpb11 and is essential for DNA replication in yeast (Kamimura et al. 1998 et al., 1998; Masumoto et al. 2002). A recent report suggests that metazoan RecQL4 protein that is mutated in Rothmund-Thomson and related syndromes may be the functional orthologue of Sld2 in animals, although homology is limited to the *N*-terminus (Sangrithi et al. 2005). Sld3 forms a complex with Cdc45 (Sld4) and is needed for the initiation of DNA replication (Kamimura et al. 2001; Nakajima et al. 2002). Sld3 is required for the association of yeast Cdc45 with the MCM complex, recruitment of Cdc45 to origins of DNA replication, and subsequent loading of RPA to the origin

(Kamimura et al. 2001; Nakajima et al. 2002). Sld5 forms the heterotetrameric, ring-shaped GINS (Go, Ichi, Nii, and San japanese for five, one, two and three; Sld5 and Psf1, Psf2 and Psf3 {Partner of Sld Five}; Table 1, Figs. 2 and 3) complex with three partner proteins (Kubota et al. 2003). In contrast to Sld2 and Sld3, the GINS complex is well conserved in higher eukaryotes (Takayama et al. 2003). Dependent on S phase Cdk activity the GINS complex binds to origins of DNA replication. Association of Dpb11/Cut5, Cdc45/Sld3 and GINS to origins appears to be mutually interdependent. The GINS complex apparently migrates with the replication fork and interacts with Cdc45 and the MCM complex (Table 1 and Fig. 3; Pacek et al. 2006).

2.3.3
Replication Protein A (RPA)

The ssDNA-binding protein RPA (replication protein A) is necessary for DNA replication, DNA repair, and DNA recombination, and has important functions in DNA damage signalling (Table 1, Figs. 2 and 3; for review Iftode et al. 1999; Nasheuer et al. 2002; Wold 1997; Zou et al. 2003). The protein complex consists of three subunits p70, p32 and p14 named according to their molecular masses of 70 kDa, 32 kDa and 14 kDa, respectively. RPA interacts with ssDNA via its four DNA binding domains DBD-A, DBD-B, DBD-C, and DBD-D. The DNA binding domains DBD-A to DBD-C are located on p70, whereas the central part of p32 comprises DBD-D. In addition to these four well established DNA binding domains, the RPA complex contains two related, structurally defined OB (oligonucleotide/oligosaccharide binding) fold domains, the *N*-terminus of p70 and one in p14. The *C*-terminal domain of p70 interacts with the p32, which also interacts with p14 (Iftode et al. 1999; Nasheuer et al. 2002; Wold 1997). Moreover, the *C*-terminus of p32 is involved in protein interactions and is required for the loading of RPA onto the origin of replication by SV40 T antigen (Arunkumar et al. 2005). For the effective coordination of its subunits and correct positioning on ssDNA, RPA requires the presence of all three subunits (Weisshart et al. 2004). These findings support the notion that the cooperation of all three RPA subunits is necessary for its functions in vivo (Wold 1997).

Recent studies have shown that RPA plays a central role in DNA damage signalling in human and yeast. RPA interacts with ATR via ATRIP (ATR Ataxia telangiectasia mutated and Rad3-related and ATR interacting protein, respectively) to signal the existence of stretches of ssDNA in the genome after DNA damage. Moreover, p32 is specifically phosphorylated during the cell cycle and after DNA damage suggesting that the RPA activities are regulated. Hyperphosphorylation of RPA causes a conformational change within the protein complex, which affects DBD-B resulting in a lower affinity of RPA for ssDNA (Liu et al. 2005). Recently, the negative charge of these phosphorylated residues was mimicked by the introduction of aspartates which interfered with the DNA unwinding activity of the RPA complex and the loading of mutant RPA onto chromatin. This change of the affinity of RPA to ssDNA is thought to impede the DNA replication function of RPA but should have no effect on its repair activities (Binz et al. 2004).

In the eukaryotic cell cycle, RPA is necessary for activation of the pre-replication complex to form the initiation complex and for the loading of Pol α complex to origins of replication. Investigations in *Xenopus* showed that Cdc45 is necessary to load RPA onto chromatin (Nasheuer et al. 2002). Moreover, direct interactions between Cdc45 and RPA p70 and p32 have been detected (Nasheuer et al. 2002). In mammals it was found that RPA interacts with ORC (Orc1, 2, 4 and 6), the MCM complex (specifically Mcm2, 4, 6 and 7), with the tumor suppressor protein p53, and Pol α (p180, p58 and p48; Kneissl et al. 2003; Nasheuer et al. 2002; Weisshart et al. 2000). To study the functions of RPA at replication forks, the SV40 DNA replication in vitro has extensively been used. In order to initiate primer synthesis, RPA, Pol α, and SV40 TAg work in concert at the SV40 origin of DNA replication. TAg interacts with the two largest subunits of RPA with the association of TAg to *C*-terminus of p32 necessary for the loading of RPA onto the origin of replication (Arunkumar et al. 2005).

2.3.4
DNA Polymerase α (Pol α)

DNA polymerase α (Pol α) is the only enzyme that can start DNA synthesis *de novo* (Table 1, Figs. 2 and 3; for a review see (Hübscher et al. 2000, 2002; Nasheuer et al. 2002, 2005) and Ramadan et al. in this book). To fulfill this task Pol α is associated with a DNA-dependent RNA polymerase – primase – synthesizing about 10 nucleotides of RNA. Then the DNA polymerase activity of the enzyme complex extends this primer to approx. 40 nucleotides and an RNA-DNA primer is formed. This process takes place at the replication origin of each leading strand and during the synthesis of each Okazaki fragment on the lagging strand. Pol α has a low processivity since it dissociates after the synthesis of each RNA-DNA primer (Hübscher et al. 2000, 2002). DNA polymerase δ and/or ε extend the RNA-DNA primers by synthesizing long stretches of deoxynucleotides.

Pol α consists of four subunits with apparent molecular masses of 180, 68 to 90, 58, and 48 to 50 kDa, which are highly conserved between all eukaryotes and which are all essential for DNA replication (Hübscher et al. 2000, 2002). The two smallest subunits, p58 and p48, form the primase. The largest subunit of Pol α, p180, carries the DNA polymerase activity. The second largest subunit, the B subunit, and p180 are phosphorylated in a cell cycle-dependent manner and have most likely regulatory functions (Hübscher et al. 2000, 2002). The ability of primase to synthesize RNA-DNA primer is the central function of Pol α. The subunit p48 carries the catalytic center of the primase which polymerizes the RNA oligonucleotide whereas no catalytic activities have been found to be associated with p58 (Arezi et al. 2000; Augustin et al. 2001; Nasheuer et al. 2002). However, in contrast to p48, p58 contains a nuclear localization sequence. Therefore, p58 is necessary for the translocation of the primase into the nucleus. Moreover, it stabilizes the enzymatic activity of p48 and controls the length of the primase products (Zerbe et al. 2002). The largest subunit of Pol α, p180, elongates the RNA primer and synthesizes a short stretch of DNA. The central part of p180 contains seven sequence motifs conserved between eukaryotic Pol α and related DNA polymerases of family B. They are required for phosphoryl transfer, Mg^{2+}, DNA, and dNTP binding activities (Nasheuer et al. 2002). The *C*-terminus is needed for the assembly of the heterotetrameric complex (Mizuno et al. 1999; Smith et al. 2002). In contrast to Pols δ and ε, Pol α does not possess $3'$ to $5'$-exonuclease activity, which is needed for proof-reading of newly synthesized DNA. It has been proposed that the nuclease activities of the interacting tumor suppressor p53, Pol δ or DNase III may proofread DNA synthesis errors of Pol α (Melle et al. 2002; Nasheuer et al. 2005).

The coordination of leading and lagging strand DNA synthesis requires the cooperation of multiple proteins. To this end, Pol α binds various cellular and viral proteins such as RPA, Pol δ, poly(ADP-ribose)polymerase (PARP), SV40 TAg, and HPV protein E2. In yeast recent results revealed that Mcm10 is responsible for an association of Pol α with chromatin (Ricke et al. 2004). In *Xenopus* biochemical analyses have shown that primase and DNA polymerase activities of Pol α are required to activate the intra-S checkpoint after DNA damage (Byun et al. 2005; Michael et al. 2000). Pol α is also involved in the maintenance of the telomere length probably via its interaction with the telomere associated proteins Cdc13 and Stn1 (Chandra et al. 2001; Grossi et al. 2004; Nasheuer et al. 2002)

2.4 From the Initiation Reaction to Processive DNA Replication

2.4.1 Proliferating Cell Nuclear Antigen (PCNA)

Proliferating cell nuclear antigen (PCNA) was initially discovered as a protein that is induced after serum stimulation and located in the nucleus (Table 1 and Fig. 3; Takasaki et al. 1981). Later it was shown to be an activator of Pol δ, which is required to replicate SV40 DNA in vitro and which is essential for cellular DNA replication (reviewed in Burgers 1998). Homotrimeric PCNA is composed of two-domain 29 kDa subunits and is loaded onto DNA by replication factor C (RF-C) (reviewed by Majka et al. 2004). The crystal structures of *S. cerevisiae* and human PCNA revealed that PCNA forms a closed ring able to encircle dsDNA and slides along it (sliding clamp), but does not have direct contacts to DNA (Gulbis et al. 1996; Krishna et al. 1994). In *E. coli* the structural and functional homologue of PCNA is the dimeric β-clamp (reviewed in Maga et al. 2003; Majka et al. 2004). Both have essentially the same three-dimensional shape and chain fold despite lacking sequence similarity.

PCNA is involved in DNA replication, DNA repair, cell cycle control, and chromatin remodeling (Maga et al. 2003). In addition to replication proteins such as RF-C, FEN-1 (flap endonuclease), DNA ligase I, Pols δ and ε, PCNA interacts with more than 30 other proteins including Pol η and the helicase Srs2 (suppressor of radiation-sensitive mutations) (Maga et al. 2003; Majka et al. 2004; Papouli et al. 2005; Pfander et al. 2005; and Ramadan et al. in this book). Interestingly, most of these proteins bind to the front side of the outer surface of the PCNA ring facing the replication primer, whereas Pol ε was shown to interact with the backside of PCNA. Simultaneous accommodation of several proteins would be possible if PCNA rings form back-to-back doublets, which has been shown to allow simultaneous binding of chromatin assembly factor CAF-1 and Pol δ (Naryzhny et al. 2005). Recently it was reported that PCNA interacts with Cdt1 and acts as a molecular switch to trigger Cdt1 proteolysis to prevent re-replication (Arias et al. 2006).

2.4.2
PCNA as a an Accessory Protein of DNA Polymerases

PCNA and RF-C provide the moving platform for Pol δ (reviewed by Hübscher et al. 2000, 2002; Nasheuer et al. 2005). PCNA associates with Pol δ at the primer-template junction and supports processive DNA synthesis, and at the same time prevents non-productive binding of Pol δ to ssDNA. Pol δ in conjunction with PCNA and RF-C is, therefore, generally referred to as the Pol δ holoenzyme (Burgers 1998). Like Pol δ, Pol ε forms a highly processive holoenzyme complex with PCNA and RF-C under physiological conditions (Burgers 1998; Hübscher et al. 2000, 2002; Nasheuer et al. 2005). However, in contrast to Pol δ, where the interaction with the interdomain loop appears to be most important, Pol ε interacts with the front side of PCNA including its *C*-terminus. A second interaction with the back is important for PCNA stimulation of the primer binding by Pol ε. Recently PCNA was described to be a co-factor of the so-called translesion DNA polymerases and Pol λ (for more details see below and the reviews Ulrich 2005 as well as Rudolph et al. and Ramadan et al. in this book).

2.4.3
PCNA Modification and Regulation

The regulation of this central factor of eukaryotic DNA metabolism only recently started to emerge when posttranslational modifications of PCNA by ubiquitin and the small ubiquitin-related modifier protein, SUMO, were detected. These modifications modulate the coordination of DNA replication, damage tolerance, and mutagenesis (for additional discussions see Ulrich 2005 and Rudolph et al. in this book). In short, after DNA damage as recently described in yeast and human cells, PCNA is ubiquitinated at the conserved lysine residue 164 in a Rad18-dependent manner (Hoege et al. 2002; Stelter et al. 2003). In human cells this ubiquitin modification is necessary for the association of PCNA with Pol η. It allows the bypass of replication-blocking lesions by the damage-tolerant Pol η in human, and Pols η as well as ζ in yeast (Garg et al. 2005; Kannouche et al. 2004; Nasheuer et al. 2005). Alternatively, even in the absence of DNA damage, PCNA is modified by SUMO during S phase. In yeast this SUMOylation is involved in the recruitment of the DNA helicase Srs2 to replication forks and regulates its physical association with the helicase, which suppresses unscheduled homologous recombination (Papouli et al. 2005; Pfander et al. 2005). This process can be summarized in a "typewriter wheel" model, in which depending on its modification PCNA serves as the landing platform and clamp for various proteins to counteract genome instability (Kannouche et al. 2004). Like the wheels of a typewriter, the various DNA replication and DNA repair proteins interact with SUMOylated and ubiquitinated PCNA, respectively, depending whether the DNA template is undamaged or damaged and duplicate chromosomal DNA or suppress homologous recombination of undamaged DNA.

2.4.4
Replication Factor C (RF-C)

Replication factor C (RF-C) loads the DNA polymerase clamp PCNA onto DNA. It is composed of a large 140 kD subunit (RFC1) and four smaller subunits with molecular masses of 37 kD (RFC2), 36 kD (RFC3), 40 kD (RFC4) and 38 kD (RFC5) (Table 1 and Fig. 3; reviewed in Bowman et al. 2005; Majka et al. 2004). RF-C is phylogenetically conserved and homologues include *E. coli* DNA polymerase III γ complex. All five subunits of RF-C are members of the AAA^+ family of ATPases and are homologous to one another.

Biochemical studies with yeast proteins and three-dimensional structures of clamp loaders from *E. coli* (γ complex and various sub-assemblies, Jeruzalmi et al. 2001a,b), *Pyrococcus furiosus* (RF-C/PCNA/DNA complex; Miyata et al. 2004), and *S. cerevisiae* (RF-C/PCNA complex; Bowman et al. 2004) have led to a model for the clamp loading. These studies might also reveal some insights in the loading of the MCM complex by ORC-Cdc6 and the function of the various RF-C-related complexes in the cells. Each subunit of RF-C has a conserved arginine residue positioned to participate in the hydrolysis of ATP bound by the neighbouring subunit. Thus, ATPase modules are organized so that the nucleotide binding sites are located at the interphases between subunits. The clamp loader assembly is held together by a circular collar formed by the *C*-terminal helical domains of the subunits into which ATPase domains are connected by flexible linkages. This organization allows the ATPase domains to alter their relative orientations and interfacial interactions in response to ATP binding and hydrolysis without disrupting the assembly of the RF-C complex. The arrangement of the subunits of RF-C is most likely RFC5 : RFC2 : RFC3 : RFC4 : RFC1, where RFC5 contributes an arginine finger to ATP in RFC2, etc. RFC5 and RFC1 come close to each other so that the ATPase domains of the subunits from RFC5 to RFC1 form a circle. It is likely that global conformational changes of RF-C are directly coupled to the binding, hydrolysis and release of nucleotide. Binding of ATP converts RF-C into a more open form, and it is able to bind on top of the PCNA ring. A conformation change in RF-C leads to opening of the PCNA ring, which in turn can then encircle the DNA. RFC1 and RFC3 mediate tight contacts of RF-C to PCNA. ATP-bound ATPase domains form a spiral arrangement above the PCNA ring matching closely the pitch of double helical DNA, whereas hydrolysis of ATP results in the release of RF-C from the clamp loaded onto DNA. The sliding clamp then recruits the replicative DNA polymerases δ and ε as well as additional proteins involved in eukaryotic DNA metabolism to DNA.

2.5
DNA Polymerase δ (Pol δ)

Mammalian Pol δ was originally identified as a DNA polymerase capable of proofreading (Lee et al. 1981). Human enzyme consists of the catalytic 125 kDa and three smaller subunits of 50, 68 and 12 kDa (subunits A, B, C and D, respectively) (Table 1 and Fig. 3; reviewed by Garg and Burgers 2005; Hübscher et al. 2002; Nasheuer et al. 2005; and additional discussions in Ramadan et al. in this book). Pol δ from *S. pombe* has a similar subunit structure, but *S. cerevisiae* enzyme may be devoid of the D subunit (Garg and Burgers 2005; Hübscher et al. 2002). Subunits A and B are highly conserved between eukaryotes. Besides DNA polymerase activity, the A subunit contains the proofreading 3′ to 5′ exonuclease activity (Garg and Burgers 2005; Hübscher et al. 2002). Subunit B is bound to the subunit A through the zinc-finger module of the latter. The B subunit does not exhibit any catalytic activity but is believed to be important for the stability of the enzyme. It belongs to the superfamily of replicative DNA polymerase B subunits conserved from archaea to human (Mäkiniemi et al. 1999). A conserved calcineurin-like phosphoesterase domain and an OB fold domain have been identified in B subunits (Aravind et al. 1998; Koonin et al. 2000).

The subunit B of Pol δ acts as a bridge between the subunit A and less conserved subunit C, which contains in its conserved regions at the *C*-terminus a consensus PCNA binding site (Gerik et al. 1998; Hughes et al. 1999; MacNeill et al. 1996; Zuo et al. 1997). Deletion of this interaction domain impairs the processivity of Pol δ in certain conditions in vitro and leads to growth defects in *S. pombe* (Bermudez et al. 2002). The consequences of these mutations are less severe in *S. cerevisiae* (Johansson et al. 2004). The two-subunit A/B form of Pol δ is strongly stimulated by PCNA, and a high affinity interaction between the two-subunit form of Pol δ and PCNA was found (Mozzherin et al. 1999). Direct interaction of PCNA with both A and B subunit has also been reported (Lu et al. 2002; Zhang et al. 1995). It seems that replicative form of Pol δ has several physical contacts with PCNA through its subunits. Besides PCNA, the C subunit of Pol δ interacts with the catalytic subunit of Pol α (Gray et al. 2004; Huang et al. 1999a). The small subunit D of human enzyme functions to stabilize the Pol δ complex (Podust et al. 2002). The elongated shape of the Pol δ complex has led to speculation that Pol δ itself associates into a dimer (Burgers et al. 1998). Evidence obtained later indicates, however, that Pol δ contains only one copy of each subunit (Bermudez et al. 2002; Johansson et al. 2001).

2.6
DNA Polymerase ε (Pol ε)

Pol ε was discovered as early as 1970 from yeast (Wintersberger et al. 1970). Human full size protein was first purified and characterized from HeLa cells (Nishida et al. 1988; Syväoja et al. 1989). The catalytic subunit A contains both DNA polymerase and proofreading 3′ to 5′ exonuclease activity, and notably, a large *C*-terminal domain of unclear function that accounts for nearly half of the large molecular weight of $\sim$ 260 kDa (Kesti et al. 1993; Morrison et al. 1990). Apart from the catalytic subunit, Pol ε contains three additional subunits, which are conserved in their primary structure from yeast to human (Table 1, Figs. 2 and 3; reviewed in (Pospiech and Syväoja, 2003) and Ramadan et al. in this book). Whereas the B subunit of 60–86 kDa is also essential in yeast the *DPB3* and *DPB4* genes code for the two smallest, nonessential subunits of budding yeast Pol ε. The latter possesses a histone-fold and forms a stable heterodimer, which interacts with the two larger subunits. The heterodimer consisting of Dpb3 and Dpb4 has affinity to dsDNA as discussed below (Tsubota et al. 2003). Moreover, Dpb4 is also a component of the chromatin remodeling complex CHRAC (Pospiech and Syväoja, 2003).

Unlike Pol δ, Pol ε binds with a high affinity both to ssDNA and dsDNA (Tsubota et al. 2003). Biophysical studies on overproduced *S. cerevisiae* enzymes indicate that Pol ε is a heterotetramer with no indication of dimerization (Chilkova et al. 2003), though evidence from yeast two-hybrid studies suggests that dimerization may occur via the *C*-terminus of the catalytic subunit (Dua et al. 1999). The globular A subunit contains a cleft that could accommodate dsDNA, and it is flexibly connected to an extended structure formed by the three smaller subunits (Asturias et al. 2006). This tail domain formed by the three small subunits could facilitate the binding of Pol ε to DNA and explain the intrinsic PCNA-independent processivity of Pol ε (Hamatake et al. 1990; Syväoja et al. 1990). Pol ε does not require PCNA to be highly active. However, PCNA reverses the inhibition of Pol ε activity by high salt concentrations (Lee et al. 1991). The PCNA interaction motif identified in the catalytic A subunit may not be necessary for DNA replication but rather for DNA repair (Dua et al. 2002).

Unexpectedly, the *C*-terminal domain of subunit A is sufficient for cell viability in yeast, whereas the DNA polymerase and exonuclease domains are dispensable (Kesti et al. 1999). However, mutant cells expressing only the *C*-terminus of subunit A suffer from severe defects including those to efficiently progress through DNA replication. It has also been shown that in healthy human cells Pol ε is responsible for the bulk of DNA synthesis (Pospiech et al. 1999; Zlotkin et al. 1996). It is obvious that in mutant yeast cells Pol δ is able to substitute Pol ε in its DNA synthesis role to the extent that the cells remain viable. The essential *C*-terminal domain of subunit A may serve a structural role in initiation and/or elongation complex. The *C*-terminus contains a zinc finger region that is particularly important, since several mutants in this domain are defective in growth and in response to DNA damages.

2.7 Flap Endonuclease 1 (FEN-1)

A flap endonuclease 1 (FEN-1) homologue was initially discovered in 1968 as a 5′ to 3′ exonuclease associated with *E. coli* DNA polymerase I that also possessed endonuclease activity, which is able to excise mismatched sequences and hydrolyze distorted regions from duplex DNA (Kelly et al. 1969). Later on the calf enzyme was found to interact functionally with Pol ε and cleaved a downstream DNA fragment in a length corresponding to the number of nucleotides incorporated at the upstream primer end (Table 1 and Fig. 3; reviewed by Liu et al. 2004). Harrington and Lieber (Harrington et al. 1994a; Harrington et al. 1994b) purified and cloned the enzyme from a mouse and found that the enzyme actually cleaves a DNA flap, a 5′-displaced single-stranded DNA from a duplex DNA, and named the enzyme FEN-1. Deletion of both copies of mouse FEN-1 genes leads to embryonic lethality and haplo-insufficiency may promote tumor progression (Liu et al. 2004). FEN-1 null blastocysts fail to maintain normal DNA replication and repair. The corresponding protein is encoded by *Rad27* gene in *S. cerevisiae* and by *rad2* in *S. pombe*. The genes in both yeasts are necessary to maintain normal cell growth rates, but are not essential for survival. The phenotypes of the mutant cells are typical for the cells having replication defects.

Eukaryotic FEN-1 consists of an *N*-terminal, an intermediate and a *C*-terminal domain (Harrington et al. 1994a). The motifs for substrate binding and catalysis reside within the *N*-terminal and intermediate domains, whereas the *C*-terminal domain is responsible for interaction with other proteins (Liu et al. 2004). A crystal structure of eukaryotic FEN-1 has not been defined but structures of homologues from bacteriophage, eubacteria, and archaebacteria have been solved. *Methanococcus jannaschii* and *Pyrococcus furiosus* FEN-1 have a preformed loop or helical clamp, respectively, above the catalytic site (Hosfield et al. 1998; Hwang et al. 1998). The structure and mutational information indicate interactions for the single- and double-stranded portions of the flap DNA substrate and suggest that DNA binding induces FEN-1 to clamp onto the cleavage junction to form the productive complex. FEN-1 loads from the 5′ end of the flap and finds the junction between the flap and downstream duplex DNA (Liu et al. 2004). Although the preformed loop would allow traversing of FEN-1 along the flap by a threading mechanism, a tracking mechanism is more likely since the enzyme is able to accommodate bulky modifications of the flap (Liu et al. 2004). This hypothesis is supported by the finding that during the catalytic process the enzyme undergoes a substrate-induced conformational change (Hosfield et al. 1998; Liu et al. 2004).

2.8 Dna2

The *DNA2* gene was identified as a DNA replication-defective mutant of *S. cerevisiae* (Table 1 and Fig. 3; Budd et al. 1995). The *S. cerevisiae* gene encodes an essential protein with a molecular mass of 172 kDa that possesses DNA-dependent ATPase, DNA helicase, and ssDNA–specific endonuclease activity (for review, Kao et al. 2003). While being absolutely essential in *S. cerevisiae* and *S. pombe*, a null mutation of *Caenorhabditis elegans* allows survival of some adults to F2 generation (Budd et al. 1995; Kang et al. 2000; Lee et al. 2003), suggesting that the Dna2 protein is essential in single cell organisms but not in metazoan organisms. The situation may be reversed regarding the requirement of FEN-1, the deletion of which results in embryonic lethality in mice; whereas, it is not essential in both yeasts. Dna2 protein interacts with FEN-1 both genetically and physically (Budd et al. 1997) and with RPA genetically and biochemically (Bae et al. 2001a) on flap substrates. Genetic interactions of Dna2 with subunits of Pol δ and DNA ligase I have also been found in *S. pombe* (Kang et al. 2000). The nature of its nuclease activity and that overproduction of FEN-1 suppresses the deletion of *Dna2* indicates that Dna2 is involved in flap removal (Bae et al. 2001a). An RPA-bound flap intermediate stimulates Dna2 activity but inhibits FEN-1 (Bae et al. 2001a; Murante et al. 1995). In summary, Dna2 is able to shorten long RPA-bound flaps that could be created by extensive strand displacement synthesis by Pol δ (Hübscher et al. 2002). Short flaps remaining could no longer bind RPA and would be cleaved by FEN-1 to allow ligation (Kao et al. 2004a; Kao et al. 2004b). Besides long flaps, Dna2 is also needed to cleave flaps containing fold-backs or repeat sequences (Kao et al. 2004a; Kao et al. 2004b). The characteristics of Dna2 suggest a supplementary function for the synthesis of the lagging strand as discussed below.

2.9 RNase H

RNases H enzymes are present in all living organisms and are distinguishable from other RNases since they hydrolyze RNA only when it is annealed to a complementary DNA (Kornberg et al. 1992). Most organisms have more than one type of RNase H. Based on amino acid sequence similarity eukaryotic RNases are classified into two distinct groups, RNase H1 and RNase H2 (Liu et al. 2004). There is no amino acid sequence homology between the two groups but they have similar three-dimensional structures and catalytic sites suggesting that they also have similar catalytic mechanisms (Liu et al. 2004). In *S. cerevisiae* RNase H2 exhibits increased activity when the gene encoding RNase H1 is deleted, probably to compensate for loss of RNase H1 activity (Arudchandran et al. 2000). This suggests that the cellular functions of the two enzymes are at least partially overlapping. Most RNase H enzymes consist of only a single polypeptide, but *S. cerevisiae* RNase H2 is composed of three distinct polypeptides (Jeong et al. 2004). Inactivation of both RNase H-functions results in only minor defects in DNA damage response in *S. cerevisiae* while deletion of the RNase H1-encoding gene results in embryonic lethality in mouse, and *Drosophila melanogaster* (Arudchandran et al. 2000; Cerritelli et al. 2003; Filippov et al. 2001). The embryonic lethality is due to inability to replicate mitochondrial rather than nuclear DNA. Although the cellular role of RNase H2 is not yet well defined, it is thought to be involved in removal of RNA primers from Okazaki fragments and misincorporated single ribonucleotides from duplex DNA (Liu et al. 2004). FEN-1 and Dna2 are clearly involved in removal of RNA primers during lagging strand synthesis, whereas RNase H2 may only aid in this process (Ayyagari et al. 2003; Bae et al. 2002).

2.10 DNA Ligase I

DNA ligases are needed for DNA replication, DNA recombination and for DNA repair to carry out the last step, sealing the nick between 3′-hydroxyl and 5′ phosphate termini. It was found as early as late 1960s that DNA ligation consists of a series of three reactions and two covalent intermediates: an enzyme-adenylate and a DNA-adenylate (Lehman 1974). DNA ligases fall in two subfamilies. Most eubacterial enzymes utilize NAD^+ as an energy source, while most eukaryotic, archaeal and bacteriophage DNA ligases utilize ATP. Vertebrate cells contain three DNA ligases – DNA ligases I, III and IV (reviewed in Martin et al. 2002; Nash et al. 1996; Timson et al. 2000). DNA ligases I and IV are conserved in all eukaryotes, with DNA ligase I being involved in chromosomal DNA replication (Table 1 and Fig. 3). In budding yeast a distinct mitochondrial form of DNA ligase I is encoded from an alternative start codon. This form of DNA ligase is involved in mitochondrial DNA replication and repair. Yeast cells harboring temperature-sensitive mutations in DNA ligase I are arrested in S-phase when shifted to the restrictive temperature, and accumulate Okazaki fragments. Unlike yeast DNA ligase I, the mammalian enzyme is not essential for viability of cultivated cells, but it is required for embryogenesis (Bentley et al. 1996, 2002).

Sequence comparisons, mutational analysis, and X-ray structures have revealed the structural and functional features of DNA ligases (Martin et al. 2002; Nash et al. 1996; Timson et al. 2000). All DNA ligases have a conserved adenylation domain (AdD) followed by an OB fold domain comprising the catalytic core of DNA ligases. These domains define a superfamily of covalent nucleotidyl-transferases. Upstream of the AdD domain all eukaryotic DNA ligases have an additional DNA binding domain. Nuclear DNA ligase I has a weakly conserved amino-terminal domain that contains nuclear localization signals and a PCNA binding motif. When binding to DNA, the DNA binding domain of DNA ligase I encircles DNA, stabilizes DNA in a distorted structure, and positions the catalytic core on the nick (Pascal et al. 2004).

3
Cooperation of Replication Factors at the DNA Replication Fork

Despite extensive research efforts and progress for several decades, several basic aspects of the eukaryotic nuclear DNA replication machinery have started to unravel only now during the recent years. Still numerous open questions require answers before the anatomy of the eukaryotic replication fork can be reconstructed (see Table 1 and Fig. 3). What is the exact nature of the replicative DNA helicase? How do the replicative Pols α, δ, and ε share the labor of DNA synthesis? How is the replication fork held together? Especially the questions of when and where different Pols synthesize DNA are a prerequisite to understand the whole replication machinery. Figure 3 presents a simplified view of our current understanding of the eukaryotic DNA replication fork.

Lagging strand DNA synthesis is the aspect of the eukaryotic replication machinery that is probably best understood (see Garg and Burgers 2005; Kao et al. 2003 for recent reviews). In the pioneering work using the SV40 DNA replication model, a switch from Pol α to Pol δ has been demonstrated as a central feature of the process (reviewed in Waga et al. 1998). The Pol α-primase complex initiates lagging strand DNA synthesis with an 8–12 nt RNA primer followed by DNA synthesis to extend the RNA-DNA primer to approximately 30–40 nt. This primer length corresponds well to the binding site of a single RPA molecule to single-stranded DNA (Wold 1997), and in fact, primer synthesis seems to be guided by displacement of RPA from the template stand (Mass et al. 2001). RPA appears therefore to play a crucial role in the coordination and regulation of the primer synthesis (Maga et al. 2001; Pestryakov et al. 2003; Yuzhakov et al. 1999).

Only a preformed RF-C-PCNA complex, but not RF-C alone, is capable of productively binding DNA and loading PCNA, suggesting that dissociation of Pol α is directly connected to the loading of PCNA (Garg and Burgers 2005). PCNA then serves as an operation platform for the following reactions at the lagging strand (reviewed in Maga et al. 2003). Pol δ is recruited by PCNA and will elongate the RNA-DNA primer until the Pol encounters a downstream Okazaki fragment. The cooperation of Pol α and Pol δ during lagging strand synthesis is also supported by the physical interaction between Pol α and Pol δ as well as by genetic data indicating that Pol δ, but not Pol ε may proofread for replication errors made by Pol α (Garg and Burgers 2005; Pavlov et al. 2006). Two competing models propose the template of the Okazaki fragments either to be covered by multiple RPA molecules or by continuously deposited RNA-DNA primers that are subsequently displaced or maturated to result in Okazaki fragments[4] (Garg and Burgers 2005; Kaufmann et al. 2004). It has been noted that the size of eukaryotic Okazaki fragments, 120–300 bp, corresponds well to the spacing of nucleosomes on DNA (Herman et al. 1981), indicating that lagging strand synthesis may be intimately linked to chromatin structure (reviewed by Kaufmann et al. 2004).

After reaching the primer end, the newly synthesized Okazaki fragment has to be joined with downstream Okazaki fragments to form a continuous DNA strand. Already reconstitution of SV40 DNA replication indicated that Pol δ cooperates with FEN-1 and RNase H to produce a ligatable nick for ligase I (Waga et al. 1998). Pol δ shows a strong coordination with FEN-1 to produce such a product (Garg and Burgers 2005). In the absence of FEN-1, Pol δ will only displace a few nucleotides of downstream DNA or RNA. Newly synthesized DNA will be degraded by the proofreading activity of the enzyme in a process called idling where no net DNA synthesis occurs. In the presence of FEN-1, Pol δ will switch to nick translation with predominantly mononucleotides released by FEN-1. In the presence of DNA ligase I, the nick will

[4] A good account of the two competing models is given by Salas et al. (1996). One model is called the "nested discontinuity" model and was first proposed by Gabriel Kaufmann (see Nethanel et al. 1992). Working with SV40 DNA replication, Nethanel et al. determined that short RNA-DNA primers accumulate prior to the appearance of Okazaki fragment sized products. These RNA-DNA primers form a closely spaced array on the template strand if maturation or ligation is prevented. Indirect support comes also from the fact that contrary to bacterial SSB (ssDNA binding protein) RPA does not seem to bind DNA cooperatively. The idea is that the lagging strand template is rapidly covered by short DNA first, and Okazaki fragments are more distributively assembled later on. This suggests that basically the whole lagging strand template is covered with short RNA/DNA pieces made by Pol α. Then most of these fragments are replaced in a displacement reaction by DNA synthesized by Pol δ, except those at the junctions of Okazaki fragments, which would mean that the lagging strand is basically synthesized twice. This is in contrast to the second more commonly considered model, which assumes synthesis of one Okazaki fragment at a time. Garg and Burgers discussed this issue also in their recent review (Garg and Burgers 2005). NB.: There is evidence for distributive Okazaki fragment synthesis in Archaea (Matsunaga et al. 2003).

be sealed rapidly, terminating the nick translation and completing Okazaki fragment maturation (Kao et al. 2003). Okazaki fragment maturation by Pol δ, FEN-1 and DNA ligase I is efficient in vitro, but the helicase and nuclease Dna2 is additionally required to complete Okazaki fragment maturation in vivo. It is proposed that Dna2 may counteract the formation of longer flaps. Flaps that are bound by RPA or other proteins, or flaps with secondary structure, which are poor substrate for FEN-1, require such a proposed backup mechanism with Dna2 nucleolytically removing flaps refractory to FEN-1 (Bae et al. 2001a; Bae et al. 2001b; Kao et al. 2004b).

Our understanding of the leading strand DNA synthesis apparatus is far less advanced, probably due to the lack of a convincing model for cellular leading strand DNA synthesis. Since Pol δ has been assigned to the lagging strand, it seems natural to place Pol ε, the second DNA polymerase implicated in synthesis of the bulk of DNA during replication, on the leading strand. This proposal was already made by Akio Sugino more than 10 years ago (Sugino 1995). Analysis of DNA replication in *Xenopus* egg extracts depleted of Pol δ or Pol ε seem to support the view that these DNA polymerases may operate on opposite arms of the replication fork (Fukui et al. 2004). Another line of support comes from genetic studies that demonstrate the presence of a strand bias in replication fidelity of proofreading-deficient Pols δ and ε yeast mutants (Shcherbakova et al. 2003). In fact, active origins establish a strand bias for replication-dependent mutagenesis in yeast, indicating an intrinsic bias for replication errors on the leading and lagging strand (Pavlov et al. 2002). Nevertheless, this strand bias of replication fidelity is not necessarily due to different DNA polymerases on the leading and the lagging strand. DNA checkpoint control and DNA repair[5] processes also contribute substantially to the strand-specific error bias (Pospiech and Syväoja 2003), and a similar bias appears to exist even in *E. coli*, where the same replicase operates on both strands (Fijalkowska et al. 1998). Moreover, it appears that DNA checkpoint control and DNA repair processes also contribute to the strand-specific error bias (Pospiech et al. 2003).

An argument against Pol ε as the leading strand replicase comes from the observation that at least yeast cells can survive without the enzymatic activity of this enzyme. Although Pol ε is essential for viability of both budding and fission yeast, its *C*-terminal checkpoint domain rather than its amino-proximal catalytic polymerase domain executes the essential function (Feng et al. 2001; Kesti et al. 1999; Morrison et al. 1990). Although cells lacking the catalytic domain of Pol ε have a severe cell growth phenotype, this is a surprising finding, considering that point mutants inactivating the DNA polymerase activity render the cells non-viable (Morrison et al. 1990). Moreover, several studies indicate that the catalytic activity of Pol ε seems to

[5] A strand bias in stalled replication forks or Synthesis Dependent Strand Annealing (SDSA), which is the mechanism of bypass repair, might be plausible as well (Lankenau and Gloor 1998; Rudolph et al. 2006).

participate in DNA replication in a number of eukaryotic models (Feng et al. 2001; Fukui et al. 2004; Pospiech et al. 1999; Zlotkin et al. 1996). This observation led to the hypothesis that the *C*-terminus of Pol ε participates as an essential component of the replication initiation complex (Garg and Burgers 2005). In addition, previous studies showed that Pol δ is sufficient to support cell-free SV40 DNA replication and that Pol ε is not necessary in this DNA replication system (Waga et al. 1998). Several studies implicate Pol ε in the initiation of replication (Aparicio et al. 1999; Aparicio et al. 1997; Masumoto et al. 2000; Mimura et al. 2000). Hiraga and coworkers (Hiraga et al. 2005) recently performed a ChIP-on-Chip approach to analyze the chromosome-wide association of Pols α, δ and ε with origin DNA during the cell cycle in yeast. Wild-type Pol ε and *pol2-16* lacking the catalytic domain both associate in early S phase with origins of replication. But whereas wild type Pol ε remains associated with DNA, the mutant is rapidly lost from the chromosomes, supporting the idea of a dual role of Pol ε during replication: An absolutely essential role during initiation that does not require DNA synthesis, and a role as replicase during elongation that can be circumvented in the absence of the catalytic domain of Pol ε.

Therefore, the apparent contradiction that DNA synthesis by an essential replicative Pol is not absolutely required may be explained by another paradox: The elongating Pol ε is loaded onto chromatin prior to and independent of RPA and the initiating Pol α (Masumoto et al. 2000; Mimura et al. 2000). In *S. cerevisiae*, the loading of Pol α actually requires Pol ε and Dpb11 (Masumoto et al. 2000). This arrangement may represent a safety mechanism guaranteeing assembly of the complete leading strand apparatus prior to initiation of DNA synthesis and also supports the potential role of Pol ε as the leading strand replicase. There is another important outcome of the study by Hiraga et al. (Hiraga et al. 2005). All three replicases associate with the same origins of replication during early S phase, indicating that all three enzymes cooperate at the same replication fork. This is not conceivable with the hypothesis that Pol ε may be responsible for the replication of heterochromatic DNA during late S phase only, as has been proposed for human Pol ε based on cell biological study (Fuss et al. 2002).

Very little is known about the additional factors involved in leading strand DNA synthesis (Table 1 and Fig. 3). As outlined above, MCM complex or a subset of it likely represents the replicative DNA helicase in eukaryotes. As reviewed by Cook (Cook 1999), the replicative DNA helicase and leading strand DNA polymerase are probably tethered together and immobilized by attachment to the nuclear matrix. Several such replication forks are concentrated to discrete sites in the nucleus called replication factories. This arrangement of DNA replication minimizes torsional stress on DNA and facilitates higher order regulation of multiple replication forks. CDC45 and the GINS complex may represent good candidates for tethering the leading strand replicase to the replicative DNA helicase. The requirement for attachment of the helicase may be the reason that has hampered the development of a functional cellular replication system in vitro. It may also represent a central difference to replication of SV40 DNA, the predominating model system for mammalian DNA replication. This mechanistic difference could at the same time provide an explanation why Pol ε is not required for replication of SV40 DNA.

One important question is how the three major replicases cooperate at the replication fork. There is no direct evidence for a dimer of replicases at the replication fork. Recent studies indicate that both Pol δ and Pol ε prevail as monomers after purification (Dua et al. 1999; Johansson et al. 2001). Except the previously mentioned non-essential interaction between Pol α and Pol δ (Gray et al. 2004; Huang et al. 1999b), there have been no reports on the direct or indirect dimerisation of eukaryotic replicases. This raises the possibility that in eukaryotes, the replication fork may not be arranged into a trombone structure familiar from the *E. coli* replisome, and the opposite polarity of the replicative DNA helicase in eukaryotes makes it well conceivable that the lagging strand synthesis may not be physically linked to the leading strand synthesis apparatus at all. For example, emetine-induced histone depletion causes preferential inhibition of the discontinuous strand, and under such conditions, uncoupling of DNA synthesis on both strands can be directly detected (Burhans et al. 1991).

Multiple processes are closely associated with the DNA replication fork. These include DNA mismatch repair, base excision repair, DNA methylation and chromatin assembly (for review see Krude et al. 2001; Kunkel et al. 2005; Maga et al. 2003; Otterlei et al. 1999), all of which are equally required on both the leading and the lagging strand. PCNA appears to be the central organizer for all these processes, and it is therefore reasonable to assume that PCNA will be present on both strands of the replication fork, independent of the requirement for PCNA by the respective replicase. However, there is to our knowledge no indication for a physical link of two DNA strands mediated by PCNA or the clamp loader RF-C, as it is discussed for *E. coli* DNA replication.

4
Outlook

DNA replication in *Escherichia coli* is the best understood model system and has served as a paradigm also for eukaryotic replication (Egel in this book). The remarkable similarity in the basic mechanisms of replication in bacteria and eukaryotes easily makes us overlook the equally striking differences in the replication apparatus itself. In particular, the replicative DNA helicase and DNA polymerases of eukaryotes show no phylogenetic relation with their bacterial counterparts (Leipe et al. 2000). This has led to the suggestion that DNA replication may have evolved twice independently (Forterre 2002; Leipe et al. 2000) and is also reflected by important mechanistic differences between the two replication systems (Kaufmann et al. 2004). Especially the opposite direction of eukaryotic-type replicative DNA helicases compared to the direction of bacterial replicative DnaB helicase means that the eukaryotic DNA helicase engulfs the leading strand template whereas the bacterial counterpart binds the lagging strand template. Moreover, eukaryotic DNA is replicated as chromatin assembled on nucleosomes, whereas such an arrangement is missing in bacteria (Kornberg et al. 1992). This is also reflected by a much slower progression of the eukaryotic replication fork compared to the bacterial counterpart (Kornberg et al. 1992).

One should, therefore, be very careful with assumptions for the eukaryotic replication apparatus based on the bacterial counterpart. Future research has to focus on the development of an in vitro model system that enables mechanistic studies of the leading strand synthesis during eukaryotic DNA replication. Only such a model would permit the unraveling of the various central questions still remaining.

References

Aparicio OM, Stout AM, Bell SP (1999) Differential assembly of cdc45p and DNA polymerases at early and late origins of DNA replication. Proc Natl Acad Sci USA 96:9130–9135

Aparicio OM, Weinstein DM, Bell SP (1997) Components and dynamics of DNA replication complexes in S. cerevisiae: redistribution of MCM proteins and Cdc45p during S phase. Cell 91:59–69

Araki H, Leem SH, Phongdara A, Sugino A (1995) Dpb11, which interacts with DNA polymerase II(epsilon) in Saccharomyces cerevisiae, has a dual role in S-phase progression and at a cell cycle checkpoint. Proc Natl Acad Sci USA 92:11791–11795

Aravind L, Koonin EV (1998) The HORMA domain: a common structural denominator in mitotic checkpoints, chromosome synapsis and DNA repair. Trends Biochem Sci 23:284–286

Arezi B, Kuchta RD (2000) Eukaryotic DNA primase. Trends Biochem Sci 25:572–576

Arias EE, Walter JC (2006) PCNA functions as a molecular platform to trigger Cdt1 destruction and prevent re-replication. Nat Cell Biol 8:84–90

Arudchandran A, Cerritelli S, Narimatsu S, Itaya M, Shin DY, Shimada Y, Crouch RJ (2000) The absence of ribonuclease H1 or H2 alters the sensitivity of Saccharomyces cerevisiae to hydroxyurea, caffeine and ethyl methanesulphonate: implications for roles of RNases H in DNA replication and repair. Genes Cells 5:789–802

Arunkumar AI, Klimovich V, Jiang X, Ott RD, Mizoue L, Fanning E, Chazin WJ (2005) Insights into hRPA32 C-terminal domain-mediated assembly of the simian virus 40 replisome. Nat Struct Mol Biol 12:332–339

Asturias FJ, Cheung IK, Sabouri N, Chilkova O, Wepplo D, Johansson E (2006) Structure of Saccharomyces cerevisiae DNA polymerase epsilon by cryo-electron microscopy. Nat Struct Mol Biol 13:35–43

Augustin MA, Huber R, Kaiser JT (2001) Crystal structure of a DNA-dependent RNA polymerase (DNA primase). Nat Struct Biol 8:57–61

Ayyagari R, Gomes XV, Gordenin DA, Burgers PM (2003) Okazaki fragment maturation in yeast. I. Distribution of functions between FEN1 AND DNA2. J Biol Chem 278:1618–1625

Bae SH, Bae KH, Kim JA, Seo YS (2001a) RPA governs endonuclease switching during processing of Okazaki fragments in eukaryotes. Nature 412:456–461

Bae SH, Kim DW, Kim J, Kim JH, Kim DH, Kim HD, Kang HY, Seo YS (2002) Coupling of DNA helicase and endonuclease activities of yeast Dna2 facilitates Okazaki fragment processing. J Biol Chem 277:26632–26641

Bae SH, Kim JA, Choi E, Lee KH, Kang HY, Kim HD, Kim JH, Bae KH, Cho Y, Park C, Seo YS (2001b) Tripartite structure of Saccharomyces cerevisiae Dna2 helicase/endonuclease. Nucleic Acids Res 29:3069–3079

Bell SP, Dutta A (2002) DNA replication in eukaryotic cells. Annu Rev Biochem 71:333–374

Bentley D, Selfridge J, Millar JK, Samuel K, Hole N, Ansell JD, Melton DW (1996) DNA ligase I is required for fetal liver erythropoiesis but is not essential for mammalian cell viability. Nat Genet 13:489–491

Bentley DJ, Harrison C, Ketchen AM, Redhead NJ, Samuel K, Waterfall M, Ansell JD, Melton DW (2002) DNA ligase I null mouse cells show normal DNA repair activity but altered DNA replication and reduced genome stability. J Cell Sci 115:1551–1561

Bermudez VP, MacNeill SA, Tappin I, Hurwitz J (2002) The influence of the Cdc27 subunit on the properties of the Schizosaccharomyces pombe DNA polymerase delta. J Biol Chem 277:36853–36862

Binz SK, Sheehan AM, Wold MS (2004) Replication Protein A phosphorylation and the cellular response to DNA damage. DNA Repair (Amst) 3:1015–1024

Blanton HL, Radford SJ, McMahan S, Kearney HM, Ibrahim JG, Sekelsky J (2005) REC, Drosophila MCM8, drives formation of meiotic crossovers. PLoS Genet 1:e40

Blow JJ (2001) Control of chromosomal DNA replication in the early Xenopus embryo. Embo J 20:3293–3297

Blow JJ, Dutta A (2005) Preventing re-replication of chromosomal DNA. Nat Rev Mol Cell Biol 6:476–486

Bose ME, McConnell KH, Gardner-Aukema KA, Muller U, Weinreich M, Keck JL, Fox CA (2004) The origin recognition complex and Sir4 protein recruit Sir1p to yeast silent chromatin through independent interactions requiring a common Sir1p domain. Mol Cell Biol 24:774–786

Bowman GD, Goedken ER, Kazmirski SL, O'Donnell M, Kuriyan J (2005) DNA polymerase clamp loaders and DNA recognition. FEBS Lett 579:863–867

Bowman GD, O'Donnell M, Kuriyan J (2004) Structural analysis of a eukaryotic sliding DNA clamp–clamp loader complex. Nature 429:724–730

Budd ME, Campbell JL (1995) A yeast gene required for DNA replication encodes a protein with homology to DNA helicases. Proc Natl Acad Sci USA 92:7642–7646

Budd ME, Campbell JL (1997) A yeast replicative helicase, Dna2 helicase, interacts with yeast FEN-1 nuclease in carrying out its essential function. Mol Cell Biol 17:2136–2142

Burgers PM (1998) Eukaryotic DNA polymerases in DNA replication and DNA repair. Chromosoma 107:218–227

Burgers PMJ, Gerik KJ (1998) Structure and processivity of two forms of saccharomyces cerevisiae DNA polymerase delta. J Biol Chem 273:19756–19762

Burhans WC, Vassilev LT, Wu J, Sogo JM, Nallaseth FS, DePamphilis ML (1991) Emetine allows identification of origins of mammalian DNA replication by imbalanced DNA synthesis, not through conservative nucleosome segregation. Embo J 10:4351–4360

Byun TS, Pacek M, Yee MC, Walter JC, Cimprich KA (2005) Functional uncoupling of MCM helicase and DNA polymerase activities activates the ATR-dependent checkpoint. Genes Dev 19:1040–1052
Cerritelli SM, Frolova EG, Feng C, Grinberg A, Love PE, Crouch RJ (2003) Failure to produce mitochondrial DNA results in embryonic lethality in RNase H1 null mice. Mol Cell 11:807–815
Chandra A, Hughes TR, Nugent CI, Lundblad V (2001) Cdc13 both positively and negatively regulates telomere replication. Genes Dev 15:404–414
Chilkova O, Jonsson BH, Johansson E (2003) The quaternary structure of DNA polymerase epsilon from Saccharomyces cerevisiae. J Biol Chem 278:14082–14086
Chou DM, Petersen P, Walter JC, Walter G (2002) Protein phosphatase 2A regulates binding of Cdc45 to the prereplication complex. J Biol Chem 277:40520–40527
Cook PR (1999) The organization of replication and transcription. Science 284:1790–1795
Costanzo V, Robertson K, Ying CY, Kim E, Avvedimento E, Gottesman M, Grieco D, Gautier J (2000) Reconstitution of an ATM-dependent checkpoint that inhibits chromosomal DNA replication following DNA damage. Mol Cell 6:649–659
Cvetic CA, Walter JC (2006) Getting a grip on licensing: mechanism of stable Mcm2–7 loading onto replication origins. Mol Cell 21:143–144
Diffley JF (2001) DNA replication: building the perfect switch. Curr Biol 11:R367-R370
Diffley JF, Labib K (2002) The chromosome replication cycle. J Cell Sci 115:869–872
Dua R, Levy DL, Campbell JL (1999) Analysis of the essential functions of the C-terminal Protein/Protein interaction domain of saccharomyces cerevisiae pol epsilon and its unexpected ability to support growth in the absence of the DNA polymerase domain. J Biol Chem 274:22283–22288
Dua R, Levy DL, Li CM, Snow PM, Campbell JL (2002) In vivo reconstitution of Saccharomyces cerevisiae DNA polymerase epsilon in insect cells. Purification and characterization. J Biol Chem 277:7889–7896
Egel R (2006) Chromosomal DNA Replication: On Replicases, Replisomes, and Bidirectional Replication Factories. In: Lankenau DH (ed) Genome Integrity: Facets and Perspectives, vol 1. Springer, Berlin Heidelberg DOI: 10.1007/7050_012
Falck J, Petrini JH, Williams BR, Lukas J, Bartek J (2002) The DNA damage-dependent intra-S phase checkpoint is regulated by parallel pathways. Nat Genet 30:290–294
Fanning E (1992) Simian virus 40 large T antigen: the puzzle, the pieces, and the emerging picture. J Virol 66:1289–1293
Feng W, D'Urso G (2001) Schizosaccharomyces pombe cells lacking the amino-terminal catalytic domains of DNA polymerase epsilon are viable but require the DNA damage checkpoint control. Mol Cell Biol 21:4495–4504
Fijalkowska IJ, Jonczyk P, Tkaczyk MM, Bialoskorska M, Schaaper RM (1998) Unequal fidelity of leading strand and lagging strand DNA replication on the Escherichia coli chromosome. Proc Natl Acad Sci USA 95:10020–10025
Filippov V, Filippov M, Gill SS (2001) Drosophila RNase H1 is essential for development but not for proliferation. Mol Genet Genomics 265:771–777
Forsburg SL (2004) Eukaryotic MCM proteins: beyond replication initiation. Microbiol Mol Biol Rev 68:109–131, table of contents
Forterre P (2002) The origin of DNA genomes and DNA replication proteins. Curr Opin Microbiol 5:525–532
Fukui T, Yamauchi K, Muroya T, Akiyama M, Maki H, Sugino A, Waga S (2004) Distinct roles of DNA polymerases delta and epsilon at the replication fork in Xenopus egg extracts. Genes Cells 9:179–191

Fuss J, Linn S (2002) Human DNA polymerase epsilon colocalizes with proliferating cell nuclear antigen and DNA replication late, but not early, in S phase. J Biol Chem 277:8658–8666

Garcia V, Furuya K, Carr AM (2005) Identification and functional analysis of TopBP1 and its homologs. DNA Repair (Amst) 4:1227–1239

Garg P, Burgers PM (2005) DNA polymerases that propagate the eukaryotic DNA replication fork. Crit Rev Biochem Mol Biol 40:115–128

Garg P, Stith CM, Majka J, Burgers PM (2005) Proliferating cell nuclear antigen promotes translesion synthesis by DNA polymerase zeta. J Biol Chem 280:23446–23450

Gerik KJ, Li X, Pautz A, Burgers PM (1998) Characterization of the two small subunits of Saccharomyces cerevisiae DNA polymerase delta. J Biol Chem 273:19747–19755

Gozuacik D, Chami M, Lagorce D, Faivre J, Murakami Y, Poch O, Biermann E, Knippers R, Brechot C, Paterlini-Brechot P (2003) Identification and functional characterization of a new member of the human Mcm protein family: hMcm8. Nucleic Acids Res 31:570–579

Gray FC, Pohler JR, Warbrick E, MacNeill SA (2004) Mapping and mutation of the conserved DNA polymerase interaction motif (DPIM) located in the C-terminal domain of fission yeast DNA polymerase delta subunit Cdc27. BMC Mol Biol 5:21

Grossi S, Puglisi A, Dmitriev PV, Lopes M, Shore D (2004) Pol12, the B subunit of DNA polymerase alpha, functions in both telomere capping and length regulation. Genes Dev 18:992–1006

Gulbis JM, Kelman Z, Hurwitz J, O'Donnell M, Kuriyan J (1996) Structure of the C-terminal region of p21(WAF1/CIP1) complexed with human PCNA. Cell 87:297–306

Hamatake RK, Hasegawa H, Clark AB, Bebenek K, Kunkel TA, Sugino A (1990) Purification and characterization of DNA polymerase II from the yeast Saccharomyces cerevisiae. Identification of the catalytic core and a possible holoenzyme form of the enzyme. J Biol Chem 265:4072–4083

Harrington JJ, Lieber MR (1994a) The characterization of a mammalian DNA structure-specific endonuclease. Embo J 13:1235–1246

Harrington JJ, Lieber MR (1994b) Functional domains within FEN-1 and RAD2 define a family of structure-specific endonucleases: implications for nucleotide excision repair. Genes Dev 8:1344–1355

Herman TM, DePamphilis ML, Wassarman PM (1981) Structure of chromatin at deoxyribonucleic acid replication forks: location of the first nucleosomes on newly synthesized simian virus 40 deoxyribonucleic acid. Biochemistry 20:621–630

Hiraga S, Hagihara-Hayashi A, Ohya T, Sugino A (2005) DNA polymerases alpha, delta, and epsilon localize and function together at replication forks in Saccharomyces cerevisiae. Genes Cells 10:297–309

Hoege C, Pfander B, Moldovan GL, Pyrowolakis G, Jentsch S (2002) RAD6-dependent DNA repair is linked to modification of PCNA by ubiquitin and SUMO. Nature 419:135–141

Hosfield DJ, Mol CD, Shen B, Tainer JA (1998) Structure of the DNA repair and replication endonuclease and exonuclease FEN-1: coupling DNA and PCNA binding to FEN-1 activity. Cell 95:135–146

Hsu HC, Stillman B, Xu RM (2005) Structural basis for origin recognition complex 1 protein-silence information regulator 1 protein interaction in epigenetic silencing. Proc Natl Acad Sci USA 102:8519–8524

Huang D, Pospiech H, Kesti T, Syväoja JE (1999a) Structural organization and splice variants of the POLE1 gene encoding the catalytic subunit of human DNA polymerase epsilon. Biochem J 339:657–665

Huang ME, Le Douarin B, Henry C, Galibert F (1999b) The Saccharomyces cerevisiae protein YJR043C (Pol32) interacts with the catalytic subunit of DNA polymerase alpha and is required for cell cycle progression in G2/M. Mol Gen Genet 260:541–550

Hübscher U, Maga G, Spadari S (2002) Eukaryotic DNA polymerases. Annu Rev Biochem 71:133–163

Hübscher U, Nasheuer HP, Syväoja J (2000) Eukaryotic DNA polymerases, a growing family. Trends Biochem Sci 25:143–147

Hughes P, Tratner I, Ducoux M, Piard K, Baldacci G (1999) Isolation and identification of the third subunit of mammalian DNA polymerase delta by PCNA-affinity chromatography of mouse FM3A cell extracts. Nucleic Acids Res 27:2108–2114

Hwang KY, Baek K, Kim HY, Cho Y (1998) The crystal structure of flap endonuclease-1 from Methanococcus jannaschii. Nat Struct Biol 5:707–713

Iftode C, Daniely Y, Borowiec JA (1999) Replication Protein A (RPA): The Eukaryotic SSB. Critical Rev Biochem Mol Biol 24:141–180

Ishimi Y (1997) A DNA helicase activity is associated with an MCM4, -6, and -7 protein complex. J Biol Chem 272:24508–24513

Jares P, Blow JJ (2000) Xenopus cdc7 function is dependent on licensing but not on XORC, XCdc6, or CDK activity and is required for XCdc45 loading. Genes Dev 14:1528–1540

Jeong HS, Backlund PS, Chen HC, Karavanov AA, Crouch RJ (2004) RNase H2 of Saccharomyces cerevisiae is a complex of three proteins. Nucleic Acids Res 32:407–414

Jeruzalmi D, O'Donnell M, Kuriyan J (2001a) Crystal structure of the processivity clamp loader gamma (gamma) complex of E. coli DNA polymerase III. Cell 106:429–441

Jeruzalmi D, Yurieva O, Zhao Y, Young M, Stewart J, Hingorani M, O'Donnell M, Kuriyan J (2001b) Mechanism of processivity clamp opening by the delta subunit wrench of the clamp loader complex of E. coli DNA polymerase III. Cell 106:417–428

Johansson E, Garg P, Burgers PM (2004) The Pol32 subunit of DNA polymerase delta contains separable domains for processive replication and proliferating cell nuclear antigen (PCNA) binding. J Biol Chem 279:1907–1915

Johansson E, Majka J, Burgers PM (2001) Structure of DNA polymerase delta from Saccharomyces cerevisiae. J Biol Chem 276:43824–43828

Kamimura Y, Masumoto H, Sugino A, Araki H (1998) Sld2, which interacts with Dpb11 in Saccharomyces cerevisiae, is required for chromosomal DNA replication. Mol Cell Biol 18:6102–6109

Kamimura Y, Tak YS, Sugino A, Araki H (2001) Sld3, which interacts with Cdc45 (Sld4), functions for chromosomal DNA replication in Saccharomyces cerevisiae. Embo J 20:2097–2107

Kang HY, Choi E, Bae SH, Lee KH, Gim BS, Kim HD, Park C, MacNeill SA, Seo YS (2000) Genetic analyses of Schizosaccharomyces pombe dna2(+) reveal that dna2 plays an essential role in Okazaki fragment metabolism. Genetics 155:1055–1067

Kannouche PL, Lehmann AR (2004) Ubiquitination of PCNA and the polymerase switch in human cells. Cell Cycle 3:1011–1013

Kao HI, Bambara RA (2003) The protein components and mechanism of eukaryotic Okazaki fragment maturation. Crit Rev Biochem Mol Biol 38:433–452

Kao HI, Campbell JL, Bambara RA (2004a) Dna2p helicase/nuclease is a tracking protein, like FEN1, for flap cleavage during Okazaki fragment maturation. J Biol Chem 279:50840–50849

Kao HI, Veeraraghavan J, Polaczek P, Campbell JL, Bambara RA (2004b) On the roles of Saccharomyces cerevisiae Dna2p and Flap endonuclease 1 in Okazaki fragment processing. J Biol Chem 279:15014–15024

Kaufmann G, Nethanel T (2004) Did an early version of the eukaryal replisome enable the emergence of chromatin? Prog Nucleic Acid Res Mol Biol 77:173–209
Kelly RB, Atkinson MR, Huberman JA, Kornberg A (1969) Excision of thymine dimers and other mismatched sequences by DNA polymerase of Escherichia coli. Nature 224:224, 495–501
Kesti T, Flick K, Keränen S, Syväoja JE, Wittenberg C (1999) DNA polymerase ε catalytic domains are dispensable for DNA replication, DNA repair, and cell viability. Mol Cell 3:679–685
Kesti T, Frantti H, Syvaoja JE (1993) Molecular cloning of the cDNA for the catalytic subunit of human DNA polymerase epsilon. J Biol Chem 268:10238–10245
Kneissl M, Putter V, Szalay AA, Grummt F (2003) Interaction and assembly of murine pre-replicative complex proteins in yeast and mouse cells. J Mol Biol 327:111–128
Koonin EV, Wolf YI, Aravind L (2000) Protein fold recognition using sequence profiles and its application in structural genomics. Adv Protein Chem 54:245–275
Kornberg A, Baker T (1992) DNA Replication, 2nd. Edition, W.H. Freeman & Company, New York
Krishna TS, Kong XP, Gary S, Burgers PM, Kuriyan J (1994) Crystal structure of the eukaryotic DNA polymerase processivity factor PCNA. Cell 79:1233–1243
Krude T, Keller C (2001) Chromatin assembly during S phase: contributions from histone deposition, DNA replication and the cell division cycle. Cell Mol Life Sci 58:665–672
Kubota Y, Takase Y, Komori Y, Hashimoto Y, Arata T, Kamimura Y, Araki H, Takisawa H (2003) A novel ring-like complex of Xenopus proteins essential for the initiation of DNA replication. Genes Dev 17:1141–1152
Kunkel TA, Erie DA (2005) DNA mismatch repair. Annu Rev Biochem 74:681–710
Lankenau DH, Gloor GB (1998) In vivo gap repair in Drosophila: a one-way street with many destinations. BioEssays 20:317–327
Laskey RA, Madine MA (2003) A rotary pumping model for helicase function of MCM proteins at a distance from replication forks. EMBO Rep 4:26–30
Lee C, Hong B, Choi JM, Kim Y, Watanabe S, Ishimi Y, Enomoto T, Tada S, Cho Y (2004) Structural basis for inhibition of the replication licensing factor Cdt1 by geminin. Nature 430:913–917
Lee JK, Hurwitz J (2001) Processive DNA helicase activity of the minichromosome maintenance proteins 4, 6, and 7 complex requires forked DNA structures. Proc Natl Acad Sci USA 98:54–59
Lee KH, Lee MH, Lee TH, Han JW, Park YJ, Ahnn J, Seo YS, Koo HS (2003) Dna2 requirement for normal reproduction of Caenorhabditis elegans is temperature-dependent. Mol Cells 15:81–86
Lee MY, Tan CK, Downey KM, So AG (1981) Structural and functional properties of calf thymus DNA polymerase delta. Prog Nucleic Acid Res Mol Biol 26:83–96
Lee SH, Pan ZQ, Kwong AD, Burgers PM, Hurwitz J (1991) Synthesis of DNA by DNA polymerase epsilon in vitro. J Biol Chem 266:22707–22717
Lehman IR (1974) T4 DNA polymerase. Methods Enzymol 29:46–53
Lei M, Tye BK (2001) Initiating DNA synthesis: from recruiting to activating the MCM complex. J Cell Sci 114:1447–1454
Leipe DD, Aravind L, Grishin NV, Koonin EV (2000) The bacterial replicative helicase DnaB evolved from a RecA duplication. Genome Res 10:5–16
Liu Y, Kao HI, Bambara RA (2004) Flap endonuclease 1: a central component of DNA metabolism. Annu Rev Biochem 73:589–615

Liu Y, Kvaratskhelia M, Hess S, Qu Y, Zou Y (2005) Modulation of replication protein A function by its hyperphosphorylation-induced conformational change involving DNA binding domain B. J Biol Chem 280:32775–32783

Lu X, Tan CK, Zhou JQ, You M, Carastro LM, Downey KM, So AG (2002) Direct interaction of proliferating cell nuclear antigen with the small subunit of DNA polymerase delta. J Biol Chem 277:24340–24345

Luciani MG, Oehlmann M, Blow JJ (2004) Characterization of a novel ATR-dependent, Chk1-independent, intra-S-phase checkpoint that suppresses initiation of replication in Xenopus. J Cell Sci 117:6019–6030

Lutzmann M, Maiorano D, Mechali M (2005) Identification of full genes and proteins of MCM9, a novel, vertebrate-specific member of the MCM2-8 protein family. Gene 362:51–56

Machida YJ, Hamlin JL, Dutta A (2005) Right place, right time, and only once: replication initiation in metazoans. Cell 123:13–24

MacNeill S, Moreno S, Reynolds N, Nurse P, Fantes P (1996) The fission yeast Cdc1 protein, a homologue of the small subunit of DNA polymerase delta, binds to Pol3 and Cdc27. EMBO J 15:4613–4628

Madine M, Laskey R (2001) Geminin bans replication licence. Nat Cell Biol 3:E49–50

Maga G, Frouin I, Spadari S, Hübscher U (2001) Replication protein A as a fidelity clamp for DNA polymerase alpha. J Biol Chem 276:18235–18242

Maga G, Hübscher U (2003) Proliferating cell nuclear antigen (PCNA): a dancer with many partners. J Cell Sci 116:3051–3060

Maiorano D, Cuvier O, Danis E, Mechali M (2005a) MCM8 is an MCM2-7-related protein that functions as a DNA helicase during replication elongation and not initiation. Cell 120:315–328

Maiorano D, Krasinska L, Lutzmann M, Mechali M (2005b) Recombinant Cdt1 induces rereplication of G2 nuclei in Xenopus egg extracts. Curr Biol 15:146–153

Majka J, Burgers PM (2004) The PCNA-RFC families of DNA clamps and clamp loaders. Prog Nucleic Acid Res Mol Biol 78:227–260

Mäkiniemi M, Hillukkala T, Tuusa J, Reini K, Vaara M, Huang D, Pospiech H, Majuri I, Westerling T, Makela TP, Syvaoja JE (2001) BRCT domain-containing protein TopBP1 functions in DNA replication and damage response. J Biol Chem 276:30399–30406

Mäkiniemi M, Pospiech H, Kilpelainen S, Jokela M, Vihinen M, Syväoja JE (1999) A novel family of DNA-polymerase-associated B subunits. Trends Biochem Sci 24:14–16

Marte B (2004) Cell division and cancer. Nature 432:293

Martin IV, MacNeill SA (2002) ATP-dependent DNA ligases. Genome Biol 3:REVIEWS 3005

Masai H, You Z, Arai K (2005) Control of DNA replication: regulation and activation of eukaryotic replicative helicase, MCM. IUBMB Life 57:323–335

Mass G, Nethanel T, Lavrik OI, Wold MS, Kaufmann G (2001) Replication protein A modulates its interface with the primed DNA template during RNA-DNA primer elongation in replicating SV40 chromosomes. Nucleic Acids Res 29:3892–3899

Masumoto H, Muramatsu S, Kamimura Y, Araki H (2002) S-Cdk-dependent phosphorylation of Sld2 essential for chromosomal DNA replication in budding yeast. Nature 415:651–655

Masumoto H, Sugino A, Araki H (2000) Dpb11 controls the association between DNA polymerases alpha and epsilon and the autonomously replicating sequence region of budding yeast. Mol Cell Biol 20:2809–2817

Matsubayashi H, Yamamoto MT (2003) REC, a new member of the MCM-related protein family, is required for meiotic recombination in Drosophila. Genes Genet Syst 78:363–371

Matsunaga F, Norais C, Forterre P, Myllykallio H (2003) Identification of short "eukaryotic" Okazaki fragments synthesized from a prokaryotic replication origin EMBO Rep 4:154–158

McGarry TJ, Kirschner MW (1998) Geminin, an inhibitor of DNA replication, is degraded during mitosis. Cell 93:1043–1053

Melle C, Nasheuer HP (2002) Physical and functional interactions of the tumor suppressor protein p53 and DNA polymerase α-primase. Nucleic Acids Res 30:1493–1499

Michael WM, Ott R, Fanning E, Newport J (2000) Activation of the DNA replication checkpoint through RNA synthesis by primase. Science 289:2133–2137

Mimura S, Masuda T, Matsui T, Takisawa H (2000) Central role for cdc45 in establishing an initiation complex of DNA replication in Xenopus egg extracts. Genes Cells 5:439–452

Miyata T, Oyama T, Mayanagi K, Ishino S, Ishino Y, Morikawa K (2004) The clamp-loading complex for processive DNA replication. Nat Struct Mol Biol 11:632–636

Mizuno T, Yamagishi K, Miyazawa H, Hanaoka F (1999) Molecular architecture of the mouse DNA polymerase alpha-primase complex. Mol Cell Biol 19:7886–7896

Montagnoli A, Bosotti R, Villa F, Rialland M, Brotherton D, Mercurio C, Berthelsen J, Santocanale C (2002) Drf1, a novel regulatory subunit for human Cdc7 kinase. Embo J 21:3171–3181

Morrison A, Araki H, Clark AB, Hamatake RK, Sugino A (1990) A third essential DNA polymerase in S. cerevisiae. Cell 62:1143–1151

Mozzherin DJ, Tan CK, Downey KM, Fisher PA (1999) Architecture of the active DNA polymerase delta.proliferating cell nuclear antigen.template-primer complex. J Biol Chem 274:19862–19867

Murante RS, Rust L, Bambara RA (1995) Calf 5′ to 3′ exo/endonuclease must slide from a 5′ end of the substrate to perform structure-specific cleavage. J Biol Chem 270:30377–30383

Nakajima R, Masukata H (2002) SpSld3 is required for loading and maintenance of SpCdc45 on chromatin in DNA replication in fission yeast. Mol Biol Cell 13:1462–1472

Naryzhny SN, Zhao H, Lee H (2005) Proliferating cell nuclear antigen (PCNA) may function as a double homotrimer complex in the mammalian cell. J Biol Chem 280:13888–13894

Nash R, Lindahl T (1996) DNA ligases. In: DePamphilis ML (ed) DNA Replication in Eukaryotic Cells. Cold Spring Harbor Laboratory Press, Cold Spring Harbor Laboratory, NY, USA, p 575–586

Nasheuer HP, Pospiech H, Syvaoja J (2005) DNA Polymerases. In: Ganten D, Ruckpaul KE (eds) Encyclopedic Reference of Genomics and Proteomics in Molecular Medicine. Springer Verlag, Berlin Heidelberg New York

Nasheuer HP, Smith R, Bauerschmidt C, Grosse F, Weisshart K (2002) Initiation of eukaryotic DNA replication: regulation and mechanisms. Prog Nucleic Acid Res Mol Biol 72:41–94

Nethanel T, Zlotkin T, Kaufmann G (1992) Assembly of simian virus 40 Okazaki pieces from DNA primers is reversibly arrested by ATP depletion. J Virol 66:6634–6640

Nishida C, Reinhard P, Linn S (1988) DNA repair synthesis in human fibroblasts requires DNA polymerase delta. J Biol Chem 263:501–510

Oehlmann M, Score AJ, Blow JJ (2004) The role of Cdc6 in ensuring complete genome licensing and S phase checkpoint activation. J Cell Biol 165:181–190
Otterlei M, Warbrick E, Nagelhus TA, Haug T, Slupphaug G, Akbari M, Aas PA, Steinsbekk K, Bakke O, Krokan HE (1999) Post-replicative base excision repair in replication foci. Embo J 18:3834–3838;3844
Pacek M, Tutter AV, Kubota Y, Takisawa H, Walter JC (2006) Localization of MCM2-7, Cdc45, and GINS to the site of DNA unwinding during eukaryotic DNA replication. Mol Cell 21:581–587
Papouli E, Chen S, Davies AA, Huttner D, Krejci L, Sung P, Ulrich HD (2005) Crosstalk between SUMO and ubiquitin on PCNA is mediated by recruitment of the helicase Srs2p. Mol Cell 19:123–133
Parrilla-Castellar ER, Karnitz LM (2003) Cut5 is required for the binding of Atr and DNA polymerase alpha to genotoxin-damaged chromatin. J Biol Chem 278:45507–45511
Pascal JM, O'Brien PJ, Tomkinson AE, Ellenberger T (2004) Human DNA ligase I completely encircles and partially unwinds nicked DNA. Nature 432:473–478
Pavlov YI, Frahm C, McElhinny SA, Niimi A, Suzuki M, Kunkel TA (2006) Evidence that errors made by DNA polymerase alpha are corrected by DNA polymerase delta. Curr Biol 16:202–207
Pestryakov PE, Weisshart K, Schlott B, Khodyreva SN, Kremmer E, Grosse F, Lavrik OI, Nasheuer HP (2003) Human replication protein A: The C-terminal RPA70 and the central RPA32 domains are involved in the interactions with the 3′-end of a primer-template DNA. J Biol Chem 278:17515–17524
Petersen BO, Lukas J, Sørensen CS, Bartek J, Helin K (1999) Phosphorylation of mammalian CDC6 by Cyclin A/CDK2 regulates its subcellular localization. Embo J 18:396–410
Pfander B, Moldovan GL, Sacher M, Hoege C, Jentsch S (2005) SUMO-modified PCNA recruits Srs2 to prevent recombination during S phase. Nature 436:428–433
Podust VN, Chang LS, Ott R, Dianov GL, Fanning E (2002) Reconstitution of human DNA polymerase delta using recombinant baculoviruses: the p12 subunit potentiates DNA polymerizing activity of the four-subunit enzyme. J Biol Chem 277:3894–3901
Pospiech H, Kursula I, Abdel-Aziz W, Malkas L, Uitto L, Kastelli M, Vihinen-Ranta M, Eskelinen S, Syväoja JE (1999) A neutralizing antibody against human DNA polymerase epsilon inhibits cellular but not SV40 DNA replication. Nucleic Acids Res 27:3799–3804
Pospiech H, Syvaoja JE (2003) DNA polymerase epsilon - more than a polymerase. Scientific World Journal 3:87–104
Prasanth SG, Prasanth KV, Siddiqui K, Spector DL, Stillman B (2004) Human Orc2 localizes to centrosomes, centromeres and heterochromatin during chromosome inheritance. Embo J 23:2651–2663
Prasanth SG, Prasanth KV, Stillman B (2002) Orc6 involved in DNA replication, chromosome segregation, and cytokinesis. Science 297:1026–1031
Ramadan K, Maga G, Hübscher U (2006) DNA Polymerases and Diseases In: Lankenau DH (ed) Genome Integrity: Facets and Perspectives, vol 1. Springer, Berlin Heidelberg DOI: 10.1007/7050_005
Randell JC, Bowers JL, Rodriguez HK, Bell SP (2006) Sequential ATP hydrolysis by Cdc6 and ORC directs loading of the Mcm2-7 helicase. Mol Cell 21:29–39
Ricke RM, Bielinsky AK (2004) Mcm10 regulates the stability and chromatin association of DNA polymerase-alpha. Mol Cell 16:173–185
Rudolph C, Schürer KA, Kramer W (2006) Facing Stalled Replication Forks: The Intricacies of Doing the Right Thing. In: Lankenau DH (ed) Genome Integrity: Facets and Perspectives, vol 1. Springer, Berlin Heidelberg DOI: 10.1007/7050_003

Saka Y, Fantes P, Yanagida M (1994) Coupling of DNA replication and mitosis by fission yeast rad4/cut5. J Cell Sci Suppl 18:57–61

Salas M, Miller JT, Leis J, DePamphilis ML (1996) Mechanisms for priming DNA synthesis. In: DePamphilis ML (ed) DNA Replication in Eukaryotic Cells. Cold Spring Harbor Laboratory Press, Cold Spring Harbor Laboratory, NY, USA, p 131–176

Sangrithi MN, Bernal JA, Madine M, Philpott A, Lee J, Dunphy WG, Venkitaraman AR (2005) Initiation of DNA replication requires the RECQL4 protein mutated in Rothmund-Thomson syndrome. Cell 121:887–898

Saxena S, Dutta A (2005) Geminin-Cdt1 balance is critical for genetic stability. Mutat Res 569:111–121

Schub O, Rohaly G, Smith RW, Schneider A, Dehde S, Dornreiter I, Nasheuer HP (2001) Multiple phosphorylation sites of DNA polymerase α-primase cooperate to regulate the initiation of DNA replication in vitro. J Biol Chem 276:38076–38083

Schwacha A, Bell SP (2001) Interactions between two catalytically distinct MCM subgroups are essential for coordinated ATP hydrolysis and DNA replication. Mol Cell 8:1093–1104

Shcherbakova PV, Bebenek K, Kunkel TA (2003) Functions of Eukaryotic DNA Polymerases. Sci. SAGE KE 2003:re3 (26 February 2003) http://sageke.sciencemag.org/cgi/content/full/sageke;2003/2008/re2003

Sherr CJ, Roberts JM (2004) Living with or without cyclins and cyclin-dependent kinases. Genes Dev 18:2699–2711

Smith RWP, Nasheuer HP (2002) Control of complex formation of DNA polymerase alpha-primase and cell-free DNA replication by the C-terminal amino acids of the largest subunit p180. FEBS Lett 527:143–146

Søe K, Rockstroh A, Grosse F (2006) Role of Human Topoisomerase I in DNA Repair and Apoptosis In: Lankenau DH (ed) Genome Integrity: Facets and Perspectives, vol 1. Springer, Berlin Heidelberg DOI: 10.1007/7050_004

Speck C, Chen Z, Li H, Stillman B (2005) ATPase-dependent cooperative binding of ORC and Cdc6 to origin DNA. Nat Struct Mol Biol

Stelter P, Ulrich HD (2003) Control of spontaneous and damage-induced mutagenesis by SUMO and ubiquitin conjugation. Nature 425:188–191

Stillman B (2005) Origin recognition and the chromosome cycle. FEBS Lett 579:877–884

Sugino A (1995) Yeast DNA polymerases and their role at the replication fork. Trends Biochem Sci 20:319–323

Sun WH, Coleman TR, DePamphilis ML (2002) Cell cycle-dependent regulation of the association between origin recognition proteins and somatic cell chromatin. Embo J 21:1437–1446

Syväoja J, Linn S (1989) Characterization of a large form of DNA polymerase delta from HeLa cells that is insensitive to proliferating cell nuclear antigen. J Biol Chem 264:2489–2497

Syväoja J, Suomensaari S, Nishida C, Godsmith JS, Chui GSJ, Jain S, Linn S (1990) Dna polymerase alpha, delta and epsilon: Three distinct enzymes from HeLa cells. Proc Natl Acad Sci USA 87:6664–6668

Tada S, Li A, Maiorano D, Mechali M, Blow JJ (2001) Repression of origin assembly in metaphase depends on inhibition of RLF-B/Cdt1 by geminin. Nat Cell Biol 3:107–113

Takahashi TS, Walter JC (2005) Cdc7-Drf1 is a developmentally regulated protein kinase required for the initiation of vertebrate DNA replication. Genes Dev 19:2295–2300

Takasaki Y, Deng JS, Tan EM (1981) A nuclear antigen associated with cell proliferation and blast transformation. J Exp Med 154:1899–1909

Takayama Y, Kamimura Y, Okawa M, Muramatsu S, Sugino A, Araki H (2003) GINS, a novel multiprotein complex required for chromosomal DNA replication in budding yeast. Genes Dev 17:1153–1165

Tanaka S, Diffley JF (2002) Interdependent nuclear accumulation of budding yeast Cdt1 and Mcm2–7 during G1 phase. Nat Cell Biol 4:198–207

Timson DJ, Singleton MR, Wigley DB (2000) DNA ligases in the repair and replication of DNA. Mutat Res 460:301–318

Tsubota T, Maki S, Kubota H, Sugino A, Maki H (2003) Double-stranded DNA binding properties of Saccharomyces cerevisiae DNA polymerase epsilon and of the Dpb3p-Dpb4p subassembly. Genes Cells 8:873–888

Tye BK (1999) MCM proteins in DNA replication. Annu Rev Biochem 68:649–686

Ulrich HD (2005) Mutual interactions between the SUMO and ubiquitin systems: a plea of no contest. Trends Cell Biol 15:525–532

Vogelauer M, Rubbi L, Lucas I, Brewer BJ, Grunstein M (2002) Histone acetylation regulates the time of replication origin firing. Mol Cell 10:1223–1233

Volkening M, Hoffmann I (2005) Involvement of human MCM8 in prereplication complex assembly by recruiting hcdc6 to chromatin. Mol Cell Biol 25:1560–1568

Waga S, Stillman B (1998) The DNA replication fork in eukaryotic cells. Annu Rev Biochem 67:721–751

Walter J, Newport J (2000) Initiation of eukaryotic DNA replication: origin unwinding and sequential chromatin association of Cdc45, RPA, and DNA polymerase alpha. Mol Cell 5:617–627

Walter JC (2000) Evidence for sequential action of cdc7 and cdk2 protein kinases during initiation of DNA replication in xenopus egg extracts. J Biol Chem 275:39773–39778

Weisshart K, Förster H, Kremmer E, Schlott B, Grosse F, Nasheuer HP (2000) Protein-protein interactions of the primase subunits p58 and p48 with simian virus 40 T antigen are required for efficient primer synthesis in a cell-free system. J Biol Chem 275:17328–17337

Weisshart K, Pestryakov P, Smith RW, Hartmann H, Kremmer E, Lavrik O, Nasheuer HP (2004) Coordinated regulation of replication protein A activities by its subunits p14 and p32. J Biol Chem 279:35368–35376

Wintersberger U, Wintersberger E (1970) Studies on deoxyribonucleic acid polymerases from yeast. 1. Parial purification and properties of two DNA polymerases from mitochondria-free cell extracts. Eur J Biochem 13:11–19

Wohlschlegel JA, Dwyer BT, Dhar SK, Cvetic C, Walter JC, Dutta A (2000) Inhibition of eukaryotic DNA replication by geminin binding to Cdt1. Science 290:2309–2312

Wold MS (1997) Replication protein A: a heterotrimeric, single-stranded DNA-binding protein required for eukaryotic DNA metabolism. Annu Rev Biochem 66:61–92

Yoshida K (2005) Identification of a novel cell-cycle-induced MCM family protein MCM9. Biochem Biophys Res Commun 331:669–674

Yuzhakov A, Kelman Z, Hurwitz J, O'Donnell M (1999) Multiple competition reactions for RPA order the assembly of the DNA polymerase delta holoenzyme. Embo J 18:6189–6199

Zerbe LK, Kuchta RD (2002) The p58 subunit of human DNA primase is important for primer initiation, elongation, and counting. Biochemistry 41:4891–4900

Zhang SJ, Zeng XR, Zhang P, Toomey NL, Chuang RY, Chang LS, Lee MY (1995) A conserved region in the amino terminus of DNA polymerase delta is involved in proliferating cell nuclear antigen binding. J Biol Chem 270:7988–7992

Zlotkin T, Kaufmann G, Jiang Y, Lee MY, Uitto L, Syväoja J, Dornreiter I, Fanning E, Nethaniel T (1996) DNA polymerase ε may be dispensable for SV40- but not cellular DNA replication. EMBO J 15:2298–2305

Zou L, Elledge SJ (2003) Sensing DNA damage through ATRIP recognition of RPA-ssDNA complexes. Science 300:1542–1548

Zuo S, Gibbs E, Kelman Z, Wang TS, O'Donnell M, MacNeill SA, Hurwitz J (1997) DNA polymerase delta isolated from Schizosaccharomyces pombe contains five subunits. Proc Natl Acad Sci USA 94:11244–11249

Genome Dyn Stab (1)
D.-H. Lankenau: Genome Integrity
DOI 10.1007/7050_005/Published online: 20 December 2005

DNA Polymerases and Diseases

Kristijan Ramadan[1,3] · Giovanni Maga[2] · Ulrich Hübscher[1] (✉)

[1]Institute of Veterinary Biochemistry and Molecular Biology, University of Zürich, Winterthurerstrasse 190, 8057 Zürich, Switzerland
kristijan.ramadan@bc.biol.ethz.ch, hubscher@vetbio.unizh.ch

[2]Istituto di Genetica Molecolare, Consiglio Nazionale delle Ricerche, via Abbiategrasso 207, 27100 Pavia, Italy
maga@igm.cnr.it

[3]*Present address:*
Institute of Biochemistry, ETH Zürich, 8093 Zürich, Switzerland
kristijan.ramadan@bc.biol.ethz.ch

Abstract In this chapter, we would like to highlight the possible connections between DNA polymerases, the main enzymes in DNA metabolism, and human diseases, also critically evaluating those cases where the experimental data are not fully convincing. To this aim, we will first give a short overview of the three main DNA metabolic events, namely replication, repair and recombination, as well as of the checkpoint pathways acting in response to DNA damage. Besides a role in replication of the genome, DNA polymerases also have fundamental functions in other aspects of DNA metabolism, such as DNA repair, DNA recombination, translesion DNA synthesis and cell cycle checkpoint. In the last 10 years, numerous novel DNA polymerases have been revealed, but their exact cellular functions still await clarification. This review summarizes the known eukaryotic DNA polymerases and their relationships with human diseases.

1 DNA Replication

DNA replication occurs in a highly organized way in vivo. All the enzymes and proteins that are involved have precise roles in a replication complex called the replisome, which is itself located in so-called "nuclear replication factories" (Hubscher et al. 2002).

An initiator protein complex known as the origin-recognition complex (ORC) is bound to an origin of replication. The ORC has to be activated by other proteins such as the minichromosome maintenance proteins (MCMs), by the cell division cycle proteins (Cdc6 and Cdc7/Dbf4) and by phosphorylation (or possibly other post-translational modifications). This leads to the formation of an initiation complex that can alter DNA structures locally, presumably by activating the intrinsic helicase activity of a MCM subcomplex, or by attracting other DNA helicases, such as MCM 8, to the replication origins (Stucki et al. 2001).

This results in single-stranded (ss)DNA, which is protected and stabilized by the ssDNA-binding protein replication protein-A (RP-A). This protein can also help to unwind DNA, either on its own or through interaction with DNA helicases, and promotes binding of DNA polymerase α/primase (pol α/prim), which acts as the initiating pol. After very limited RNA and DNA synthesis, a DNA polymerase switch occurs, from pol α/prim to the processive pol δ or pol ε holoenzyme (composed of pol δ or pol ε, proliferating cell nuclear antigen (PCNA) and RF-C), probably mediated by the pol auxiliary protein replication factor C (RF-C) (Shevelev and Hubscher 2002).

DNA synthesis on the leading and lagging strands is coordinated by dimerization of two pol holoenzymes (pol δ and pol ε), either through direct physical interaction or through a clamp factor. Thus, mammalian chromosomal replication requires at least the three pols: pol α, pol δ and pol ε (Stucki et al. 2001).

After full replication, the two DNA strands are separated by DNA topoisomerase II (Hubscher et al. 2000).

DNA replication has to be performed in an accurate manner. To achieve this, the fidelity of DNA synthesis is controlled by several mechanisms: (i) by the intrinsic mechanisms of steric and energetic exclusion of non-complementary base pairing (base selection) of the pols that operate at the fork, (ii) by the proofreading activity of the $3' \rightarrow 5'$ exonucleases (exos) of pol δ and pol ε or possibly by autonomous exos and (iii) by postreplication DNA mismatch repair (Shevelev and Hubscher 2002).

2
DNA Repair

The DNA in the cell is constantly exposed to various harmful factors from the environment (e.g. sunlight, cigarette smoke, pollution) or from the organism itself (different metabolic intermediates and products). DNA damage is a common event in the life of a cell, and this may lead to mutations, cancer or death. Damage to DNA induces several cellular responses that enable the cell either to eliminate or tolerate the damage or to activate a programmed cell death process, presumably to eliminate cells with potentially catastrophic mutations (Radman 2001). DNA damage response reactions include: (i) DNA repair itself, with the removal of DNA damage and restoration of the continuity of the DNA duplex; (ii) activation of a DNA damage checkpoint, which arrests cell cycle progression to allow DNA repair and prevention of the transmission of damaged or incompletely replicated chromosomes; (iii) an SOS type of response, which allows DNA replication to proceed through damaged DNA, at the expense of generating mutations (translesion synthesis, TLS); (iv) a transcriptional response, which causes changes in the transcription profile

Table 1 DNA polymerases and diseases[1]

DNA polymerase	Possible physiological functions	Knock out or mutation phenotype	Diseases related to DNA polymerase
α	– initiation of Okazaki fragments (synthesis of a short RNA primers in DNA synthesis)	– Unknown	– Unknown
β	– Short patch BER (dRP-lyase activity and gap filling synthesis)	– Embryos are not viable – Cells (MEF) = hypersensitive to alkylating agents	– Overexpressed in many tumors (prostate, breast, colon, ovarian, chronic myeloid leukemia)
δ	– DNA synthesis of the lagging strand – DNA repair (long patch BER, MMR, TLS)	– Mouse pol δ oxo – /– = cancer susceptibility – Yeast POL3 exo – /– = mutator phenotype	– Cancer
ε	– DNA synthesis of leading strand – DNA repair (long patch BER and NER)		– Unknown
γ	– Mitochondrial DNA replication and repair	– Double mutant in exo I and II motif in *Drosophila* = generation phenotype	– Autosomal dominant or recessive progressive external ophthalmoplegia – Parkinsons disease – Aging, premature menopause – Alpers syndrome
TdT	– V(D)J recombination	– Unknown	– Immunodeficiency, and overexpressed in several acute leukemia cells: its overexpression correlated with poor prognosis
θ	– DNA crosslinking repair and AP site translesion synthesis	– Mutation in *Drosophila* showed sensitivity to DNA crosslinking agents, elevated frequency of chromosomal aberration and altered DNA metabolism	– Chromosome instability with potential cancer susceptibility

Table 1 (continued)

DNA polymerase	Possible physiological functions	Knock out or mutation phenotype	Diseases related to DNA polymerase
ν	– Unknown	– Unknown	– Unknown
ζ	– Mitosis/meiosis (spindle checkpoint) – Somatic hypermutation of IgG – TLS-like a "mismatch extender" – Mouse embryonic development	– *S. cerevisiae* – /– = sensitive to various DNA damage agents (UV-light, MMS, cisplatin and IR) – Apoptosis, cell proliferation and control of cell cycle	– Might be involved mammalian untargeted mutation
η	– TLS over TT dimer and the flanking bases – IgG hypermutation	– Mutation in human = XPV = skin cancer	– Skin cancer
κ	– TLS synthesis of various distort helix geometry	– Overexpression cause DNA breaks and stimulates DNA exchange	– Up-regulated in lung cancer with aneuploidia – Reduce expression in mammary carcinoma – Excess cause proliferation of solid tumors in immunodeficient mice
ι	– Incorporation of nucleotides opposite DNA lesion	– Overexpression in breast cancer and significant decreasing in DNA replication fidelity, consequently contributing to the accumulation of genetic damage	– Up- or down-regulated in different cancer cell types
REV1	– TLS synthesis of AP site	– Mice – /– are normal – Cells – /– are sensitive to DNA damage agents	– Unknown

Table 1 (continued)

DNA polymerase	Possible physiological functions	Knock out or mutation phenotype	Diseases related to DNA polymerase
λ	– DSBR (NHEJ)	– Mice – /– = immotile cilia syndrome – Cells – /– = no sensitivity to DNA damage agents	– Immotile cilia syndrome?
μ	– V(D)J recombination	– Mice – /– = defect in light chain of IgG	– Non-Hodgkin's lymphoma
σ and 2	– Sister chromatid cohesion	– Cells with double mutant die and cannot complete S phase of the cell cycle	Unknown
telomerase	– Maintenance of physical integrity of chromosomal ends		– Overexpressed in 90% tumor – Aging – Werner syndrome

[1] For references see text

that may be beneficial to the cell; and (v) apoptosis, which eliminates heavily damaged or seriously deregulated cells (Sancar et al. 2004).

The major DNA repair mechanisms include base excision repair (BER), nucleotide excision repair (NER), non-homologous end-joining (NHEJ), homologous recombination (HR), and mismatch repair (MMR) (reviewed in Christmann et al. 2003).

3 DNA-damage Checkpoints

DNA-damage checkpoints (Sancar et al. 2004) are biochemical pathways that arrest or delay cell-cycle progression in response to DNA damage, thereby providing the opportunity for DNA repair pathways to function. DNA-damage checkpoints are specific points in the cell cycle when the integrity of the DNA is examined, before allowing progression through the following phases. The main checkpoints are: (i) the G1 checkpoint, which prevents entry into S phase with damaged DNA, (ii) the intra-S-phase checkpoint, which is activated by genotoxic insults, causes transient but reversible delays in cell-cycle progression and (iii) the G2/M checkpoint, which prevents cells from initiating mitosis when damaged DNA is detected during the G2 phase of the cell cycle. The key components of the sophisticated network of mammalian checkpoint pathways can be divided into five categories: (i) the sensors, (ii) the mediators, (iii) the apical signal-transducing kinases, (vi) the distal signal-transducing kinases, and (v) the effector proteins (Li and Zou 2005).

4 DNA Polymerases

DNA polymerases (pol) are template dependent enzymes that catalyze the transfer of deoxyribonucleoside triphosphates (dNTPs) to the 3′OH group of a primer. They have the ability to synthesize long polymers of nucleoside monophosphates, whose linear spatial deposition is directed by the sequence of the complementary template DNA strand (Kelman and O'Donnell 1994). In addition, some of them contain a $3' \rightarrow 5'$ exo proofreading activity, which prevents misincorporations from occurring, thus improving fidelity.

Pols can be classified into six groups, based upon the phylogenetic relationships with *E. coli* pol I (class A), *E. coli* pol II (class B), *E. coli* pol III (class C), euryarchaeotic pol II (class D), human pol β (class X), and *E. coli* UmuC/DinB and eukaryotic RAD30/xeroderma pigmentosum variant (class Y). All known eukaryotic pols are class A, class B, class X, or class Y enzymes. No eukaryotic homologues of class C or class D pols have been detected so far, despite detailed sequence searches using the PSI-BLAST program (Burgers et al. 2001).

Albeit pols from different families are structurally quite dissimilar, several common features have emerged. Six highly conserved regions, termed I—VI, have been identified in eukaryotic, prokaryotic and viral pols. Their relative position along the primary sequence is also conserved; region IV is at the N terminus, followed by regions II, VI, III, I and V. (Joyce and Steitz 1994; Brautigam and Steitz 1998). The high degree of structural and sequence conservation of these domains between eukaryotic, prokaryotic and viral pols suggested that these pols derived from a common ancestor gene.

5 Eukaryotic DNA Polymerases and Diseases

Since the discovery of pol α in 1957, the number of eukaryotic pols identified has grown to at least 19, including the terminal transferase, telomerase, and several new pols discovered in the last few years (Hubscher et al. 2002). In the early 1970s, pol β and γ were discovered, leading to the simple concept that pol α was the enzyme responsible for DNA replication, pol β for DNA repair and pol γ for mitochondrial DNA replication. The discovery of pol δ and pol ε in the 1980s and at least 10 novel pols at the end of the 1990s led to the hypothesis that a particular pol might have more than one functional task, and that a particular DNA synthetic event may require more than one pol, a feature that is especially evident for the variety of most recently found pols. Table 1 summarizes the known eukaryotic DNA polymerases and their relationships with human diseases.

5.1 The Classical DNA Polymerases

The classical pols have been isolated from different tissues before the human genome was sequenced. All of them are now very well biochemically and genetically characterized and their main molecular functions in the cell are known. They are: pol α, pol β, pol γ, pol δ and pol ε. Here, we will review their roles, either directly or indirectly, in the onset of human diseases.

DNA polymerase α: pol α-prim has an essential role in DNA replication (Foiani et al. 1997). In all eukaryotic organisms pol α-prim complex consists of four subunits, with molecular masses of 165 kDa, 67–86 kDa, 58 kDa and 48 kDa. The p48 and p58 subunits form the heterodimeric primase (Hubscher et al. 2002). Pol α-prim also plays an important role in coordinating DNA replication, DNA repair and cell-cycle checkpoints (Foiani et al. 1997). It couples mitosis to the completion of DNA replication and to the repair of DNA damage, and is essential for the arrest of yeast cells in G1 prior to the start of S phase (Marini et al. 1997). Furthermore, cyclin dependent kinases (Cdk's) phosphorylate and regulate pol α-prim during the cell cycle. A direct role of pol α-prim in DNA repair and DNA recombination is still under discussion (Burgers 1998), but published data suggested that the replicative pols, pol α, pol δ and pol ε, are required for double-strand breaks (DSBR) in yeast by homologous recombination (Holmes and Haber 1999). Thus, even though non-direct experimental evidence has directly implicated pol α-prim in human diseases, so far it is likely that even minor misregulations of its activity or response to checkpoint controls might have a profound influence in, for example, the development of cancer. Indeed, altered catalytic properties have been reported for pol α-prim isolated from a hepatoma cell line.

DNA polymerase β: pol β is a 39 kDa single chain polypeptide folded into distinct domains each associated with a specific functional activity. The amino-terminus (8 kDa) is connected to the polymerase domain (31 kDa) by a protease sensitive hinge region (Prasad et al. 1998). The 8 kDa domain has a lyase activity that removes the 5′-deoxyribose phosphate (dRP) generated after incision by an apurinic/apyrimidinic (AP) endonulcease in BER (Matsumoto and Kim 1995). The nucleotidyl transferase reaction is catalysed by the 31 kDa polymerase domain. Pol β deficient cells have a reduced BER, which suggests a role of pol β in this pathway in vivo (Dianov et al. 1999). Pol β is very well suited to its role in BER: (i) its DNA synthesis specificity for short gaps; (ii) its associated dRPlyase activity; and (iii) its ability to associate with other BER enzymes, including DNA ligase I, AP endonuclease and XRCC1-DNA ligase III (Idriss et al. 2002). Beside its role in BER, pol β is involved in translesion synthesis. Pol β efficiently bypasses cisplatin and oxaliplatin adducts (Hoffmann et al. 1995; Vaisman and Chaney 2000). Moreover, pol β can synthesize up to 8-mer long RNA and can also efficiently incorporate ribonucleotides into DNA in the absence of specific deoxyribonucleotide (Bergoglio et al. 2003). All four rNTPs can be incorporated opposite the CPD lesion, and this process of ribonucleotide incorporation into DNA may contribute to the high frequency of mutagenesis observed in cells that overproduce pol β (Bergoglio et al. 2003). Lesion bypass by pol β occurs predominantly by "skipping over" the lesion, by insertion of a nucleotide complementary to an adjacent downstream template site. Misalignment incorporation for pol β occurs by a "dNTP-stabilized" mechanism resulting in both deletion and base substitution errors (Efrati et al. 1997).

Pol β –/– mouse embryos are not viable, suggesting its role in embryogenesis and development (Gu et al. 1994). Pol β null (–/–) embryonic cells survive in culture but are severely compromised in their ability to carry out short patch BER resulting in their hypersensitivity to alkylating agents (Sobol et al. 1996). Interestingly, high levels of pol β have been detected at the transcriptional and protein levels in many cancer tissues, mostly solid tumors (prostate, breast, colon, ovarian) as well as in chronic myeloid leukemia (Louat et al. 2001). Its up-regulation could contribute to enhancing chromosome instability and tumorigenesis when overexpressed by only 2-fold in cells, suggesting that a rigorous regulation of its expression may be essential in vivo (Bergoglio et al. 2004; Albertella et al. 2005). Pol β causes chromosomal instability, probably either by competing with replicative pols or by translesion synthesis over DNA lesions. Cells overexpressing pol β are much more sensitive to ionizing radiation (IR) treatments by increasing apoptosis, and hypermutator phenotype in surviving cells. These data indicate that pol β strengthens both cell death and genetic changes associated with a malignant phenotype (Frechet et al. 2002). Also, overexpression of pol β strengthens the mutagenicity of oxidative damages, concomitantly with a higher cellular sensitivity and increased apoptosis (Frechet et al. 2001). Regulation of pol β overexpression could be fundamental in some cancer treatment, e.g. tumor cell resistant to cisplatin drugs. Finally, finding of molecules which specifically inhibit pol β and are not harmful for the cell has to be one of the next steps in cancer treatment.

DNA polymerase γ: the mitochondrial (mt) pol γ is a heterodimeric protein composed of a large catalytic subunit, and a small accessory subunit. The large subunit of human pol γ has a molecular mass of 140 kDa (p140) and possesses polymerase, and both $5' \rightarrow 3'$ and $3' \rightarrow 5'$ exo activities (Graves et al. 1998). Pol γ is additionally able to catalyze the removal of a dRP (Carrodeguas et al. 2001). The small p55 subunit stimulates both pol and exo activities of p140. Pol γ participates in mt DNA replication as well as in DNA repair. In *Drosophila* mt DNA mutagenesis during development and aging showed that constitutive, low-level expression of pol γ carrying a double substitution (D185A/D263A) in the exo I and II motif results in a so-called *generational phenotype*. The phenotype progresses from normal development, reproductive capacity, and life span in the first generation to substantial developmental delays, developmental arrest at the pupal stage, severe loss of reproductive capacity, and reduction in adult life span from 3 weeks to 10 days in the 6th generation (Kaguni 2004).

Point mutations and deletions of mtDNA accumulate in a variety of tissues during aging in mammals (Corral-Debrinski et al. 1992). These mutations are unevenly distributed and can accumulate in certain cells, causing pattern of respiratory chain deficiency in tissues such as heart, skeletal muscle and brain (Cottrell et al. 2001). Mutations in the gene POLG1, encoding catalytic pol γ have been associated with a mitochondrial disease – autosomal dominant or recessive progressive external ophthalmoplegia – and multiple deletions of mtDNA (Graziewicz et al. 2004). Mitochondrial dysfunction is also suspected to participate in the pathogenesis of Parkinson's disease. Dysfunction of POLG1 causes a severe progressive disorder, including parkinsonism and premature menopause, which are not typical symptoms of mitochondrial disease. Cosegregation of parkinsonism and POLG1 mutations in all tested families suggests that when defective, this gene can underlie mendelian transmission of parkinsonism (Luoma et al. 2004; Mancuso et al. 2004). Alpers' syndrome is a fatal neurogenetic disorder first described more than 70 years ago. It is an autosomal recessive, developmental mitochondrial DNA depletion disorder characterized by deficiency in pol γ catalytic activity, refractory seizures, neurodegeneration, and liver disease. Infant patients with a combination of progressive neurological and hepatic failure were affected by Alpers' hepatopathic poliodystrophy or severe floppy infant syndrome associated with liver failure. Analysis of POLG1 revealed that all the patients carried different allelic mutations in this gene, suggesting that POLG1 is the major disease gene in mitochondrial disorders (Ferrari et al. 2005). Knock-in mice that express a proof-reading deficient version of pol γ in the second exo domain (D257A) develop an mtDNA mutator phenotype with a 3-fold to 5-fold increase in the levels of point mutations, as well as increased amounts of deleted mtDNA. This increase in somatic mtDNA mutations in mice is associated with reduced lifespan and premature onset of aging-related phenotypes (weight loss, reduced subcutaneous fat, alopecia, osteoporosis, anemia, reduced fertility and heart enlargement) (Trifunovic et al. 2004).

DNA polymerase δ: Mammalian pol δ exists as a heterotetrameric enzyme with subunits of 125, 67, 50 and 12 kDa (Burgers 1998). Pol δ possesses a wide range of functions: it is needed (i) in DNA replication (leading and lagging strand) (Burgers 1998), (ii) in several DNA repair events (long patch BER, NER, MMR) (Aboussekhra et al. 1995; Longley et al. 1997), (iii) in translesion synthesis, besides the pivotal roles of the translesion pols (pol ζ, pol η and Rev 1) (Giot et al. 1997), (iv) possibly in cell cycle control, because different temperature-sensitive mutants in the large 125-kDa subunit of *Schizosaccharomyces pombe* (*S. pombe*) pol δ exhibited a typical "cell division cycle" terminal phenotype (Francesconi et al. 1993), and finally (v) in meiotic recombination, where a novel allele of the *S. cerevisiae POL3* gene (pol δ) was identified, and the mutant strain during meiotic recombination produces shorter strand exchange intermediates and fewer crossover products (Maloisel et al. 2004).

Beside polymerase activity, pol δ possesses an intrinsic $3' \rightarrow 5'$ exo activity in the catalytic subunit (p125) (Byrnes et al. 1976). Yeast POL3 mutant strains that lack the exo activity of the pol δ have a strong mutator phenotype. POL3 mutations in regions outside of the exo domain elevate the frequency of deletions likely reflecting an increased frequency of DNA polymerase slippage. Moreover, reduction in the expression level of wild-type pol δ results in a similar mutator phenotype and in increased sensitivity to the DNA-damaging agent methyl methane sulfonate (MMS). These results might suggest that both the quantity and the quality of pol δ is important in ensuring genome stability (Kokoska et al. 2000). The $3' \rightarrow 5'$ exo activity of pol δ, beside proofreading activity, has additional biological roles in Okazaki fragment maturation and MMR (Jin et al. 2005).

Defective pol δ proofreading causes cancer susceptibility in mice (Goldsby et al. 2001). Single point mutation inactivates the $3' \rightarrow 5'$ exo of pol δ and causes a mutator and cancer phenotype, indicating that pol δ proofreading suppresses spontaneous cancer development and suggest that unrepaired pol δ errors contribute to carcinogenesis (Goldsby et al. 2002).

DNA polymerase ε: pol ε is composed of four subunits in mammalian and fission yeast cells. Pol ε is essential for cell viability and is required for chromosomal DNA replication in *S. cerevisiae.* In addition to DNA replication, pol ε is thought to be involved in DNA repair and cell-cycle checkpoint control (Burgers 1998). Pol ε from *S. cerevisiae* consists of four subunits (Pol2, Dpb2, Dpb3, and Dpb4) and the genes encoding the 256-kDa catalytic and 79-kDa subunit of *S. cerevisiae* pol ε are essential for growth (Sugino 1995), as is the catalytic subunit of *S. pombe* pol ε (D'Urso and Nurse 1997). Requirement of the catalytic subunit has been taken to reflect the need for polymerase activity of pol ε in replication in yeast cells. However, the precise role of pol ε both in yeast and in mammalian cells still remains to be determined. The question concerning the role of pol ε in replication has become even more intriguing because the catalytic polymerase domain within the 256-kDa polypeptide is apparently not essential in budding yeast (Kesti et al. 1999). In addition to DNA replication and viability, cells that lack the catalytic polymerase domain are also proficient in DNA repair and recombination. Although genes encoding both the 256-kDa and the 79-kDa subunits are required for *S. cerevisiae*, the pol ε catalytic domain is not. The C-terminal portion of the 256-kDa catalytic subunit is both necessary and sufficient for the essential functions of *S. cerevisiae* pol ε (Kesti et al. 1999). Besides DNA replication, it is likely that pol ε plays a role in NER, since it is required in the DNA synthesis process of reconstituted NER (Aboussekhra et al. 1995). Using soluble yeast repair extracts from mutant budding yeast strains, it was suggested that pol ε rather than pol δ is required for BER (Wang et al. 1993).

No experimental evidence of direct involvement of pol ε in humans diseases has been reported so far.

Terminal deoxynucleotidyl transferase (TdT): TdT exist only in mammalian cells and belongs to the pol X family (Holm and Sander 1995). TdT elongates DNA strands in a template-independent manner (Bollum et al. 1974) and can incorporate both ribo- and deoxyribonucleotides in vitro (Roychoudhury 1972; Boule et al. 2001) as well as a large array of unnatural nucleoside triphosphates (Semizarov et al. 1997; Krayevsky et al. 2000). TdT has so far only been identified in vertebrates. Despite a rather low sequence identity of 22–24%, the C-terminal part of TdT (C-TdT) has striking structural homology with pol β. TdT is responsible for the addition of non-templated nucleotides (N addition). Two splice variants of TdT have been identified in mice: a short form (TdTS) contains 509 amino acids and a long form (TdTL) contains 529 amino acids. Both TdTS and TdTL are encoded by a single gene. TdTS and TdTL share all domains that are essential for TdT binding to DNA, nucleotide and cations. TdTS is a transferase that catalyzes the non-template addition of nucleoside triphosphate to coding ends during V(D)J recombination, but lacking exo activities. In contrast, TdTL exhibited $3' \rightarrow 5'$ exo activity with no *N* addition (transferase) activity. TdTS and TdTL are expressed simultaneously during B cell development, and it was shown that TdTL modulated the N addition by TdTS (Thai et al. 2002).

Mutation experiments indicated that the exo and the transferase activities are interindependent. In vivo and in vitro, *N* addition and nucleotide deletion were modulated by the simultaneous expression of TdTS and TdTL. Thus, TdTL is a lymphoid-specific enzyme with exo activity, and its expression is restricted to lymphoid precursor cells (Thai and Kearney 2004).

Human and bovine TdT have three splice variants. The presence of the conserved exo core motifs in human and bovine TdTL sequenced suggests that these isoforms may also function as exo during V(D)J recombination. *N*-terminal part of TdT via BRCT-like domain interacts with Ku70/80 and, Pso4 proteins which function in DNA repair and recombination (Grey et al. 1996; Mahajan et al. 1999).

TdT is overexpressed in several acute leukemia cells, and its overexpression correlated with poor prognosis and low response to chemotherapy. Recently, it has been shown that the nucleoside analogue cordycepine, in combination with the antitumor drug coformycine, inhbits TdT and displays selective toxicity against TdT^+ leukemic cells, suggesting an important functional role of TdT in the cancerous phenotype (Kodama et al. 2000).

5.2 The Novel DNA Polymerases

Thanks to the wealth of genetic information coming from the human, mouse and yeast genome projects, at least 10 novel putative pols have been identified, which belong to the pol B, X and Y families (Hubscher et al. 2002).

5.2.1 DNA Polymerase A Family

DNA polymerase θ: The human POLQ gene shares a high degree of homology with the MUS308 gene of *Drosophila melanogaster (D. melanogaster)*. The *D. melanogaster* MUS308 gene is unusual in encoding both a family A pol domain and a DNA/RNA helicase domain. The human POLQ gene was mapped to chromosome 3, encoding the pol θ protein of 2592 amino acids (Sharief et al. 1999; Seki et al. 2003). In a survey of human and mouse tissues, pol θ expression was highest in testis. Immunoblotting with pol θ antibodies detected a protein of > 250 kDa in HeLa cells extracts (Seki et al. 2003). Full-length recombinant human pol θ showed pol activity on nicked double-stranded DNA and on a singly primed DNA template. In addition pol θ in *N*-terminal region contains an ATPase-helicase domain, and exhibited a single-stranded DNA-dependent ATPase activity (Seki et al. 2003).

The natural pol θ from HeLa cells was devoid of detectable helicase activity, possessed a $3' \rightarrow 5'$ exo activity and showed biochemical properties clearly distinct from any other eukaryotic pol. Misincorporation and fidelity assays showed that human pol θ was able to catalyze efficiently DNA synthesis past an AP site displayed high fidelity (Maga et al. 2002a). In contrast to the natural enzyme, recombinant human pol θ is protein with low fidelity and no exo activity or conserved exo motifs in the sequence (Seki et al. 2004). These data suggest that natural pol θ was probably isolated in the complex with an autonomous exo.

Moreover, pol θ can perform translesion DNA synthesis at an AP site and a thymine glycol. Unusually, pol θ both inserts a base opposite an AP site and efficiently extends the misincorporated nucleotide, making it the most efficient known polymerase for AP-site bypass. dAMP is the most efficiently incorporated base opposite an AP site and a primer ending with an dAMP residue opposite an AP site is the best primer extended by pol θ. It is interesting that pol θ is extremely poor at incorporating dCMP opposite an AP site, completely opposite to the preferred reaction of REV1 at an AP site. Moreover, pol θ poorly extends a primer end having dCMP opposite an AP site. This could ensure that pol θ works for bypass of AP sites in situations quite separate from those where REV1 may be employed (Seki et al. 2004).

There are no yeast homologs of pol θ, but flies carrying a mutation in this gene showed sensitivity to DNA crosslinking agents, elevated frequency of chromosomal aberration and altered DNA metabolism (Boyd et al. 1990). Mus308 apparently functions in a pathway of DNA crosslink repair or tolerance because fly MUS308 mutants are sensitive to DNA crosslinking agents (Harris et al. 1996).

The small scale ENU mutagenesis screen for chromosome instability in the mouse genome identified a mutation named *chaos1* (*c*hromosome *a*berration *o*ccurring *s*pontaneously 1) and mapped it to a region of the chromosome 16 where the POLQ gene is located (Shima et al. 2003). The identity of pol θ and *chaos1* has been confirmed by direct disruption of POLQ in the mouse and by correction of the phenotype with the POLQ gene. This result provides evidence that pol θ is potentially involved in cancer susceptibility.

DNA polymerase ν: the human POLN gene encodes a new mammalian pol, also related to MUS308, as the above described pol θ. The human cDNA encodes a protein of 900 amino acid residues. Pol ν is expressed in human cell lines with numerous alternatively spliced transcripts, and a full-length human coding region that comprises 24 exons within 160 kilobases of genomic DNA. Expression analysis by northern blotting and in situ hybridization showed highest expression of full-length pol ν in human and mouse testis, but also in hematopoetic tissues, implying that pol ν may have roles in tissues other than testis. Pol ν has polymerase activity on activated DNA as well as on a singly primed template (Marini et al. 2003). Its biological function and influence on diseases is still not known.

5.2.2 DNA Polymerase B Family

DNA polymerase ζ: pol ζ belongs to the pol B family (Lawrence and Hinkle 1996). Pol ζ of *S. cerevisiae* and human is composed of the catalytic subunit Rev3 and the stimulating subunit Rev7 (Murakumo 2002). Rev3 (980 tyrosine) is the polymerase active site. Human Rev3 amino acid residues 1776–2195 contain a Rev7 binding domain, and Rev7 amino acid residues 1–211 the binding domain for Rev1, Rev3 and Rev7. Pol ζ possesses several physiological functions: (i) Rev7 interacts with hMAD2 and therefore might function in cell cycle control by affecting the activation of anaphase promoting complex (APC) (Murakumo et al. 2000); (ii) it is involved in most spontaneous mutations, lesion-targeted mutation via TLS and somatic hypermutation of immunoglobulin genes (Harfe and Jinks-Robertson 2000; Poltoratsky et al. 2001; Murakumo 2002); (iii) pol ζ acts as a "mismatch extender" in TLS with combination of other pols, such as pol η (Prakash and Prakash 2002); (iv) in mouse embryonic development (Esposito et al. 2000), since mice lacking pol ζ are embryonic lethal (Lawrence 2002; Murakumo 2002).

REV3 –/– cells (*S. cerevisiae*) are sensitive to various DNA-damaging agents, such as UV light, MMS, cisplatin and IR. Moreover, they show reduced gene targeting efficiencies and a significant increase in chromosomal breaks in the subsequent M phase after IR in G2 suggesting an involvement of Rev3 in HR-mediated DSBR. REV3 –/– cells showed a significant increase in sister chromatid exchange and chromosomal breaks even in the absence of exogenous genotoxic stress (Sonoda et al. 2003). Therefore, pol ζ plays a central role in apoptosis, cell proliferation, the control of cell cycle by protein-protein interaction and embryonic development.

In mouse T-lymphoma cells, stress response induced by DNA damage agents (8-methoxy-psoralen or UV-A) leads to specific, delayed and untargeted mutations (UTM) (Boesen et al. 1992). It has been found that low concentration N-methyl-N'-nitro-N-nitrosoguanidine (MNNG), a carcinogen which can induce gastric cancer, could induce mammalian UTM (Zhang et al. 1994). However, it is not clear which factor capable of inhibiting fidelity can be induced or activated. More recently, it was found that pol ζ might be involved in the mammalian UTM induced by MNNG. The transcriptional level of REV3 gene is upregulated when human cells are treated by low concentration MNNG (Zhu et al. 2003). Finally, human cells, in which the function of pol ζ is inhibited by antisense REV3 RNA, display characteristics of both UTM and targeted mutation (Zhu and Zhang 2003).

5.2.3
DNA polymerase Y family

Sequence comparisons allowed the identification of a completely new pol family, called family Y, whose members have been shown to be present in procaryotes, eukaryotes and archaea (Goodman 2002). In human cells, Y-family members are pol ι, pol κ, pol η and Rev1 (Kunkel et al. 2003). Biochemical studies indicated that each of the Y-family pols has a preference for catalysing DNA synthesis across certain kind of lesions, raising the interesting hypothesis that lesion-specific DNA damage response pathways might operate in human cells. Given the highly mutagenic potential of these pols, it is of no surprise that their expression has been investigated, in recent times, in the context of tumor cells. In fact, the so-called "mutator hypothesis" for cancer (reviewed in Loeb and Loeb 2000) suggests that an early event in tumorigenesis is a mutation that inactivates a gene that normally functions to maintain genome stability, resulting in an elevated mutation rate. This elevated mutation rate in turn results in additional mutations in other genes associated with multistage carcinogenesis, which then confer selective advantages that allow mutated cells to expand and achieve clonal dominance.

DNA polymerase η: Human pol η is encoded by the PolH (xeroderma pigmentosum variant, XPV) gene on chromosome 6 and contains 713 amino acids. Its discovery generated particular interest because humans with mutations in the PolH gene (Johnson et al. 1999; Masutani et al. 1999) suffer from xeroderma pigmentosum, a disease characterized, among other phenotypes, by an increased susceptibility to sunlight-induced skin cancer (Kunkel et al. 2003). The loss of pol η results in a reduced efficiency to copy DNA containing a cis-syn cyclobutane dimer (CPDs) (Johnson et al. 1999; Masutani et al. 2000), one of several major lesions generated by exposure to sunlight. Pol η copies thymine-thymine (TT) dimers and the flanking bases with higher processivity than it copies undamaged DNA, and then switches to less processive synthesis. This ability of pol η to sense the dimer location as synthesis proceeds may facilitate polymerase switching before and after lesion bypass. Pol η bypasses a dimer with low fidelity and with higher error rates at the 3′ thymine than at the 5′ thymine (McCulloch et al. 2004). A similar bias was observed with *Sulfolobus solfataricus* DNA pol 4, which forms a Watson-Crick base pair at the 3′ thymine of a dimer but a Hoogsteen base pair at the 5′ thymine. Ultraviolet-induced mutagenesis is also higher at the 3′ base of dipyrimidine sequences. Thus, in individuals with NER-defective xeroderma pigmentosum who accumulate dimers, errors made by pol η during dimer bypass could contribute to mutagenesis and skin cancer. The current hypothesis (reviewed in Friedberg et al. 2002; Prakash and Prakash 2002) is that, unlike certain other UV photoproducts, CPDs are only slowly removed by excision repair and are therefore encountered during DNA replication. Although CPDs block DNA synthesis by pols such as pol α, δ, ε, λ, pol η efficiently bypasses them. In addition to efficient bypass of TT dimers, pol η has been shown to bypass other lesions that distort the DNA helix, albeit with varying efficiencies. Human pol η copies DNA with low processivity and lacks a $3' \rightarrow 5'$ exo activity, so that mistakes have to be proofread by autonomous exos or exos from the classical pols (Bebenek et al. 2001).

Analysis of DNA from lymphocytes of XPV patients showed that pol η generates hypermutation in the μ and γ switch regions of immunoglobulin genes (Zeng et al. 2004). In addition, in vitro it was shown that human pol η and the two other Y-family enzymes pols κ and ι, both of which are dispensable for somatic hypermutation, possess reverse transcriptase activity. It is thus feasible that pol η may act as both an RNA-dependent and a DNA-dependent pol in somatic hypermutation (Franklin et al. 2004).

DNA polymerase κ: Human pol κ is encoded by the POLK gene on chromosome 5 and contains 870 amino acid residues. Like pol η, pol κ also lacks an intrinsic $3' \rightarrow 5'$ exo. Like pol η, pol κ can copy DNA containing lesions that substantially distort the DNA helix geometry. However, its lesion bypass specificity is different from that of pol η. Pol κ does not bypass a TT dimer but it does bypass other lesions, including a benzo[a]pyrene diol epoxide (BPDE) adduct on the N2 of guanine (Rechkoblit et al. 2002; Suzuki et al. 2002). Bypass involves insertion of dCMP, an event that would avoid benzo[a]pyrene-induced mutations. This is particularly interesting because expression of the mouse pol κ gene is under the control of the arylhydrocarbon receptor, a crucial factor for the activation of benzo[a]pyrene into BPDE in mammalian cells (Ogi et al. 2001). This has led to the suggestion that pol κ may participate in the bypass of lesions generated by polycyclic aromatic hydrocarbons (PAH) in a manner that avoids PAH-induced mutations.

Some evidences also exist about a role of pol κ in tumorigenesis. Bavoux et al. showed that ectopic expression of pol κ promoted DNA strand breaks, aneuploidy as well as tumorigenesis in nude mice (Bavoux et al. 2005). Additionally, of eight non-squamos cell lung carcinoma biopsies overexpressing pol κ, seven displayed losses of heterozygosity compared with adjacent non-tumoral tissues. Taken together, these data suggest that misregulation of pol κ can promote the emergence of a large spectrum of genetic disorders associated with a malignant phenotype. However, conflicting results have been published about pol κ, as was the case with pol ι. Pan et al. and Okochi et al. found a reduced expression of pol κ in rat mammary carcinoma cell lines and primary mammary carcinomas in comparison with that of the normal tissues (Okochi et al. 2002; Pan et al. 2005). In addition, Wang et al. reported that pol κ is regulated by p53 in human cells, as well as in murine cells (Wang et al. 2004b). The functional loss of p53 by mutation resulting in the up-regulation of pol κ in human lung cancer tissues (O-Wang et al. 2001; Wang et al. 2004b). However, the results from other investigators indicated that mouse, but not the human, pol κ was primarily regulated by p53 and was up-regulated in response to exposure to various DNA-damaging agents in a p53 dependent manner (Velasco-Miguel et al. 2003). In sum, these results point to the complexity of the regulation of pol κ.

DNA polymerase ι: Human pol ι is encoded by the POLI gene on chromosome 18 and comprises of 715 amino acid residues (McDonald et al. 1999), has a low processivity and lacks an intrinsic $3' \rightarrow 5'$ exo. It can insert nucleotides opposite lesions present in DNA (Johnson et al. 2000; Tissier et al. 2000b). However, pol ι does not extend the resulting template/primers efficiently, such that complete TLS may require extension by a second pol, e.g. pol ζ (Tissier et al. 2000b; Guo et al. 2001; Prakash and Prakash 2002). Pol ι physically interacts with and is stimulated by PCNA (Haracska et al. 2001a). It also physically interacts with pol η and co-localizes with pol η and in a pol η dependent manner following UV irradiation (Kannouche et al. 2003). The fidelity of nucleotide incorporation opposite the different template residues varies in a similar manner. The fidelity of pol ι is high opposite template dAMP, with error frequencies of 10^{-4} to 10^{-5}. Opposite templates dGMP and dCMP, pol ι as a relatively low fidelity, with error frequencies ranging from 10^{-1} to 10^{-2}. Pol ι has an unprecedented low fidelity opposite template dTMP, with error frequencies ranging from 10^{+1} to 10^{-1}. The fidelity opposite template dTMP is so poor that pol ι preferentially inserts the incorrect dGMP 10-fold more efficiently than the correct dAMP (Tissier et al. 2000a; Zhang et al. 2000; Haracska et al. 2001b). Recent data might explain pol ι fidelity differences for nucleotides. In the early stage base pair is accommodated differently in the active site of pol ι, depending on whether dAMP or dTMP is the templating base. When the dTMP is the templating base, the polymerase active site better accommodates the Wobble base pair rather than the Watson-Crick base pair (Washington et al. 2004).

A growing body of evidences point to the involvement of pol ι in lung cancer progression. In one study, Wang et al. performed a systematic candidate gene analyses of the pulmonary adenoma resistance 2 locus. Differential gene expression in lung tissues and nucleotide polymorphisms analysis revealed that the gene encoding pol ι contained 25 nucleotide polymorphisms in its coding region between A/J and BALB/cJ mice, resulting in a total of ten amino acid changes. Purified BALB/cJ and A/J pol ι were shown to differ in substrate discrimination in vitro. Moreover, altered expression of pol ι protein and an amino acid-changing nucleotide polymorphism were observed in human lung cancer cells, suggesting a possible role in the development of lung cancer (Wang et al. 2004a). In another case-control study, Lee et al. identified a single nucleotide polymorphism in pol ι (Thr706Ala), which correlated with a significantly higher risk of lung adenocarcinoma and squamous cell carcinoma (Lee et al. 2003). A third study provided a potential link between altered pol ι expression and mutagenesis in breast cancer. Yang et al. found that pol ι expression was elevated in breast cancer cells and this correlated with a significant decrease in DNA replication fidelity (Yang et al. 2004). Interestingly, after immunodepletion of pol ι from nuclear extracts of the cells, a reduction in mutation frequency was found, as measured in vitro. These data suggest that pol ι may play a role in the high mutation frequencies observed in breast cancer cells. Another conflicting study, however, questioned the existence of a link between pol ι and cancer. Pan et al., measured the transcripts of pol κ, η, ι, and ζ in 131 self-paired cancerous and non-tumor samples, including 23 lung cancers, 49 stomach cancers, and 59 colorectal cancers (Pan et al. 2005). Their results indicated that, except pol η in colorectal cancers, all other enzymes were significantly down-regulated in human lung, stomach, and colorectal cancers. However, this study still does not rule out the possibility that the variation in the expression levels might be a function of the tumour stage or cell-type specific.

Rev1: The human REV1 gene is localized on chromosome 2 and codes for a protein of 1251 amino acid residues with a calculated molecular weight of 138 kDa. The REV1 gene is ubiquitously expressed in various human tissues (Lin et al. 1999). Yeast and human Rev1 protein is a terminal nucleotidyltransferase that inserts dCMP opposite template dGMP, dAMP and AP sites (Lawrence and Hinkle 1996) and is a dGMP template-specific pol (Haracska et al. 2002; Masuda and Kamiya 2002). In *S. cerevisiae* REV 1 belongs to RAD6 repair pathway and depends on the functions of the REV3 and REV7 genes (subunits of pol ζ, that appears to carry out translesion synthesis). In vivo, Pol-dependent translesion replication also requires the activity of the Rev1 protein (Lawrence and Christensen 1976). These results suggest that the Rev1 transferase may play a critical role during mutagenic translesion DNA synthesis, bypassing a template AP site in the cells, and thus might play an important role in regulating the enzymatic activities of hRev1 and hRev3 for mutagenesis in response to DNA damage (Murakumo et al. 2001). Rev1 interacts independently with Rev7 and with other Y-family pols, pol η pol κ and pol ι. These observations suggest that Rev1 plays a role(s) in mediating protein-protein interactions among pols required for TLS. The precise function(s) of these interactions during TLS remains to be determined (Guo et al. 2003). Cellular localization studies in human cells demonstrated that Rev1 is present, with pol η in replication factories at stalled replication forks and is tightly associated with nuclear structures. This Rev1 nuclear localization occurs independently of the presence of pol η. Taken together, a central role is proposed for Rev1 as a scaffold that recruits pols involved in TLS (Tissier et al. 2004).

Rev1-deficient chicken DT40 cells display reduced viability and are sensitive to a wide range of DNA-damaging agents (Simpson and Sale 2003). "Knock down" of REV1 mRNA in human cells results in a hypomutable phenotype after UV treatment (Clark et al. 2003). In addition to its catalytic domain, Rev1 possesses a so-called BRCA1 C-terminal (BRCT) domain. Mice containing a targeted deletion of this domain are healthy, fertile and display normal somatic hypermutation. *Rev1BRCT1 –/–* cells display an elevated spontaneous frequency of intragenic deletions at Hprt. In addition, these

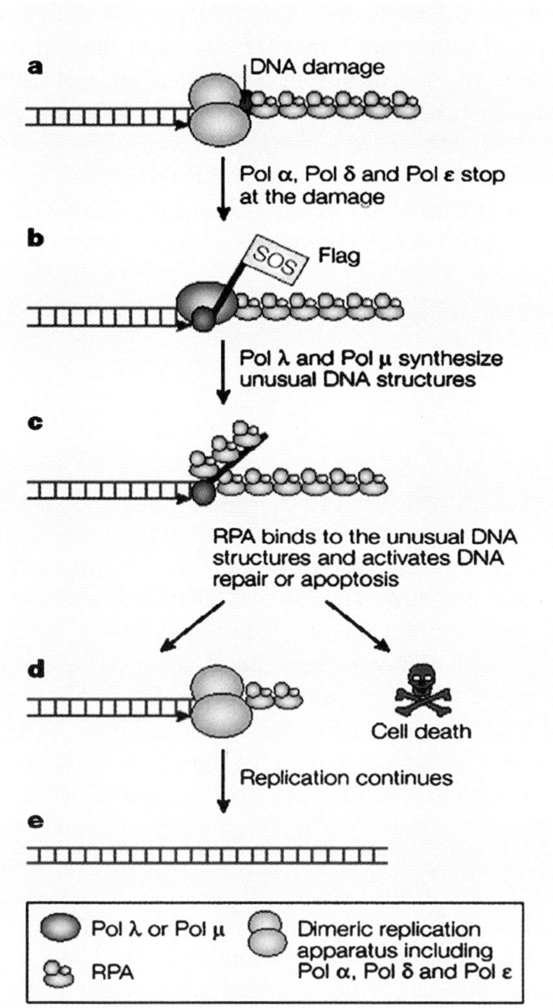

Fig. 1 "Flag waving" model for Pol X family. Our hypothetical "flag waving" model is based on specific properties of Pol λ, Pol μ or TdT. **a** The DNA-replication machinery meets a lesion on the DNA (*black sphere*). Pols α, δ and ε are unable to bypass DNA lesions and their arrest causes stalling of the replication fork, which might be rescued by switching between replicative DNA polymerases and Pol λ and/or Pol μ. **b** As a consequence, these two pol X family members might then synthesize unusual DNA structures, thereby creating the alarm signal "flag" **c** Replication protein A (RPA) binds to unusual DNA structures, which might provide the signal at the sites of DNA damage that activates the initiation of certain DNA-repair and/or DNA-damage-checkpoint pathways. **d** In case that the damaged DNA was not properly repaired, the final consequences of the DNA-damage-checkpoint pathways are further DNA repair or induction of apoptosis. **e** In the event of successful DNA repair, DNA replication continues rendering the correct information to the daughter cell (from Ramadan et al. 2004a). (Courtesy of Nature Reviews Molecular Cell Biology, www.nature.com/reviews)

cells were sensitized to exogenous DNA damages. UV-C light induced a delayed progression through late S and G2 phases of the cell cycle and many chromatid aberrations, specifically in a subset of mutant cells, but not enhanced sister chromatid exchanges. UV-C-induced mutagenesis was reduced and mutations at TT dimers were absent in *Rev1BRCT1* −/− cells, the opposite phenotype of UV-C-exposed cells from XP-V patients, lacking pol η. This suggested that the enhanced UV-induced mutagenesis in XP-V patients may depend on error-prone Rev1-dependent TLS. These data indicate a regulatory role of the Rev1 BRCT domain in TLS of a limited spectrum of endogenous and exogenous nucleotide damages during a defined phase of the cell cycle (Jansen et al. 2005).

5.2.4 DNA Polymerase X Family

Beside pol β and TdT, which have been described above, novel members of the pol family X were identified in mammalian cells and were called pol λ (67 kDa) and pol μ (55 kDa). According to the size and the amino acid sequence motifs they are more similar to TdT and *S. cerevisiae* pol IV than to pol β. They both contain on the *N*-terminal part a BRCT domain and a proline/serine rich motif not found in pol β. Genome database searching for homologous of pol X, orthologs of the pol X family members were identified in archaea and bacteria (for details, see Ramadan et al. 2004a and citations therein). Moreover, the three members of the pol X family pol λ, pol μ and TdT were found to synthesize de novo DNA of unusual structure (Ramadan et al. 2004b). Figure 1 represents a hypothetical model how pol X family pols might sense damaged DNA, and thus prevent pathological consequences for the cell.

DNA polymerase λ: The POLL gene, is localized on chromosome 10 in human (Aoufouchi et al. 2000; Garcia-Diaz et al. 2000). Human pol λ contains 575 amino acid residues resulting in a molecular weight of 67–70. Structural modelling suggested that pol λ contains the palm, finger, thumb and 8 kDa domains present in pol β, as well as an additional domain not occurring in pol β. The first 230 amino acids contain the BRCT and proline/serine rich domains. The C-terminal part of pol λ bears 32–33% sequence identity to the corresponding region of pol β. Pol λ is expressed at the highest level in testis, ovary (Garcia-Diaz et al. 2000) and fetal liver (Aoufouchi et al. 2000). Pol λ possess DNA template dependent, terminal deoxynucleotidyl transferase activitiy and can synthesize DNA de novo (Garcia-Diaz et al. 2002; Shimazaki et al. 2002; Ramadan et al. 2003, 2004b). Moreover, pol λ efficiently elongates DNA from an RNA primer (Ramadan et al. 2003). The terminal deoxynucleotidyl transferase activity of pol λ is sequence specific and it preferentially incorporates pyrimidine nucleotides (Ramadan et al. 2003). Naturally isolated pol λ can preferentially synthesize over apurinic DNA, suggesting a role in

DNA repair (Ramadan et al. 2002), and this translesion synthesis is stimulated by PCNA (Maga et al. 2002b). Moreover, dRP-lyase activity suggests a role in BER. Indeed pol λ can efficiently repair uracil-containing DNA in an in vitro reconstituted BER reaction (Garcia-Diaz et al. 2001). Pol λ might also participate in DSBR, since immunodepletion of a human cell extract indicated that pol λ might be the primary gap filling pol for accurate NHEJ, and that the BRCT domain might be required for this activity (Lee et al. 2004). Pol λ synthesizes DNA in a distributive manner, and its interaction with PCNA increases the processivity in normal and translesion synthesis (Maga et al. 2002b). A remarkable feature of pol λ is its preference for Mn^{2+}, over Mg^{2+}, and this might be related to its ability to bypass an abasic site (Blanca et al. 2003).

Upon treatment of mammalian cells with DNA damaging agents (UV, γ-irradiation and H_2O_2) mRNA of pol λ is downregulated (Aoufouchi et al. 2000). Knock out mouse pol λ cells were obtained from two groups (Bertocci et al. 2002; Kobayashi et al. 2002). Kobayashi et al. showed no differences between mouse embryonic fibroblast (MEF) pol $\lambda^{-/-}$ and pol $\lambda^{+/+}$ cells to various DNA damaging agents. They were also successful in obtaining pol $\lambda^{-/-}$ mice. The surviving males, but not the females were sterile and this is a result of spermatozoal immobility. The main phenotypes were hydrocephalus, situs inversus, and chronic sinusitis resulting in *the immotile cilia syndrome.* There is, however, suspicion regarding this strong phenotype of pol $\lambda^{-/-}$ mice, and the current belief is that Kobayashi et al. knocked out another closer gene responsible for *the immotile cilia syndrome* (L. Blanco and C-A. Reynaud, personal communications). The group of Reynaud obtained knock-out pol $\lambda^{-/-}$ mice in which the males were fertile, and homozygous breeding has been performed up to the third generation without a noticeable problem (Bertocci et al. 2002), suggesting that this enzyme is dispensable for mouse development. They also showed that pol λ is dispensable for Ig gene hypermutation.

DNA polymerase μ: Pol μ has among the pol X family the strongest homology to TdT with 42% amino acid identity (Aoufouchi et al. 2000; Dominguez et al. 2000). The human POLM gene is located on chromosome 7 and encodes a protein of 494 amino acid (55 kDa). Pol μ is predominantly expressed in peripheral lymphoid tissues, although basal levels were detected in other tissues. In situ analysis of mRNA pol μ in human tissues section showed an abundant association with follicular areas of secondary lymphoid tissues (Ruiz et al. 2001). Pol μ possesses both DNA template dependent, template independent (terminal transferase) activities and can perform de novo DNA synthesis. All of these activities prefer Mn^{2+} to Mg^{2+} as a divalent cation. The terminal transferase of pol μ preferentially incorporates pyrimidine, with the highest affinity for TTP over dCTP > dGTP > dATP. In the presence of Mn^{2+}, pol μ behaves as a strong mutator, lacking base discrimination during nucleotide insertion. In rare cases, pol μ preferentially inserts a dGMP in front of its complementary dCMP (Dominguez et al. 2000). Pol μ

can perform translesion synthesis. It can efficiently extend different small DNA lesions such as single *N*-2-acetyl-aminofluorene (AAF), 8-oxoguanine, AP site and 1,N^6-ethenoadenine as well as bulky DNA lesions such as AAF, + and – trans-anti-benzo[α]pyrene-N^2-dG and TT dimer (Duvauchelle et al. 2002; Zhang et al. 2002).

A novel bypass mechanism over AP lesions by pol μ was described and is probably due to template-dependent but sequence independent nucleotidyl transferase activity (Covo et al. 2004). Thereby pol μ would extend via its nucleotidyl transferase activity primers whose 3′-terminal nucleotides are located opposite an AP site, which does not depend on the sequence of the template. This is not due to simple terminal transferase activity, since pol μ is unable to add dNTPs under standard conditions to an single stranded oligo(dT) primer or to a blunt-end duplex oligonucleotide. The AP is probably skipped over, leading to a minus one deletion at the lesion site. Pol μ, can finally form a protein complex with the end-joining factors Ku and XRCC4-Ligase IV thus further supporting a role in DSBR (Mahajan et al. 2002). In addition, the biochemical properties of pol μ suggest a function in NHEJ and V(D)J recombination. Pol μ can incorporate both dNTPs and rNTP into DNA and thus contributes to genetic variability and DNA repair (Ruiz et al. 2003).

Treating mammalian cells with DNA damaging agents (UV, γ-irradiation and H_2O_2) caused down-regulation of pol μ mRNA (Aoufouchi et al. 2000). Knock-out pol μ did not show a specific defect in lymphoid tissues. In contrast, about half of the mice had a marked depletion of B cells in peripheral lymphoid organs (Bertocci et al. 2002). Pol μ mice have a specific alteration in the $IgM^{-/-}$ to $IgM^{+/+}$ transition in bone marrow. Immunoglobulin light chain gene rearrangement was impaired at the levels of Vκ-Jκ and Vλ-Jλ junctions. These alterations lead to a profound defect in the peripheral B cell compartment, which causes 40% reduction in the splenic B cell fractions. Pol μ appears, therefore, as a key factor contributing to the relative homogeneity in size of light chain CDR3 and takes part in immunoglobulin gene rearrangement at a stage where TdT is no longer expressed (Bertocci et al. 2003). The close association of pol μ with cells of the germinal centres cells, its error-prone nature and its strong expression by postgerminal centres in non-Hodgkin's lymphoma (NHL) B cells suggested that pol μ is a candidate to be involved in this cancer. Moreover, a close association was established between pol μ expression and B cells NHL (Chiu et al. 2002).

DNA polymerase σ 1(Trf4) and 2 (Trf): The pol σ 1 gene (TRF4) was first identified in a genetic screen for a function redundant with DNA topoisomerase I (Sadoff et al. 1995), and showed that together with its close homolog, TRF5, it is essential for chromosomal segregation (Castano et al. 1996a,b), and DNA damage repair (Walowsky et al. 1999). The TFR family is represented by multiple members in all eukaryotes. In yeast *S. cerevisiae* the name of the protein encoded by TRF gene is Trf and in mammalian cells pol σ. The Trf5 (pol σ) protein is 55% identical and 72% similar to Trf4. Trf4 is able to extend a primer hybridized to a template in a distributive manner, and its polymerization reaction is template dependent and directed. Deletion of the *N*-terminal part inactivates the Trf4 pol.

A *S. cerevisiae* TRF4/5 double mutant is hypersensitive to the anti-microtubule agent thiabendazole, suggesting that Trf4 and Trf5 functions are required at the time of mitosis (Castano et al. 1996a). Trf4 was proposed to be a linker between DNA replication and cohesion between sister chromatids. This event occurs during DNA replication and is crucial for accurate segregation at anaphase. Trf4 genetically and physically interacts with Smc1, a protein required for sister chromatin cohesion. TRF4 mutant cells showed defective cohesion. Cells with a double mutant TRF4/5 cannot completely replicate the genome. The double mutant TRF4/5 cells rapidly die and are unable to complete the S phase. The Trf4 pol activity is required for proper cohesion and also for overlapping essential functions of Trf4 and Trf5 (Wang et al. 2000). Genetic analysis in *S. cerevisiae* suggested that regions of structural similarities to the pol β are important for the Trf/pol σ function. Recently, it was found that the large subunit of *S. cerevisiae* pol ε interacts with pol σ. This interaction significantly stimulated pol ε, suggesting that pol ε might form a link between the replication machinery and sister chromatid cohesion (Edwards et al. 2003). The model proposes that the Trf4/pol σ functions to establish cohesion through a pol switch mechanism in which the replicative pols are used to replicate from an origin of replication to a cohesion site. Finally, when the replicative pols encounters a cohesion site, a switch to pol σ might occur (Carson and Christman 2001).

The fission yeast (*S. pombe*) genome contains 6 TRF4/5 related genes: cid1, cid11, cid12, cid13, cid14 and cid16. It has not yet been shown that they are DNA polymerases, but Cid1 and Cid 13 are cytoplasmic poly(A) polymerases (Read et al. 2002; Saitoh et al. 2002). Cid1 and Cid13 constitutively localize to the cytoplasm, suggesting that their targets are also cytoplasmic (Read et al. 2002). Moreover, Cid13 is a cytoplasmic enzyme that specifically polyadenylates suc22 mRNA, which encodes a subunit of ribonucleotide reductase (RNR) (Saitoh et al. 2002). Cid13 mutants have reduced dNTP pools and are sensitive to hydroxyurea, an RNR inhibitor (Saitoh et al. 2002). Probably Cid13 functions during difficult conditions in the cell when rapid consumption of dNTPs is needed for DNA replication such as DNA damage or heat shock.

5.3
Telomerases

A specialized pol, called telomerase, is responsible for synthesis of telomeres (chromosomal ends) in eukaryotic cells. Telomeres are essential for genome stability. Changes in telomere functions and the associated chromosomal abnormalities have been implicated in human aging and cancer. Maintenance of the physical integrity of telomeres is critical event coupled to DNA replication. Telomeres are composed of repetitive sequences that can be maintained by telomerase, a complex containing a reverse transcriptase (TERT in humans and Est2 in budding yeast), a template RNA (TERC in humans and Tlc1 in yeast), and accessory factors (the Est1 protein and dyskerin in humans and Est1, Est3, and Sm in budding yeast). Telomerase is regulated by proteins that bind to telomeric DNA. This regulation can take place at the telomere terminus, involving a single-stranded DNA-binding protein (POT1 in humans and Cdc13 in budding yeast), which have been proposed to contribute to the recruitment of telomerase and may also regulate the extent or frequency of elongation. In addition, an active role of DNA replication enzymes was postulated in controlling telomere length (Adams Martin et al. 2000). Some indications exist that pol α/prim has a direct role in synthesis of telomere (Qi and Zakian 2000). Telomeric DNA synthesis by telomerase should be tightly co-regulated with the production of the opposite strand to prevent telomerase from generating excessively long single-strand tails, which may be deleterious to chromosome stability.

Telomerase reactivation is the most common tumour marker, since more than 90% express the enzyme. Evidence indicates that telomerase plays roles in tumorigenesis besides telomere maintenance (Chang and DePinho 2002; Chang et al. 2004). The protection against apoptosis may be one of the ways by which telomerase favours tumour development. The mechanisms underlying this protection could be of great significance for the development of new cancer therapies. The knowledge that anti-cancer drugs can sensitize tumour cells by inhibiting telomerase can be used to develop new protocols for chemosensitizing tumors. Moreover, because telomeres can be long in cancer cells, and because a long lag time may be required for a cytotoxic effect, the efficacy of treatments could be improved with agents that simultaneously target telomeres and telomerase, such as G-quadruplex ligands. Interestingly, it has been reported that, in human embryonic kidney cells and in endothelial cells, the regulation of telomerase activity and its anti-apoptotic effect are post-transcriptionally modulated by phosphorylation and by the interaction with binding proteins such as HSP90 and Akt (Haendeler et al. 2003). By interfering with telomerase, specific kinases inhibitors could provide another option for treating human malignancy. Although the idea of interfering with the metabolism of telomeres to kill tumour cells is very attractive, but there are two issues to be considered. First, telomerase activation is not the only mechanism by which cells stabilize their telomeres. Up to 10% of human tumour cells that lack detectable telomerase activity are able to maintain the length of their telomeres with ALT (alternative lengthening of telomeres), a telomerase-independent mechanism that may be mediated by recombination (Dunham et al. 2000). Therefore, specific treatments to block the ALT pathway have to be developed. Second, recent data indicate that telomerase is also expressed in normal cells (Masutomi et al. 2003); more investigation is required to determine the consequences of telomerase inhibition in normal cells.

The mouse model showed direct involvement of telomerase in aging. The varied and complex cellular phenotypes of Werner syndrome are precipitated by exhaustion of telomere reserves in mice. Telomere dysfunction elicits a classical Werner-like premature aging syndrome typified by premature death, hair greying, alopecia, osteoporosis, type II diabetes and cataracts. This mouse model also showed accelerated replicative senescence and accumulation of DNA-damage foci in cultured cells, as well as increased chromosomal instability and cancer, particularly non-epithelial malignancies typical for Werner syndrome. These genetic data indicate that the delayed manifestation of the complex pleiotropic of Werner deficiency relates to telomere shortening (Chang et al. 2004).

6 Conclusion

For 50 years DNA polymerases have been studied for their biochemical (in vitro) properties and as results, some of them are today used in biotechnology (e.g. used for PCR or Tunel assay). Although many researchers work with mammalian diseases, especially human, very few are focused on enzymes that act as key players in DNA metabolism. In the last 10 years at least 10 novel mammalian DNA polymerases have been identified, arguing their important role in the cell. Why the nature creates so huge diversity of DNA polymerases in the cell? Genetic or epigenetic factors involved in misregulation of the DNA polymerases could present new prognostic, diagnostic and therapeutic tools in medicine. Understanding the molecular details of these processes will be the challenge for the near future.

References

Aboussekhra A et al. (1995) Mammalian DNA nucleotide excision repair reconstituted with purified protein components. Cell 80:859–868

Adams Martin A, Dionne I, Wellinger RJ, Holm C (2000) The function of DNA polymerase alpha at telomeric G tails is important for telomere homeostasis. Mol Cell Biol 20:786–796

Albertella MR, Lau A, O'Connor MJ (2005) The overexpression of specialized DNA polymerases in cancer. DNA Repair (Amst) 4:583–593

Aoufouchi S et al. (2000) Two novel human and mouse DNA polymerases of the polX family. Nucleic Acids Res 28:3684–3693

Bavoux C et al. (2005) Up-regulation of the error-prone DNA polymerase $\{\kappa\}$ promotes pleiotropic genetic alterations and tumorigenesis. Cancer Res 65:325–330

Bebenek K, Matsuda T, Masutani C, Hanaoka F, Kunkel TA (2001) Proofreading of DNA polymerase eta-dependent replication errors. J Biol Chem 276:2317–2320

Bergoglio V, Ferrari E, Hubscher U, Cazaux C, Hoffmann JS (2003) DNA Polymerase beta can Incorporate Ribonucleotides during DNA Synthesis of Undamaged, CPD-damaged DNA. Mol J Biol 331:1017–1023

Bergoglio V et al. (2004) Evidence of finely tuned expression of DNA polymerase beta in vivo using transgenic mice. FEBS Lett 566:147–150

Bertocci B, De Smet A, Berek C, Weill JC, Reynaud CA (2003) Immunoglobulin kappa light chain gene rearrangement is impaired in mice deficient for DNA polymerase mu. Immunity 19:203–211

Bertocci B et al. (2002) Cutting edge: DNA polymerases mu and lambda are dispensable for Ig gene hypermutation. J Immunol 168:3702–3706

Blanca G et al. (2003) Human DNA polymerase lambda diverged in evolution from DNA polymerase beta toward specific Mn(++) dependence: a kinetic and thermodynamic study. Biochemistry 42:7467–7476

Boesen JJ et al. (1992) Stress response induced by DNA damage leads to specific, delayed and untargeted mutations. Mol Gen Genet 234:217–227

Bollum FJ, Chang LM, Tsiapalis CM, Dorson JW (1974) Nucleotide polymerizing enzymes from calf thymus gland. Methods Enzymol 29:70–81

Boule JB, Rougeon F, Papanicolaou C (2001) Terminal deoxynucleotidyl transferase indiscriminately incorporates ribonucleotides and deoxyribonucleotides. J Biol Chem 276:31388–31393

Boyd JB, Sakaguchi K, Harris PV (1990) mus308 mutants of Drosophila exhibit hypersensitivity to DNA cross-linking agents and are defective in a deoxyribonuclease. Genetics 125:813–819

Brautigam CA, Steitz TA (1998) Structural and functional insights provided by crystal structures of DNA polymerases and their substrate complexes. Curr Opin Struct Biol 8:54–63

Burgers PM (1998) Eukaryotic DNA polymerases in DNA replication, DNA repair. Chromosoma 107:218–227

Burgers PM et al. (2001) Eukaryotic DNA polymerases: proposal for a revised nomenclature. J Biol Chem 276:43487–43490

Byrnes JJ, Downey KM, Black VL, So AG (1976) A new mammalian DNA polymerase with 3′ to 5′ exonuclease activity: DNA polymerase delta. Biochemistry 15:2817–2823

Carrodeguas JA, Theis K, Bogenhagen DF, Kisker C (2001) Crystal structure and deletion analysis show that the accessory subunit of mammalian DNA polymerase gamma, Pol gamma B, functions as a homodimer. Mol Cell 7:43–54

Carson DR, Christman MF (2001) Evidence that replication fork components catalyze establishment of cohesion between sister chromatids. Proc Natl Acad Sci USA 98:8270–8275

Castano IB, Brzoska PM, Sadoff BU, Chen H, Christman MF (1996) Mitotic chromosome condensation in the rDNA requires TRF4, DNA topoisomerase I in *Saccharomyces cerevisiae*. Genes Dev 10:2564–2576

Castano IB, Heath-Pagliuso S, Sadoff BU, Fitzhugh DJ, Christman MF (1996) A novel family of TRF (DNA topoisomerase I-related function) genes required for proper nuclear segregation. Nucleic Acids Res 24:2404–2410

Chang S, DePinho RA (2002) Telomerase extracurricular activities. Proc Natl Acad Sci USA 99:12520–12522. Epub 12002 Sep 12523

Chang S et al. (2004) Essential role of limiting telomeres in the pathogenesis of Werner syndrome. Nat Genet 36:877–882. Epub 2004 Jul 2004

Chiu A et al. (2002) DNA polymerase mu gene expression in B-cell non-Hodgkin's lymphomas: an analysis utilizing in situ hybridization. Am J Pathol 161:1349–1355

Christmann M, Tomicic MT, Roos WP, Kaina B (2003) Mechanisms of human DNA repair: an update. Toxicology 193:3–34

Clark DR, Zacharias W, Panaitescu L, McGregor WG (2003) Ribozyme-mediated REV1 inhibition reduces the frequency of UV-induced mutations in the human HPRT gene. Nucleic Acids Res 31:4981–4988

Corral-Debrinski M, Horton T, Lott MT, Shoffner JM, Beal MF, Wallace DC (1992) Mitochondrial DNA deletions in human brain: regional variability and increase with advanced age. Nat Genet 2:324–329

Cottrell DA, Blakely EL, Johnson MA, Ince PG, Borthwick GM, Turnbull DM (2001) Cytochrome c oxidase deficient cells accumulate in the hippocampus and choroid plexus with age. Neurobiol Aging 22:265–272

Covo S, Blanco L, Livneh Z (2004) Lesion bypass by human DNA polymerase mu reveals a template-dependent, sequence-independent nucleotidyl transferase activity. J Biol Chem 279:859–865

D'Urso G, Nurse P (1997) Schizosaccharomyces pombe cdc20+ encodes DNA polymerase epsilon and is required for chromosomal replication but not for the S phase checkpoint. Proc Natl Acad Sci USA 94:12491–12496

Dianov GL, Prasad R, Wilson SH, Bohr VA (1999) Role of DNA polymerase beta in the excision step of long patch mammalian base excision repair. J Biol Chem 274:13741–13743

Dominguez O et al. (2000) DNA polymerase mu (Pol mu), homologous to Td T, could act as a DNA mutator in eukaryotic cells. Embo J 19:1731–1742

Dunham MA, Neumann AA, Fasching CL, Reddel RR (2000) Telomere maintenance by recombination in human cells. Nat Genet 26:447–450

Duvauchelle JB, Blanco L, Fuchs RP, Cordonnier AM (2002) Human DNA polymerase mu (Pol mu) exhibits an unusual replication slippage ability at AAF lesion. Nucleic Acids Res 30:2061–2067

Edwards S, Li CM, Levy DL, Brown J, Snow PM, Campbell JL (2003) *Saccharomyces cerevisiae* DNA polymerase epsilon and polymerase sigma interact physically and functionally, suggesting a role for polymerase epsilon in sister chromatid cohesion. Mol Cell Biol 23:2733–2748

Efrati E, Tocco G, Eritja R, Wilson SH, Goodman MF (1997) Abasic translesion synthesis by DNA polymerase beta violates the A-rule. Novel types of nucleotide incorporation by human DNA polymerase beta at an abasic lesion in different sequence contexts. J Biol Chem 272:2559–2569

Esposito G, Godindagger I, Klein U, Yaspo ML, Cumano A, Rajewsky K (2000) Disruption of the Rev3l-encoded catalytic subunit of polymerase zeta in mice results in early embryonic lethality. Curr Biol 10:1221–1224

Ferrari G et al. (2005) Infantile hepatocerebral syndromes associated with mutations in the mitochondrial DNA polymerase-gamma. Brain A 128:723–731

Foiani M, Lucchini G, Plevani P (1997) The DNA polymerase alpha-primase complex couples DNA replication, cell-cycle progression, DNA-damage response. Trends Biochem Sci 22:424–427

Francesconi S, Park H, Wang TS (1993) Fission yeast with DNA polymerase delta temperature-sensitive alleles exhibits cell division cycle phenotype. Nucleic Acids Res 21:3821–3828

Franklin A, Milburn PJ, Blanden RV, Steele EJ (2004) Human DNA polymerase-eta, an A-T mutator in somatic hypermutation of rearranged immunoglobulin genes, is a reverse transcriptase. Immunol Cell Biol 82:219–225

Frechet M, Canitrot Y, Bieth A, Dogliotti E, Cazaux C, Hoffmann JS (2002) Deregulated DNA polymerase beta strengthens ionizing radiation-induced nucleotidic and chromosomal instabilities. Oncogene 21:2320–2327

Frechet M, Canitrot Y, Cazaux C, Hoffmann JS (2001) DNA polymerase beta imbalance increases apoptosis and mutagenesis induced by oxidative stress. FEBS Lett 505:229–232

Friedberg EC, Wagner R, Radman M (2002) Specialized DNA polymerases, cellular survival, and the genesis of mutations. Science 296:1627–1630

Garcia-Diaz M, Bebenek K, Kunkel TA, Blanco L (2001) Identification of an intrinsic 5′-deoxyribose-5-phosphate lyase activity in human DNA polymerase lambda: a possible role in base excision repair. J Biol Chem 276:34659–34663

Garcia-Diaz M et al. (2002) DNA polymerase lambda, a novel DNA repair enzyme in human cells. J Biol Chem 277:13184–13191

Garcia-Diaz M et al. (2000) DNA polymerase lambda (Pol lambda), a novel eukaryotic DNA polymerase with a potential role in meiosis. J Mol Biol 301:851–867

Giot L, Chanet R, Simon M, Facca C, Faye G (1997) Involvement of the yeast DNA polymerase delta in DNA repair in vivo. Genetics 146:1239–1251

Goldsby RE et al. (2002) High incidence of epithelial cancers in mice deficient for DNA polymerase delta proofreading. Proc Natl Acad Sci USA 99:15560–15565

Goldsby RE et al. (2001) Defective DNA polymerase-delta proofreading causes cancer susceptibility in mice. Nat Med 7:638–639

Goodman MF (2002) Error-prone repair DNA polymerases in prokaryotes and eukaryotes. Annu Rev Biochem 71:17–50

Graves SW, Johnson AA, Johnson KA (1998) Expression, purification, and initial kinetic characterization of the large subunit of the human mitochondrial DNA polymerase. Biochemistry 37:6050–6058

Graziewicz MA, Longley MJ, Bienstock RJ, Zeviani M, Copeland WC (2004) Structure-function defects of human mitochondrial DNA polymerase in autosomal dominant progressive external ophthalmoplegia. Nat Struct Mol Biol 11:770–776

Grey M, Dusterhoft A, Henriques JA, Brendel M (1996) Allelism of PSO4, PRP19 links pre-mRNA processing with recombination and error-prone DNA repair in Saccharomyces cerevisiae. Nucleic Acids Res 24:4009–4014

Gu H, Marth JD, Orban PC, Mossmann H, Rajewsky K (1994) Deletion of a DNA polymerase beta gene segment in T cells using cell type-specific gene targeting. Science 265:103–106

Guo C et al. (2003) Mouse Rev1 protein interacts with multiple DNA polymerases involved in translesion DNA synthesis. Embo J 22:6621–6630

Guo D, Wu X, Rajpal DK, Taylor JS, Wang Z (2001) Translesion synthesis by yeast DNA polymerase zeta from templates containing lesions of ultraviolet radiation and acetylaminofluorene. Nucleic Acids Res 29:2875–2883

Haendeler J, Hoffmann J, Rahman S, Zeiher AM, Dimmeler S (2003) Regulation of telomerase activity and anti-apoptotic function by protein-protein interaction and phosphorylation. FEBS Lett 536:180–186

Haracska L et al. (2001a) Targeting of human DNA polymerase iota to the replication machinery via interaction with PCN. Proc Natl Acad Sci USA 98:14256–14261

Haracska L, Prakash S, Prakash L (2002) Yeast Rev1 protein is a G template-specific DNA polymerase. J Biol Chem 277:15546–15551

Haracska L et al. (2001b) Roles of yeast DNA polymerases delta and zeta and of Rev1 in the bypass of abasic sites. Genes Dev 15:945–954

Harfe BD, Jinks-Robertson S (2000) DNA mismatch repair and genetic instability. Annu Rev Genet 34:359–399

Harris PV, Mazina OM, Leonhardt EA, Case RB, Boyd JB, Burtis KC (1996) Molecular cloning of Drosophila mus308, a gene involved in DNA cross-link repair with homology to prokaryotic DNA polymerase I genes. Mol Cell Biol 16:5764–5771

Hoffmann JS, Pillaire MJ, Maga G, Podust V, Hubscher U, Villani G (1995) DNA polymerase beta bypasses in vitro a single d(GpG)-cisplatin adduct placed on codon 13 of the HRAS gene. Proc Natl Acad Sci USA 92:5356–5360

Holm L, Sander C (1995) DNA polymerase beta belongs to an ancient nucleotidyltransferase superfamily. Trends Biochem Sci 20:345–347

Holmes AM, Haber JE (1999) Double-strand break repair in yeast requires both leading and lagging strand DNA polymerases. Cell 96:415–424

Hubscher U, Maga G, Spadari S (2002) Eukaryotic DNA polymerases. Annu Rev Biochem 71:133–163

Hubscher U, Nasheuer HP, Syvaoja JE (2000) Eukaryotic DNA polymerases, a growing family. Trends Biochem Sci 25:143–147

Idriss HT, Al-Assar O, Wilson SH (2002) DNA polymerase beta. Int J Biochem Cell Biol 34:321–324

Jansen JG et al. (2005) The BRCT domain of mammalian Rev1 is involved in regulating DNA translesion synthesis. Nucleic Acids Res 33:356–365. Print 2005

Jin YH et al. (2005) The multiple biological roles of the 3′–5′ exonuclease of *Saccharomyces Cerevisiae* DNA polymerase delta require switching between the polymerase and exonuclease domains. Mol Cell Biol 25:461–471

Johnson RE, Kondratick CM, Prakash S, Prakash L (1999) hRAD30 mutations in the variant form of xeroderma pigmentosum. Science 285:263–265

Johnson RE, Washington MT, Haracska L, Prakash S, Prakash L (2000) Eukaryotic polymerases iota and zeta act sequentially to bypass DNA lesions. Nature 406:1015–1019

Joyce CM, Steitz TA (1994) Function and structure relationships in DNA polymerases. Annu Rev Biochem 63:777–822

Kaguni LS (2004) DNA polymerase gamma, the mitochondrial replicase. Annu Rev Biochem 73:293–320

Kannouche P et al. (2003) Localization of DNA polymerases eta and iota to the replication machinery is tightly co-ordinated in human cells. Embo J 22:1223–1233

Kelman Z, O'Donnell M (1994) DNA replication: enzymology and mechanisms. Curr Opin Genet Dev 4:185–195

Kesti T, Flick K, Keranen S, Syvaoja JE, Wittenberg C (1999) DNA polymerase epsilon catalytic domains are dispensable for DNA replication, DNA repair, and cell viability. Mol Cell 3:679–685

Kobayashi Y et al. (2002) Hydrocephalus, situs inversus, chronic sinusitis, and male infertility in DNA polymerase lambda-deficient mice: possible implication for the pathogenesis of immotile cilia syndrome. Mol Cell Biol 22:2769–2776

Kodama EN, McCaffrey RP, Yusa K, Mitsuya H (2000) Antileukemic activity and mechanism of action of cordycepin against terminal deoxynucleotidyl transferase-positive (TdT+) leukemic cells. Biochem Pharmacol 59:273–281

Kokoska RJ, Stefanovic L, DeMai J, Petes TD (2000) Increased rates of genomic deletions generated by mutations in the yeast gene encoding DNA polymerase delta or by decreases in the cellular levels of DNA polymerase delta. Mol Cell Biol 20:7490–7504

Krayevsky AA, Victorova LS, Arzumanov AA, Jasko MV (2000) Terminal deoxynucleotidyl transferase. catalysis of DNA (oligodeoxynucleotide) phosphorylation. Pharmacol Ther 85:165–173

Kunkel TA, Pavlov YI, Bebenek K, Rogozin IB, Matsuda T (2003) Functions of human DNA polymerases eta, kappa and iota suggested by their properties, including fidelity with undamaged DNA templates. DNA Repair (Amst) 2:135–149

Lawrence CW (2002) Cellular roles of DNA polymerase zeta, Revl protein. DNA Repair (Amst) 1:425–435

Lawrence CW, Christensen R (1976) UV mutagenesis in radiation-sensitive strains of yeast. Genetics 82:207–232

Lawrence CW, Hinkle DC (1996) DNA polymerase zeta and the control of DNA damage induced mutagenesis in eukaryotes. Cancer Surv 28:21–31

Lee GH, Nishimori H, Sasaki Y, Matsushita H, Kitagawa T, Tokino T (2003) Analysis of lung tumorigenesis in chimeric mice indicates the pulmonary adenoma resistance 2 (Par2) locus to operate in the tumor-initiation stage in a cell-autonomous manner: detection of polymorphisms in the Poli gene as a candidate for Par2. Oncogene 22:2374–2382

Lee JW et al. (2004) Implication of DNA polymerase lambda in alignment-based gap filling for nonhomologous DNA end joining in human nuclear extracts. J Biol Chem 279:805–811

Li L, Zou L (2005) Sensing, signaling, and responding to DNA damage: organization of the checkpoint pathways in mammalian cells. J Cell Biochem 94:298–306

Lin W, Xin H, Zhang Y, Wu X, Yuan F, Wang Z (1999) The human REV1 gene codes for a DNA template-dependent dCMP transferase. Nucleic Acids Res 27:4468–4475

Loeb KR, Loeb LA (2000) Significance of multiple mutations in cancer. Carcinogenesis 21:379–385

Longley MJ, Pierce AJ, Modrich P (1997) DNA polymerase delta is required for human mismatch repair in vitro. J Biol Chem 272:10917–10921

Louat T et al. (2001) Antitumor activity of 2′,3′-dideoxycytidine nucleotide analog against tumors up-regulating DNA polymerase beta. Mol Pharmacol 60:553–558

Luoma P et al. (2004) Parkinsonism, premature menopause, and mitochondrial DNA polymerase gamma mutations: clinical and molecular genetic study. Lancet 364:875–882

Maga G, Shevelev I, Ramadan K, Spadari S, Hubscher U (2002) DNA polymerase theta purified from human cells is a high-fidelity enzyme. J Mol Biol 319:359–369

Maga G et al. (2002) Human DNA polymerase lambda functionally and physically interacts with proliferating cell nuclear antigen in normal and translesion DNA synthesis. J Biol Chem 3:3

Mahajan KN et al. (1999) Association of terminal deoxynucleotidyl transferase with Ku. Proc Natl Acad Sci USA 96:13926–13931

Mahajan KN, Nick McElhinny SA, Mitchell BS, Ramsden DA (2002) Association of DNA polymerase mu (pol mu) with Ku and ligase IV: role for pol mu in end-joining double-strand break repair. Mol Cell Biol 22:5194–5202

Maloisel L, Bhargava J, Roeder GS (2004) A role for DNA polymerase delta in gene conversion and crossing over during meiosis in *Saccharomyces cerevisiae*. Genetics 167:1133–1142

Mancuso M, Filosto M, Oh SJ, DiMauro S (2004) A novel polymerase gamma mutation in a family with ophthalmoplegia, neuropathy, Parkinsonism. Arch Neurol 61:1777–1779

Marini F, Kim N, Schuffert A, Wood RD (2003) POLN, a nuclear PolA family DNA polymerase homologous to the DNA cross-link sensitivity protein Mus308. J Biol Chem 278:32014–32019

Marini F et al. (1997) A role for DNA primase in coupling DNA replication to DNA damage response. Embo J 16:639–650

Masuda Y, Kamiya K (2002) Biochemical properties of the human REV1 protein. FEBS Lett 520:88–92

Masutani C, Kusumoto R, Iwai S, Hanaoka F (2000) Mechanisms of accurate translesion synthesis by human DNA polymerase eta. Embo J 19:3100–3109

Masutani C et al. (1999) The XPV (xeroderma pigmentosum variant) gene encodes human DNA polymerase eta. Nature 399:700–704

Masutomi K et al. (2003) Telomerase maintains telomere structure in normal human cells. Cell 114:241–253

Matsumoto Y, Kim K (1995) Excision of deoxyribose phosphate residues by DNA polymerase beta during DNA repair. Science 269:699–702

McCulloch SD, Kokoska RJ, Masutani C, Iwai S, Hanaoka F, Kunkel TA (2004) Preferential cis-syn thymine dimer bypass by DNA polymerase eta occurs with biased fidelity. Nature 428:97–100

McDonald JP et al. (1999) Novel human and mouse homologs of *Saccharomyces cerevisiae* DNA polymerase eta. Genomics 60:20–30

Murakumo Y (2002) The property of DNA polymerase zeta: REV7 is a putative protein involved in translesion DNA synthesis and cell cycle control. Mutat Res 510:37–44

Murakumo Y et al. (2001) Interactions in the error-prone postreplication repair proteins hREV1, hREV3, and hREV7. J Biol Chem 276:35644–35651

Murakumo Y et al. (2000) A human REV7 homolog that interacts with the polymerase zeta catalytic subunit hREV3 and the spindle assembly checkpoint protein hMAD2. J Biol Chem 275:4391–4397

O-Wang J et al. (2001) DNA polymerase kappa, implicated in spontaneous, DNA damage-induced mutagenesis, is overexpressed in lung cancer. Cancer Res 61:5366–5369

Ogi T, Mimura J, Hikida M, Fujimoto H, Fujii-Kuriyama Y, Ohmori H (2001) Expression of human and mouse genes encoding polkappa: testis-specific developmental regulation, AhR-dependent inducible transcription. Genes Cells 6:943–953

Okochi E, Ichimura S, Sugimura T, Ushijima T (2002) The absence of Mth1 inactivation, DNA polymerase kappa overexpression in rat mammary carcinomas with frequent A:T to C:G transversions. Jpn J Cancer Res 93:501–506

Pan Q, Fang Y, Xu Y, Zhang K, Hu X (2005) Down-regulation of DNA polymerases kappa, eta, iota, and zeta in human lung, stomach, and colorectal cancers. Cancer Lett 217:139–147

Poltoratsky V, Woo CJ, Tippin B, Martin A, Goodman MF, Scharff MD (2001) Expression of error-prone polymerases in BL2 cells activated for Ig somatic hypermutation. Proc Natl Acad Sci USA 98:7976–7981

Prakash S, Prakash L (2002) Translesion DNA synthesis in eukaryotes: a one- or two-polymerase affair. Genes Dev 16:1872–1883

Prasad R, Beard WA, Strauss PR, Wilson SH (1998) Human DNA polymerase beta deoxyribose phosphate lyase. Substrate specificity and catalytic mechanism. J Biol Chem 273:15263–15270

Qi H, Zakian VA (2000) The *Saccharomyces* telomere-binding protein Cdc13p interacts with both the catalytic subunit of DNA polymerase alpha and the telomerase-associated est1 protein. Genes Dev 14:1777–1788

Radman M (2001) Fidelity and infidelity. Nature 413:115

Ramadan K, Maga G, Shevelev IV, Villani G, Blanco L, Hubscher U (2003) Human DNA polymerase lambda possesses terminal deoxyribonucleotidyl transferase activity and can elongate RNA primers: implications for novel functions. J Mol Biol 328:63–72

Ramadan K, Shevelev IV, Hubscher U (2004a) The DNA-polymerase-X family: controllers of DNA quality? Nat Rev Mol Cell Biol 5:1038–1043

Ramadan K, Shevelev IV, Maga G, Hubscher U (2002) DNA polymerase lambda from calf thymus preferentially replicates damaged DNA. J Biol Chem 277:18454–18458

Ramadan K, Shevelev IV, Maga G, Hubscher U (2004b) De novo DNA synthesis by human DNA polymerase lambda, DNA polymerase mu and terminal deoxyribonucleotidyl transferase. J Mol Biol 339:395–404

Read RL, Martinho RG, Wang SW, Carr AM, Norbury CJ (2002) Cytoplasmic poly(A) polymerases mediate cellular responses to S phase arrest. Proc Natl Acad Sci USA 99:12079–12084

Rechkoblit O et al. (2002) Trans-lesion synthesis past bulky benzo[a]pyrene diol epoxide N2-dG, N6-dA lesions catalyzed by DNA bypass polymerases. J Biol Chem 277:30488–30494

Roychoudhury R (1972) Enzymic synthesis of polynucleotides. Oligodeoxynucleotides with one 3′-terminal ribonucleotide as primers for polydeoxynucleotide synthesis. J Biol Chem 247:3910–3917

Ruiz JF, Dominguez O, Lain de Lera T, Garcia-Diaz M, Bernad A, Blanco L (2001) DNA polymerase mu, a candidate hypermutase? Philos Trans R Soc Lond B Biol Sci 356:99–109

Ruiz JF et al. (2003) Lack of sugar discrimination by human Pol mu requires a single glycine residue. Nucleic Acids Res 31:4441–4449

Sadoff BU, Heath-Pagliuso S, Castano IB, Zhu Y, Kieff FS, Christman MF (1995) Isolation of mutants of *Saccharomyces cerevisiae* requiring DNA topoisomerase. Genetics I 141:465–479

Saitoh S, Chabes A, McDonald WH, Thelander L, Yates JR, Russell P (2002) Cid13 is a cytoplasmic poly(A) polymerase that regulates ribonucleotide reductase mRNA. Cell 109:563–573

Sancar A, Lindsey-Boltz LA, Unsal-Kaccmaz K, Linn S (2004) Molecular mechanisms of mammalian DNA repair and the DNA damage checkpoints. Annu Rev Biochem 73:39–85

Seki M, Marini F, Wood RD (2003) POLQ (Pol theta), a DNA polymerase, DNA-dependent ATPase in human cells. Nucleic Acids Res 31:6117–6126

Seki M et al. (2004) High-efficiency bypass of DNA damage by human DNA polymerase Q. EMBO J 23:4484–4494. Epub 2004 Oct 4421

Semizarov DG et al. (1997) Stereoisomers of deoxynucleoside 5′-triphosphates as substrates for template-dependent and -independent DNA polymerases. J Biol Chem 272:9556–9560

Sharief FS, Vojta PJ, Ropp PA, Copeland WC (1999) Cloning and chromosomal mapping of the human DNA polymerase theta (POLQ), the eighth human DNA polymerase. Genomics 59:90–96

Shevelev IV, Hubscher U (2002) The 3′ 5′ exonucleases. Nat Rev Mol Cell Biol 3:364–376

Shima N, Hartford SA, Duffy T, Wilson LA, Schimenti KJ, Schimenti JC (2003) Phenotype-based identification of mouse chromosome instability mutants. Genetics 163:1031–1040

Shimazaki N, Yoshida K, Kobayashi T, Toji S, Tamai K, Koiwai O (2002) Over-expression of human DNA polymerase lambda in *E. coli* and characterization of the recombinant enzyme. Genes Cells 7:639–651

Simpson LJ, Sale JE (2003) Rev1 is essential for DNA damage tolerance and non-templated immunoglobulin gene mutation in a vertebrate cell line. EMBO J 22:1654–1664

Sobol RW et al. (1996) Requirement of mammalian DNA polymerase-beta in base-excision repair. Nature 379:183–186

Sonoda E et al. (2003) Multiple roles of Rev3, the catalytic subunit of pol zeta in maintaining genome stability in vertebrates. Embo J 22:3188–3197

Stucki M, Stagljar I, Jonsson ZO, Hubscher U (2001) A coordinated interplay: proteins with multiple functions in DNA replication, DNA repair, cell cycle/checkpoint control, and transcription. Prog Nucleic Acid Res Mol Biol 65:261–298

Sugino A (1995) Yeast DNA polymerases and their role at the replication fork. Trends Biochem Sci 20:319–323

Suzuki N et al. (2002) Translesion synthesis by human DNA polymerase kappa on a DNA template containing a single stereoisomer of dG-(+)- or dG-(−)-anti-N(2)-BPDE (7,8-dihydroxy-anti-9,10-epoxy-7,8,9,10-tetrahydrobenzo[a]pyrene). Biochemistry 41:6100–6106

Thai TH, Kearney JF (2004) Distinct and opposite activities of human terminal deoxynucleotidyltransferase splice variants. J Immunol 173:4009–4019

Thai TH, Purugganan MM, Roth DB, Kearney JF (2002) Distinct and opposite diversifying activities of terminal transferase splice variants. Nat Immunol 3:457–462

Tissier A, Frank EG, McDonald JP, Iwai S, Hanaoka F, Woodgate R (2000) Misinsertion and bypass of thymine-thymine dimers by human DNA polymerase iota. Embo J 19:5259–5266

Tissier A, Kannouche P, Reck MP, Lehmann AR, Fuchs RP, Cordonnier A (2004) Co-localization in replication foci and interaction of human Y-family members, DNA polymerase pol eta, REVl protein. DNA Repair (Amst) 3:1503–1514

Tissier A, McDonald JP, Frank EG, Woodgate R (2000) poliota, a remarkably error-prone human DNA polymerase. Genes Dev 14:1642–1650

Trifunovic A et al. (2004) Premature ageing in mice expressing defective mitochondrial DNA polymerase. Nature 429:417–423

Vaisman A, Chaney SG (2000) The efficiency and fidelity of translesion synthesis past cisplatin and oxaliplatin GpG adducts by human DNA polymerase beta. J Biol Chem 275:13017–13025

Velasco-Miguel S et al. (2003) Constitutive and regulated expression of the mouse Dinb (Polkappa) gene encoding DNA polymerase kappa. DNA Repair (Amst) 2:91–106

Walowsky C, Fitzhugh DJ, Castano IB, Ju JY, Levin NA, Christman MF (1999) The topoisomerase-related function gene TRF4 affects cellular sensitivity to the antitumor agent camptothecin. J Biol Chem 274:7302–7308

Wang M et al. (2004) Pol iota is a candidate for the mouse pulmonary adenoma resistance 2 locus, a major modifier of chemically induced lung neoplasia. Cancer Res 64:1924–1931

Wang Y et al. (2004) Elevated expression of DNA polymerase kappa in human lung cancer is associated with p53 inactivation: negative regulation of POLK promoter activity by p53. Int J Oncol 25:161–165

Wang Z, Castano IB, De Las Penas A, Adams C, Christman MF (2000) Pol kappa: A DNA polymerase required for sister chromatid cohesion. Science 289:774–779

Wang Z, Wu X, Friedberg EC (1993) DNA repair synthesis during base excision repair in vitro is catalyzed by DNA polymerase epsilon and is influenced by DNA polymerases alpha and delta in *Saccharomyces cerevisiae*. Mol Cell Biol 13:1051–1058

Washington MT, Johnson RE, Prakash L, Prakash S (2004) Human DNA polymerase iota utilizes different nucleotide incorporation mechanisms dependent upon the template base. Mol Cell Biol 24:936–943

Yang J, Chen Z, Liu Y, Hickey RJ, Malkas LH (2004) Altered DNA polymerase iota expression in breast cancer cells leads to a reduction in DNA replication fidelity and a higher rate of mutagenesis. Cancer Res 64:5597–5607

Zeng X, Negrete GA, Kasmer C, Yang WW, Gearhart PJ (2004) Absence of DNA polymerase eta reveals targeting of C mutations on the nontranscribed strand in immunoglobulin switch regions. J Exp Med 199:917–924. Epub 2004 Mar 2029

Zhang X, Yu Y, Chen X (1994) Evidence for nontargeted mutagenesis in a monkey kidney cell line and analysis of its sequence specificity using a shuttle-vector plasmid. Mutat Res 323:105–112

Zhang Y et al. (2002) Lesion bypass activities of human DNA polymerase mu. J Biol Chem 277:44582–44587

Zhang Y, Yuan F, Wu X, Wang Z (2000) Preferential incorporation of G opposite template T by the low-fidelity human DNA polymerase iota. Mol Cell Biol 20:7099–7108

Zhu F, Jin CX, Song T, Yang J, Guo L, Yu YN (2003) Response of human REV3 gene to gastric cancer inducing carcinogen *N*-methyl-*N′*-nitro-*N*-nitrosoguanidine and its role in mutagenesis. World J Gastroenterol 9:888–893

Zhu F, Zhang M (2003) DNA polymerase zeta: new insight into eukaryotic mutagenesis and mammalian embryonic development. World J Gastroenterol 9:1165–1169

Part B
Replicative Double-Strand Break Repair

Genome Dyn Stab (1)
D.-H. Lankenau: Genome Integrity
DOI 10.1007/7050_003/Published online: 18 November 2005

Facing Stalled Replication Forks: The Intricacies of Doing the Right Thing

Christian Rudolph[1] · K. Anke Schürer[2] · Wilfried Kramer[3] (✉)

[1]Institute of Genetics, University of Nottingham, Queen's Medical Centre, Nottingham NG7 2UH, UK
c.rudolph@gmx.de

[2]Leibniz-Institut für Altersforschung/Fritz-Lipmann-Institut, Beutenbergstr. 11, 07745 Jena, Germany
aschuer@fli-leibniz.de

[3]Institut für Mikrobiologie und Genetik, Abt. Molekulare Genetik und Präparative Molekularbiologie, Grisebachstr. 8, 37077 Göttingen, Germany
wkramer@gwdg.de

Abstract Replication forks stalled at DNA damage or other obstacles can pose a serious problem for cells. Besides error-prone mechanisms that rescue stalled replication forks *via* translesion synthesis at the expense of replicational fidelity, error-free bypass mechanisms are available in all organisms analyzed so far to reinstall replication without corrosive effects on the integrity of the genetic information. Reasonable models for error-free replication restart include excision of the lesion, interaction with the sister chromatid to gather the correct information both with and without homologous recombination, and generation of double stranded ends with subsequent recombination with the sister chromatid. Cells need to decide which of the reinitiation mechanisms to employ. In this review, we will place special emphasis on what is known so far on these decision processes, in particular for the SOS response in *Escherichia coli*, the modification of PCNA in *Saccharomyces cerevisiae* and the eukaryotic DNA damage checkpoint.

1 Introduction

Replication of DNA is one of the fundamental processes in all living cells. The result of replication are two – within the limits of replicational fidelity – identical copies of the genetic material. To stably maintain genetic information over many generations, replication has to be very accurate. A second, much more immediate requirement, however, is that replication has to be completed. At first glance, this statement seems to be trivial, but replication forks can encounter many obstacles preventing their progression. Such obstacles include DNA lesions such as base modifications and strand breaks, secondary structures, interstrand crosslinks and proteins tightly bound to DNA.

In this review, we will discuss cellular mechanisms, first in *E. coli* and then in eukaryotes, that allow resumption of replication after an interfering event with a focus on the regulation of these pathways and factors influencing the decision, which pathway to use, if alternatives are available. Many conceptual problems apply to both bacteria and eukaroytes. Since most of the results were initially obtained in *E. coli*, these problems are discussed in that section with occasional reference to eukaroytes. Apparent differences are specifically mentioned in the section on eukaryotes, which will focus mainly on yeasts.

2
Reinitiation in *E. coli*

2.1
A Blocked Replication Fork is Particularly Dangerous in *E. coli*

Replication in *E. coli* is initiated at a single origin of replication (*oriC*). Binding of DnaA in its ATP-bound form results in local melting of the DNA duplex. At this site, two hexameric rings of the replicative helicase DnaB are loaded (with the help of the helicase loader DnaC), each encircling one of the strands. One primer is synthesized by the primase DnaG on each strand. Elongation of the primers by DNA polymerase III will start leading strand syntheses. DNA polymerase III holoenzyme consists of two catalytic Pol III core subunits which are connected *via* the γ-complex and two τ-proteins. The γ-complex is the so-called clamp loader, which loads the processivity factor β onto DNA. Two β subunits form a ring-shaped homodimer with a central opening large enough to accommodate double stranded DNA. (In eukaroytes, a homotrimer of PCNA is the functional analog, which has a very similar overall shape (Krishna et al. 1994).) After loading of the β dimer onto DNA it forms a sliding clamp that increases the processivity of the DNA polymerase *via* protein-protein interactions. In the normal elongation mode DnaB helicase, a 5′-3′ helicase, which is associated with and stimulated by the Pol III holoenzyme, unwinds the parental strands. Whereas the leading strand is made continuously by one catalytic subunit of the Pol III holoenzyme, the lagging strand has to be synthesized in fragments (Okazaki fragments), since DNA polymerases elongate only in 5′-3′ direction. Once a sufficient stretch of DNA is unwound by DnaB, primase will synthesize a new primer for the next Okazaki and the other catalytic subunit of Pol III holoenzyme will elongate this primer. A sliding clamp is loaded by γ for each Okazaki fragment. Thus, in the current picture leading and lagging strand synthesis are coupled and carried out by one Pol III holoenzyme. The loop formed by the lagging strand template will continuously shrink and increase like the slide of a trombone, why this model is also called the "trombone model" (for a more detailed description of basic replication processes see article of R. Egel in this book). For eukaryotes, an alternative model has also been proposed, where the leading strand polymerases of both replications forks and the lagging strand polymerases of both forks are coupled, but not the leading and lagging strand polymerases (Falaschi 2000).

The processes sketched above mean that the *E. coli* genome is basically replicated by two replication forks, starting at the origin and moving in opposite directions around the circular chromosome, although it should be noted that present evidence is strongly in favor of a stationary replisome, where the DNA moves with respect to immobile replication enzymes (for review see, e.g. Sawitzke and Austin 2001). Somewhat more than halfway in each direction are specific termination sites (*terA–terF*) that unidirectionally inactivate replication forks by the antihelicase action of the Tus protein bound to these sites (Hill and Marians 1990; Khatri et al. 1989). Therefore, if one replication fork arrests between the origin and the *ter*-sites due to an unscheduled replication block, the unreplicated portion cannot be replicated by the other fork, since this terminates at the *ter*/Tus complex. If the arrested fork cannot be reactivated, the cell will be left with a partially replicated chromosome and probably die. In eukaryotes, which have several origins of replication per chromosome, a single blocked forked is presumably less deleterious, since the unreplicated portion can be duplicated by forks that have started at adjacent origins.

2.2 Arrests of DNA Polymerases at DNA Lesions

One critical determinant for the high accuracy of replicative DNA polymerases is their tight binding pocket (Friedberg et al. 2001; Kool 2002). This in turn prevents copying of damaged nucleotides in the template strand, which results in stalling of the polymerase at a lesion. One option for reinitiation in this case is always translesion synthesis (TLS). TLS is carried out by specialized DNA polymerases that are less accurate than replicative polymerases, but in turn are able to copy damaged templates. The arrested DNA polymerase needs to be removed and the translesion polymerase has to elongate the primer past the lesion, before the replicative DNA polymerase can take over again. Deployment of TLS polymerases, however, is always on the expense of replicational fidelity. Since TLS polymerases are covered in the article by U. Hübscher in this book, we will concentrate mainly on error-free pathways and discuss TLS in more detail only in cases, where it is necessary to position the different cellular responses with respect to each other.

2.3 Removal of Lesions to Prevent DNA Polymerase Arrests

One way to prevent DNA polymerase arrests is, of course, the removal of potential blocks by repair before an arrest has actually occurred. DNA is a reactive chemical compound that is vulnerable to both endogenous and environmental damaging agents. All living cells have developed a network of repair system for the continuous replacement of damaged or miscoding parts of the DNA (Friedberg et al. 1995; Lindahl and Wood 1999). During normal growth conditions a basal level of a variety of different repair proteins are present in *E. coli* cells to cope with modifications by reactive metabolites like the methylating agent SAM or reducing sugars (Bucala et al. 1984, 1985; Rydberg and Lindahl 1982), reactive oxygen species, UV light, etc. Besides occasional direct reversion of the modification, most lesions are removed by base or nucleotide excision repair (for overview see e.g. Friedberg et al. 1995). Interestingly, there are indications that excessive repair activity can be harmful to cells (Glassner et al. 1998). This may be the reason, why many repair systems can be induced, if the respective damage density increases (Volkert and Landini 2001).

One of the most extensively studied system is the SOS response (Crowley and Courcelle 2002; Janion 2001), which is induced after a variety of physical and chemical treatments such as UV light, the cross-linking agent mitomycin C, the methylating agent methyl methanesulfonate (MMS), 4-nitroquinoline 1-oxide (4-NQO), which causes bulky DNA lesions etc. (Janion 2001). However, different mutagens seem to have different effects. Whereas SOS is induced strongly and fast after exposure to even comparatively low UV doses (Ishioka et al. 1997), induction by 4-NQO, even though it is sometimes referred to as UV-mimetic, since the bulky lesions are repaired mainly by NER (Friedberg et al. 1995), requires a higher dose regarding cell survival and also a longer time (C.R. and R.G. Lloyd, unpublished observations).

In contrast to other inducible repair systems, the SOS response is not triggered by the harmful components directly but rather by persistent ssDNA in the cell, as, for example, it occurs after arrest of a replicative DNA polymerase at a lesion. In the presence of NTP (usually ATP or dATP) RecA protein binds with high affinity to ssDNA and forms a RecA-DNA filament, which leads to activation of RecA to RecA* (reviewed in Bianco and Kowalczykowski 1999). RecA* stimulates autocatalytic cleavage of the LexA repressor protein, which inhibits transcription of the SOS regulated genes. RecA* catalysed auto cleavage of LexA thus leads to transcriptional activation of the LexA regulated genes as well as transcriptional down-regulation or specific degradation of transcripts (Crowley and Courcelle 2002; Janion 2001). The SOS regulated proteins comprise repair enzymes as well as many proteins required for reinitiation of stalled replication forks (see Sect. 2.9). Thus, if a cell is exposed to a DNA damaging agent, SOS induction by stalled replication forks allows for reinitiation but at the same time the repair capacity is increased to clear the future path of the DNA polymerases, either by direct induction of the repair system or as part of the SOS induction as in the case of NER. Coupling of the latter to the SOS response may be due to the fact that NER recognizes a broad spectrum of diverse lesions with ill-defined causative agents.

2.4 Forks Stalled at DNA Polymerase Blocking Lesions

If a lesion blocks a replicative DNA polymerase, the consequences depend on whether the lesion is in the leading or the lagging strand template. A lesion in the lagging strand template can block synthesis of one Okazaki fragment but should not impede the replisome as a whole. Lagging strand synthesis can proceed simply by elongating the primer for the next Okazaki fragment (Cordeiro-Stone et al. 1999; Higuchi et al. 2003). The remaining gap in the lagging strand (Svoboda and Vos 1995) can subsequently be filled *via* recombination or TLS.

For the leading strand, however, primer synthesis at the *oriC* is strictly regulated to once per cell cycle (Katayama 2001) and it was held as a dogma that resumption of replication after a damage-induced DNA polymerase arrest cannot be achieved by synthesis of a new primer downstream of the lesion (see, e.g. McGlynn 2004). However, recent biochemical experiments on the assembly of the restart replisome at various replication fork-like substrates raised the possibility that the primase DnaG is able to create a new primer on the leading strand template. The replisome assembly protein PriC, which seems to be involved in one distinct sub-pathway for reinitiation together with the key reinitiation protein PriA (Sandler et al. 2001), is required for this reaction (Heller and Marians 2005; Lovett 2005). Intriguingly, these findings open the possibility that reinitiation of replication can be achieved in a similar manner as described above for the lagging strand.

At any rate, arrest of the leading strand DNA polymerase does not cause an immediate arrest of the complete replication fork. Synthesis of the lagging strand can still continue (Cordeiro-Stone et al. 1999; Higuchi et al. 2003; Pagés and Fuchs 2003; Svoboda and Vos 1995), producing a structure where the newly synthesized lagging strand is longer than the leading strand. The majority of models for error-free reinitiation mechanisms build upon this structure as the initial situation.

2.5 Reinitiation by Lesion Removal or Error-free Bypass

2.5.1 Excision Repair of DNA Lesions at Stalled Forks

A replication-blocking lesion dictates that the lesion must be either removed or bypassed. Since the parental strands are separated, the damage cannot be repaired by excision repair, which requires a complementary strand for gap filling after excision. However, the fork can move backwards (hence called replication fork regression: RFR, or fork regression) to allow reannealing of the parental strands, forming a four-way Holliday junction (Fig. 1b). Thus,

Fig. 1 Some models for error-free reinitiation of replication forks stalled at a damage on the leading strand template. The damage is indicated by a *grey rectangle*. Newly synthesized DNA is shown as a *grey line*. (i) The terminated leading strand may invade the sister chromatid to form intermediate (**a**). Elongation of the 3′-end of the leading strand on the lagging strand as template could lead to intermediate (**d**). This may be converted to intermediate (**g**) by hybridization of the displaced lagging strand template with the leading strand template. Elongation of the invaded 3′-end in intermediate (**d**) until the end of the lagging strand would yield intermediate (**e**). Disruption of the D-loop by, e.g. a helicase leads to intermediate (**h**), where the lesions is bypassed in an error-free manner and a 3′-end of the leading strand is created, that can be elongated by replicative DNA polymerases. Basically the same intermediates (with the exception of intermediate (**e**)) can occur if the damage is on the lagging strand template. (ii) Fork regression leads to intermediate (**b**), which can be converted after a polymerase strand switch to the lagging strand as template into intermediate (**e**). Both in (**b**) and (**e**) the damage can also be removed by excision repair. Intermediate (**e**) can be converted to intermediate (**h**) (*see above*) by reverse branch migration, or by cleavage *via* resolvases with subsequent processing by recombination enzymes to produce a single stranded 3′ overhang into structures similar to intermediate (**f**) (*see below*). Depending on the direction of cleavage the damage may still reside in the DNA or may be excised during end processing as described below. (iii) Direct cleavage of the leading strand template will yield intermediate (**c**). Resection of the 5′-end of the resulting end by recombination enzymes to create a 3′ overhang will remove the lesion as shown for intermediate (**f**), where also the gap or nick in the sister chromatid has been sealed. Invasion of the 3′-overhang into the sister chromatid will yield intermediate (**j**). For processing of this structure refer to Fig. 2

the lesion again resides in a double stranded DNA, where excision repair can take place.

Several mechanisms have been discussed for RFR. Positive supercoiling in front of the replication fork can lead to spontaneous RFR (Postow et al. 2001). Additionally, several proteins such as RecG and RecA have been suggested to be involved in fork reversal (McGlynn and Lloyd 2002b; Michel et al. 2004). To convert a replication fork into a Holliday junction, at least three steps are necessary: 1) the leading and lagging strands, which are hybridized to the parental strands, have to be unwound; 2) the leading and lagging strands have to anneal, forming a double strand; 3) the parental strands have to re-anneal. Insights into how these steps can be carried out simultaneously by a RecG monomer have come from the crystal structure of the *Thermotoga maritima* protein in complex with a partial fork substrate (Singleton et al. 2001). The protein contains two helicase domains, which are linked *via* a long α-helix to a wedge domain providing specificity for binding branched DNA (Briggs et al. 2004; Singleton et al. 2001). Two channels flank either side of the wedge, which are too small to accommodate double stranded DNA. It is thought that RecG helicase acts as an ATP-dependent dsDNA translocase, pulling the parental strands through these channels, stripping off the nascent leading and lagging strands. As the parental strands re-anneal on one side of the wedge, the nascent strands are thought to re-anneal on the opposite side such that a Holliday junction is formed around the wedge (Briggs

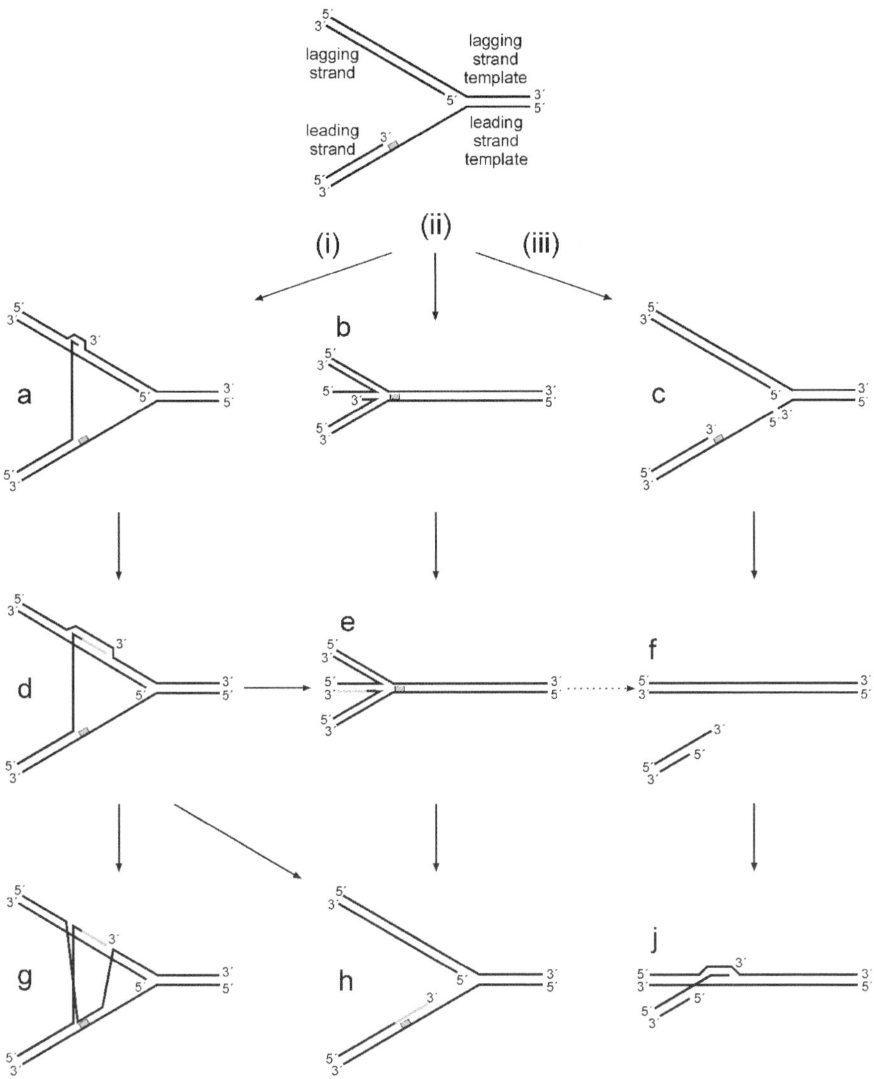

et al. 2004; Singleton et al. 2001). Given that the sequence of *E. coli* RecG is essentially identical to that of *Thermotoga maritima* RecG, it seems likely that both proteins have similar structures (Briggs et al. 2004). Interestingly, the postulated unwinding mechanism of RecG *via* translocation on double stranded DNA differs from all other helicases characterized so far (Mahdi et al. 2003).

After the lesions has been excised, several reinitiation pathways are possible after Holliday junction formation. The Holliday junction can be resolved by branch migration in direction of replication, which would basically reform the same structure as before RFR (Fig. 1b back to starting structure, but without lesion), where the reinitiation primosome can assemble. Another way (not shown in Fig. 1) would be degradation of the free arm of the Holliday junction by exonuclease V (RecBCD) (Seigneur et al. 1998). This pathway would allow the nascent leading strand end to be brought into register with the junction point, in cases where lagging strand synthesis has proceeded and differs significantly in length from the leading strand (Higuchi et al. 2003; Pagés and Fuchs 2003; Svoboda and Vos 1995) since it is unclear whether reinitiation can take place at the leading strand when the lagging strand covers several 100 bp of the lagging strand template. Furthermore, it has been shown that RecJ exonuclease degrades the lagging strand with the aid of RecQ helicase in *E. coli* cells irradiated with UV. It has been suggested that due to this action ssDNA is created which can be bound by RecA for stabilization of the stalled fork (Courcelle and Hanawalt 1999). Clearly, this process would also serve to equalize the length of leading and lagging strand in a similar manner to RecBCD degradation of an extruded end.

Additionally, forks can be re-established by recombinational processes. The Holliday junction can be cleaved by the Holliday junction resolvase RuvABC, resulting in a double stranded end (not shown; compare Figs. 1e and f). This can be processed by RecBCD and invade the sister chromatid after a RecA filament has been assembled (Figs. 1f and j). The resulting D-loop is an efficient substrate for the restart primosome (reviewed in Cox 2001; McGlynn and Lloyd 2002b). Resolution of the Holliday junction can also take place in a different order if there is a χ site between the end of the leading and lagging strand hybrid and the four-way junction. Degradation stops and a single stranded 3′-overhang is created by RecBCD, which can invade the parental duplex before the junction is cleaved by RuvABC (Hyrien 2000; McGlynn and Lloyd 2002a; Seigneur et al. 1998). Processing in this order would prevent formation of a double stranded end, which might be dangerous to the cell.

2.5.2
Error-free Bypass of Lesions

If the lesion is not removed, the Holliday junction intermediate offers several possibilities for error-free bypass of the damaged base. If the lagging strand is elongated beyond the lesion, the lagging strand can serve as template for elongation of the terminated leading strand in the Holliday junction intermediate (Figs. 1b and e). For this, a polymerase switch to the lagging strand as template has to take place. If such a Holliday junction resolves by branch migration into replication direction (Fig. 1h), the restart primosome can assemble at the leading strand that is now elongated beyond the lesion (Fujiwara and Tatsumi 1976; Higgins et al. 1976; reviewed in Cox 2001). Cleavage of the Holliday junction by RuvABC with subsequent strand invasion as discussed above would also lead to bypass of the damage, since invasion would occur past the lesion. However, degradation of the leading-lagging strand hybrid by RecBCD would result in a structure, where the leading strand ends before the lesion, requiring an additional attempt for bypass or repair.

2.6
Forks Stalled at Helicase Blocking Complexes or Lesions

Lesions, which interfere with strand separation like interstrand crosslinks, protein-DNA complexes, or specific types of secondary structures are likely to block progression of the replicative helicases. The best studied examples of replisomes blocked at a protein-DNA complex are forks stalled at natural replication fork barriers (RFBs). A fork blocked at the *ter*/Tus-complex has a leading strand 50–70 nt longer than the lagging strand (Hill and Marians 1990) and is thus different from a fork, where the leading strand DNA polymerase has been arrested. However, it has to be stressed that at the *ter*/Tus-complex replication is blocked due to the specific anti-helicase function of the Tus protein (Sahoo et al. 1995). Thus it might be that the structures of replication forks blocked at naturally occurring RFBs differ significantly from forks, which are stopped due to, e.g. stable repressor-operator complexes or stalled RNA polymerases.

2.7
Removal of Protein-DNA Complexes

Rep helicase has been implicated in the removal of proteins bound to DNA as well as DNA secondary structures (reviewed in Michel 2000). This conjecture is based on observations of Michel and colleagues that replication mutants can be classified into two groups: The first group requires both RecA and RecBCD for restart, whereas the second group requires only RecBCD, but not RecA. DNA polymerase I as well as ligase are in class I, whereas *rep* (coding for a 3′-5′ helicase; see below) and *holD* mutants are in class II (Flores et al. 2001; Seigneur et al. 1998; reviewed in Michel et al. 2001). *holD* codes for the ψ subunit of the Pol III holoenzyme. The ψ subunit is part of the γ complex clamp loader. *holD recBCD* mutants are synthetically lethal whereas *holD recA* double mutants are not (Flores et al. 2001). It has been shown that the Rep helicase is able to remove proteins bound to DNA (reviewed in Michel 2000) and replication is slowed down significantly in *rep* mutants (Lane and Denhardt 1975), which lead to the above mentioned hypothesis on the role of Rep helicase.

The synthetic lethality of a *rep recBCD* mutant (Uzest et al. 1995) seems to indicate that blockage of replication forks at secondary structures and stable protein-DNA complexes can induce recombination events. (However, it has to be stressed that Rep helicase plays a role in replication restart together with PriC; Sandler 2000). Therefore. the question arises, how homologous recombination might help the cell to clear protein-DNA complexes. In gyrase (DNA topoisomerase II) mutants the additional deletion of *priA* is synthetically lethal (Grompone et al. 2003), which seems to indicate that replication is impaired. However, the amounts of linear chromosomes are not increased, which indicates that no fork reversal occurs. It has been postulated that replication is stopped in gyrase mutants due to positive supercoiling of the DNA in front of the replication fork and that reinitiation can happen directly and without regression at those arrested forks (reviewed in Michel et al. 2004).

2.7.1 Replication Arrests Induced by Stalled RNA Polymerases

The question remains why in some cases forks reverse after arrests at DNA-protein complexes. Since neither of the strands is elongated beyond the obstacle, the latter cannot simply be bypassed, but has to be removed. There is accumulating evidence that stalled RNA polymerase-DNA complexes cause severe problems in DNA replication (Hyrien 2000; McGlynn and Lloyd 2000; Trautinger and Lloyd 2002). RNA polymerases (RNAP) can be blocked spontaneously or by DNA lesions and undergo reverse translocation. This so-called backtracking displaces the 3′-end of the growing transcript from the active site, resulting in a highly stable inactivated complex (Park et al. 2002). For resumption of transcription, factors like Mfd and GreA are necessary. GreA induces cleavage of the transcript, resulting in a structure where the 3′-end again is located in the active center of the RNA polymerase (Perederina et al. 2004; Stebbins et al. 1995). However, Mfd is thought to push stalled RNA polymerases forward and therefore the 3′-end of the transcript can be elongated (Park et al. 2002). Furthermore, in case of a stalled RNAP at a DNA lesion, Mfd can recruit NER factors to repair the damage and release the RNAP molecule as well as the transcript (Park et al. 2002; Selby and Sancar 1993; reviewed in Roberts and Park 2004). Given that Mfd needs at least 25 nucleotides upstream of the stalled RNAP-DNA complex to modulate the complex (Park et al. 2002), one possible explanation for why replication fork reversal would facilitate replication restart after hitting a RNAP-DNA complex would be that due to the fork reversal, a platform for a protein removing factor like Mfd is created which is then able to remove the stalled complex, recruit excision repair and clear the path for replication. For a gene with a high level of transcription, arrest of the first RNAP at a lesion might cause a pile-up due to repeated loading of RNA polymerases at the promoter, leading to an array of stalled ternary transcription complexes. It is difficult to see how Mfd might cope with such arrays.

2.7.2
Forks Stalled at Natural Replication Fork Barriers

Beside stable protein-DNA complexes, which represent obstacles to DNA replication, there are also blocks, where replication is "deliberately" terminated. As already mentioned, such a replication fork barrier is represented in *E. coli* by the *ter*/Tus complex. Initiation of recombination events by the *ter*/Tus complex has been used for a long time as an argument that blockage of the replication fork can cause recombination. However, recent results indicate that the mechanism differs significantly from that of replication forks blocked at, e.g. stalled RNAP complexes. Replication is blocked due to the specific anti-helicase function of the Tus protein (Hill and Marians 1990; Khatri et al. 1989; Sahoo et al. 1995). Genetic data in a system where an ectopic *ter* site was used suggested that no fork regression occurred. The complex seems to be stable and the problem arises, when a second replication fork runs into the complex, causing the formation of double strand breaks (Bidnenko et al. 2002). Those double strand breaks apparently have to be repaired *via* homologous recombination, accounting for the observation of recombination induction. This is in contrast to, e.g. stalled RNAP complexes where the involvement of homologous recombination does not seem to be a necessity.

2.8
Fork Collapse at Single Strand Breaks

If the template for the leading strand contains a single strand break, the fork can collapse. The replicative helicase DnaB, which encircles the lagging strand template (Yuzhakov et al. 1996), will probably continue unwinding the parental strand. As soon as the replicative DNA polymerase reaches the nick, it will release the DNA and the leading strand DNA polymerase can create a double stranded DNA end by run-off synthesis (Cox 2001, see also Strumberg et al. 2000). Reinitiation in this case can be achieved by a very similar mechanism as after RuvABC-mediated cleavage of a Holliday junction (see Sect. 2.5). For a nick on the lagging strand template, however, it seems plausible that the replicative helicase simply falls off the DNA and the parental strands cannot be unwound further. Since both DNA polymerases are behind the helicase, it seems likely that they still hold onto the DNA, preventing fork collapse and that the template strand is single stranded from the Okazaki fragment until the break site.

2.9
Transcriptional Hierarchy of Repair Pathways after SOS Induction

As we have tried to illustrate above, stalled or collapsed forks can result from very diverse lesions or obstacles, which may result in different structures of the stalled forks. Apparently, many different pathways are involved in the reinitation of these forks. In the following, we want to discuss what is known about the factors that may influence the decision as to which pathway to use.

It has already been mentioned that presumably the structure of a stalled fork can trigger different responses. The structures of forks blocked at RFBs indicate that the amount of ssDNA is comparatively small when a fork is blocked, due to failure in unwinding the parental strands. Hence, it seems logical that such a structure does not lead to a strong induction of the SOS response in the first place, since expression of NER proteins or specialized DNA polymerases are not useful to remove a protein block in front of a stalled replication fork. Recruitment of factors such as Mfd or Rep helicase seems to be a more appropriate measure to deal with this problem. In apparent contradiction to this idea it was observed that an ectopic *ter*-site (which reflects a protein block, which arrests the replicative helicase) induces SOS (Taki and Horiuchi 1999). However, as mentioned in Sect. 2.7, forks arrested at *ter*-sites are quite stable and double strand ends are created by additional rounds of replication started at *oriC*. A double strand end should be processed rapidly by RecBCD and a single stranded 3′-end is created, which is probably responsible for induction of the SOS response.

In cases of blocked DNA polymerases and nicks on the template strand, the level of ssDNA increases as described in Sect. 2.3. SOS induction in these cases seems very reasonable, since the necessary functions – increase in NER, TLS and homologous recombination capacity – are all provided after induction of the SOS response. The SOS response involves the coordinated expression of at least 43 genes (reviewed in Sweasy 2005) and has been divided into a fast error-free part and a late error-prone response (reviewed in Tippin et al. 2004). Genes that are early expressed include *uvrA, uvrB* and *uvrD* (involved in NER), *polB* (DNA polymerase II), *lexA, dinI* (blocks RecA mediated autocleavage of LexA and UmuD; see below) and *ruvAB* (reviewed in Janion 2001). The up-regulation of the *uvr* genes, which are necessary for NER, is quite fast. An increase in the cellular amount of UvrA and UvrB is detectable already 10 min after induction and reaches the maximum after 40 min, where roughly 80% of the lesions have been repaired (Crowley and Courcelle 2002).

If the fast up-regulation of NER after SOS induction, especially after arrest of replication at small, replication blocking lesions, is considered to be important, mutants defective in SOS induction or NER should have severe problems regarding reinitiation of replication after UV irradiation. Indeed, it has been shown that *uvr* mutants (Courcelle et al. 1999; reviewed in Smith 2004) as well as mutants in which no LexA cleavage occurs (Waleh and Stocker 1979), are very sensitive to UV irradiation. Furthermore, reinitiation of replication seems to be severely hindered in *uvrA* mutants after UV irradiation (Courcelle et al. 1999; Otsuji et al. 1974). In the absence of *uvr*, recombination is required for survival (reviewed in Smith 2004) and the lesions remain in the DNA (Bridges and Munson 1968; Courcelle et al. 1999; Ganesan 1974).

All these facts seem to indicate that NER is one of the key players for lesion removal and reinitiation after UV irradiation.

Furthermore, expression of *polB* and *dinI* is up-regulated. *polB* codes for DNA polymerase II, which seems to be involved in the polymerase strand switch in error-free bypass (Rangarajan et al. 1999). DinI is involved in controlling the level of RecA and hence the activity of UmuD, which is one component of the translesion polymerase V (see below).

After this first stage of the SOS responses, recombination genes such as *recA* and *recN* are expressed. RecA levels in the cell reach a maximum after 1 h; the protein level is $\sim$ 20-fold increased. Between 4–6 h after induction of the SOS response the RecA concentration drops to a normal level (Bianco and Kowalczykowski 1999). Thus, induction of recombination – according to transcriptional activation – seems to be the second stage in the SOS response.

The late stage leads to expression of *sfiA*, whose gene product inhibits septum formation and leads to filamentation of the cells (Huisman et al. 1984), and *umuD* and *umuC* (reviewed in Janion 2001; Tippin et al. 2004), which can form the translesion polymerase V. The carboxy terminal domain of UmuD can be removed in an auto-cleavage reaction stimulated by a RecA::ssDNA nucleoprotein filament, leading to the activated form UmuD′. UmuD tends to form dimers and the dimers can interact with UmuC. Thus, after activation, $UmuD_2C$, UmuDD′C and $UmuD'_2C$ complexes are formed (Ferentz et al. 2001). Only the $UmuD'_2C$ complex is active in translesion synthesis. This activity is stimulated by binding of RecA, although it is inhibited by a RecA filament on the template (Pham et al. 2002; Schlacher et al. 2005). UmuD is cleaved at a comparatively slow rate and UmuD predominates over UmuD′ for 20 min (Opperman et al. 1999), which allows NER to remove the majority of the lesions (reviewed in Crowley and Courcelle 2002).

2.10 Interaction of Repair Pathways

The SOS response creates a "transcriptional hierarchy" where transcription of NER genes predominates at the beginning. Then, elevated transcription of recombinational repair genes supports the process of damage removal and reinitiation of replication. At the end, error-prone processes have their chance to deal with remaining obstacles, which still hinder replication. However, the network of repair processes is more complicated, since beside this transcriptional regulation, some of the repair systems interact specifically with other pathways, leading to activation or inhibition. For instance, under normal growth conditions, homologous and especially homeologous recombination is suppressed by MutS (reviewed in Schofield and Hsieh 2003), ensuring that the potential danger of homologous recombination, such as strand invasion at the wrong place, is minimized or can even be avoided if a less dangerous and more accurate mechanism can be used. As mentioned in Sect. 2.9, *uvrD* is an

SOS regulated gene, but apparently increased levels of UvrD are not necessary for an increase in NER capacity (Crowley and Hanawalt 2001). The function of increased levels of UvrD seems to be rather the repression of recombination (Petranovic et al. 2001; Veaute et al. 2005), which is a further indication that recombinational repair is called upon if NER alone is not sufficient to solve the problem.

Since TLS is error-prone (Tippin et al. 2004), it is not very surprising that homologous recombination inhibits TLS in vitro as well as in vivo (Berdichevsky et al. 2002). Interestingly, inhibition seems to be mutual, since Pol V inhibits RecA-dependent recombination in vivo (Boudsocq et al. 1997; Sommer et al. 1993) as well as in vitro (Rehrauer et al. 1998). The in vitro data imply that active $Umu(D')_2C$ complexes bind the RecA filament and inhibit homologous recombination (Rehrauer et al. 1998). Therefore, it seems once the cell has decided to use an error-prone process like TLS, this predominates over other processes.

Besides the inhibition of TLS by homologous recombination another control mechanisms is processing of UmuD to UmuD′ (see Sect. 2.9). The kinetics of this reaction is slow, allowing the removal of the majority of lesions by NER (Crowley and Courcelle 2002), before active Pol V is available. Pol V formation is further retarded by, e.g. the early expression of *dinI*, which slows UmuD processing by inhibiting the ability of RecA to mediate auto-cleavage (Yasuda et al. 1998). These regulatory mechanisms suggest that the cells avoid error-prone mechanisms such as translesion synthesis as long as possible (Berdichevsky et al. 2002; McGlynn 2004).

On the other hand, uncleaved UmuD seems to have an important checkpoint function. UmuD interacts directly with Pol III holoenzyme and it has been suggested that this interaction slows down DNA synthesis by Pol III (Opperman et al. 1999). In addition, it has been suggested that replication is slowed down by competition of Pol III and Pol V for the sliding clamp (Lopéz de Saro et al. 2003). Since this regulation does not affect initiation of replication directly, it seems to be comparable with the intra-S checkpoint in eukaryotes (Opperman et al. 1999). Even more, it has been reported that breakage of unresolved chromosomal dimers by septum formation leads to DNA degradation (comparable with apoptosis in mammalian cells) and growth inhibition, implying the presence of a checkpoint which is able to recognize failures of chromosomal segregation (Cairns 2002; Hendricks et al. 2000).

Strikingly, although initiation of replication seems to be retarded in cells irradiated with low levels of UV (also resulting in filamentation; see Sect. 2.9), replication at the *oriC* undergoes multiple reinitiation events after a certain period of time. Between 1 and 2 h after UV irradiation the cells contain multiple replicating chromosomes, which are finally segregated, leading to break down of the filaments (C.R. and R.G. Lloyd, unpublished). This seems to imply that, although there is an intra-S checkpoint in *E. coli*, no prolonged inhibition of DNA synthesis happens after the DNA has been damaged.

As important as tightly regulated activation of the SOS response seems to be that – after the signal has been removed, indicating that damage is repaired – the SOS response is downregulated. If, for instance NER is able to remove almost all lesions that have blocked progression of replication, TLS activation must be avoided. This can be achieved mainly by inactivation of RecA. Once the amount of ssDNA drops to normal levels, the activated form of RecA decreases. As a result, LexA auto-cleavage is no longer stimulated and expression of the SOS inducible genes is shut down (Crowley and Courcelle 2002; Janion 2001). Furthermore, the *dinI* gene product is involved in the down-regulation by titrating the RecA protein (reviewed in Crowley and Courcelle 2002). Unprocessed UmuD protein as well as UmuC is degraded by the Lon protease, whereas processed UmuD′ is relatively stable as a homodimer (Frank et al. 1996). If the active form of RecA is no longer present, UmuD remains in an uncleaved form (Schlacher et al. 2005). This leads to the accumulation of UmuD′D heterodimers, which apparently inhibit TLS (Battista et al. 1990). UmuD′ present in those dimers is rapidly degraded by the protease ClpXP in vitro (Gonzalez et al. 2000) as well as in vivo (Frank et al. 1996). These mechanisms appear to keep the level of *E. coli* UmuD, UmuD′ and UmuC to a minimum in wild type cells.

3 Error-free Mechanisms for Reinitiation of Stalled Replication Forks in Eukaryotes

3.1 Intermediates after Replication Arrests in Eukaryotes

Eukaroytes certainly have to deal with basically the same problems after replication arrest as prokaryotes, although in contrast to *E. coli* a few stalled forks may be tolerable to the cells, since the unreplicated portion of the genome could be replicated by forks started at adjacent origins. However, if damage density increases resulting in many stalled forks, the lesions have to be somehow bypassed to provide a primer past the damage that can be elongated and would thus allow continuation of replication. Error-free bypass of such a lesion requires that the information regarding which nucleotide to insert opposite the lesion is gathered from an undamaged strand containing the same information. In replicating diploid cells the information could either be provided by the homolog or the sister chromatid that results from copying the complementary strand. Genetic evidence in yeast indicates that predominantly the sister chromatid is used and not the homolog (Kadyk and Hartwell 1992; Kadyk and Hartwell 1993).

Three principle ways that are schematically shown in Fig. 1 are presently discussed in eukaryotes that would allow elongation of the terminated

3′-end: (i) invasion of the terminated strand into the sister chromatid duplex with elongation of the 3′-end, which is similar in mechanism to synthesis-dependent strand annealing (SDSA; Nassif et al. 1994; for review, see, e.g. Paques and Haber 1999; Symington 2002); (ii) in case of a leading strand arrest, regression of the fork resulting in a Holliday junction-like structure (often called "chicken foot") as was discussed in Sect. 2.5; (iii) cleavage of the leading strand template after leading strand arrest. The first step for reinstalling the replication fork in this case would be invasion of the resulting double stranded end into the sister chromatid. To achieve this, the 5′ strand has to be resected to create a 3′ single stranded overhang suitable for Rad51 (the eukaryotic RecA homolog) filament formation. The lesion would be removed in the process of generating this recombination intermediate and replication would resume upstream of the position, where the lesion had been. (Proteins of homologous recombination in yeast will not be discussed in detail in this review, since several excellent reviews are available; Krogh and Symington 2004; Paques and Haber 1999; Symington 2002.)

For most of the intermediates, several different processing pathways are plausible. The chicken foot may be resolved as in prokaryotes *via* reverse branch migration (Figs. 1e and h) or *via* cleavage by a Holliday junction resolvase. (It should be noted, however, that so far no nuclear resolvase has been firmly identified in eukaroytes, although biochemically such an activity has been detected in human cell extracts (Liu et al. 2004).) For the double stranded end resulting from resolvase cleavage, a similar downstream pathway as for the end created by direct fork cleavage should be necessary (Figs. 1f and j; see also Fig. 2). The strand invasion listed under (i) above would result in a D-loop, which may move during elongation of the 3′-end and be resolved by disruption of the D-loop by a helicase (Figs. 1d and h). Alternatively, the D-loop may be extended during 3′-end elongation and eventually hybridize with the complementary strand of the sister chromatid forming a Holliday junction (Figs. 1d and g). Placing the end of the invaded strand back on its original template would yield a double Holliday junction. If the leading strand invades, elongation towards the end of the lagging strand would produce a chicken foot (Figs. 1d and e). Invasion of the end created by either direct leading strand template or chicken foot cleavage should also form a D-loop that could either be cut in the strand hybridized to the invading strand, or the displaced strand can be degraded to restore the replication fork (Figs. 2 A and B). Alternatively, this D-loop may be extended on the side where the invading strand enters, which eventually would result in formation of a Holliday junction, if the displaced strand hybridizes with the 5′ resected strand of the original double stranded end (Fig. 2C). Cleavage of this Holliday junction by a resolvase would also restore the replication fork. More alternatives branching off at different intermediates are conceivable that will not be mentioned here.

As can be seen from these brief sketches, Holliday junction-like structures may result from any of the principle pathways delineated above, and there-

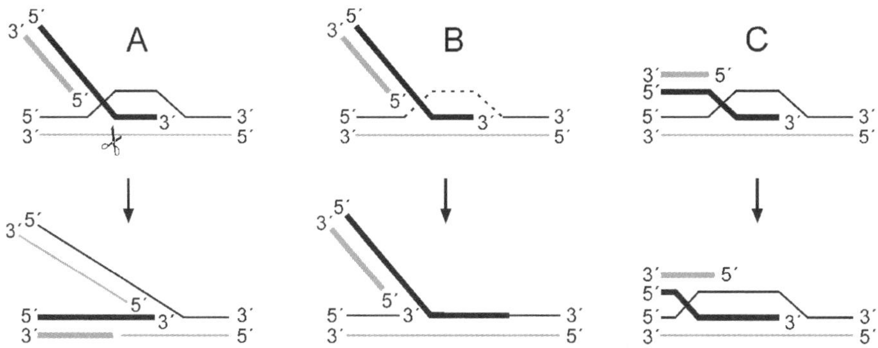

Fig. 2 Some possibilities for resolutions of D-loops resulting from invasion of a processed double stranded end into duplex DNA. The processed double stranded end is represented by *thick lines*, the duplex DNA by *thin lines*. Strands with 5′-3′ polarity (*from left to right*) are in *black*, strands with 3′-5′ polarity in *grey*. **A** If the D-loop is cleaved by an endonuclease at the position indicated by the scissors, a structure will be formed that is suitable to reassemble a replication fork. **B** Nucleolytic degradation of the D-loop and closing the gap will also result in a structure suitable for reinitiation. Note that in this case in a collapsed replication fork the identity of the strands will be switched, i.e. the former leading strand will be connected to the lagging strand template, the former leading strand template will become the lagging strand and the former lagging strand template will be the leading strand. **C** Extension of the D-loop in direction of the invading strand allows formation of a Holliday junction. This may be resolved with or without a sister chromatid exchange (not shown)

fore detection of such structures by electron microscopy in damage exposed cells from higher eukaryotes (Fujiwara and Tatsumi 1976; Higgins et al. 1976; Inman 1984) do not allow a decision as to whether all of these pathways are realized by cells and whether one is preferred over another. Similarly, genetic evidence such as the rescue of certain mutant phenotypes in *S. pombe* by production of the bacterial resolvase RusA also suffers from the same uncertainties regarding the possible pathways (Doe et al. 2002). In *S. cerevisiae* Holliday junction formation could only be demonstrated in *rad53* mutants, but not in wild type cells (Sogo et al. 2002). It should be noted in this context that during repair of double strand breaks resulting from excision of P-elements in *Drosophila* SDSA seems to be the more appropriate model to explain the genetic phenomena observed, rather than the "classical" double strand break repair model (Szostak et al. 1983) involving formation of a double Holliday junction (Nassif et al. 1994; reviewed in Lankenau 1995).

Another method to detect unusual intermediates arising during replication is 2D agarose gel electrophoresis (Brewer and Fangman 1987). In *S. cerevisiae*, molecules in the so-called X-spike, where also Holliday junctions migrate (Brewer and Fangman 1988), have been observed at the Fob1 protein bound to RFB (Brewer and Fangman 1988; Brewer et al. 1992; Kobayashi and Horiuchi 1996) that blocks replication forks at the 3′-end of the rDNA locus to prevent collision of transcription and replication. Formation of these molecules was dependent on Rad52, but not on Rad51 (Zou and Rothstein 1997). However, as already discussed in Sects. 2.6 and 2.7 for *E. coli*, blocks at proteins are probably different from arrests of replicative DNA polymerases (Gruber et al. 2000). Furthermore, since Fob1 has evolved to block replication forks, it may act differently from spurious protein-induced arrests. Molecules migrating in the X-spike of 2D gels have been detected in *S. pombe* and *S. cerevisiae* associated with replication origins in cells with presumably unperturbed replication (Lopes et al. 2003; Segurado et al. 2002). However, whereas in *S. pombe* formation of these molecules was dependent on Rad22, Rhp51 and Rhp54 (the homologs of the *S. cerevisiae* proteins Rad52, Rad51 and Rad54, respectively), they were found to be independent of Rad52 and Rad51 in *S. cerevisiae*. Characterization of these molecules in *S. cerevisiae* lead to the hypothesis that they may represent hemicatenanes that may serve a function of sister chromatid cohesion (Lopes et al. 2003). In an *sgs1* mutant (see Sect. 3.2), a significant increase in molecules migrating in the X-spike have been observed, whose formation is dependent on function of homologous recombination (Ira et al. 2003; Liberi et al. 2005). However, resistance of these molecules to the Holliday junction resolvases RuvC and T4 endo VII strongly suggest that they are not classical Holliday junction but rather pseudo-double Holliday junctions, where only two strands in between the junctions are hybridized, whereas the other two are not.

In summary, at present it is not clear whether in eukaryotes Holliday junctions – which are thought to be crucial intermediates in prokaryotes – are formed in response to DNA damage at replication forks and, if yes, which pathway would form these junctions and how they may be processed.

3.2 Pathways and Factors Involved in Reinitiation in Eukaryotes

3.2.1 Post-replicative Repair (PRR)

The first repair pathway that was implicated in the reinitiation of stalled replication forks is the so-called post-replicative repair pathway (PRR). PRR was initially defined in *E. coli* by the observation that newly synthesized chromosomal DNA becomes fragmented after UV irradiation, but is converted into high molecular weight DNA upon post-irradiation incubation (Rupp and Howard-Flanders 1968). Subsequently, a similar observation has also been made in *S. cerevisiae* and the *RAD6* gene has been identified to be required for this repair (Jachymczyk et al. 1977; Prakash 1981). Further analysis of this pathway revealed that it consists of an error-prone and an error-free branch, both requiring Rad6 function. Recent experiments have suggested that post-translational modification of the homotrimeric eukaryotic sliding clamp PCNA (proliferating cell nuclear antigen), which is the functional analog of the prokaryotic β-clamp (see Sect. 2.1), is an important determinant for the switch between error-prone and error-free pathways as shown schematically in Fig. 3; it is described in more detail below.

Genes are classified into error-prone PRR, if the mutants, where those genes are inactivated, fulfill the following two criteria: First, the mutants have a defect in UV-induced mutagenesis, indicating that UV damages cannot be processed by error-prone translesion synthesis. Secondly, the mutants are hypostatic to *rad6*, meaning that the respective double mutant with *rad6* is no more sensitive to DNA damage than the *rad6* single mutant. Error-prone PRR comprises DNA-polymerase ζ, which consist of the two subunits Rev3 (Morrison et al. 1989) and Rev7, and the Rev1 protein, that seems to be a dCMP transferase (Haracska et al. 2001, 2002; Nelson et al. 1996; Washington et al. 2004). Another translesion polymerase in budding yeast is Pol η (encoded by

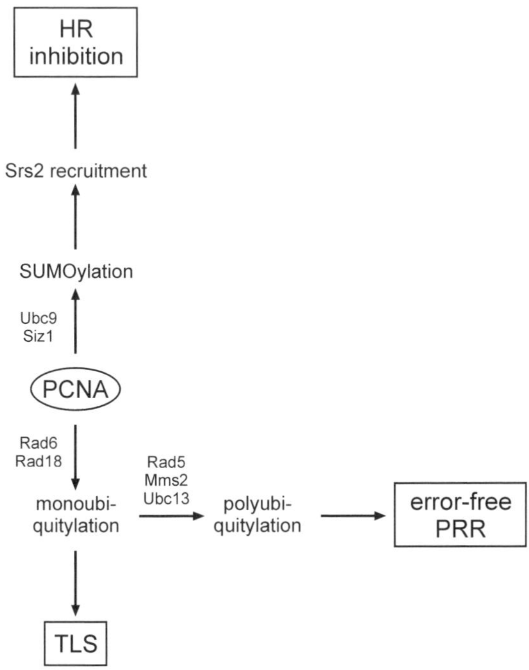

Fig. 3 Posttranslational modification of PCNA as a molecular switch for different reinitiation pathways. Lysine 164 of PCNA can be monoubiquitylated in response to DNA damage by the Rad6/Rad18 heterodimer. This modification is a prerequisite for translesion synthesis (TLS). The monoubiquitylated form of PCNA can be polyubiquitylated by the heterotrimer Rad5/Mms2/Ubc13 *via* unusual lysine 63 linkages between the ubiquitin moieties. This modification is probably a signal for error-free post-replicative repair (error-free PRR). Instead of being ubiquitylated, lysine 164 of PCNA can alternatively be SUMOylated by the SUMO-conjugating enzyme Ubc9 and the SUMO ligase Siz1. This modification recruits Srs2, which inhibits homologous recombination (HR)

the *RAD30* gene), which incorporates two A's opposite a T<>T dimer and is thus error-free for this type of lesion (Washington et al. 1999; Washington et al. 2000). Therefore, it cannot be neatly classified by the criteria given above as belonging to error-prone PRR.

The error-free branch comprises genes, where the respective mutants are hypostatic to *rad6* (and thus belong to the *rad6* epistasis group), but have no defect in UV-induced mutagenesis. On the contrary, most of the mutants are mutators by themselves. This can be interpreted such that DNA-lesions not processed by error-free PRR are channeled into the error-prone pathway instead. Three genes from this group will be considered in more detail in the following: *RAD5*, *MMS2* and *UBC13*.

At least one reason for the central role of *RAD6*, defining the epistasis group, has recently been discovered. It was already known since 1987 that Rad6 is a ubiquitin conjugating enzyme (Jentsch et al. 1987) that is probably recruited to DNA by interaction with Rad18 (Bailly et al. 1994, 1997). Lysine 164 of the sliding clamp PCNA (encoded by the *POL30* gene) has been identified as a target for ubiquitylation by Rad6 (Hoege et al. 2002). As indicated in Fig. 3, this modification is apparently a prerequisite for TLS by both Pol η and Pol ζ, since a PCNA mutant that cannot ubiquitylated at K164 has the same DNA damage sensitivity as a *rev3 rad30* double mutant (if polyubiquitylation and SUMOylation is excluded; see below) (Stelter and Ulrich 2003). Once monoubiquitylated by Rad6, an unusual K63-linked polyubiquitin-chain can be assembled on PCNA by the Rad5/Mms2/Ubc13 complex (Hoege et al. 2002) as shown schematically in Fig. 3. The finding that monoubiquitylation is required for TLS and for polyubiquitylation, which might well be the signal for activation of error-free PRR, suggests that TLS is employed prior to error-free repair attempts *via* this pathway. This is in stark contrast to *E. coli* (see Sect. 2.9), where SOS induced TLS *via* Pol V occurs comparably late after a replication arrest. But as discussed below, there seems to be an error-free mechanism also in eukaryotes that acts prior to TLS.

The picture sketched above for regulating the switch between TLS and error-free PRR is certainly oversimplified. K164 is not only subject to ubiquitylation but can also be modified by the small ubiquitin-like modifier SUMO (see Fig. 3). This SUMOylation is carried out by Ubc9 and Siz1, where the former has been shown to physically interact with PCNA, Rad18 and Rad5. In addition to K164, SUMO can also be transferred to K127. In contrast to K164, K127 is not conserved in PCNA of higher eukaryotes. Modification of K127 seems to inhibit repair, since a *pol30* K127R K164R double mutant is more resistant to DNA damaging agents than a *pol30* K164R single mutant (Hoege et al. 2002).

SUMO modification is apparently present in cells not challenged with a DNA damaging agent. Treatment with low doses of MMS resulted in appearance of mono- and polyubiquitylated PCNA alongside with SUMO-modified PCNA. Upon high doses of MMS, ubiquitylated PCNA disappeared and high-levels of SUMO-modified PCNA could be detected instead (Hoege et al. 2002).

These results suggest that SUMO-modification plays a role both in undamaged and in heavily damaged cells.

From the statement made above, that monoubiquitylation of PCNA by Rad6/Rad18 is required for TLS, it seems surprising that both *rad6* and *rad18* mutants exhibit a largely *REV3*-dependent spontaneous mutator phenotype (Quah et al. 1980; Roche et al. 1995). However, the observation that a PCNA K127R/K164R double mutant has a spontaneous mutator phenotype that is lower than that of a PCNA K164R single mutant would suggest that also K127 modification plays a role in control of spontaneous mutagenesis. From the suppression of the mutator phenotype of *rad18* by a *siz1* mutation one would conclude that this is due to the absence of SUMOylation. From the mutants analyzed it cannot be decided whether the latter phenotype is due to K164 or K127 modification (Stelter and Ulrich 2003), but the results with the PCNA mutants would point rather to K127.

The modification of PCNA apparently controls both branches of PRR, i.e. *REV3*- and *RAD30* dependent TLS as well as *RAD5/MMS2/UBC13*-dependent error-free bypass, since a mutant carrying a K164R mutation in PCNA is epistatic to *rev3*, *rad30* and *rad5* (Haracska et al. 2004; Stelter and Ulrich 2003). From this one would expect that the K164R mutant has the same sensitivity as a *rad6* mutant, since Rad6 is the enzyme that controls the modification. However, *rad6* is more sensitive than the PCNA mutant and actually the *rad6* sensitivity is partially suppressed by introduction of a K164R mutation. This would suggest that in the K164R mutant an alternative pathway can be activated, which cannot operate in the *rad6* mutant due to the presence of K164 in PCNA. The alternative pathway is probably dependent on functions of homologous recombination, since a *rad52* mutation, that confers only mild sensitivity to UV-light as a single mutation, is strongly synergistic with the K164R mutation and the respective double mutant is identical in sensitivity to a *rad6 rad52* double mutant. This demonstrates that in a *rad52* background the K164R mutation is equivalent to *rad6*. A very likely scenario to explain these results is that SUMOylation of K164 inhibits homologous recombination (see following section on Srs2). In a *rad6* mutant, where all branches of PRR are defective, also the rescue pathway *via* homologous recombination would be unavailable, since PCNA would be SUMO-modified at K164, inhibiting homologous recombination. In the K164R mutant, however, neither ubiquitin modification required for PRR nor SUMO-modification repressing homologous recombination would be possible, which would effectively result in an increase in homologous recombination (Haracska et al. 2004), which has been demonstrated to contribute to post-replicative repair (Prakash 1981). Also consistent with this idea is the hyperrecombination phenotype both for ectopic recombination and for sister chromatid recombination exhibited by *rad5* and *rad18* mutants after UV irradiation in logarithmically growing cells (Friedl et al. 2001). Lesions normally handled by PRR would then be processed by recombination with the sister chromatid or also with another homologous copy of the respective DNA sequence. This would imply that error-free PRR most likely requiring interaction of the sister chromatid is non-recombinogenic in the experimental systems required to detect sister chromatid recombination. The most obvious mechanism for this would be chicken foot formation by fork regression followed by a polymerase strand switch as described in Sect. 2.5 for *E. coli*. However, this is a reasonable mechanism for lesions on the leading strand template only, which would then mean that lesions on the lagging strand template should preferably be handled by TLS or HR. Thus, at present the mechanism by which the error-free branch of PRR contributes to damage tolerance is still unclear.

3.2.2 Srs2

The *SRS2* gene (*s*uppressor of *r*ad *s*ix) has originally been identified by the observation that respective mutants are able to suppress the UV-sensitivity of *rad6* and *rad18* mutants (Lawrence and Christensen 1979). The same gene has been found independently as a mutant with increased intrachromosomal gene conversion called *hpr5* (Aguilera and Klein 1988) and as a mutant increasing radiation sensitivity called *radH* (Aboussekhra et al. 1989). It could be demonstrated that the suppression of *rad6* by *srs2* is dependent on functional homologous recombination (Schiestl et al. 1990) and that the suppression acts specifically on the *RAD5/MMS2/UBC13* dependent error-free branch of PRR (Ulrich 2001). The protein is a 3′ to 5′ helicase (Rong and Klein 1993) that was recently shown to be able to disrupt presynaptic Rad51-filaments (Krejci et al. 2003; Veaute et al. 2003) and may thus have a similar function as UvrD in *E. coli* (Veaute et al. 2005; see also Sect. 2.9). This biochemical activity apparently provides a reasonable explanation for the hyperrecombination phenotype of *srs2* mutants and the suppression of the UV-sensitivity of *rad6* and of mutants in error-free PRR assuming that homologous recombination can handle the lesions instead of PRR. In a recent publication it was demonstrated that Srs2 interacts preferentially with the SUMOylated form of PCNA. Furthermore, mutations in *SIZ1*, coding for the SUMO ligase necessary for PCNA K164-SUMOylation, suppress *rad6* and *rad18* mutation as efficiently as *srs2*ΔC mutations, where a truncated version of Srs2 is made lacking the C-terminal part of the protein, which is required for interaction of Srs2 with SUMO-PCNA. Epistatic relationships of *siz1*, *srs2*ΔC and *pol30 K164R K127R* further support the conclusion that Srs2 is recruited by SUMO-PCNA to counteract homologous recombination (see Fig. 3). The stronger suppression of *rad6* and *rad18* by an *srs2* null allele, however, indicates that Srs2 acts also *via* additional mechanisms (Pfander et al. 2005).

It should, however, not go unnoticed that some genetic observations cannot be satisfactorily be explained by such a model. First, a mutation in *srs2* reduces the hyperrecombination phenotype of *rad5* and *rad18* mutant in UV-induced ectopic recombination. Secondly, *srs2* mutants display less damage induced sister chromatid recombination than wild type cells (Friedl et al. 2001, C.R., A.S. and W.K. unpublished observations), which is opposite to the behaviour in spontaneous recombination (Pfander et al. 2005). From an inhibitor of homologous recombination one would have expected a hyper-recombination phenotype in these assays, since all models for recombinogenic bypass of replication arresting lesion invoke an interaction with the sister chromatid. Therefore, Srs2 may also be a necessary factor for sister chromatid interactions.

3.2.3 Mph1

A recombination dependent pathway for error-free bypass of DNA polymerase arresting lesions that is apparently distinct from PRR is defined by the *MPH1* gene (*m*utator *ph*enotype). This gene was originally identified in a genomic screen to define the functions of unknown open reading frames discovered in the genome sequencing project by virtue of the spontaneous mutator phenotype of respective mutants (Entian et al. 1999). The mutator phenotype is dependent on functional TLS and *mph1* mutants are sensitive to several DNA damaging agents (Scheller et al. 2000). Epistasis analyses demonstrated that *mph1* is hypostatic to mutants from homologous recombination but with respect to DNA damage sensitivity not to mutants from PRR (Schürer et al. 2004). The Mph1 protein belongs to the DEAH family (Scheller et al. 2000) and possesses a 3′-5′ helicase activity (Prakash et al. 2005). Homologous proteins can be found in archaea and in other eukaryotes including humans (Komori et al. 2002; Scheller et al. 2000). The archaeal homologs have an additional C-terminal domain that carries sequence motifs of structure specific endonucleases like XPF and Mus81 (Enzlin and Schärer 2002; Komori et al. 2002). It was recently demonstrated that the N-terminal helicase stimulates the endonuclease activity residing in the C-terminal domain (Komori et al. 2004). Intriguingly, for the human gene a cDNA can be predicted from the chromosomal sequence that in addition to the N-terminal putative Mph1 homolog codes for a protein with a large C-terminal extension that also carries the conserved endonuclease motifs. It is therefore tempting to speculate that one function of yeast Mph1 may be the stimulation of Mms4/Mus81 endonuclease (see below). However, this cannot be the sole cellular function of Mph1, since *mms4* is not epistatic to *mph1* (C.R., A.S. and W.K., unpublished results).

3.2.4 Mms4/Mus81 (Eme1/Mus81)

The structure specific heterodimeric endonuclease Mms4/Mus81 in *S. cerevisiae* and its *S. pombe* counterpart Eme1/Mus81 are presently studied extensively for their role in reinitiation of stalled replication forks and in meiosis. For the latter, the reader may refer to a recent review by Hollingsworth and Brill (Hollingsworth and Brill 2004). This review also presents a comprehensive summary of the substrate specificity of these endonucleases.

The *MMS4* gene has been identified in 1977 by virtue of the MMS sensitivity of respective mutants (Prakash and Prakash 1977) and their damage sensitivity phenotypes characterized in more detail (Xiao et al. 1998). Mus81 (*M*ethyl methansulfonate, *UV* *s*ensitive) has been found in *S. cerevisiae* as a protein interacting with the recombination protein Rad54 (Interthal and Heyer 2000) and in *S. pombe* as a protein interacting with the checkpoint kinase Cds1 (Boddy et al. 2000), the *S. pombe* homolog of Chk2 (see Sect. 4.2). Subsequently, in *S. pombe* Eme1 was found as an interaction partner of Mus81 (Boddy et al. 2001). The finding that overproduction of the bacterial Holliday junction resolvase RusA partially rescued the meiotic defects associated with a *mus81* mutation in both S. *pombe* (Boddy et al. 2000) and S. *cerevisiae* (Interthal and Heyer 2000) as well as the finding that a partially purified protein cleaved a synthetic Holliday junction lead to the hypothesis that Eme1/Mus81 may constitute the long sought after eukaryotic Holliday junction resolvase (Boddy et al. 2001).

Further biochemical characterization of purified Eme1/Mus81 and Mms4/ Mus81 heterodimers, however, suggested that this endonuclease may rather cleave stalled replication forks in the leading strand template after an arrest of the leading strand DNA polymerase (Doe et al. 2002; Kaliraman et al. 2001; Whitby et al. 2003). Another function of Mms4/Mus81 could be removal of 3′ flaps (Bastin-Shanower et al. 2003) as they may result during filling of gaps in the lagging strand (Fabre et al. 2002). A third function of this endonuclease as demonstrated for Eme1/Mus81 is probably the cleavage of D-loops in the strand that hybridizes with the invading strand, which could explain several of the meiotic phenotypes of the respective mutants (Osman et al. 2003). The substrate specificity in general can probably be characterized by preferentially cleaving structures that have a single strand close to a branch point and an adjacent 5′-end (Fricke et al. 2005). During reinitiation of stalled replication forks, cleavage of D-loops would be useful after fork collapse, resulting in a double stranded end at the duplex of the leading strand/leading strand template. Such a double stranded end could, e.g. be created by Mms4/Mus81 cleavage of the leading strand template itself (see above) or by polymerase run-off at a nick in the leading strand template. (Run-off at the lagging strand template would probably be repaired as a normal double strand break.) In addition, treatment with drugs inhibiting topoisomerase I such as camptothecin would result in formation of double stranded ends on the leading strand template (Strumberg et al. 2000). Reinitiation of replication would probably require resection of the 5′-end to create a single stranded 3′ overhang, that could invade the sister chromatid forming a D-loop. Cleavage of the D-loop in the strand hybridized to the invading strand as shown in Fig. 2A, which would be one of the Mms4/Mus81 activities described above, reforms a forked DNA structure that seems suitable for replication restart.

Alternative resolution pathways may include the transformation of the D-loop into a Holliday junction, which could be achieved by extending the D-loop to the side where the invading strand enters until the displaced strand can hybridize with the recessed 5′-end of the double strand end. Cleavage of the D-loop may also be exerted by other structure specific endonucleases in the cell like Rad1/Rad10 (Bardwell et al. 1994; Davies et al. 1995; de Laat et al. 1998) or Slx1/Slx4 (Fricke and Brill 2003) or by degradation of the displaced strand as shown in Fig. 2B. Also here other structure-specific endonucleases may play a role as well as the 5′ flap endonuclease Rad27, which could be one reason for the synthetic lethality of *rad27* and *mus81* (Tong et al. 2001).

3.2.5 Sgs1

Another helicase that is involved in the reinitiation of stalled replication forks is Sgs1. *sgs1* mutants (*s*low *g*rowth *s*uppressor) were originally identified as suppressors of the slow growth phenotype of *top3* mutants (Gangloff et al. 1994). Sgs1 is the only member of the RecQ-family in yeast and mutations in several family members in humans are responsible for hereditary diseases associated with cancer susceptibility and genome instability, namely Bloom's, Werner and Rothmund-Thompson syndromes (Hickson 2003; Oakley and Hickson 2002). Sgs1 physically interacts with topoisomerase III (Bennett et al. 2000; Fricke et al. 2001) and this interaction seems to be at least functionally conserved during evolution (Wu et al. 2000). A characteristic feature of Bloom's syndrome is a high number of sister chromatid exchanges (Chaganti et al. 1974), which is also observed in yeast *sgs1* mutants (Onoda et al. 2000; Onoda et al. 2004). One biochemical activity of BLM is branch migrating Holliday junctions (Karow et al. 2000). By moving each junction of a double Holliday junction towards the other, the double Holliday junction would be converted into a hemicatenane, which could then be resolved by the action of topoisomerase III. This so-called Holliday junction dissolution can explain the increase of sister chromatid exchanges in Bloom's syndrome, since it would result in the resolution of Holliday junctions without the possibility of generating a cross-over (Wu and Hickson 2003). This would also explain the increase in X-structures observed in *sgs1* mutants on 2D-gels mentioned in Sect. 3.1 (Liberi et al. 2005).

sgs1 mutants display a variety of synthetically lethal or synthetically sick interactions. One of these synthetic interactions is observed with *srs2*, which can be suppressed by mutations in homologous recombination (Gangloff et al. 2000). This could suggest that the lack of suppression of homologous recombination *via* Srs2 would induce too many double Holliday junctions that in the absence of the dissolution process may be lethal to the cells, either as such or by the high number of cross-overs. A variety of other mutations have been found to synthetically interact with *sgs1* (Ooi et al. 2003). They include *mms4*, *mus81*, *slx1* and *slx4* (Mullen et al. 2001). Also with mutants that probably affect progression of the replication fork like *mgs1* (Hishida et al. 2001, 2002) or *rrm3* (Ivessa et al. 2002, 2003) synthetic interaction are observed (Branzei et al. 2002; Torres et al. 2004). Thus, Sgs1 apparently is a very versatile helicase that seem to be involved in a variety of cellular processes not only linked to replication restart (Watt et al. 1995).

3.3
Interaction and Hierarchy of the Different Pathways

Having a plethora of different systems that all seems to be involved in the reinitiation of stalled replication forks the question immediately comes to mind, how the cell decides, which of the systems should be employed. There are two extreme views: Either there is a strict hierarchy of the different pathways, in that one pathway acts on the top and if it fails for any reason, the next pathway is activated, or all the different pathway coexist and are employed more or less on a random basis. Given the high degree of organization for most cellular processes, the latter view seems very unlikely. TLS can probably serve as a suitable reference point in reasoning on a possible hierarchy of the different pathways. With respect to this reference point one would place error-free PRR certainly after TLS, since monoubiquitylation of PCNA is a prerequisite for both TLS and PCNA polyubiquitylation, which probably is one signal for error-free PRR (Hoege et al. 2002; Stelter and Ulrich 2003). Thus, error-free PRR can only take place in a situation, where TLS could have worked already. The mutator phenotype of *rad5*, *mms2* and *ubc13* (Broomfield et al. 1998; Cejka et al. 2001; Schürer et al. 2004), however, indicates that error-free bypass is not always acting after TLS has failed, but that the step of TLS is often skipped in favor of error-free processing. The observation that *rev3* mutants are not very sensitive to DNA damage would argue that at least in the absence of TLS most lesions can be processed by alternative, presumably error-free pathways, since the sensitivity is synergistic with mutations in error-free pathways (Broomfield et al. 1998; Cejka et al. 2001; Schürer et al. 2004). Nevertheless, spontaneous lesions are to a certain extent processed by TLS, since the spontaneous mutator phenotype of a *rev3* mutant is lower than that of wild type (Quah et al. 1980). From these considerations one would conclude that in contrast to bacteria, where TLS seems to be the last resort (see Sect. 2.9), in eukaryotes TLS is probably employed more regularly, even if not all other resources have been exploited.

As indicated by the suppression of sensitivity of mutants in error-free PRR by an *srs2* mutation (Ulrich 2001), which is dependent on functional HR (Schiestl et al. 1990), one would suspect that HR could in principle substitute for error-free PRR, but that its action is prevented by Srs2 (Krejci et al. 2003; Veaute et al. 2003; Pfander et al. 2005). Most of the intermediates that are part of the models sketched at the beginning of this section can in principle be arrived at by the action of HR. With the *MPH1*-HR pathway a function of HR in error-free bypass could be defined that is distinct from PRR (Schürer et al. 2004). We would tentatively place this pathway upstream of TLS, since mutants that are defective in this pathway have a spontaneous *REV3*-dependent mutator phenotype. Furthermore, deletion of the *APN1* and *APN2* genes, coding for the two known AP-endonucleases in yeast, does not result in an increase of the spontaneous mutator phenotype. Additional deletion of *MPH1*, however, leads to a strong synergistic increase of the mutator phenotype in the triple mutant. This indicates that the increase of AP-sites can be handled by the *MPH1*-HR pathway and would also imply that AP sites are not normally processed by TLS (Schürer et al. 2004).

Placement of the *MMS4/MUS81* pathway may be aided by the fact that neither of the respective mutants has a spontaneous mutator phenotype in the canavanine forward mutation system (C.R., A.S. and W.K., unpublished results), although for a reversion assay of *trp1-289* a mutator phenotype of *mms4* mutants has been reported (Xiao et al. 1998). From the lack of a mutator phenotype one would conclude that Mms4/Mus81 acts at a point, where TLS is not a valid option. This may mean that Mms4/Mus81 works on a structure, which in case of failure to be processed cannot be reverted to an intermediate, where TLS can occur. Alternatively, Mms4/Mus81 may work on replication arrest that for intrinsic reason are not amenable to TLS. This could be, e.g. protein induced arrests or adducts that are too large even for a translesion polymerase. However, at least in the case of protein induced fork stalling, it is hardly conceivable how fork cleavage could alleviate the problem. Therefore, probably resolution of recombination intermediates like D-loops is the more attractive option, where Mms4/Mus81 would then be placed after homologous recombination and would be mainly required for situations, where a double stranded end is produced. In favor of this hypothesis would be the interaction of Mus81 with Rad54 that originally led to the identification of Mus81 (Interthal and Heyer 2000). In addition, the epistasis of *rad52* and *mms4* seems to point into the same direction (Odagiri et al. 2003). It was shown in *S. pombe* that after camptothecin, MMS and HU treatment Mus81 is required for Rhp51-independent (Rhp51 is the *S. pombe* homolog of Rad51), but Rad22-dependent (Rad22 is the Rad52 homolog in *S. pombe*) events (Doe et al. 2004). The above assertion for Mus81 function would then require that D-loops can be formed by Rad52-homologs alone, which in vitro could be demonstrated (Doe et al. 2004; Kagawa et al. 2001).

4 The Eukaroytic DNA Damage Checkpoint

4.1 Architecture of the DNA Damage Checkpoint

The major control and coordination center for dealing with DNA damage in eukaroytic cells is the DNA damage checkpoint, which is actually subdivided

Table 1 Homologous checkpoint proteins from yeasts and mammals

Group	Function	*S. cerevisiae*	*S. pombe*	Mammals
Sensors	PI3-like kinase	Mec1	Rad3	ATR
	PI3-like kinase	Tel1	Tel1	ATM
	interaction with ATR-like proteins	Ddc2/Lcd1	Rad26	ATRIP
	DSB recognition	Mre11	Mre11	MRE11
		Rad50	Rad50	RAD50
		Xrs2	Nbs1	NBS1
	RFC-like clamp loader, interaction with Rfc2–5	Rad24	Rad17	RAD17
	PCNA-like clamp, 9-1-1-complex; associates with Mec3 and Rad17	Ddc1	Rad9	RAD9
	PCNA-like clamp, 9-1-1-complex; associates with Mec3 and Ddc1	Rad17	Rad1	RAD1
	PCNA-like clamp, 9-1-1-complex; associates with Rad17 and Ddc1	Mec3	Hus1	HUS1
Mediators	mediates Chk1/2 activation mainly in G1 and G2, BRCT domains	Rad9	Crb2/Rhp9	BRCA1? 53BP?
	mediates Chk1/2 activation, S-phase checkpoint, BRCT domains	Mrc1	Mrc1	Claspin
Transducers	effector S/T kinases	Rad53	Cds1	CHK2
	effector S/T kinases	Chk1	Chk1	CHK1
	effector S/T kinases	Dun1		
Others (involved in bypass of stalled replication forks)	XPF-like endonuclease, complex with Eme1/Mms4^{SC}	Mus81	Mus81	MUS81
	helicase of the RecQ family, interaction with Top3	Sgs1	Rqh1	BLM, WRN, RECQ4L
	helicase, regulates negatively homologous recombination	Srs2	Srs2	

into several subpathways as described below. Besides delay in cell cycle transitions in order to coordinate DNA replication with DNA repair, the DNA damage checkpoint controls damage-induced transcription, deoxynucleoside triphosphate metabolism and possibly other physiological responses to genotoxic stress. In G1 cells the G1/S checkpoint can prevent entry into S-phase. The G2/M checkpoint can delay the onset of mitosis. The re-establishment of normal cell cycle is defined as recovery, which is associated with checkpoint inactivation (Pellicioli et al. 1999; Sandell and Zakian 1993). Unlike the G1/S or G2/M checkpoint, in which the cell cycle progression is blocked in response to DNA damage, the S-phase checkpoint only slows the rate of replication. The so-called intra-S-phase and replication checkpoints are involved in stabilization of stalled replication forks in a replication-competent state, inhibition of late origin firing and prevention of entry into mitosis. Moreover, a process called adaptation ensures that DNA replication is completed even in the presence of DNA damage (Sandell and Zakian 1993; Toczyski et al. 1997). The significance of these checkpoint responses is emphasized by the evolutionary conservation of the proteins involved from yeast to human. Homologous checkpoint components from budding and fission yeasts and mammals are listed in Table 1. For simplification, only the name of the human protein is mentioned in the following, if common aspects are discussed.

The DNA damage checkpoint pathway is divided into signaling factors that recognize and bind to sites of DNA damage (sensors) and into signal transducers, which initiate the action of the effectors in response to obstacles that interfere with the ongoing cell cycle. The key players will be described relatively briefly; for a more detailed overview see, e.g. (Boddy and Russell 2001; Carr 2002; Longhese et al. 2003; McGowan 2003).

4.2 DNA Damage Recognition by the Checkpoint and its Activation

4.2.1 Activation of Sensor Kinases

It is still quite enigmatic how the S-phase DNA damage checkpoint is activated, especially in response to a variety of different obstacles and DNA lesions. However, it is clear that replication must be initiated for the S-phase checkpoint to be activated (Labib et al. 2001; Sidorova and Breeden 2003). During regular DNA repair abnormal structures and DNA lesions are recognized *via* specific DNA-protein interactions (various DNA glycosylases for different DNA lesions in base excision repair, MutS in mismatch repair, UvrA (*E. coli*) and XPC (human) in nucleotide excision repair, etc.). Many of those proteins, in addition to ATM, BLM, MRN and RFC subunits (see below) are found in a complex called BASC (*B*RCA1-*a*ssociated genome *s*urveillance *c*omplex), which is found in S-phase irradiated cells in nuclear foci that also contain PCNA (Wang et al. 2000). On the other hand, a unique intermediate that often occurs after replication arrests is single stranded DNA, which is in vivo coated by RPA, the eukaryotic single stranded binding protein. Indeed, it could be shown that an ssDNA-RPA complex is bound by ATRIP (Carr 2003; Cortez et al. 2001; Zou and Elledge 2003). ATRIP is associated with the checkpoint kinase ATR, which belongs to the phosphatidylinositol 3-kinase (PI-3 kinase) family (Cortez et al. 2001; Paciotti et al. 2000; Rouse and Jackson 2002). A PI-3 kinase functionally distinguished from ATR, but with overlapping function, is ATM. In fact, ATR and its yeast orthologs respond to DNA polymerase blocking DNA lesions and stalled replication forks, whereas ATM and its orthologs respond primarily to double strand breaks (DSBs) (for review see, e.g. Abraham 2001; Nyberg et al. 2002; Shiloh 2003). Whereas ATM plays a key role in responding to DSB in humans, the yeast homolog Tel1 has just a secondary role, apparently only in the absence of Mec1 (Usui et al. 2001). The ATM/Tel1-dependent checkpoint involves additional proteins, namely a trimeric complex known as Mre11/Rad50/Xrs2 (MRX) in yeasts and Mre11/Rad50/Nbs1 (MRN) in mammals (D'Amours and Jackson 2002; Petrini 2000). MRE11 and NBS1 are phosphorylated in response to DNA damage in an ATM dependent manner (D'Amours and Jackson 2002; Grenon et al. 2001; Lim et al. 2000). The activation of the PI 3-kinases, probably induced by binding to DNA, occurs by a change in the dimer conformation of the kinase with subsequent autophosphorylation, as was shown for ATM (Bakkenist and Kastan 2003). This enables them to phosphorylate their accessory factor and many other substrates. In order to phosphorylate their other substrates, loading of the 9-1-1-complex (see below), which occurs independently of ATR/ATM, is required. This indicates that two DNA damage recognition events are required to activate the checkpoint, which may safeguard the cell from inappropriate checkpoint activation (Caspari and Carr 2002; Kondo et al. 2001; Melo et al. 2001).

4.2.2
The 9-1-1-Complex and Alternative Clamp Loaders

During normal replication, the processivity factor PCNA is loaded onto DNA by the clamp loader replication factor C (RFC). A heterotrimeric PCNA-like structure, the so called 9-1-1-complex composed of Rad9, Rad1 and Hus1 in fission yeast and mammals and of Ddc1, Rad17 and Mec1 in budding yeast, respectively, is loaded onto DNA in case of damage (for review see, e.g. Longhese et al. 1998; Majka and Burgers 2003). Recruitment of this complex to damaged sites depends upon an RFC-like complex (Zou et al. 2002). The checkpoint-specific clamp loader is composed of the four small subunits of RFC, Rfc2–5, and hRAD17/*sp*Rad17/*sc*Rad24, which shares homology to the large subunit of RFC, Rfc1 (Green et al. 2000). Recently, also other RFC-like complexes were identified in *S. cerevisiae* and in other organisms, which consist of Rfc2–5, but contain Elg1 or Ctf18 instead of Rfc1. Elg1 (*e*nhanced *l*evels of *g*enome instability) seems to play an important role in limiting deleterious recombination (Ben-Aroya et al. 2003; Kanellis et al. 2003). Ctf18, identified in a screen for yeast mutants that exhibit elevated rates of mitotic chromosome loss (Spencer et al. 1990), loads PCNA onto gapped DNA and is implicated into sister chromatid cohesion (Bermudez et al. 2003; Mayer et al. 2001). It seems that those different RFC-like complexes work in separate but redundant pathways, probably by loading PCNA or PCNA-related complexes onto DNA in order to recruit specific DNA polymerases and contribute to maintenance of genomic integrity (Aroya and Kupiec 2005; Majka and Burgers 2004).

4.2.3 Effector Kinases

The main effectors of the checkpoint pathway are the CHK1 and CHK2 serine/threonine protein kinases, which also undergo an ATM/ATR dependent phosphorylation and activation. Despite being well conserved at the sequence level, they do not have equivalent function in all species. The Chk2 homolog from budding yeast, Rad53, is an effector of both the G1/S and G2/M DNA damage checkpoint as well as the S-phase checkpoint. Its function in replication resumption can be explained by preventing formation of abnormal structures at stalled replication forks (Sogo et al. 2002; Tercero and Diffley 2001). In addition, Rad53 actively inhibits firing of late origin of replication (Santocanale and Diffley 1998; Shirahige et al. 1998). *RAD53* and *MEC1* are essential genes. Their essential role, however, is not due to their role in the DNA damage checkpoint but rather due to the inhibition of Sml1, a suppressor of ribonucleotide reductase Rnr1, which regulates the dNTP pool in the cell. Deletion of *SML1* allows the generation of *rad53* and *mec1* null mutants (Zhao et al. 2000; Zhao et al. 1998). In contrast to *sc*Rad53, the fission yeast homolog Cds1 is not activated outside S-phase (Brondello et al. 1999). The human homolog CHK2 is activated after ionizing radiation at any stage of the cell cycle, but is poorly activated in response to replicational blocks (Blasina et al. 1999; Matsuoka et al. 1998). Conversely, CHK1 appears to be the primary effector for the replication checkpoint in humans, which is supported by the finding that disruption of CHK1 in mice is embryonically lethal. hCHK1 is phosphorylated and activated in response to incomplete replication, but it also seems to be involved in both G1 and G2 DNA damage checkpoint (Takai et al. 2000; Zhao and Piwnica-Worms 2001; Zhao et al. 2002). By contrast, *sc*Chk1 functions in parallel to Rad53, mainly to prevent mitosis (Sanchez et al. 1999). *sp*Chk1 has also a function in the G2 DNA damage checkpoint (Rhind and Russell 2000). In fission yeast, neither Cds1 nor Chk1 are essential.

4.2.4
Activation of Effector Kinases

In order for these kinases to be activated by ATR- and ATM like kinases, other factors in addition to RFC and alternative clamp loaders are required, usually called mediators or adaptors. The mediator proteins, characterized by a large, acidic BRCT domain (*BR*AC1 *c*arboxy-*t*erminal), are weakly conserved. The founding member of this group is budding yeast Rad9, with homologs in *S. pombe*, Crb2, and humans, where BRCA1 and/or 53BP1 (p*53* *b*inding *p*rotein) are discussed to be the mediators. BRCA1, a protein mutated in 50% of familiar hereditary breast cancer (Easton et al. 1993), is phosphorylated by both ATM and ATR in response to DNA damage (Cortez et al. 1999; Tibbetts et al. 2000), colocalizes in ATR foci (Cortez et al. 1999; Tibbetts et al. 2000) and is found in association with hCHK2 (Lee et al. 2000). Likewise, 53BP1 forms discrete foci in response to ionizing radiation presumably at sites of DSBs and is phosphorylated in an ATM-dependent manner (Anderson et al. 2001; Schultz et al. 2000). The mediator function is best studied for *S. cerevisiae* Rad9. After DNA damage, but not in response to replication stress, Rad9 is phosphorylated in a Mec1- or Tel1-dependent manner, which triggers its oligomerization *via* the BRCT domain (Soulier and Lowndes 1999; Vialard et al. 1998). Rad9 can now act as scaffold for Rad53. This interaction occurs *via* the *f*ork-*h*ead *a*ssociated domain (FHA) of Rad53, a domain found in many proteins, which mediates specific phosphorylation dependent protein-protein interactions. The result is an autophosphorylation of Rad53 in *trans*, followed by dissociation of Rad9 (Gilbert et al. 2001; Sun et al. 1998). The Rad9 counterpart in Rad53 activation for responses to replicational stress is *sc*/*sp*Mrc1 (Alcasabas et al. 2001; Tanaka and Russell 2001), for which a similar mechanism is proposed. Mrc1 is hyperphosphorylated in response to replication stress (HU treatment) in budding and fission yeast (Osborn and Elledge 2003). Similarly, a protein homologous to *sp*Mrc1, Claspin, is required for Chk1 activation during replication block in higher eukaryotes (Kumagai and Dunphy 2000; Lee et al. 2003). Another S/T effector kinase, *sc*Dun1, homologous to Cds1 and Chk2, was first isolated as an essential factor for damage induced transcription of *RNR* genes (Zhou and Elledge 1993). Its FHA domain is required for transient interaction with Rad53, which leads to Rad53 hyperphosphorylation and activation. Dun1 functions primarily in DNA damage checkpoint pathways controlling G2/M arrest (Bashkirov et al. 2003). Once activated, the effector kinases CHK1, CHK2 and *sc*Dun1 phosphorylate a variety of target proteins that mediate, e.g. delay in cell cycle and mitosis (p53, MDM2, Cdc25C, Pds1, Swi6, Dbf4), and/or induce repair/bypass mechanisms (BRCA1, NBS1, Rad55, RPA; for review see, e.g. Foiani et al. 2000; Zhou and Elledge 2000).

4.3 Links between Intra-S-Phase Checkpoint and Replication, Repair and Recombination

The intra-S-phase checkpoint plays an important role in maintenance of the replication fork arrested at a DNA lesion. In particular, direct interaction of Tof1 and Mrc1 with the replication fork seems to be necessary for maintaining a stable replication complex (Katou et al. 2003; Lopes et al. 2001). Tof1, which was identified by virtue of its interaction with topoisomerase I (Park and Sternglanz 1999), seems to have a mediator-like function during S-phase to link Mec1 with Rad53 (Foss 2001). In addition, the fission yeast homolog Swi1, a specific mediator of Cds1 activation, is required in complex with Swi3 to prevent fork collapse (Noguchi et al. 2003; Noguchi et al. 2004).

Other proteins that are presumably involved in reinitiation of replication like primase, Rad55, RPA, Srs2, Sgs1 and Mus81 also show links to the DNA damage checkpoint. In yeast, Rad53 and Mec1 are involved in regulation of these proteins, probably by direct modification. For example, phosphorylation of polα-primase prevents accumulation of ssDNA at stalled replication forks (Pellicioli et al. 1999; Sogo et al. 2002). RPA is also phosphorylated by Mec1 in yeast and is required for efficient checkpoint function, although this interaction is not well understood (Brush and Kelly 2000; Brush et al. 1996; Kim and Brill 2003; Longhese et al. 1996). RPA is likely to function with an alternative RFC complex (*sc*Rad24/RFC2–5; Kim and Brill 2001) as sensor for the checkpoint pathway and could serve to recruit other proteins to stalled replication forks. Rad55, a protein involved in Rad51 filament formation (Sung 1997), is phosphorylated by Dun1 in response to MMS treatment, UV light and gamma radiation. This supports the idea that the DNA damage checkpoint activates repair and possibly affects the balance between different competing DNA repair pathways (Bashkirov et al. 2000).

Dun1 also interacts with Sgs1 and Top3 in a two hybrid system (Hazbun et al. 2003). This interaction might be a link to *sc*Sgs1/*sp*Rqh1/hBLM, all homologues of the RecQ-helicase from *E. coli*, and its role in the intra-S-phase checkpoint (for other Sgs1 functions confer to Sect. 3.2). Consistent with this hypothesis, Sgs1 protein levels, like those of BLM, peak in S-phase (Dutertre et al. 2000; Frei and Gasser 2000). Furthermore, Sgs1 co-localizes with components of the replication machinery, e.g. Orc2, topoisomerase I and II, PCNA, RPA and polδ, polε and polα (Bjergbaek et al. 2002; Cobb et al. 2003). Sgs1 may be a key element of the intra-S-phase checkpoint to stabilize stalled replication forks, keeping proteins of the replication machinery in a replication proficient state to ensure resumption of elongation, if the obstacle is bypassed or repaired. On the other hand, *sgs1* cells have only a minor defect in intra-S-phase response after HU treatment, which is synergistic with the RFC like clamp loader mutant *rad24*. Whereas both single mutants are proficient for Rad53 phosphorylation, it is strongly reduced in the *sgs1 rad24* double mutant. This indicates that Sgs1 and Rad24 act indepently upstream of Rad53 (Frei and Gasser 2000). *S. pombe rqh1*$^{-}$ cells recover poorly after HU treatment (Stewart et al. 1997). The latter findings raise the question if *sc*Sgs1/*sp*Rqh1 has a function in the DNA damage checkpoint as a sensor in parallel to *sc*Rad24 to activate *sc*Rad53. Recently, a function of Sgs1 in checkpoint activation by binding to Rad53, which was independent of its helicase activity, was described (Bjergbaek et al. 2005). BLM/Sgs1 show a conserved physical interaction with topoisomerase type IA hTop3α/*sc*Top3 to dissolve double Holliday junctions without cross-overs (see Sect. 3.2).

Besides its functions described in Sect. 3.2, also for Srs2 a connection to the checkpoint is postulated. As for Sgs1, Srs2 expression peaks shortly before S-phase and can be induced by DNA damage (Heude et al. 1995). Srs2 is phosphorylated by Dun1 or Cdk1, which is dependent on functional Mec1 and Rad53 (Liberi et al. 2000). These findings suggest that Srs2 activity is regulated and that the protein is targeted to stalled replication forks by checkpoint functions. Srs2 may be involved in turning off the DNA damage checkpoint, since *srs2* cells do not recover from a checkpoint arrest induced by a DSB (Vaze et al. 2002). On the other hand, Srs2 may also be required for proper activation of the checkpoint, since after MMS treatment of an *srs2* mutant, Rad53 activation was decreased, which lead to the assumption that Srs2 may generate an intermediate that contributes to checkpoint activation (Liberi et al. 2000). This is supported by the finding that Srs2 can prevent Rad51 filament formation (Krejci et al. 2003; Veaute et al. 2003) and in *srs2* cells recruitment of recombination factors might mask the checkpoint signal.

The link of Mus81 (see Sect. 3.2 for other functions) to the DNA damage checkpoint came from the observation that it binds to the FHA domain of Cds2, the *sc*Rad53 and hCHK2 homolog from *S. pombe* (Boddy et al. 2000) and that it is phosphorylated by Cds1, which might control its biochemical activity directly or its chromatin-binding ability. This might prevent deletion mutations and thus regulate recombination-dependent restart of replication.

In addition to the mechanisms discussed in Sect. 3.2, also the checkpoint seems to control TLS. Expression of DinB from fission yeast, the homologue of human polκ, is regulated by *sp*Rad17: when replication is perturbed, DinB binds to chromatin and is up-regulated both at RNA and protein level (Kai and Wang 2003a). This does not require *sp*Chk1 and Cds1, but rather a functional *sp*Rad17. In addition, DinB can physically associate with *sp*Hus1 and *sp*Rad1 from the 9-1-1-complex. These data suggest that after replication perturbations, e.g. in a replication mutant or by DNA damage, the *sp*Rad17/Rfc2-5 complex loads the 9-1-1-complex onto DNA which subsequently recruits the translesion polymerase DinB to chromatin for mutagenic bypass (Kai and Wang 2003b).

5 Conclusion

The considerations outlined in the three sections above should have demonstrated that a plethora of different mechanisms are available for reinitation of stalled replication forks. There are certainly more to be discovered as is indicated by some recent publications. For many of the genes involved in reinitiation, the molecular function is still quite obscure, which makes a coherent reconstruction of the network impossible at the moment. However, from what is known to date it is clear that there is extensive regulation and crosstalk between different pathways. Regulatory mechanisms include transcriptional regulation, posttranslational modifications like protein cleavage, ubiquitylation and phosphorylation, protein-protein interactions and conformational changes. Although in some cases like the SOS response from *E. coli* the outline of the picture appears to be quite complete, it is still rather sketchy in eukaroytes. Future experiments should be able to fill more and more blank spots in the drawing of these vital processes, which have an immediate impact on the understanding of the etiology of human cancer.

Acknowledgement The authors thank T. Moore for critical comments on the manuscript.

References

Aboussekhra A, Chanet R, Zgaga Z, Cassier-Chauvat C, Heude M, Fabre F (1989) *RADH*, a gene of *Saccharomyces cerevisiae* encoding a putative DNA helicase involved in DNA repair. Characteristics of *radH* mutants and sequence of the gene. Nucleic Acids Res 17:7211–7219

Abraham RT (2001) Cell cycle checkpoint signaling through the ATM and ATR kinases. Genes Dev 15:2177–2196

Aguilera A, Klein HL (1988) Genetic control of intrachromosomal recombination in *Saccharomyces cerevisiae.* I. Isolation and genetic characterization of hyper-recombination mutations. Genetics 119:779–790

Alcasabas AA et al. (2001) Mrc1 transduces signals of DNA replication stress to activate Rad53. Nat Cell Biol 3:958–965

Anderson L, Henderson C, Adachi Y (2001) Phosphorylation and rapid relocalization of 53BP1 to nuclear foci upon DNA damage. Mol Cell Biol 21:1719–1729

Aroya SB, Kupiec M (2005) The Elg1 replication factor C-like complex: a novel guardian of genome stability. DNA Repair (Amst) 4:409–417

Bailly V, Lamb J, Sung P, Prakash S, Prakash L (1994) Specific complex formation between yeast RAD6 and RAD18 proteins: a potential mechanism for targeting RAD6 ubiquitin-conjugating activity to DNA damage sites. Genes Dev 8:811–820

Bailly V, Lauder S, Prakash S, Prakash L (1997) Yeast DNA repair proteins Rad6 and Rad18 form a heterodimer that has ubiquitin conjugating, DNA binding, and ATP hydrolytic activities. J Biol Chem 272:23360–23365

Bakkenist CJ, Kastan MB (2003) DNA damage activates ATM through intermolecular autophosphorylation and dimer dissociation. Nature 421:499–506

Bardwell AJ, Bardwell L, Tomkinson AE, Friedberg EC (1994) Specific cleavage of model recombination and repair intermediates by the yeast Rad1-Rad10 DNA endonuclease. Science 265:2082–2085

Bashkirov VI, Bashkirova EV, Haghnazari E, Heyer WD (2003) Direct kinase-to-kinase signaling mediated by the FHA phosphoprotein recognition domain of the Dun1 DNA damage checkpoint kinase. Mol Cell Biol 23:1441–1452

Bashkirov VI, King JS, Bashkirova EV, Schmuckli-Maurer J, Heyer WD (2000) DNA repair protein Rad55 is a terminal substrate of the DNA damage checkpoints. Mol Cell Biol 20:4393–4404

Bastin-Shanower SA, Fricke WM, Mullen JR, Brill SJ (2003) The mechanism of Mus81-Mms4 cleavage site selection distinguishes it from the homologous endonuclease Rad1-Rad10. Mol Cell Biol 23:3487–3496

Battista JR, Ohta T, Nohmi T, Sun W, Walker GC (1990) Dominant negative *umuD* mutations decreasing RecA-mediated cleavage suggest roles for intact UmuD in modulation of SOS mutagenesis. Proc Natl Acad Sci USA 87:7190–7194

Ben-Aroya S, Koren A, Liefshitz B, Steinlauf R, Kupiec M (2003) *ELG1*, a yeast gene required for genome stability, forms a complex related to replication factor C. Proc Natl Acad Sci USA 100:9906–9911

Bennett RJ, Noirot-Gros MF, Wang JC (2000) Interaction between yeast Sgs1 helicase and DNA topoisomerase III. J Biol Chem 275:26898–26905

Berdichevsky A, Izhar L, Livneh Z (2002) Error-free recombinational repair predominates over mutagenic translesion replication in E. coli. Mol Cell 10:917–924

Bermudez VP, Maniwa Y, Tappin I, Ozato K, Yokomori K, Hurwitz J (2003) The alternative Ctf18-Dcc1-Ctf8-replication factor C complex required for sister chromatid cohesion loads proliferating cell nuclear antigen onto DNA. Proc Natl Acad Sci USA 100:10237–10242

Bianco PR, Kowalczykowski SC (1999) RecA protein. In: Encyclopedia of Life Sciences. Nature Publishing Group, London

Bidnenko V, Ehrlich SD, Michel B (2002) Replication fork collapse at replication terminator sequences. EMBO J 21:3898–3907

Bjergbaek L, Cobb JA, Gasser SM (2002) RecQ helicases and genome stability: lessons from model organisms and human disease. Swiss Med Wkly 132:433–442

Bjergbaek L, Cobb JA, Tsai-Pflugfelder M, Gasser SM (2005) Mechanistically distinct roles for Sgs1p in checkpoint activation and replication fork maintenance. EMBO J 24:405–417

Blasina A, de Weyer IV, Laus MC, Luyten WH, Parker AE, McGowan CH (1999) A human homologue of the checkpoint kinase Cds1 directly inhibits Cdc25 phosphatase. Curr Biol 9:1–10

Boddy MN, Gaillard PH, McDonald WH, Shanahan P, Yates JR 3rd, Russell P (2001) Mus81-Eme1 are essential components of a Holliday junction resolvase. Cell 107:537–548

Boddy MN, Lopez-Girona A, Shanahan P, Interthal H, Heyer WD, Russell P (2000) Damage tolerance protein Mus81 associates with the FHA1 domain of checkpoint kinase Cds1. Mol Cell Biol 20:8758–8766

Boddy MN, Russell P (2001) DNA replication checkpoint. Curr Biol 11:R953–956

Boudsocq F, Campbell M, Devoret R, Bailone A (1997) Quantitation of the inhibition of Hfr x F-recombination by the mutagenesis complex UmuD'C. J Mol Biol 270:201–211

Branzei D, Seki M, Onoda F, Yagi H, Kawabe Y, Enomoto T (2002) Characterization of the slow-growth phenotype of *S. cerevisiae* Whip/Mgs1 Sgs1 double deletion mutants. DNA Repair (Amst) 1:671–682

Brewer BJ, Fangman WL (1987) The localization of replication origins on ARS plasmids in *S. cerevisiae*. Cell 51:463–471

Brewer BJ, Fangman WL (1988) A replication fork barrier at the 3′ end of yeast ribosomal RNA genes. Cell 55:637–643

Brewer BJ, Lockshon D, Fangman WL (1992) The arrest of replication forks in the rDNA of yeast occurs independently of transcription. Cell 71:267–276

Bridges BA, Munson RJ (1968) The persistence through several replication cycles of mutation-producing pyrimidine dimers in a strain of *Escherichia coli* deficient in excision-repair. Biochem Biophys Res Commun 30:620–624

Briggs GS, Mahdi AA, Weller GR, Wen Q, Lloyd RG (2004) Interplay between DNA replication, recombination and repair based on the structure of RecG helicase. Philos Trans R Soc Lond B Biol Sci 359:49–59

Brondello JM, Boddy MN, Furnari B, Russell P (1999) Basis for the checkpoint signal specificity that regulates Chk1 and Cds1 protein kinases. Mol Cell Biol 19:4262–4269

Broomfield S, Chow BL, Xiao W (1998) *MMS2*, encoding a ubiquitin-conjugating-enzyme-like protein, is a member of the yeast error-free postreplication repair pathway. Proc Natl Acad Sci USA 95:5678–5683

Brush GS, Kelly TJ (2000) Phosphorylation of the replication protein A large subunit in the *Saccharomyces cerevisiae* checkpoint response. Nucleic Acids Res 28:3725–3732

Brush GS, Morrow DM, Hieter P, Kelly TJ (1996) The ATM homologue *MEC1* is required for phosphorylation of replication protein A in yeast. Proc Natl Acad Sci USA 93:15075–15080

Bucala R, Model P, Cerami A (1984) Modification of DNA by reducing sugars: a possible mechanism for nucleic acid aging and age-related dysfunction in gene expression. Proc Natl Acad Sci USA 81:105–109

Bucala R, Model P, Russel M, Cerami A (1985) Modification of DNA by glucose 6-phosphate induces DNA rearrangements in an *Escherichia coli* plasmid. Proc Natl Acad Sci USA 82:8439–8442

Cairns J (2002) A DNA damage checkpoint in *Escherichia coli.* DNA Repair 1:699–701

Carr AM (2002) DNA structure dependent checkpoints as regulators of DNA repair. DNA Repair 1:983–994

Carr AM (2003) Molecular biology. Beginning at the end. Science 300:1512–1513

Caspari T, Carr AM (2002) Checkpoints: how to flag up double-strand breaks. Curr Biol 12:R105–107

Čejka P, Vondrejs V, Storchova Z (2001) Dissection of the functions of the *Saccharomyces cerevisiae RAD6* postreplicative repair group in mutagenesis and UV sensitivity. Genetics 159:953–963

Chaganti RS, Schonberg S, German J (1974) A manyfold increase in sister chromatid exchanges in Bloom's syndrome lymphocytes. Proc Natl Acad Sci USA 71:4508–4512

Cobb JA, Bjergbaek L, Shimada K, Frei C, Gasser SM (2003) DNA polymerase stabilization at stalled replication forks requires Mec1 and the RecQ helicase Sgs1. EMBO J 22:4325–4336

Cordeiro-Stone M, Makhov AM, Zaritskaya LS, Griffith JD (1999) Analysis of DNA replication forks encountering a pyrimidine dimer in the template to the leading strand. J Mol Biol 289:1207–1218

Cortez D, Guntuku S, Qin J, Elledge SJ (2001) ATR and ATRIP: partners in checkpoint signaling. Science 294:1713–1716

Cortez D, Wang Y, Qin J, Elledge SJ (1999) Requirement of ATM-dependent phosphorylation of BRCA1 in the DNA damage response to double-strand breaks. Science 286:1162–1166

Courcelle J, Crowley DJ, Hanawalt PC (1999) Recovery of DNA replication in UV-irradiated *Escherichia coli* requires both excision repair and recF protein function. J Bacteriol 181:916–922

Courcelle J, Hanawalt PC (1999) RecQ and RecJ process blocked replication forks prior to the resumption of replication in UV-irradiated *Escherichia coli*. Mol Gen Genet 262:543–551

Cox MM (2001) Recombinational DNA repair of damaged replication forks in *Escherichia coli*: questions. Annu Rev Genet 35:53–82

Crowley DJ, Courcelle J (2002) Answering the call: coping with DNA damage at the most inopportune time. J Biomed Biotechnol 2:66–74

Crowley DJ, Hanawalt PC (2001) The SOS-dependent upregulation of *uvrD* is not required for efficient nucleotide excision repair of ultraviolet light induced DNA photoproducts in *Escherichia coli*. Mutat Res 485:319–329

D'Amours D, Jackson SP (2002) The Mre11 complex: at the crossroads of DNA repair and checkpoint signalling. Nat Rev Mol Cell Biol 3:317–327

Davies AA, Friedberg EC, Tomkinson AE, Wood RD, West SC (1995) Role of the Rad1 and Rad10 proteins in nucleotide excision repair and recombination. J Biol Chem 270:24638–24641

de Laat WL, Appeldoorn E, Jaspers NG, Hoeijmakers JH (1998) DNA structural elements required for ERCC1-XPF endonuclease activity. J Biol Chem 273:7835–7842

Doe CL, Ahn JS, Dixon J, Whitby MC (2002) Mus81-Eme1 and Rqh1 involvement in processing stalled and collapsed replication forks. J Biol Chem 277:32753–32759

Doe CL, Osman F, Dixon J, Whitby MC (2004) DNA repair by a Rad22-Mus81-dependent pathway that is independent of Rhp51. Nucleic Acids Res 32:5570–5581

Dutertre S et al. (2000) Cell cycle regulation of the endogenous wild type Bloom's syndrome DNA helicase. Oncogene 19:2731–2738

Easton DF, Bishop DT, Ford D, Crockford GP (1993) Genetic linkage analysis in familial breast and ovarian cancer: results from 214 families. The Breast Cancer Linkage Consortium. Am J Hum Genet 52:678–701

Entian KD et al. (1999) Functional analysis of 150 deletion mutants in *Saccharomyces cerevisiae* by a systematic approach. Mol Gen Genet 262:683–702

Enzlin JH, Schärer OD (2002) The active site of the DNA repair endonuclease XPF-ERCC1 forms a highly conserved nuclease motif. EMBO J 21:2045–2053

Fabre F, Chan A, Heyer WD, Gangloff S (2002) Alternate pathways involving Sgs1/Top3, Mus81/Mms4, and Srs2 prevent formation of toxic recombination intermediates from single-stranded gaps created by DNA replication. Proc Natl Acad Sci USA 99:16887–16892

Falaschi A (2000) Eukaryotic DNA replication: a model for a fixed double replisome. Trends Genet 16:88–92

Ferentz AE, Walker GC, Wagner G (2001) Converting a DNA damage checkpoint effector (UmuD2C) into a lesion bypass polymerase (UmuD'2C). EMBO J 20:4287–4298

Flores MJ, Bierne H, Ehrlich SD, Michel B (2001) Impairment of lagging strand synthesis triggers the formation of a RuvABC substrate at replication forks. EMBO J 20:619–629

Foiani M et al. (2000) DNA damage checkpoints and DNA replication controls in *Saccharomyces cerevisiae*. Mutat Res 451:187–196

Foss EJ (2001) Tof1p regulates DNA damage responses during S phase in *Saccharomyces cerevisiae*. Genetics 157:567–577

Frank EG, Ennis DG, Gonzalez M, Levine AS, Woodgate R (1996) Regulation of SOS mutagenesis by proteolysis. Proc Natl Acad Sci USA 93:10291–10296

Frei C, Gasser SM (2000) The yeast Sgs1p helicase acts upstream of Rad53p in the DNA replication checkpoint and colocalizes with Rad53p in S-phase-specific foci. Genes Dev 14:81–96

Fricke WM, Bastin-Shanower SA, Brill SJ (2005) Substrate specificity of the *Saccharomyces cerevisiae* Mus81-Mms4 endonuclease. DNA Repair (Amst) 4:243–251

Fricke WM, Brill SJ (2003) Slx1-Slx4 is a second structure-specific endonuclease functionally redundant with Sgs1-Top3. Genes Dev 17:1768–1778

Fricke WM, Kaliraman V, Brill SJ (2001) Mapping the DNA topoisomerase III binding domain of the Sgs1 DNA helicase. J Biol Chem 276:8848–8855

Friedberg EC, Fischhaber PL, Kisker C (2001) Error-prone DNA polymerases: novel structures and the benefits of infidelity. Cell 107:9–12

Friedberg EC, Walker GC, Siede W (1995) DNA repair and mutagenesis. ASM Press, Washington D.C.

Friedl AA, Liefshitz B, Steinlauf R, Kupiec M (2001) Deletion of the *SRS2* gene suppresses elevated recombination and DNA damage sensitivity in *rad5* and *rad18* mutants of *Saccharomyces cerevisiae.* Mutat Res 486:137–146

Fujiwara Y, Tatsumi M (1976) Replicative bypass repair of ultraviolet damage to DNA of mammalian cells: caffeine sensitive and caffeine resistant mechanisms. Mutat Res 37:91–110

Ganesan AK (1974) Persistence of pyrimidine dimers during post-replication repair in ultraviolet light-irradiated *Escherichia coli* K12. J Mol Biol 87:103–119

Gangloff S, McDonald JP, Bendixen C, Arthur L, Rothstein R (1994) The yeast type I topoisomerase Top3 interacts with Sgs1, a DNA helicase homolog: a potential eukaryotic reverse gyrase. Mol Cell Biol 14:8391–8398

Gangloff S, Soustelle C, Fabre F (2000) Homologous recombination is responsible for cell death in the absence of the Sgs1 and Srs2 helicases. Nat Genet 25:192–194

Gilbert CS, Green CM, Lowndes NF (2001) Budding yeast Rad9 is an ATP-dependent Rad53 activating machine. Mol Cell 8:129–136

Glassner BJ, Rasmussen LJ, Najarian MT, Posnick LM, Samson LD (1998) Generation of a strong mutator phenotype in yeast by imbalanced base excision repair. Proc Natl Acad Sci USA 95:9997–10002

Gonzalez M, Rasulova F, Maurizi MR, Woodgate R (2000) Subunit-specific degradation of the UmuD/D′ heterodimer by the ClpXP protease: the role of trans recognition in UmuD' stability. EMBO J 19:5251–5258

Green CM, Erdjument-Bromage H, Tempst P, Lowndes NF (2000) A novel Rad24 checkpoint protein complex closely related to replication factor C. Curr Biol 10:39–42

Grenon M, Gilbert C, Lowndes NF (2001) Checkpoint activation in response to double-strand breaks requires the Mre11/Rad50/Xrs2 complex. Nat Cell Biol 3:844–847

Grompone G, Ehrlich SD, Michel B (2003) Replication restart in *gyrB Escherichia coli* mutants. Mol Microbiol 48:845–854

Gruber M, Wellinger RE, Sogo JM (2000) Architecture of the replication fork stalled at the 3′ end of yeast ribosomal genes. Mol Cell Biol 20:5777–5787

Haracska L, Prakash S, Prakash L (2002) Yeast Rev1 protein is a G template-specific DNA polymerase. J Biol Chem 277:15546–15551

Haracska L, Torres-Ramos CA, Johnson RE, Prakash S, Prakash L (2004) Opposing effects of ubiquitin conjugation and SUMO modification of PCNA on replicational bypass of DNA lesions in *Saccharomyces cerevisiae.* Mol Cell Biol 24:4267–4274

Haracska L et al. (2001) Roles of yeast DNA polymerases δ and ζ and of Rev1 in the bypass of abasic sites. Genes Dev 15:945–954

Hazbun TR et al. (2003) Assigning function to yeast proteins by integration of technologies. Mol Cell 12:1353–1365

Heller RC, Marians KJ (2005) The disposition of nascent strands at stalled replication forks dictates the pathway of replisome loading during restart. Mol Cell 17:733–743

Hendricks EC, Szerlong H, Hill T, Kuempel P (2000) Cell division, guillotining of dimer chromosomes and SOS induction in resolution mutants (*dif*, *xerC* and *xerD*) of *Escherichia coli*. Mol Microbiol. 36:973–981

Heude M, Chanet R, Fabre F (1995) Regulation of the *Saccharomyces cerevisiae* Srs2 helicase during the mitotic cell cycle, meiosis and after irradiation. Mol Gen Genet 248:59–68

Hickson ID (2003) RecQ helicases: caretakers of the genome. Nat Rev Cancer 3:169–178

Higgins NP, Kato K, Strauss B (1976) A model for replication repair in mammalian cells. J Mol Biol 101:417–425

Higuchi K, Katayama T, Iwai S, Hidaka M, Horiuchi T, Maki H (2003) Fate of DNA replication fork encountering a single DNA lesion during *oriC* plasmid DNA replication *in vitro*. Genes Cells 8:437–449

Hill TM, Marians KJ (1990) *Escherichia coli* Tus protein acts to arrest the progression of DNA replication forks *in vitro*. Proc Natl Acad Sci USA 87:2481–2485

Hishida T, Iwasaki H, Ohno T, Morishita T, Shinagawa H (2001) A yeast gene, *MGS1*, encoding a DNA-dependent AAA(+) ATPase is required to maintain genome stability. Proc Natl Acad Sci USA 98:8283–8289

Hishida T, Ohno T, Iwasaki H, Shinagawa H (2002) *Saccharomyces cerevisiae MGS1* is essential in strains deficient in the *RAD6*-dependent DNA damage tolerance pathway. EMBO J 21:2019–2029

Hoege C, Pfander B, Moldovan GL, Pyrowolakis G, Jentsch S (2002) *RAD6*-dependent DNA repair is linked to modification of PCNA by ubiquitin and SUMO. Nature 419:135–141

Hollingsworth NM, Brill SJ (2004) The Mus81 solution to resolution: generating meiotic crossovers without Holliday junctions. Genes Dev 18:117–125

Huisman O, D'Ari R, Gottesman S (1984) Cell-division control in *Escherichia coli*: specific induction of the SOS function SfiA protein is sufficient to block septation. Proc Natl Acad Sci USA 81:4490–4494

Hyrien O (2000) Mechanisms and consequences of replication fork arrest. Biochimie 82:5–17

Inman RB (1984) Methodology for the study of the effect of drugs on development and DNA replication in *Drosophila melanogaster* embryonic tissue. Biochim Biophys Acta 783:205–215

Interthal H, Heyer WD (2000) *MUS81* encodes a novel helix-hairpin-helix protein involved in the response to UV- and methylation-induced DNA damage in *Saccharomyces cerevisiae*. Mol Gen Genet 263:812–827

Ira G, Malkova A, Liberi G, Foiani M, Haber JE (2003) Srs2 and Sgs1-Top3 suppress crossovers during double-strand break repair in yeast. Cell 115:401–411

Ishioka K, Iwasaki H, Shinagawa H (1997) Roles of the *recG* gene product of *Escherichia coli* in recombination repair: effects of the *ΔrecG* mutation on cell division and chromosome partition. Genes Genet Syst 72:91–99

Ivessa AS, Lenzmeier BA, Bessler JB, Goudsouzian LK, Schnakenberg SL, Zakian VA (2003) The *Saccharomyces cerevisiae* helicase Rrm3p facilitates replication past nonhistone protein-DNA complexes. Mol Cell 12:1525–1536

Ivessa AS, Zhou JQ, Schulz VP, Monson EK, Zakian VA (2002) *Saccharomyces* Rrm3p, a 5′ to 3′ DNA helicase that promotes replication fork progression through telomeric and subtelomeric DNA. Genes Dev 16:1383–1396

Jachymczyk WJ, Chlebowicz E, Swietlinska Z, Zuk J (1977) Alkaline sucrose sedimentation studies of MMS-induced DNA single-strand breakage and rejoining in the wild type and in UV-sensitive mutants of *Saccharomyces cerevisiae*. Mutat Res 43:1–10

Janion C (2001) Some aspects of the SOS response system – a critical survey. Acta Biochim Pol 48:599–610

Jentsch S, McGrath JP, Varshavsky A (1987) The yeast DNA repair gene *RAD6* encodes a ubiquitin-conjugating enzyme. Nature 329:131–134

Kadyk LC, Hartwell LH (1992) Sister chromatids are preferred over homologs as substrates for recombinational repair in *Saccharomyces cerevisiae*. Genetics 132:387–402

Kadyk LC, Hartwell LH (1993) Replication-dependent sister chromatid recombination in *rad1* mutants of *Saccharomyces cerevisiae*. Genetics 133:469–487

Kagawa W, Kurumizaka H, Ikawa S, Yokoyama S, Shibata T (2001) Homologous pairing promoted by the human Rad52 protein. J Biol Chem 276:35201–35208

Kai M, Wang TS (2003a) Checkpoint activation regulates mutagenic translesion synthesis. Genes Dev 17:64–76

Kai M, Wang TS (2003b) Checkpoint responses to replication stalling: inducing tolerance and preventing mutagenesis. Mutat Res 532:59–73

Kaliraman V, Mullen JR, Fricke WM, Bastin-Shanower SA, Brill SJ (2001) Functional overlap between Sgs1-Top3 and the Mms4-Mus81 endonuclease. Genes Dev 15:2730–2740

Kanellis P, Agyei R, Durocher D (2003) Elg1 forms an alternative PCNA-interacting RFC complex required to maintain genome stability. Curr Biol 13:1583–1595

Karow JK, Constantinou A, Li JL, West SC, Hickson ID (2000) The Bloom's syndrome gene product promotes branch migration of Holliday junctions. Proc Natl Acad Sci USA 97:6504–6508

Katayama T (2001) Feedback controls restrain the initiation of *Escherichia coli* chromosomal replication. Mol Microbiol 41:9–17

Katou Y et al. (2003) S-phase checkpoint proteins Tof1 and Mrc1 form a stable replication-pausing complex. Nature 424:1078–1083

Khatri GS, MacAllister T, Sista PR, Bastia D (1989) The replication terminator protein of *E. coli* is a DNA sequence-specific contra-helicase. Cell 59:667–674

Kim HS, Brill SJ (2001) Rfc4 interacts with Rpa1 and is required for both DNA replication and DNA damage checkpoints in *Saccharomyces cerevisiae*. Mol Cell Biol 21:3725–3737

Kim HS, Brill SJ (2003) *MEC1*-dependent phosphorylation of yeast RPA1 *in vitro*. DNA Repair (Amst) 2:1321–1335

Kobayashi T, Horiuchi T (1996) A yeast gene product, Fob1 protein, required for both replication fork blocking and recombinational hotspot activities. Genes Cells 1:465–474

Komori K, Fujikane R, Shinagawa H, Ishino Y (2002) Novel endonuclease in Archaea cleaving DNA with various branched structure. Genes Genet Syst 77:227–241

Komori K, Hidaka M, Horiuchi T, Fujikane R, Shinagawa H, Ishino Y (2004) Cooperation of the N-terminal Helicase and C-terminal endonuclease activities of Archaeal Hef protein in processing stalled replication forks. J Biol Chem 279:53175–53185

Kondo T, Wakayama T, Naiki T, Matsumoto K, Sugimoto K (2001) Recruitment of Mec1 and Ddc1 checkpoint proteins to double-strand breaks through distinct mechanisms. Science 294:867–870

Kool ET (2002) Active site tightness and substrate fit in DNA replication. Annu Rev Biochem 71:191–219

Krejci L et al. (2003) DNA helicase Srs2 disrupts the Rad51 presynaptic filament. Nature 423:305–309

Krishna TS, Kong XP, Gary S, Burgers PM, Kuriyan J (1994) Crystal structure of the eukaryotic DNA polymerase processivity factor PCNA. Cell 79:1233–1243

Krogh BO, Symington LS (2004) Recombination proteins in yeast. Annu Rev Genet 38:233–271

Kumagai A, Dunphy WG (2000) Claspin, a novel protein required for the activation of Chk1 during a DNA replication checkpoint response in *Xenopus* egg extracts. Mol Cell 6:839–849

Labib K, Kearsey SE, Diffley JF (2001) MCM2–7 proteins are essential components of pre-replicative complexes that accumulate cooperatively in the nucleus during G1-phase and are required to establish, but not maintain, the S-phase checkpoint. Mol Biol Cell 12:3658–3667

Lane HE, Denhardt DT (1975) The rep mutation. IV. Slower movement of replication forks in Escherichia coli rep strains. J Mol Biol 97:99–112

Lankenau D-H (1995) Genetics of genetics in *Drosophila*: P elements serving the study of homologous recombination, gene conversion and targeting. Chromosoma 103:659–668

Lawrence CW, Christensen RB (1979) Metabolic suppressors of trimethoprim and ultraviolet light sensitivities of *Saccharomyces cerevisiae rad6* mutants. J Bacteriol 139:866–876

Lee J, Kumagai A, Dunphy WG (2003) Claspin, a Chk1-regulatory protein, monitors DNA replication on chromatin independently of RPA, ATR, and Rad17. Mol Cell 11:329–340

Lee JS, Collins KM, Brown AL, Lee CH, Chung JH (2000) hCds1-mediated phosphorylation of BRCA1 regulates the DNA damage response. Nature 404:201–204

Liberi G et al. (2000) Srs2 DNA helicase is involved in checkpoint response and its regulation requires a functional Mec1-dependent pathway and Cdk1 activity. EMBO J 19:5027–5038

Liberi G et al. (2005) Rad51-dependent DNA structures accumulate at damaged replication forks in *sgs1* mutants defective in the yeast ortholog of BLM RecQ helicase. Genes Dev 19:339–350

Lim DS et al. (2000) ATM phosphorylates p95/nbs1 in an S-phase checkpoint pathway. Nature 404:613–617

Lindahl T, Wood RD (1999) Quality control by DNA repair. Science 286:1897–1905

Liu Y, Masson JY, Shah R, O'Regan P, West SC (2004) RAD51C is required for Holliday junction processing in mammalian cells. Science 303:243–246

Longhese MP, Clerici M, Lucchini G (2003) The S-phase checkpoint and its regulation in *Saccharomyces cerevisiae*. Mutat Res 532:41–58

Longhese MP, Foiani M, Muzi-Falconi M, Lucchini G, Plevani P (1998) DNA damage checkpoint in budding yeast. EMBO J 17:5525–5528

Longhese MP, Neecke H, Paciotti V, Lucchini G, Plevani P (1996) The 70 kDa subunit of replication protein A is required for the G1/S and intra-S DNA damage checkpoints in budding yeast. Nucleic Acids Res 24:3533–3537

Lopes M, Cotta-Ramusino C, Liberi G, Foiani M (2003) Branch migrating sister chromatid junctions form at replication origins through Rad51/Rad52-independent mechanisms. Mol Cell 12:1499–1510

Lopes M et al. (2001) The DNA replication checkpoint response stabilizes stalled replication forks. Nature 412:557–561

Lopéz de Saro FJ, Georgescu RE, Goodman MF, O'Donnell M (2003) Competitive processivity-clamp usage by DNA polymerases during DNA replication and repair. EMBO J 22:6408–6418

Lovett ST (2005) Filling the gaps in replication restart pathways. Mol Cell 17:751–752

Mahdi AA, Briggs GS, Sharples GJ, Wen Q, Lloyd RG (2003) A model for dsDNA translocation revealed by a structural motif common to RecG and Mfd proteins. EMBO J 22:724–734

Majka J, Burgers PM (2003) Yeast Rad17/Mec3/Ddc1: a sliding clamp for the DNA damage checkpoint. Proc Natl Acad Sci USA 100:2249–2254

Majka J, Burgers PM (2004) The PCNA-RFC families of DNA clamps and clamp loaders. Prog Nucleic Acid Res Mol Biol 78:227–260

Matsuoka S, Huang M, Elledge SJ (1998) Linkage of ATM to cell cycle regulation by the Chk2 protein kinase. Science 282:1893–1897

Mayer ML, Gygi SP, Aebersold R, Hieter P (2001) Identification of RFC (Ctf18p, Ctf8p, Dcc1p): an alternative RFC complex required for sister chromatid cohesion in *S. cerevisiae*. Mol Cell 7:959–970

McGlynn P (2004) Links between DNA replication and recombination in prokaryotes. Curr Opin Genet Dev 14:107–112

McGlynn P, Lloyd RG (2000) Modulation of RNA polymerase by (p)ppGpp reveals a RecG-dependent mechanism for replication fork progression. Cell 101:35–45

McGlynn P, Lloyd RG (2002a) Genome stability and the processing of damaged replication forks by RecG. Trends Genet 18:413–419

McGlynn P, Lloyd RG (2002b) Recombinational repair and restart of damaged replication forks. Nat Rev Mol Cell Biol 3:859–870

McGowan CH (2003) Running into problems: how cells cope with replicating damaged DNA. Mutat Res 532:75–84

Melo JA, Cohen J, Toczyski DP (2001) Two checkpoint complexes are independently recruited to sites of DNA damage *in vivo*. Genes Dev 15:2809–2821

Michel B (2000) Replication fork arrest and DNA recombination. Trends Biochem Sci 25:173–178

Michel B, Flores MJ, Viguera E, Grompone G, Seigneur M, Bidnenko V (2001) Rescue of arrested replication forks by homologous recombination. Proc Natl Acad Sci USA 98:8181–8188

Michel B, Grompone G, Flores MJ, Bidnenko V (2004) Multiple pathways process stalled replication forks. Proc Natl Acad Sci USA 101:12783–12788

Morrison A et al. (1989) *REV3*, a *Saccharomyces cerevisiae* gene whose function is required for induced mutagenesis, is predicted to encode a nonessential DNA polymerase. J Bacteriol 171:5659–5667

Mullen JR, Kaliraman V, Ibrahim SS, Brill SJ (2001) Requirement for three novel protein complexes in the absence of the Sgs1 DNA helicase in *Saccharomyces cerevisiae*. Genetics 157:103–118

Nassif N, Penney J, Pal S, Engels WR, Gloor GB (1994) Efficient copying of nonhomologous sequences from ectopic sites via P-element-induced gap repair. Mol Cell Biol 14:1613–1625

Nelson JR, Lawrence CW, Hinkle DC (1996) Deoxycytidyl transferase activity of yeast Rev1 protein. Nature 382:729–731

Noguchi E, Noguchi C, Du LL, Russel P (2003) Swi1 prevents replication fork collapse and controls checkpoint kinase Cds1. Mol Cell Biol 23:7861–7874

Noguchi E, Noguchi C, McDonald WH, Yates JR 3rd, Russel P (2004) Swi1 and Swi3 are components of a replication fork protection complex in fission yeast. Mol Cell Biol 24:8342–8355

Nyberg KA, Michelson RJ, Putnam CW, Weinert TA (2002) Toward maintaining the genome: DNA damage and replication checkpoints. Annu Rev Genet 36:617–656

Oakley TJ, Hickson ID (2002) Defending genome integrity during S-phase: putative roles for RecQ helicases and topoisomerase III. DNA Repair (Amst) 1:175–207

Odagiri N, Seki M, Onoda F, Yoshimura A, Watanabe S, Enomoto T (2003) Budding yeast mms4 is epistatic with *rad52* and the function of Mms4 can be replaced by a bacterial Holliday junction resolvase. DNA Repair (Amst) 2:347–358

Onoda F, Seki M, Miyajima A, Enomoto T (2000) Elevation of sister chromatid exchange in *Saccharomyces cerevisiae sgs1* disruptants and the relevance of the disruptants as a system to evaluate mutations in Bloom's syndrome gene. Mutat Res 459:203–209

Onoda F, Seki M, Wang W, Enomoto T (2004) The hyper unequal sister chromatid recombination in an *sgs1* mutant of budding yeast requires *MSH2*. DNA Repair (Amst) 3:1355–1362

Ooi SL, Shoemaker DD, Boeke JD (2003) DNA helicase gene interaction network defined using synthetic lethality analyzed by microarray. Nat Genet 35:277–286

Opperman T, Murli S, Smith BT, Walker GC (1999) A model for a *umuDC*-dependent prokaryotic DNA damage checkpoint. Proc Natl Acad Sci USA 96:9218–9223

Osborn AJ, Elledge SJ (2003) Mrc1 is a replication fork component whose phosphorylation in response to DNA replication stress activates Rad53. Genes Dev 17:1755–1767

Osman F, Dixon J, Doe CL, Whitby MC (2003) Generating crossovers by resolution of nicked Holliday junctions: a role for Mus81-Eme1 in meiosis. Mol Cell 12:761–774

Otsuji N, Iyehara H, Hideshima Y (1974) Isolation and characterization of an *Escherichia coli ruv* mutant which forms nonseptate filaments after low doses of ultraviolet light irradiation. J Bacteriol 117:337–344

Paciotti V, Clerici M, Lucchini G, Longhese MP (2000) The checkpoint protein Ddc2, functionally related to *S. pombe* Rad26, interacts with Mec1 and is regulated by Mec1-dependent phosphorylation in budding yeast. Genes Dev 14:2046–2059

Pagés V, Fuchs RP (2003) Uncoupling of leading- and lagging-strand DNA replication during lesion bypass *in vivo*. Science 300:1300–1303

Paques F, Haber JE (1999) Multiple pathways of recombination induced by double-strand breaks in *Saccharomyces cerevisiae*. Microbiol Mol Biol Rev 63:349–404

Park H, Sternglanz R (1999) Identification and characterization of the genes for two topoisomerase I-interacting proteins from *Saccharomyces cerevisiae*. Yeast 15:35–41

Park JS, Marr MT, Roberts JW (2002) *E. coli* transcription repair coupling factor (Mfd protein) rescues arrested complexes by promoting forward translocation. Cell 109:757–767

Pellicioli A et al. (1999) Activation of Rad53 kinase in response to DNA damage and its effect in modulating phosphorylation of the lagging strand DNA polymerase. EMBO J 18:6561–6572

Perederina A et al. (2004) Regulation through the secondary channel–structural framework for ppGpp-DksA synergism during transcription. Cell 118:297–309

Petranovic M, Zahradka K, Zahradka D, Petranovic D, Nagy B, Salaj-Smic E (2001) Genetic evidence that the elevated levels of *Escherichia coli* helicase II antagonize recombinational DNA repair. Biochimie 83:1041–1047

Petrini JH (2000) The Mre11 complex and ATM: collaborating to navigate S phase. Curr Opin Cell Biol 12:293–296

Pfander B, Moldovan G-L, Sacher M, Hoege C, Jentsch S (2005) SUMO-modified PCNA recruits Srs2 to prevent recombination during S phase. Nature 436:428–433

Pham P et al. (2002) Two distinct modes of RecA action are required for DNA polymerase V-catalyzed translesion synthesis. Proc Natl Acad Sci USA 99:11061–11066

Postow L, Ullsperger C, Keller RW, Bustamante C, Vologodskii AV, Cozzarelli NR (2001) Positive torsional strain causes the formation of a four-way junction at replication forks. J Biol Chem 276:2790–2796

Prakash L (1981) Characterization of postreplication repair in *Saccharomyces cerevisiae* and effects of *rad6, rad18, rev3* and *rad52* mutations. Mol Gen Genet 184:471–478

Prakash L, Prakash S (1977) Isolation and characterization of MMS-sensitive mutants of *Saccharomyces cerevisiae*. Genetics 86:33–55

Prakash R, Krejci L, Van Komen S, Schürer KA, Kramer W, Sung P (2005) *Saccharomyces cerevisiae MPH1* gene, required for homologous recombination-mediated mutation avoidance, encodes a 3′ to 5′ DNA helicase. J Biol Chem 280:7854–7860

Quah SK, von Borstel RC, Hastings PJ (1980) The origin of spontaneous mutation in *Saccharomyces cerevisiae*. Genetics 96:819–839

Rangarajan S, Woodgate R, Goodman MF (1999) A phenotype for enigmatic DNA polymerase II: a pivotal role for pol II in replication restart in UV-irradiated *Escherichia coli*. Proc Natl Acad Sci USA 96:9224–9229

Rehrauer WM, Bruck I, Woodgate R, Goodman MF, Kowalczykowski SC (1998) Modulation of RecA nucleoprotein function by the mutagenic UmuD'C protein complex. J Biol Chem 273:32384–32387

Rhind N, Russell P (2000) Chk1 and Cds1: linchpins of the DNA damage and replication checkpoint pathways. J Cell Sci 113:3889–3896

Roberts J, Park JS (2004) Mfd, the bacterial transcription repair coupling factor: translocation, repair and termination. Curr Opin Microbiol 7:120–125

Roche H, Gietz RD, Kunz BA (1995) Specificities of the *Saccharomyces cerevisiae rad6*, *rad18*, and *rad52* mutators exhibit different degrees of dependence on the *REV3* gene product, a putative nonessential DNA polymerase. Genetics 140:443–456

Rong L, Klein HL (1993) Purification and characterization of the Srs2 DNA helicase of the yeast *Saccharomyces cerevisiae*. J Biol Chem 268:1252–1259

Rouse J, Jackson SP (2002) Lcd1p recruits Mec1p to DNA lesions *in vitro* and *in vivo*. Mol Cell 9:857–869

Rupp WD, Howard-Flanders P (1968) Discontinuities in the DNA synthesized in an excision-defective strain of *Escherichia coli* following ultraviolet irradiation. J Mol Biol 31:291–304

Rydberg B, Lindahl T (1982) Nonenzymatic methylation of DNA by the intracellular methyl group donor S-adenosyl-L-methionine is a potentially mutagenic reaction. EMBO J 1:211–216

Sahoo T, Mohanty BK, Lobert M, Manna AC, Bastia D (1995) The contrahelicase activities of the replication terminator proteins of *Escherichia coli* and *Bacillus subtilis* are helicase-specific and impede both helicase translocation and authentic DNA unwinding. J Biol Chem 270:29138–29144

Sanchez Y et al. (1999) Control of the DNA damage checkpoint by Chk1 and Rad53 protein kinases through distinct mechanisms. Science 286:1166–1171

Sandell LL, Zakian VA (1993) Loss of a yeast telomere: arrest, recovery, and chromosome loss. Cell 75:729–739

Sandler SJ (2000) Multiple genetic pathways for restarting DNA replication forks in *Escherichia coli* K-12. Genetics 155:487–497

Sandler SJ, McCool JD, Do TT, Johansen RU (2001) *priA* mutations that affect PriA-PriC function during replication restart. Mol Microbiol 41:697–704

Santocanale C, Diffley JF (1998) A Mec1- and Rad53-dependent checkpoint controls late-firing origins of DNA replication. Nature 395:615–618

Sawitzke J, Austin S (2001) An analysis of the factory model for chromosome replication and segregation in bacteria. Mol Microbiol 40:786–794

Scheller J, Schürer A, Rudolph C, Hettwer S, Kramer W (2000) *MPH1*, a yeast gene encoding a DEAH protein, plays a role in protection of the genome from spontaneous and chemically induced damage. Genetics 155:1069–1081

Schiestl RH, Prakash S, Prakash L (1990) The *SRS2* suppressor of *rad6* mutations of *Saccharomyces cerevisiae* acts by channeling DNA lesions into the *RAD52* DNA repair pathway. Genetics 124:817–831

Schlacher K, Leslie K, Wyman C, Woodgate R, Cox MM, Goodman MF (2005) DNA polymerase V and RecA protein, a minimal mutasome. Mol Cell 17:561–572

Schofield MJ, Hsieh P (2003) DNA mismatch repair: molecular mechanisms and biological function. Annu Rev Microbiol 57:579–608

Schultz LB, Chehab NH, Malikzay A, Halazonetis TD (2000) p53 binding protein 1 (53BP1) is an early participant in the cellular response to DNA double-strand breaks. J Cell Biol 151:1381–1390

Schürer KA, Rudolph C, Ulrich HD, Kramer W (2004) Yeast *MPH1* gene functions in an error-free DNA damage bypass pathway that requires genes from homologous recombination, but not from postreplicative repair. Genetics 166:1673–1686

Segurado M, Gomez M, Antequera F (2002) Increased recombination intermediates and homologous integration hot spots at DNA replication origins. Mol Cell 10:907–916

Seigneur M, Bidnenko V, Ehrlich SD, Michel B (1998) RuvAB acts at arrested replication forks. Cell 95:419–430

Selby CP, Sancar A (1993) Molecular mechanism of transcription-repair coupling. Science 260:53–58

Shiloh Y (2003) ATM and related protein kinases: safeguarding genome integrity. Nat Rev Cancer 3:155–168

Shirahige K et al. (1998) Regulation of DNA-replication origins during cell-cycle progression. Nature 395:618–621

Sidorova JM, Breeden LL (2003) Precocious G1/S transitions and genomic instability: the origin connection. Mutat Res 532:5–19

Singleton MR, Scaife S, Wigley DB (2001) Structural analysis of DNA replication fork reversal by RecG. Cell 107:79–89

Smith KC (2004) Recombinational DNA repair: the ignored repair systems. Bioessays 26:1322–1326

Sogo JM, Lopes M, Foiani M (2002) Fork reversal and ssDNA accumulation at stalled replication forks owing to checkpoint defects. Science 297:599–602

Sommer S, Bailone A, Devoret R (1993) The appearance of the UmuD'C protein complex in *Escherichia coli* switches repair from homologous recombination to SOS mutagenesis. Mol Microbiol 10:963–971

Soulier J, Lowndes NF (1999) The BRCT domain of the *S. cerevisiae* checkpoint protein Rad9 mediates a Rad9-Rad9 interaction after DNA damage. Curr Biol 9:551–554

Spencer F, Gerring SL, Connelly C, Hieter P (1990) Mitotic chromosome transmission fidelity mutants in *Saccharomyces cerevisiae*. Genetics 124:237–249

Stebbins CE, Borukhov S, Orlova M, Polyakov A, Goldfarb A, Darst SA (1995) Crystal structure of the GreA transcript cleavage factor from *Escherichia coli*. Nature 373:636–640

Stelter P, Ulrich HD (2003) Control of spontaneous and damage-induced mutagenesis by SUMO and ubiquitin conjugation. Nature 425:188–191

Stewart E, Chapman CR, Al-Khodairy F, Carr AM, Enoch T (1997) *rqh1*$^{+}$, a fission yeast gene related to the Bloom's and Werner's syndrome genes, is required for reversible S phase arrest. EMBO J 16:2682–2692

Strumberg D, Pilon AA, Smith M, Hickey R, Malkas L, Pommier Y (2000) Conversion of topoisomerase I cleavage complexes on the leading strand of ribosomal DNA into 5′-phosphorylated DNA double-strand breaks by replication runoff. Mol Cell Biol 20:3977–3987

Sun Z, Hsiao J, Fay DS, Stern DF (1998) Rad53 FHA domain associated with phosphorylated Rad9 in the DNA damage checkpoint. Science 281:272–274

Sung P (1997) Yeast Rad55 and Rad57 proteins form a heterodimer that functions with replication protein A to promote DNA strand exchange by Rad51 recombinase. Genes Dev 11:1111–1121

Svoboda DL, Vos JM (1995) Differential replication of a single, UV-induced lesion in the leading or lagging strand by a human cell extract: fork uncoupling or gap formation. Proc Natl Acad Sci USA 92:11975–11979

Sweasy JB (2005) RecA kicks Pol V into gear. Nat Struct Mol Biol 12:215–216
Symington LS (2002) Role of *RAD52* epistasis group genes in homologous recombination and double-strand break repair. Microbiol Mol Biol Rev 66:630–670, table of contents
Szostak JW, Orr-Weaver TL, Rothstein RJ, Stahl FW (1983) The double-strand-break repair model for recombination. Cell 33:25–35
Takai H et al. (2000) Aberrant cell cycle checkpoint function and early embryonic death in $Chk1^{(-/-)}$ mice. Genes Dev 14:1439–1447
Taki K, Horiuchi T (1999) The SOS response is induced by replication fork blockage at a ter site located on a pUC-derived plasmid: dependence on the distance between *ori* and *ter* sites. Mol Gen Genet 262:302–309
Tanaka K, Russell P (2001) Mrc1 channels the DNA replication arrest signal to checkpoint kinase Cds1. Nat Cell Biol 3:966–972
Tercero JA, Diffley JF (2001) Regulation of DNA replication fork progression through damaged DNA by the Mec1/Rad53 checkpoint. Nature 412:553–557
Tibbetts RS et al. (2000) Functional interactions between BRCA1 and the checkpoint kinase ATR during genotoxic stress. Genes Dev 14:2989–3002
Tippin B, Pham P, Goodman MF (2004) Error-prone replication for better or worse. Trends Microbiol 12:288–295
Toczyski DP, Galgoczy DJ, Hartwell LH (1997) *CDC5* and *CKII* control adaptation to the yeast DNA damage checkpoint. Cell 90:1097–1106
Tong AH et al. (2001) Systematic genetic analysis with ordered arrays of yeast deletion mutants. Science 294:2364–2368
Torres JZ, Schnakenberg SL, Zakian VA (2004) *Saccharomyces cerevisiae* Rrm3p DNA helicase promotes genome integrity by preventing replication fork stalling: viability of *rrm3* cells requires the intra-S-phase checkpoint and fork restart activities. Mol Cell Biol 24:3198–3212
Trautinger BW, Lloyd RG (2002) Modulation of DNA repair by mutations flanking the DNA channel through RNA polymerase. EMBO J 21:6944–6953
Ulrich HD (2001) The *srs2* suppressor of UV sensitivity acts specifically on the *RAD5*- and *MMS2*-dependent branch of the *RAD6* pathway. Nucleic Acids Res 29:3487–3494
Usui T, Ogawa H, Petrini JH (2001) A DNA damage response pathway controlled by Tel1 and the Mre11 complex. Mol Cell 7:1255–1266
Uzest M, Ehrlich SD, Michel B (1995) Lethality of *rep recB* and *rep recC* double mutants of *Escherichia coli*. Mol Microbiol 17:1177–1188
Vaze MB et al. (2002) Recovery from checkpoint-mediated arrest after repair of a double-strand break requires Srs2 helicase. Mol Cell 10:373–385
Veaute X et al. (2005) UvrD helicase, unlike Rep helicase, dismantles RecA nucleoprotein filaments in *Escherichia coli*. EMBO J 24:180–189
Veaute X, Jeusset J, Soustelle C, Kowalczykowski SC, Le Cam E, Fabre F (2003) The Srs2 helicase prevents recombination by disrupting Rad51 nucleoprotein filaments. Nature 423:309–312
Vialard JE, Gilbert CS, Green CM, Lowndes NF (1998) The budding yeast Rad9 checkpoint protein is subjected to Mec1/Tel1-dependent hyperphosphorylation and interacts with Rad53 after DNA damage. EMBO J 17:5679–5688
Volkert MR, Landini P (2001) Transcriptional responses to DNA damage. Curr Opin Microbiol 4:178–185
Waleh NS, Stocker BA (1979) Effect of host *lex, recA, recF*, and *uvrD* genotypes on the ultraviolet light-protecting and related properties of plasmid R46 in *Escherichia coli*. J Bacteriol 137:830–838

Wang Y, Cortez D, Yazdi P, Neff N, Elledge SJ, Qin J (2000) BASC, a super complex of BRCA1-associated proteins involved in the recognition and repair of aberrant DNA structures. Genes Dev 14:927–939

Washington MT, Johnson RE, Prakash S, Prakash L (1999) Fidelity and processivity of *Saccharomyces cerevisiae* DNA polymerase η. J Biol Chem 274:36835–36838

Washington MT, Johnson RE, Prakash S, Prakash L (2000) Accuracy of thymine-thymine dimer bypass by *Saccharomyces cerevisiae* DNA polymerase η. Proc Natl Acad Sci USA 97:3094–3099

Washington MT et al. (2004) Efficient and error-free replication past a minor-groove N2-guanine adduct by the sequential action of yeast Rev1 and DNA polymerase ζ. Mol Cell Biol 24:6900–6906

Watt PM, Louis EJ, Borts RH, Hickson ID (1995) Sgs1: a eukaryotic homolog of *E. coli* RecQ that interacts with topoisomerase II in vivo and is required for faithful chromosome segregation. Cell 81:253–260

Whitby MC, Osman F, Dixon J (2003) Cleavage of model replication forks by fission yeast Mus81-Eme1 and budding yeast Mus81-Mms4. J Biol Chem 278:6928–6935

Wu L et al. (2000) The Bloom's syndrome gene product interacts with topoisomerase III. J Biol Chem 275:9636–9644

Wu L, Hickson ID (2003) The Bloom's syndrome helicase suppresses crossing over during homologous recombination. Nature 426:870–874

Xiao W, Chow BL, Milo CN (1998) Mms4, a putative transcriptional (co)activator, protects *Saccharomyces cerevisiae* cells from endogenous and environmental DNA damage. Mol Gen Genet 257:614–623

Yasuda T, Morimatsu K, Horii T, Nagata T, Ohmori H (1998) Inhibition of *Escherichia coli* RecA coprotease activities by DinI. EMBO J 17:3207–3216

Yuzhakov A, Turner J, O'Donnell M (1996) Replisome assembly reveals the basis for asymmetric function in leading and lagging strand replication. Cell 86:877–886

Zhao H, Piwnica-Worms H (2001) ATR-mediated checkpoint pathways regulate phosphorylation and activation of human Chk1. Mol Cell Biol 21:4129–4139

Zhao H, Watkins JL, Piwnica-Worms H (2002) Disruption of the checkpoint kinase 1/cell division cycle 25A pathway abrogates ionizing radiation-induced S and G2 checkpoints. Proc Natl Acad Sci USA 99:14795–14800

Zhao X et al. (2000) Mutational and structural analyses of the ribonucleotide reductase inhibitor Sml1 define its Rnr1 interaction domain whose inactivation allows suppression of *mec1* and *rad53* lethality. Mol Cell Biol 20:9076–9083

Zhao X, Muller EG, Rothstein R (1998) A suppressor of two essential checkpoint genes identifies a novel protein that negatively affects dNTP pools. Mol Cell 2:329–340

Zhou BB, Elledge SJ (2000) The DNA damage response: putting checkpoints in perspective. Nature 408:433–439

Zhou Z, Elledge SJ (1993) *DUN1* encodes a protein kinase that controls the DNA damage response in yeast. Cell 75:1119–1127

Zou H, Rothstein R (1997) Holliday junctions accumulate in replication mutants via a RecA homolog-independent mechanism. Cell 90:87–96

Zou L, Cortez D, Elledge SJ (2002) Regulation of ATR substrate selection by Rad17-dependent loading of Rad9 complexes onto chromatin. Genes Dev 16:198–208

Zou L, Elledge SJ (2003) Sensing DNA damage through ATRIP recognition of RPA-ssDNA complexes. Science 300:1542–1548

Genome Dyn Stab (1)
D.-H. Lankenau: Genome Integrity
DOI 10.1007/7050_019/Published online: 12 September 2006

Germline Double-Strand Break Repair and Gene Targeting in *Drosophila*: A Trajectory System throughout Evolution

D.-H. Lankenau

Hinterer Rindweg 21, 68526 Ladenburg, Germany
d.lankenau@t-online.de

Abstract Replicative double-strand break (DSB) repair has been predominantly studied in single-celled organisms such as bacteria and fungi who lack the germline of metazoans. In contrast, most studies concerning DNA repair in mammals focused on somatic cells while only few authors engaged in embryonic stem cells. *Drosophila* represents a field of in vivo biochemistry systems linking DNA repair research with the power of classical genetics and modern developmental, molecular and cellular biology. In addition to its biomedical relevance, the *Drosophila* system helped—until recently—to form evolutionary theory over the past hundred years. There is reasonable evidence that the DSB repair mechanism termed Synthesis-Dependent Strand Annealing (SDSA), represents a candidate primordial repair apparatus opening a better understanding of genome evolution for all organisms. On the basis of in vivo gene knockout targeting experiments exemplified here, recombination tract data reflect the relatedness of the DSB repair mechanics of SDSA, Break Induced Replication (BIR), Single Strand Annealing (SSA), and the involvement of Holliday junctions (HJs) and non-DSB repair pathways.

Abbreviations

BIR	break induced replication
DHJ	double Holliday junction
DSB	double-strand DNA break
ds(ss)DNA	double-strand (single-strand) DNA
HR	homologous recombination
Nap1	nucleosome assembly protein 1 gene
NHEJ	non-homologous DNA end joining
RT	reverse transcriptase
SDSA	synthesis-dependent strand annealing
SSA	single strand annealing

1
Introduction

Most modern geneticists will agree that gene[1] *mutations form the chief basis of organic evolution, and therefore of most complexities of living things.*

(Muller 1927a)

Beginning with Thomas Hunt Morgan's trailblazing genetic research (Morgan et al. 1915) *Drosophila*[2] is a metazoan model system which successfully served the past hundred years to focus efforts at understanding the nature of the gene along with its mutations. Furthermore, it fundamentally helped to grasp how genes control the development of an embryo (Nüsslein-Volhard 2004). Pushing the frontier further, for the past 16 years *Drosophila* opened experimental strategies of in vivo *biochemistry* otherwise only known from microbial systems where the repair of unique DSBs was physically monitored after DNA break-induction by site-specific endonucleases (Engels et al. 1990; Haber 1995; Lankenau 1995; Lankenau and Gloor 1998; Rio 2002; Rong et al. 2002; Szostak et al. 1983). These studies provided keen molecular-genetic insight into the world of mitotic recombinational DNA repair within the *Drosophila* germ line cells while the establishment of targeting techniques were just a convenient by-product.

Simultaneously to mouse research (Capecchi 1989a,b), the system led to two gene targeting tactics now being routinely applied in *Drosophila* genetics. In the first system the repair of a DSB in a target gene occurs mainly by replicating sequence from a transgenic template located anywhere in the respective germline nucleus. The process is called Synthesis-Dependent Strand Annealing (SDSA) (Lankenau 1995; Lankenau and Gloor 1998; Nassif et al. 1994). The second targeting system (Rong and Golic 2000; Rong et al. 2002) uses a transgenic, operationally excised episome with an artificially introduced precise DSB. It takes advantage of two alternative DSB repair pathways, i.e. Break Induced Replication (BIR) and, after production of a knockin mutation, Single Strand Annealing (SSA) (Engels 2000; Lankenau et al. 2003; Rong and Golic 2001). The classical double Holliday junction (DHJ) pathway of DSB repair (Szostak et al. 1983) here appears to be a topological feature which may or may not be a mechanistic consequence of BIR and the timing of independent strand invasion events (Kraus et al. 2001; Lankenau 1995; Lankenau et al. 2003). In summary, the *Drosophila* in vivo *biochemistry* system was the first of its kind in a metazoan organism and with SDSA it promises to open a molecu-

[1] The term "*gene*" is understood here in its broadest definition which can be traced back to G.C. Williams: "*A gene is defined as any portion of chromosomal material that potentially lasts for enough generations to serve as a unit of natural selection.*" Dawkins R (1976) The selfish gene. Oxford University Press, Oxford.

[2] The official, taxonomic nomenclature for the "laboratory-fruitfly" as biological species is: *Drosophila melanogaster* Meigen, 1830

lar window to the deep time of ancient life, its evolution and contemporary genomic function.

This review concerns the fruit fly's role in understanding the nature of *mutation* from a historical perspective and our current insight into the ubiquitous DSB repair pathways, i.e. SDSA, BIR, SSA and non-homologous DNA end joining (NHEJ), where most of the latest experiments were carried out jointly with germline gene knockout targeting approaches (Gloor et al. 1991; Rong and Golic 2000). On the basis of combined gene-knockin experiments, and experiments physically monitoring recombination induced by site-specific endonucleases (Lankenau et al. 2003) an example of the making of bulk single-copy knockout flies (with the nucleosome assembly protein (*Nap1*) gene as the target locus) is presented. The latter procedure especially involved SSA which has now been incorporated into another intelligent, in-vivo recombination-monitoring system (Preston et al. 2006).

2 DNA Double-Chain Break Repair, Evolution and the Origin of Life

The common roots of replication and DNA repair alike date back to the most ancient forms of life at least 3.5 billion years ago (Summons et al. 1999). This review argues for SDSA representing a replicative DSB repair mechanism (Lankenau 1995; Lankenau and Gloor 1998; Nassif et al. 1994) as a candidate-primordial DSB repair apparatus. To this end, let's take the hypothesis for granted that SDSA represented a realistic primordial replicative DSB repair pathway.

First, let us travel back in time to the imaginary origins of life[3]. The reliable replication of nucleotide chains in combination with high error rates were both a fundamental prerequisite of life. Replication error rates established themselves near an error-threshold where the best conditions for evolution exist (Eigen 1992). The emergence of life was chemically complex with carbon fixation as the most fundamental reaction (Hazen 2005). Then, at some point, prebiotic metabolism-like chemistry arrived at the manifestation of the more stable ribonucleotide chains (RNA world hypothesis) and only later deoxyribonucleotide replicators evolved. There is general agreement that RNA replicators came prior to DNA replicating systems, because, in contrast to deoxyribose, ribose is promptly formed by aldol condensation, and the nucleophilicities of the $2'$, $3'$ hydroxyl groups of ribonucleotides are significantly

[3] To comprehend the beginnings of life requires that we explain the origin of replication as well as of metabolism synergistically (Maynard Smith J, Szathmary E 1997). The genetic aspect of the modern definition of life as first proposed by Muller in 1966: "*It is to define as alive any entities that have the properties of multiplication, variation and heredity*" (Muller HJ 1966). While metabolism supplies the monomers from which the replicators (i.e. genes) are made, replicators alter the kinds of chemical reactions occurring in metabolism. Only then can natural selection, acting on replicators, power the evolution of metabolism.

higher than the one of the corresponding 3′ hydroxyl group of deoxyribonucleotides (Lohrmann and Orgel 1976). Also, the basepair interactions of ribonucleotides are much stronger than those of DNA. While DNA exists almost exclusively as a double-helix, RNA typically is a single-stranded polymer with an enormous range of secondary structures. Further, in all modern cells, the building blocks of DNA are synthesized from those of RNA, and DNA replication still depends on primase-driven RNA primer synthesis during chain initiation in replication forks (Egel 2006; Nasheuer et al. 2006, in this volume).

Prebiotic evolution may have produced small nucleotide chains spontaneously. But replication fidelity without proteins was low and did not allow synthesis of long genes. A recent model suggests the origin of life in sea ice with abiotically synthesized RNA molecules of lengths as long as 400 nucleotides (Trinks et al. 2005). But even if RNA molecules of this length were synthesized abiotically and spontaneously there was a discrepancy which is known as Eigen's paradox[4]. Only the fusion of such molecules and catalytically interacting ribozyme activities in a hypercycle might have led to larger genes and chromosomes by ligase-like fusion of RNA chains (Bartel and Szostak 1993; Eigen 1992; Eigen and Schuster 1978). The involved molecular reactions may have resembled in many ways modern mechanistics of DSB repair (e.g. tripartite SDSA repair) as observed in *Drosophila* and in yeast. One key transition protein between an early RNA-world and the DNA-world was reverse transcriptase [RT] encoded in the genome of retroelements (Temin and Mizutani 1970). RT is a versatile enzyme capable of transcribing RNA into DNA and replicating DNA. While in many organisms, a specialized RT—termed telomerase—acts to elongate the ends of linear chromosomes, in *Drosophila* the telomere maintenance is accomplished via the RT of non-LTR retrotransposons (Ahmad and Golic 1998; Biessmann et al. 1992a,b; Melnikova et al. 2005; Melnikova and Georgiev 2002; Pardue 1996). Consequently, from a contemporary point of view, the three "Rs" of DNA/RNA metabolism, i.e. *replication, recombination* and *repair*, represent the most indispensable processes in the evolution of life and continue to be the universal ingredients of all present-day organisms[5].

Once DNA replicators had evolved, there must have been mechanisms simultaneously stabilizing the chemical integrity of DNA chains. In comparison to RNA, DNA molecules are highly susceptible to decay and DSBs are the most devastating occurrences. They can arise for a variety of reasons such as replication fork collapse, the action of chemical agents, radiation or endonucleases. In eukaryotic cells a major source of endogenous mutagens are oxygen radicals leaking from mitochondrial respiratory chains. For instance, spontaneous

[4] Eigen's paradox: *No enzymes without a large genome, and no genomes without enzymes* (Eigen M 1992; Maynard Smith J, Szathmary E 1997)

[5] The viewpoint of carbon fixation as "the" most fundamental prebiotic and biological reaction (Hazen RM 2005) is completely compatible and a prerequisite to DNA/RNA replicators as essentials of life.

strand breaks can be triggered by a sequence of successive chemical reactions which are temperature-dependent: Depurination and depyrimidination leading to loss of bases occur spontaneously with high frequency in every genome. There is approximately one purine lost per *Escherichia coli* genome per generation at 37 °C while 300 purines are lost per *Thermus thermophilus* generation at 85 °C. 5000 depurination reactions per day occur in a mouse cell. While bases are lost, the sugar backbone of single DNA strands remains intact. However, at sites of base loss the deoxyribose residues exist in equilibrium between the closed furanose form and the open aldehyde form. In the open aldehyde form the 3′ phosphodiester bonds are readily hydrolyzed by β-elimination such that, partly based on this mechanism, 10 000 strand breaks occur per day in every human cell (Friedberg et al. 1995). Strand breaks, and especially DSBs were detrimental to life at all times. Therefore, it is likely that the earliest DNA replicators not only evolved efficient strategies for DNA-replication but that the establishment of replication-compatible DNA repair machineries was a precondition of and coincided with the evolution of a DNA-based Last Universal Common Cellular Ancestor (LUCA) (see also Maynard Smith and Szathmary 1997). Such replication-compatible DNA repair mending activities likely were ancestral to all modern cells. One possible relic of the transition from an RNA-world to the DNA-world is the *thumb, palm* and *finger*-domain structure of RT/RNaseH which indeed resembles modern proofreading DNA polymerases (Arnold et al. 1992; Kohlstaedt 1992). Also the *E. coli* DNA polymerase I enzyme encompasses an N-terminal, 5′-3′exonuclease domain orthologous to flap structure-specific endonucleases (FEN), which represents one of the most conserved DNA repair proteins. FENs are a central component of Okazaki fragment maturation in eukaryotes further underlining their primordial ancestry.

Hence, the contemporary biochemical pathways of DNA maintenance are likely to have been transmitted from a deep past (billions of years ago) opening a practical window for research into the replicative, molecular processes that always have been central to life (compare Aravind et al. 1999). Of course, we have to be aware that life had a history of at least 3.5 billion years before about 550 million years ago multicellular diversity emerged in some kind of punctuated equilibrium during the Cambrian explosion (Bottjer 2005; Briggs et al. 1994; Chen et al. 2004; Conway Morris 1994, 1998, 2000; Gould 1989; Hou 2004; Mattick 2004). This means that most of the molecular evolution of DNA-repair related proteins took place during the vast time before metazoans originated. Proteins involved in DNA-repair of today's organisms are likely to have evolved non-ancestral features that made them compatible to modern cellular and metabolic functions. Therefore, not every function encountered in DNA-repair proteins is ancestral as we have to be aware of functional innovations which Stephen Jay Gould calls *quirky functional shifts* (Gould 2002). The use of novel protein domains as established by the shuffling of functional catalytic domains between DNA repair relevant proteins, i.e. *exaptations* (Brosius 2005; Gould 2002) represents such quirky shifts. Exaptations may practically impede accurately retracing the origins of life but still the ancestral functions of all DNA-repair domains will serve as models for the emergence of life in the RNA/DNA world and the transition from acellular to cellular life forms (Aravind et al. 1999).

In agreement with the hypothesis that SDSA is an ancient replicative DSB repair pathway, it appears to be universally conserved in bacteria (Formosa and Alberts 1986), yeasts (Paques and Haber 1999; Paques et al. 1998) fungi (Ferguson and Holloman 1996), plants (Puchta 2005) and metazoans (Johnson and Jasin 2000; Lankenau 1995). SDSA also represents a mechanism acting within stalled replication forks bypassing single-strand DNA lesions (Rudolph et al., 2006, in this volume). This could be an evolutionarily primordial function, as replication forks are primordial. Even though repair systems appear to have undergone distinct, divergent evolutionary histories in the three superkingdoms of life (Egel, 2006, in this volume), there is a core of universally conserved DNA repair protein domains (Aravind et al. 1999). The most prominent is the RecA recombinase whose relatives may play a central role in the annealing step of SDSA (Aguilera 2001). The most notable variation on the main theme of RecA during evolution was its fusion with a modified nucleic acid-binding HhH[6] domain as in the Rad51 ATPase. The HhH module actually appears to be the most common DNA-binding domain in repair systems (e.g. in the ERCC4 nuclease/helicase family, the archaeal and eukaryotic members of the RadA/RecA family and the UmuC/REV1 superfamily, the RNA-binding protein NusA of *E. coli* and *Chlamydia*). It also has been detected by iterative database searches in the DNA ligases (Aravind et al. 1999). Recently the *Bacilus subtilis* sporulation preventing and DNA integrity scanning A (DisA) protein was reported to bind non-specifically DNA forming a single focus which moves rapidly along the chromosomes. Upon encountering a lesion, DisA delays entry into sporulation until DNA repair is completed. The domain responsible for the scanning process has been predicted to be a HhH domain at the C terminus of DisA (Bejerano-Sagie et al. 2006). As mentioned above, another universally conserved domain is the FLAP[7] nuclease which fulfills different functions in bacteria[8] on the one hand and archaea and eukaryotes[9] on the other hand (Aravind et al. 1999).

Only a small number of repair proteins are conserved in most representatives of the three superkingdoms of life although they are not universally represented. With regard to the ancestry-hypothesis of SDSA, a most remarkable family seems to be the Rad50/SMC-like[10] ATPases and associated nucleases

---

[6] Helix-hairpin-Helix

[7] FLAP is the term for a flap structure in DNA. Flaps exist at the ends of Okazaki fragments in lagging strand DNA synthesis and as intermediate structures of SSA DNA repair (see Figs. 5A and 6A).

[8] N-terminal, 5′-3′ exonuclease domain of DNA polymerase I

[9] FLAP nuclease cleaves overlapping 5′-flaps at sites of nicks. They also have 5′-3′ exonuclease activity possibly involved in the hydrolysis of flaps.

[10] structural maintenance of chromosome

(Hopfner, 2005, in this volume). This ATPase family is conserved in almost all completely sequenced genomes, and in eukaryotes, it is involved in ATP-dependent modifications of the chromatin structure and plays a significant role in chromosome segregation. The Rad50/SMC family forms multiprotein complexes with the equally well-conserved calcineurin-like phosphoesterase superfamily. The MRN[11] complex mediates telomere maintenance in addition to a variety of functions in response to DSBs, including NHEJ and activation of DNA damage checkpoints. It was suggested that the MRN complex protects *Drosophila* telomeres by facilitating recruitment of two proteins, HOAP[12] and heterochromatin protein one (HP1) at chromosome ends (Ciapponi et al. 2004). MRN may indeed be also highly relevant for SDSA: One key-aspect of DSB repair appears to be the "biochemical decision" of whether a DSB is closed by NHEJ or if its termini are processed by exonuclease activities. When the ends are processed exonucleolytically the biochemical specifics of their degradation determine the length of conversion tracts in SDSA (Gloor et al. 1991) and/or whether the related pathways NHEJ, BIR or SSA are utilized. In agreement with this, biochemically, the MRN complex is an ATP stimulated nuclease that acts endo- and exonucleolytically on ssDNA and hairpins, and exonucleolytically on various types of dsDNA ends. Mre11 by itself only degrades ssDNA while Mre11/Rad50 proteins degrade dsDNA in 3′-> 5′ direction (Hopfner, 2005, in this volume). Hopfner argues that this proofreading-like nuclease activity of MRN may not directly generate formation of 3′ tails in recombination, but more likely processes DNA secondary structures or otherwise misfolded or damaged DNA ends for subsequent repair steps (Hopfner, 2005, in this volume). Finally, clean 3′-OH termini will result which seem to be the required substrates for the subsequent SDSA reactions. However, this assumption requires future biochemical and genetic experiments to be verified. Also, the RecQ helicase/RNaseD-related family of repair proteins represents a window into ancient biochemical reactions. While RNaseD is required for tRNA trimming through dsRNA degradation this truly ancient protein shares its 3′-> 5′ exonuclease domain and its Helicase and RNAseD (HRD) nucleic acid binding-domain with two aging relevant DNA repair genes (BLM, WRN) mutated in Bloom syndrome and Werner syndrome. The BLM RecQ helicase preferentially unwinds or migrates recombination intermediates, including Holliday junctions (HJ) and D loops which both play an important role in BIR and SDSA, respectively, and have been experimentally associated with these DSB repair mechanisms (Figs. 2 and 4) (Adams et al. 2003; Kusano et al. 1999, 2001; McVey et al. 2004; Morozov et al. 1997). BLM and WRN both physically interact with p53 (Gatz and Wiesmüller, 2005, in this volume), a classical tumor suppressor gene under investigation in *Drosophila* as well (Brodsky et al. 2004).

[11] Mre11-Rad50-Nbs1

[12] (HP1)/origin recognition complex-associated protein.

Yet another feature of SDSA is its contributing to within-species homogeneity and between-species heterogeneity for a family of repeated genes such as histone gene clusters or rRNA gene-repeats (Paques et al. 1998). These genomic homogenization effects are termed "*concerted evolution*" (Dover et al. 1982). Further, the eukaryotic Mre11-like nucleases and the bacterial SbcD-like proteins belong to the same superfamily. Aravind et al. noted that SbcD and Mre11 were vertically inherited in all life forms with loss in only a few lineages (Aravind et al. 1999). Finally, it needs to be mentioned that both essential functions, i.e. DNA repair and replication, must be also compatible with present genome architectures dominated by chromatin, non-coding DNA and redundant sequences (Mattick 2003, 2004; Friedl 2005, in this volume). If a repetitive sequence which occurs in various patches of clusters at various chromosome loci requests replicative DSB repair the use of alternative templates creates the risk of crossover products that lead to scrambled chromosome rearrangements. SDSA overcomes such a problem (Lankenau et al. 2000).

3
Drosophila: Historical Background

Recently, Karl Woese stated that biology today has a choice to make, between the comfortable path of continuing to follow molecular biology's lead or the more invigorating one of seeking a new and inspiring vision of the living world (Woese 2004). *Drosophila* as a role model system of metazoan genetics incorporates both aspects, with gene targeting as a pragmatic engineering discipline, and DSB-repair research addressing more fundamental questions along with physics probing and defining reality.

Gene targeting is an instrument to modify the function and the structure of selected genes within the germline of a living organism in a very precise, stable and predictable molecular manner. The initial biochemical targeting reaction takes place in vivo within the nucleus of a cell. It is intimately associated with cell-endogenous key-processes maintaining genome stability such as replication, repair and recombination. As genomes grew in size during early evolution, these chromosome- and chromatin-associated processes became fundamental to life in general. As a tool, gene targeting serves the various aims of many different interests by experimentally altering gene functions in vivo. As a biochemical and cell physiological reaction it gave insight into the chromosome maintenance programs of metazoan germline cells. This review is a brief trajectory through the history of genetics culminating in the establishment of gene targeting in *Drosophila*. Recent progress in *Drosophila*-based research, mainly the employment of site specific endonucleases, gave insight into the underlying molecular DNA-maintenance machineries in vivo. As the combination of molecular and classical genetics of *Drosophila* provide a powerful tool to study the basics of DNA repair the approach is called *in vivo biochemistry* of repair, recombination and replication in analogy to fungal genetics (Haber 1995).

3.1
Drosophila as a Role Model Organism

When the primary research goal is to understand broad biological principles—without underplaying or denying the contingency of evolutionary change—the organism chosen for study is called a model organism. As a most typical model organism, *Drosophila* represents a metazoan pilot system which is unparalleled up to now. Its break through came as early as 1916 with Bridges' paper on the proof of the chromosome theory of inheritance (Bridges 1916). Since then, it served in numerous ways our modern understanding of the multiple facets of evolution. The fly was mainly popularized by Thomas Hunt Morgan as an experimental organism serving to map genes on chromosomes (Morgan et al. 1915). Morgan and his colleagues Sturtevant, Muller and Bridges measured the frequencies of crossing over in order to chart the distances between new discovered genes. In 1931, Curt Stern, a German geneticist working with the fly at last confirmed genetic recombination (*Faktorenaustausch*) by means of cytological analysis (Stern 1931)[13]. Crossing over created variety in the offspring, and Theodosius Dobzhansky realizing this fact, integrated the Mendelian experimental world with natural history and established chromosome-inversion polymorphisms as intrinsic properties of natural populations (Dobzhansky 1937). The fusion of genetics with Darwinian evolutionary biology established the "*Modern Synthesis*" of evolution (Dobzhansky 1937; Fisher 1930; Haldane 1932; Huxley 1942; Mayr 1942; Rensch 1947; Simpson 1944; Wright 1931). In X-ray radiation experiments, *Drosophila* served to establish that mutation was linearly related to dose (Muller 1927a). This made mutation a researchable subject, and Muller notes:

"*Perhaps the most hopeful feature of the present data is that they show that mutation is indeed capable of being influenced 'artificially'—that it does not stand as an unreachable god playing its pranks upon us from some impregnable citadel in the germ plasm.*"

(Muller 1928a).

After a long time of uncertainty about how to link evolutionary theory and genetics with the world of physics, these radiation experiments after all convinced the leading physicists of the physical nature of the gene:

"*And the gene is most certainly not just a homogeneous drop of liquid.*"

(Schrödinger 1944).

[13] Independently, in the same year, Creighton and McClintock had established priority on this subject with maize (Creighton HB, McClintock B 1931)

This statement is from Erwin Schrödinger "*What is Life*". It is now widely believed that this book had a significant impact on the discovery of the structure of DNA.

Another frontier of *Drosophila*-based research was opened at the end of the 1970s, when saturation mutagenesis experiments established the hierarchy of genes responsible for early embryonic development of the fly. As this knowledge could be transferred to the early development of the human embryo and to other vertebrates it accomplished a link to the biomedical realm. In 1995 Edward B. Lewis, Christiane Nüsslein-Volhard and Eric F. Wieschaus won the Nobel Prize in Physiology or Medicine for their discoveries concerning "the genetic control of early embryonic development". It was the third Nobel Prize for *Drosophila*-research with the two others awarded to Morgan in 1933 and Muller in 1946 (Nüsslein-Vollhard 1995)[14]. Then, only a few years later, sequencing the entire *Drosophila* genome served as a practical warm-up effort for the Human Genome Project (Adams et al. 2000).

3.2
Drosophila and the Discovery of Irradiation-Induced Gene Mutations

In 1927 and 1928 Muller published a series of key papers establishing that genes can be mutated by means of X-ray radiation (Muller 1927a,b, 1928a,b; 1946). In a few months Muller found more mutants than all *Drosophila* researchers up to that time. Muller had been a member of Thomas Hunt Morgan's laboratory (*the Fly Room*) and he now worked at the Department of Zoology of the University of Texas[15]. Several prior and contemporary studies addressed the influences of radiation on chromosomes, but none had revealed conclusive results establishing the material (i.e. not spiritual) nature of the "*gene*" fitting the terms of physics (Altenburg 1928; Dognon 1925; Gager and Blakeslee 1927; Goodspeed and Olson 1928; Harrison and Garrett 1926; Holthusen 1920; Mohr 1919; Stadler 1928a,b; Stein 1922; Whiting 1928). Muller comments on this issue:

"It has been repeatedly reported that germinal changes, presumably mutational, could be induced by X or radium rays, but ... the work has been done in such a way that the meaning of the data, as analyzed from a modern genetic standpoint, has been highly disputatious at best; ..."

(Muller 1927a).

Muller's quantitative, genetic studies of X-ray irradiated *Drosophila* flies represent the turning point in the relationship between biosciences and physics. Figure 1A shows the critical experiment in 1926, where Muller irradi-

[14] http://nobelprize.org/medicine/laureates/1995/nusslein-volhard-lecture.pdf
[15] Stadler confirmed his experiments independently with barley and other plants at about the same time (Stadler 1928a,b).

ated wild-type male flies and mated them to females carrying a "*balancer*"[16] X-chromosome. In a historical account, Ernst Peter Fischer notes the tight interdependence of *quantum mechanics*, radiation and classical genetics during the 1930s (Fischer 1988): In physics, wave-particle duality holds that light (i.e. radiation) and matter simultaneously exhibit properties of waves and of particles (or photons). The 1900 rediscovery of Mendel's laws of inheritance established classical genetics as an independent science. It was known that genes do exist and are lined up like beads on a string on microscopically visible chromosomes. But genes were invisible in the microscope and nothing was known about the physical nature of genes. And thus, a connection between "*genes*" and wave-particles seemed illusive. Muller's experiments showed that genes were by no means non-material abstractions without any relationship to the physical system of measurement, but his data indicated for the first time that each gene was indeed a macromolecule (Peter Starlinger, cited by Fischer 1988).

Thus, Muller's data overturned the skepticism concerning the physical reality of genes. For the achievement of altering genes in the laboratory he received the Nobel Prize in 1946 (Muller 1946). Utilizing *Drosophila melanogaster* as a genetic system, and irradiating whole flies, Muller quantified mutations particularly in germline cells ("*germinal changes*"). Others studied X-ray induced mutations in *Drosophila* somatic cells as well (Patterson 1928).

3.3
Initial Stages of DNA Repair Research in *Drosophila*

The molecular concepts on DNA recombination and repair rooted mainly in bacteria and lower-order fungi as model organisms (Friedberg 1997). *Drosophila*, however, as an eukaryotic organism combines many of the genetic tools available for fungi. Like mammals, it is a multicellular metazoan with distinct germ cell and somatic cell lineages. In addition to its genetic strengths, *Drosophila* offers several unique experimental advantages for the molecular analysis of DNA repair. For many years, Boyd and co-workers promoted the utilization of cultured *Drosophila* cells to investigate DNA repair pathways. A number of cell lines were permanently established, but also the use of short cultures in vitro of dissected larval brain ganglia became prominent (Echalier 1997). However, the use of germline cell cultures as a DNA repair research tool might be a goal to strive at. Indeed, embryonic pole cells were cultured early on in the 1970s (Allis et al. 1979). Another advantage is that the first 13 nuclear divisions of the early embryo last together

[16] Balancer chromosomes are genetic tools unique to *Drosophila*. A balancer chromosome contains at least one recessive lethal gene mutation and several chromosome inversions and translocations such that nearly no meiotic crossing over occurs with the wild-type homolog. Typically, it also contains a couple of other mutant genes serving as phenotypic markers.

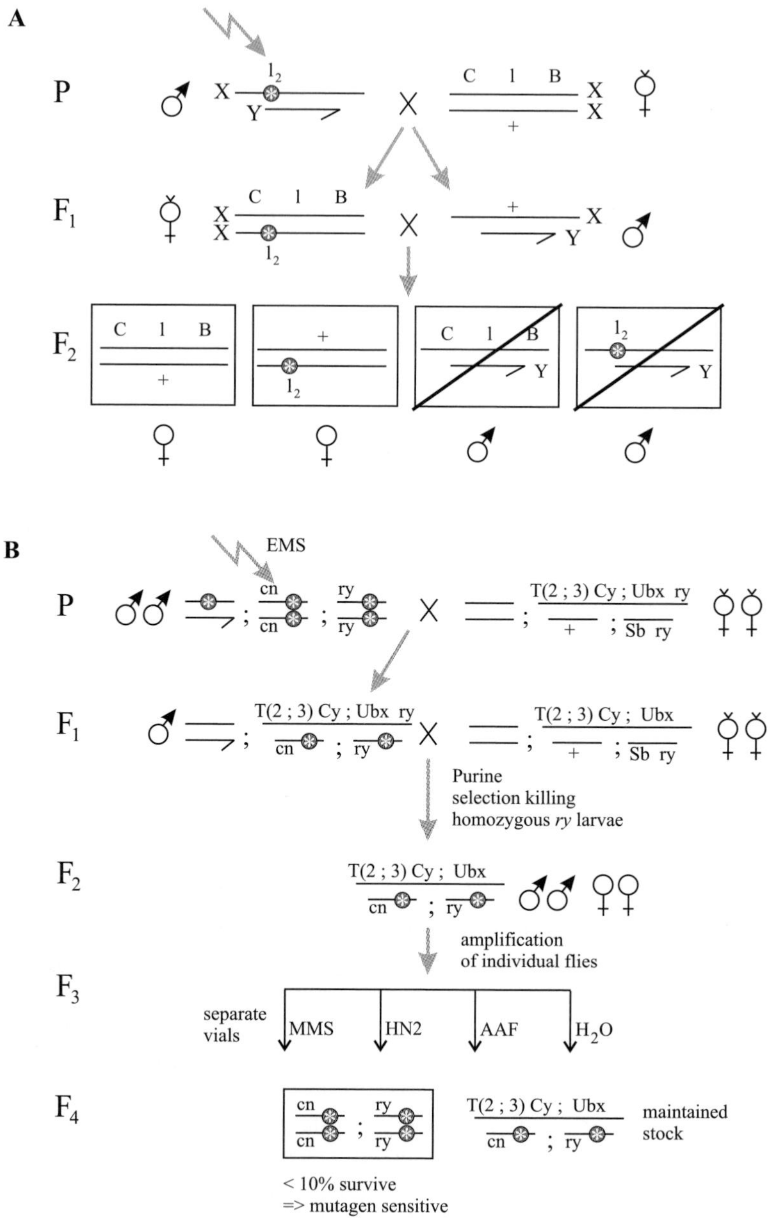

only about 130 minutes each skipping the G1 or G2 phase of the cell cycle. Compared to cell cycle 14 which alone lasts at least 115 minutes, and includes G1 and G2 phases of the cell cycle (Foe et al. 1993) this is a very short time. It made the early embryo an extraordinary source of DNA metabolic enzymes and provided an entry for in vitro experiments (Gaillard et al.

◄ **Fig. 1** Classical genetic experiments with *Drosophila* identifying the physical character of the *gene*, and identifying DNA-repair relevant mutations. **A** Muller's 1926 two-generation cross where males were X-ray irradiated (*gray arrow*) causing multiple mutations. Lethal mutations on the X-chromosome (l_2, *circle marked with star*) were genetically captured in the next generation (F_1) by crossing irradiated parental (P) males to non-irradiated virgin females containing the C l B crossover-suppression (C) chromosome, where l is a homozygous lethal mutation and Bar (B) is a dominant phenotypic eye marker mutation. By crossing F_1 females to wild-type males the mutated X chromosome carrying the X-ray induced l_2 mutation would be passed on to half of the offspring males while the other half would carry the l mutation of the balancer chromosome. All individual crosses without male F_2 offspring contained an X ray-induced lethal mutation. The cross allowed quantitation of X-ray dose to numbers of induced mutations. **B** Simultaneous third- and second-chromosome mutagen-sensitive screen (Boyd et al. 1981). Males homozygous for marker-mutations cn and ry were fed EMS (ethane methyl sulfonate) for the induction of point mutations. After mass-mating to virgin females carrying a T(2;3) translocation balancer chromosome F_1 males carrying the indicated genetic marker combination and the potential mutagen-induced DNA-repair gene mutations (*circle filled with a star*) were mated individually to virgin females. After three days, F1 parents were removed and purine solution was added to kill homozygous *ry* larvae. Surviving F_2 males and females were crossed to expand the stock and to identify recessive lethals. F_3 flies were treated with various mutagens testing for mutagen sensitivity. If DNA-repair relevant genes were captured (*circle filled with star*), less than 10% of homozygous cn/cn; ry/ry flies would be found (*boxed genotype*). For gene nomenclature refer to: http://flybase.bio.indiana.edu/

1999). A further advantage is the lack of chromosomal recombination in the male germline during meiosis, permitting the experimental distinction between meiotic and mitotic DNA repair. Aiming at the discovery of DNA-repair genes, the genetic strengths of *Drosophila* was first employed during the 1970s and early 1980s by hunting directly for mutagen-sensitive mutants (Boyd et al. 1981, 1987). Mutants were genetically selected that were conditionally lethal upon exposure to moderate doses of mutagens such as ethyl methane sulphonate (EMS) or low doses of X-rays. A classical genetic screen for autosomal mutagen sensitive genes is shown in Fig. 1B. Together with the complete sequence of the *Drosophila* genome, this approach has led to an extensive database of *Drosophila* repair genes with the long-term goal to have at least one mutant allele and DNA clone for each *Drosophila* repair gene (http://www.dmrepair.ucdavis.edu/).

4
The Roots of Gene Targeting in *Drosophila*

Our sophisticated, present-day techniques of gene knockout targeting and associated DNA-repair research in the *Drosophila* germline rooted in a seemingly different branch of scientific interest. This was the discovery of transposable elements in metazoan eukaryotes.

During the 1960s population geneticists became interested in the question how chromosomes affect the variation of fitness (Hiraizumi and Crow 1960). When wild-caught *Drosophila* flies were crossed with genetically marked laboratory stocks it resulted in a phenomenon called *hybrid dysgenesis*, encompassing high sterility and mutation rates, chromosomal segregation distortion, and chromosome nondisjunction (Engels 1989; Kidwell et al. 1977; Sved 1976). At least one study was in apparent contrast to Morgan's 1914 observation that recombination was always absent in male *Drosophila*: Hiraizumi and Crow noticed that the males of some crosses transmitted recombinant progeny (Hiraizumi 1971). Today, this inconsistency is known to be triggered by a transposable element, the P element. The first cloned P element was an incomplete element, identified by virtue of having disrupted the famous *Drosophila white*[17] gene. The P element fragment was then used as a molecular probe to isolate a complete element. After injecting the full P element as an integral part of a plasmid into the germline of *Drosophila* embryos, it transposed from the plasmid into the *Drosophila* chromosomes (Rubin and Spradling 1982; Spradling and Rubin 1982). This transposon produced DSBs upon excision which are subsequently repaired in a gene-conversion-like process involving SDSA (for a review see Gloor and Lankenau 1998; Lankenau 1995).

Together with the culturing techniques of embryonic stem cells (ESC) as a prerequisite for the targeted modification of mouse genes (Capecchi 1989a; Thomas and Capecchi 1987), the recent establishment of gene knockout targeting within the *Drosophila* germline (Gloor et al. 1991; Rong and Golic 2000) represents a new conceptual brick in the "*Modern Synthesis of Evolution*". Here, the concepts of developmental biology, the molecular genetics of DNA repair, replication, recombination and aspects of epigenetics and other areas fuse with the classical *Synthesis*.

One obvious aspect of gene targeting is to facilitate functional analysis of genes in vivo which are of developmental and physiological importance. However, there is far less awareness that these same techniques employ real cellular pathways of DNA/RNA metabolism that could not be analyzed and tackled before the establishment of targeting as a tool in metazoans (Lankenau and Gloor 1998).

Therefore, the coupling of targeted gene modification to the analysis of double-strand break (DSB) repair now promises to link developmental pattern formation programs with DNA maintenance networks.

4.1 P Transposable Elements Transpose via a Cut-and-Paste Mechanism

P transposable elements are among the most-studied DNA transposons of eukaryotes (Engels 1989; Rio 2002). A P element (w^{hd}) insertion within the

[17] The *white* gene was Thomas Hunt Morgan's first discovered mutation at the beginning of the twentieth century.

Drosophila white locus causes loss of the red eye color phenotype. Taking advantage of these *white* eye mutants W. R. Engels and colleagues measured the rates of precise P element excisions using the backmutation to red eye color as a comfortable phenotypic marker. Their experiments indicated that "precise" P excisions were found at a rate hundreds of times greater if a homologous chromosome carried a wild-type *white* sequence or just a corresponding functional *white* fragment not containing the P insertion anywhere in the genome. Their findings led to the conclusion that P elements excise via DSB formation and that these breaks can be repaired either by non-homologous end joining (NHEJ) leaving a footprint of the P or by a replicative DSB repair mechanism termed SDSA (Fig. 2) (Engels et al. 1990, 1994; Gloor and Lankenau 1998; Gloor et al. 1991; Lankenau 1995; Lankenau and Gloor 1998; Nassif et al. 1994). These principals were subsequently verified for other gene loci (i.e. *forked*, *white*^{eh}) as well (Lankenau 1995; Lankenau et al. 1996, 2000).

4.1.1 Non-Homologous End Joining (NHEJ)

The experimental manipulation of genomes in living organisms is tightly associated with transposable elements and becomes increasingly important also in vertebrates (Ivics et al. 2004; Miskey et al. 2005). While class I elements transpose via an RNA intermediate[18], class II elements such as the P element, do it without RNA using a self-encoded transposase (Finnegan 1989). The conservative excision-insertion mechanism of class II elements is also called cut-and-paste transposition. Among all eukaryotic transposable elements, the 2.9 kb P element of *Drosophila* is most heavily exploited as an experimental tool (Engels 1989; Kaiser et al. 1995). It is a class II element better known as DNA transposon. P elements transpose via a cut-and-paste mechanism (Engels et al. 1990). Upon excision as a double stranded DNA fragment, the P element leaves a DSB at the original location of the transposon. This DSB is subsequently healed by the DNA-repair machinery (Sentry and Kaiser 1992). The DSB is produced by a transposon-encoded transposase which recognizes about 20 nucleotides at the ends of the transposon's terminal inverted repeats. The meganuclease activity of the P element transposase introduces two DSBs each with a 17 nt single strand overhang (Beall and Rio 1997). After the main body of a P element has left its original location, 17 nt ssDNA flank each side of the DSB gap. When this DNA is not fully removed by exonucleolytic activity (presumably involving the MRN complex; Hopfner 2005) the remaining microhomologies lead to non-homologous, illegitimate annealing resulting in a non-homologous end joining (NHEJ) product. The ensuing footprint of the P element is on average 30 nt long and is some-

[18] class I elements also classify as retroelements

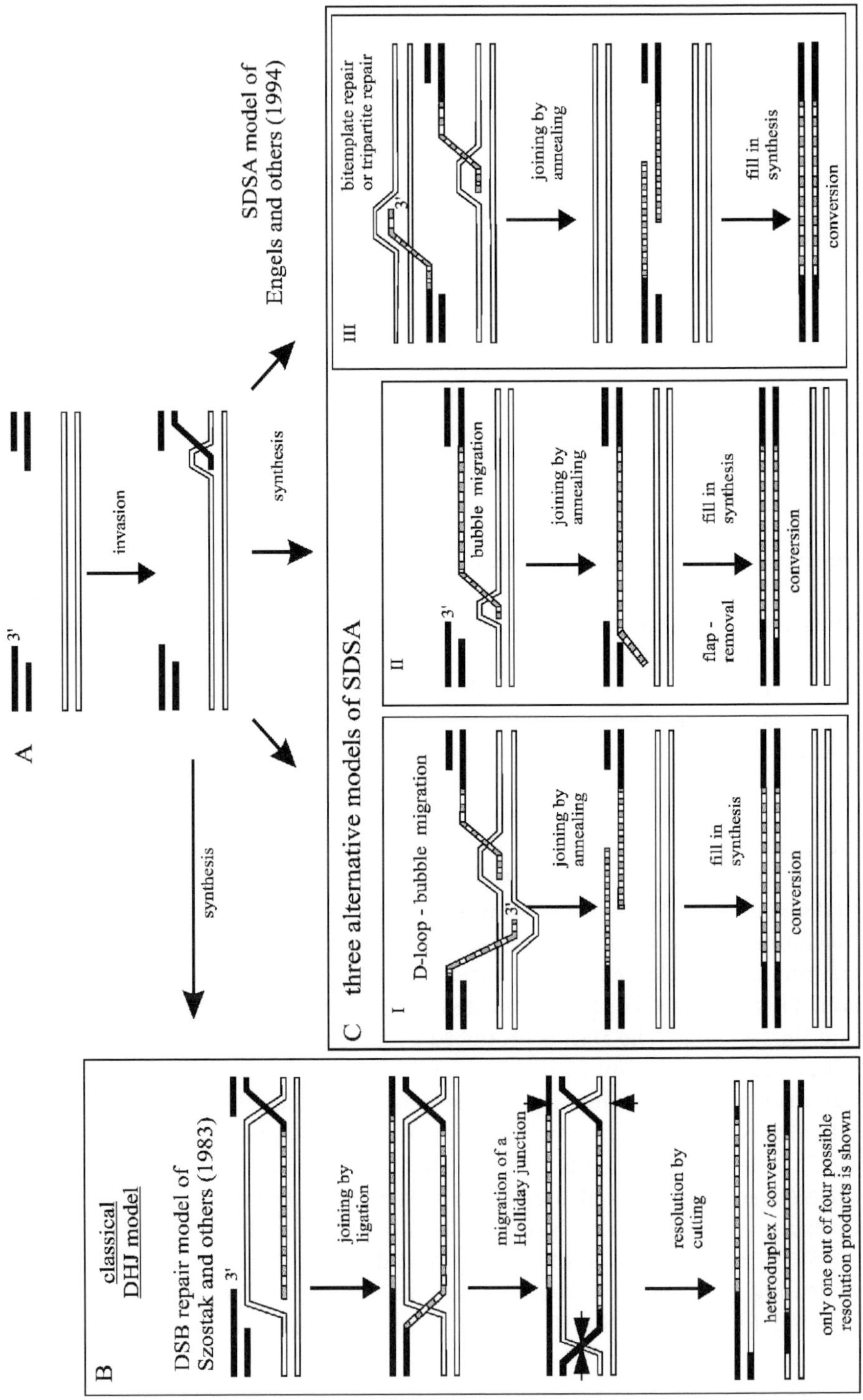
A
3'
invasion
synthesis
synthesis
SDSA model of
Engels and others (1994)
B
classical
DHJ model
DSB repair model of
Szostak and others (1983)
3'
joining by
ligation
migration of a
Holliday junction
resolution by
cutting
heteroduplex / conversion
only one out of four possible
resolution products is shown
C three alternative models of SDSA
I
D-loop - bubble migration
3'
joining by
annealing
fill in
synthesis
conversion
II
3'
bubble migration
joining by
annealing
flap -
removal
fill in
synthesis
conversion
III
bitemplate repair
or tripartite repair
3'
joining by
annealing
fill in
synthesis
conversion

◀ **Fig. 2** Comparison of the classical DHJ-DSB repair model of Szostak et al. (1983), and variants of the SDSA model of Engels et al. (Nassif et al. 1994). **A** Replicative repair is initiated in both cases when homology is found between the broken end and a complementary template molecule. **B** In the DSB repair model DNA synthesis proceeds across the template and the newly synthesized DNA is finally ligated to the opposite broken end. Two Holliday junctions (HJ) exist simultaneously and are resolved by endonucleolytic cleavage (only one product of four possible is shown). **C** In the SDSA model, the newly synthesized DNA is completely displaced from the template leaving the template always unaltered, and the newly synthesized DNA recognizes the opposite end by annealing. Annealing can take place either with two newly synthesized strands (C_{I}) or between the new strand and the old strand at the opposite end of the break (C_{II}). The independence of the two ends during SDSA repair has been experimentally proven in bitemplate repair experiments, this alternative is shown in panel C_{III}

times, and misleadingly, termed to be the result of an imprecise excision event (O'Brochta et al. 1991). Significantly larger (e.g. 300 nt) but still internally truncated P element-remnants are not the products of NHEJ but of incomplete homologous recombination (HR) activities resulting from DSB repair. The contributions from *Drosophila* complement other NHEJ systems rooted in well-established model organisms. For example, this volume contains a summary of experimental aspects concerning in vitro assays of NHEJ (Pfeiffer et al. 2005).

4.1.2
SDSA

In contrast to single-strand DNA break repair mechanisms such as methyl-directed mismatch repair (Modrich 1994), NER[19] and BER[20], the repair of DSBs has been described as too complicated to be deciphered exclusively by means of biochemical analysis (Kuzminov 1996). Mechanistic models of recombinational DSB repair are mainly derived from genetic systems, and they have gone through a long history over the past hundred years. Kuzminov underscores that two-strand DNA repair is an intricate multistage process in which individual components are often part of a bigger multiprotein-complex machine. Many of the participating enzymes are specifically maintained at very low concentrations because they are poisonous for the cell, and their abundance changes during the cell cycle (Spellman et al. 1998) and in different tissues (Christmann et al. 2006). Fortunately, this is exactly the situation where genetic analysis in *E. coli*, *Neurospora*, *Saccharomyces*, maize and *Drosophila*, the latter engaging P elements, feels at ease. But also progress using transposable elements in vertebrates as DSB-inducers has been made (Miskey et al. 2005; Walisko et al. 2006).

[19] NER = nucleotide excision repair
[20] BER = base excision repair

These applications of genetics are reflected by the history of recombinational repair (RR) and HR models which have been carefully detailed by Kuzminov (1996; p 8–22). On the basis of microscopic observations Janssens first proposed in 1909 a "chiasmatype" model for crossing over (Janssens 1909). Formal physical "proof" for his hypothesis came in 1931 from Barbara McClintock and Harriet B. Creighton (Creighton and McClintock 1931; McClintock 1987). Several additional major researchers (e.g. *Drosophila* geneticist Alfred Sturtevant in Tom Hunt Morgan's lab) were involved in break-join, copy-choice, and break-copy-type models (Leach 1996). But only after the structure of duplex DNA had been resolved by Watson and Crick (1953a,b) and the semiconservative nature of DNA replication had been experimentally demonstrated in the famous "Meselson and Stahl experiment" (Meselson and Stahl 1958) did the molecular rules for designing recombination models become straight forward. Several early models (Barricelli 1956; Meselson and Weigle 1961; Stahl 1956) assumed that, in order to recombine, DNA has to be damaged, and a broken end invades the intact homologue to start DNA synthesis (replication). Other models followed such as the first "molecular" break-join model of crossing over (Whitehouse 1963). Then, in 1964 the first of three fundamental models were developed: the contemporary four-way junction model of Robin Holliday (1964, 1968) although it was similar to that proposed by Meselson and Weigle (1961). Meselson and Radding then published the second major model which was also similar to the Holliday model but with more flexibility concerning the symmetry of the recombination products (compare Fig. 1 in Lankenau 1995).

In the end, models of HR converged upon those for RR. On the one hand, HR models were designed mainly to explain reciprocal exchanges (crossovers) and associated non-reciprocal events (conversions) observed during meiosis. On the other hand, RR models were proposed to explain repair of double-strand lesions such as those produced after irradiation or endonuclease activity, for example, during transposon motion. These two separate lineages of understanding were united in 1983 by the DSB repair model for meiotic recombination (Fig. 2B) (Szostak et al. 1983). The DSB repair model involves DNA synthesis, two migrating Holliday junctions and D-loop formation. In 1986 Formosa and Alberts used bacteriophage T4 and proposed that DNA synthesis takes place in a rather small replication-bubble instead of extensive D-loop formation. This model was the basis to explain the P element induced gap repair data of *Drosophila* leading to the SDSA model of mitotic DSB repair (Gloor et al. 1991; Nassif et al. 1994). A number of observations from the *Drosophila* P element gap repair-experiments were not compatible with the classical DSB repair model. Most prominent was the repair of the gap from ectopic templates located on non-homologous chromosomes and the absence of evidence for sequence changes within template DNA. Three-partner or bitemplate- experiments in yeast subsequently confirmed the observations at the *white* and *forked* loci and proposed SDSA to be a major mechanism of mating type switching in *Saccharomyces cerevisiae* (Paques and Haber 1999; Paques et al. 1998; Watson et al. 2004).

5
Gene Targeting and the Mechanism of HR in *Drosophila*: Recent Experiments

In vivo modification of a gene in its original chromosomal position is the ultimate, technical goal of genetic experimentation. P element-induced gap repair involving SDSA is clearly a straight forward approach to introduce every mutation desired to study the function of a gene, including insertions, deletions and point mutations. However, there is always the need to have an appropriate P element insertion at the right place within or near the gene of interest. Only a maximum of 1 kb left and right to the P insertion can be modified in the targeting experiment (Gloor et al. 1991). Also the targeting frequencies depend on presence and absence of homologous sequences on the homologous chromosome and the sister chromatids. Even though, the *Drosophila* stock centers provide a large number of P insertions aiming at a genomic density of one P element in 100 kilobases, this is not sufficient to target every nucleotide in the genome. Particularly the growing interest for non-protein-coding genes which represents a significant fraction of genome function (Taft and Mattick 2003) desires high-performance targeting tools. For this reason Rong and Golic developed a universal targeted gene replacement technique which uses P elements not as DSB-inducing agents but as vectors containing a modified template sequence and two vector-encoded endonucleases whose expression can be conditionally triggered, for example, by a temperature shock to the transgenic flies. The P element vector also contains the *white* gene as a positive selection marker (Rong 2002; Rong and Golic 2000, 2001; Rong et al. 2002; Xie and Golic 2004) (Fig. 3A).

5.1
BIR and Knockin Targeting of the *Nap1* Gene

In a recent effort to understand replication-related chromatin assembly in vivo the *Drosophila* nucleosome assembly protein 1 (*Nap1*) gene was targeted by means of Rong's and Golic's targeting system (Lankenau et al. 2003). Simultaneously, four artificial recombination tract markers were placed within the targeting vector in order to examine the DNA repair tracts within the recombination products (Fig. 3A, open circles). Upon homologous recombination between the donor vector and the native *Nap1* gene, the four tract markers doubled in number to the amount of eight in the two targeted *Nap1* knockin products, each representing a chimera consisting of outer wild-type *Nap1* sequence (Fig. 3C, marked as solid circles), and central donor-vector

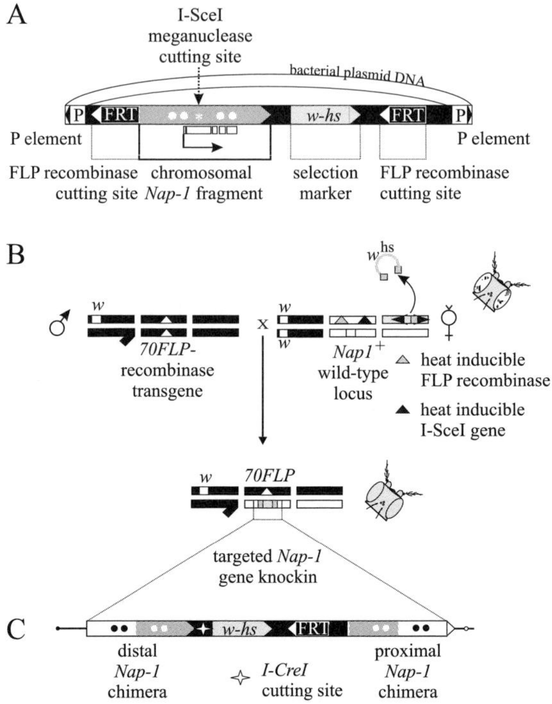

Fig. 3 Strategy for gene knockin targeting exemplified for *Nap1* (Lankenau et al. 2003). **A** Structure of a plasmid donor targeting vector as developed by Rong and Golic (Rong et al. 2002) encompassing in vitro-mutagenized sequences serving as recombination tract markers (*white circles*). The rest of the genomic *Nap1* fragment otherwise consists of a wild-type sequence. The plasmid contains a non-autonomous P transposable element. The P vector carries the *white-hs* (*w-hs*) marker gene, two FRT sites in direct orientation for FLP-mediated episomal excision, and the in vitro mutagenized *Nap1* gene. The intron-exon structure and its transcriptional orientation of *Nap1* is as indicated. The mutagenized center of *Nap1* contains an *I-SceI* meganuclease cutting site (*white star*). **B** Genetic cross to generate a targeted gene knockin at *Nap1*. The third chromosome of a virgin female fly contains the transgenic *Nap1* P element donor construct. The wild-type *Nap1* gene (*Nap1*$^{+}$) is located on both 2nd chromosome homologs (*only one copy is shown as a rectangle*). One of these homologs contains two heat shock-inducible transgenes: 1. FLP-recombinase (*gray triangle*), 2. *I-SceI* endonuclease (*filled triangle*). The X chromosomes are homozygous for a *white* (w) mutation. Upon heat shock during larval development FLP and *I-SceI* produce the extra chromosomal targeting molecule (*bent arrow*). The resulting mosaic eye phenotype serves monitoring the reaction (*upper right fly-head symbol*) and demonstrates loss of the episome in somatic eye cell clusters. The female virgin fly is crossed to a transgenic male with strong constitutive expression of FLP-recombinase (*70FLP, white triangle*). **C** Anticipated chimeric *Nap1* knockin targeting duplication. Mutagenized *Nap1* sequences (*gray rectangle*) are marked after in vitro mutagenesis by restriction site sequences (*white circles*). The reciprocal wild-type sites are shown as *filled circles*. The *outlined star* represents the *I-CreI* meganuclease cutting site for DSB induction to perform the SSA reaction required to remove vector sequences and for the production of a true single-copy *Nap1* knockout

derived *Nap1* sequences (Fig. 3C, white circles). In the knockin flies the conglomerated *Nap1* sequences flank the targeting vector sequence within the native *Nap1*-site on chromosome 2 encompassing the *white* gene as the phenotypic selection marker. Four independent germline events, i.e. four knockin-flies with different recombination tract-marker patterns were recovered in the targeting screen (Fig. 3B and C: one example shown) (Lankenau et al. 2003). The question was asked as to which DSB repair mechanism underlies the germline targeting reactions in vivo. NHEJ and SSA are DSB repair mechanisms that cannot explain the products of the targeting reaction. However, the sequence structure of targeting products and the particular distribution of tract markers allows for distinguishing between the remaining two alternative mechanisms of HR, i.e. SDSA versus BIR. SDSA does not explain the tract data of the knockins (Lankenau et al. 2003). However, BIR, a replicative DSB repair pathway mechanistically related to SDSA, seems to be the mechanism of choice, best explaining the recombination tract data and all knockin products (Fig. 4) (Engels 2000; Lankenau et al. 2003; Rong and Golic 2001). BIR reactions begin as one-ended recombination events, either because there is only one free DNA end available or, as perhaps in the case of a broken episome, only one of the two ends of the DSB succeeds in strand invasion at a time. This again is similar to the independent strand invasions preceding SDSA as suggested first for P-induced gap repair products (Nassif et al. 1994) and experimentally shown in bitemplate-repair experiments in *Saccharomyces* (Paques and Haber 1999; Paques et al. 1998). In budding yeast two or more alternative BIR mechanisms have been proposed (Kraus et al. 2001). Unlike the direct rivalry between the SDSA model (Lankenau 1995; Lankenau and Gloor 1998) and the classical DSB repair model (Szostak et al. 1983) where SDSA does not involve HJs while the DSB repair model is based on the formation and resolution of DHJs, BIR alternatively may or may not involve the formation and resolution of HJs (Fig. 4B versus 4C) (Kraus et al. 2001). Like in SDSA, in both alternative BIR mechanisms, the 3′ ends of the episome opened by *I-SceI* (Figs. 3A,B and 4A) are thought to be resected in the 5′ to 3′ direction exonucleolytically (possibly, but experimentally not yet established, involving the MRN complex; Hopfner 2005). The 3′-OH overhang of one end interacts with a number of recombination proteins and invades the homologous target strands in the native chromosome (Fig. 4A). One can imagine various ways in which BIR might occur—although they have not been distinguished yet experimentally: (1) Like in SDSA, the 3′ end of the invading strand initiates replication (a feature absent from the DSB repair model; Szostak et al. 1983). This leads to a migrating D-loop "bubble" as first suggested by Formosa and Alberts for prokaryotes (1986). The newly synthesized DNA will be displaced from the template strand and subsequently made double-stranded (Fig. 4B). In this process the episome may actually not be integrated into the chromosome. (2) Strand invasion sets up a true replication fork and branch migration proteins melt and displace both newly synthesized

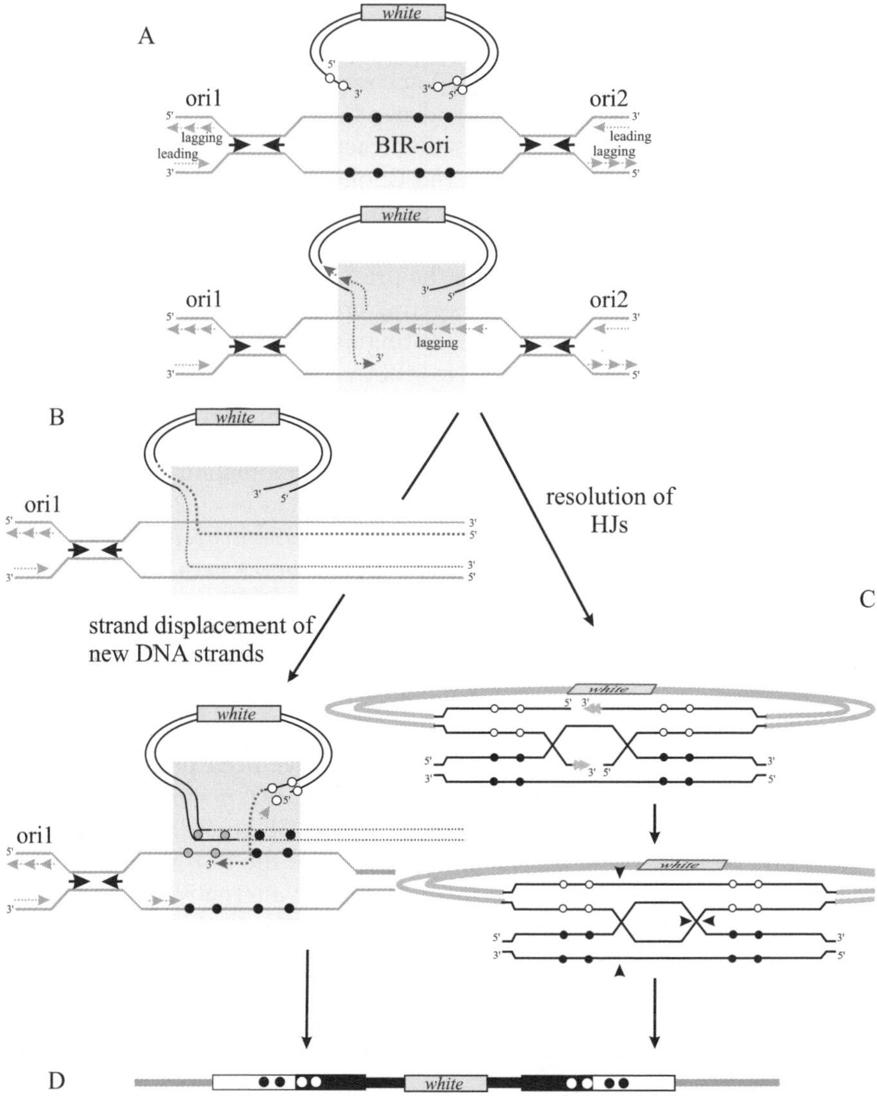

lagging and leading DNA strands. When strand displacement accompanies DNA-synthesis throughout an entire chromosome, and the reaction involved both broken ends of the episome, the knockin-chimera becomes an integrated part of the chromosome. Here, DNA synthesis is conservative (Fig. 4B). An argument in favor of this process was given by an experiment demonstrating that BIR could occur in G1 cells (Bosco and Haber 1998). (3) A third alternative is when strand invasion sets up a true replication fork and DNA will be synthesized semiconservatively in two directions involving both processed episome ends one at a time. When the synthesis takes place during

◀ **Fig. 4** BIR with and without involvement of Holliday junctions: The mechanism of Rong and Golic's ends-in knockout targeting. **A** The ends of a broken episome (Fig. 3) will be resected by 5′ to 3′ exonuclease activities, allowing each 3′ end to interact independently with various proteins (i.e. possibly: MRN-complex, Rad51, Rad52, Rad54, Rad55, Rad57) to carry out strand invasion. This triggers DNA replication at a BIR induced-origin of replication (BIR-ori). As this reaction may or may not occur simultaneously with normal, cell cycle dependent ORI activities two endogenous replication forks are indicated (ori1 and 2) which may finally merge with one another involving topoisomerases. Tract markers are as in Fig. 3. Leading and lagging strand synthesis is indicated. DNA synthesis into the episome involves lagging strand displacement followed by alternative template-use similar to reactions at stalled replication forks (Rudolph et al. 2006; this book) and subsequent fill-in synthesis. **B** Endogenous ORIs one and two merge with BIR-ori. Branch migration enzymes such as BLM and WRN displace both newly synthesized DNA strands as the proteins migrate over the length of an entire chromosome. This leads to a conservative replication product. *Gray filled circles* indicate tract markers involved in transient heteroduplex formation. **C** BIR intermediates resolved by resolution of HJs. Strand displacements over lengths of megabases in eukaryotes have so far been demonstrated only at the termini of yeast chromosomes (Kraus et al. 2001). *Drosophila* chromosomes encompass a much longer DNA chain. It therefore seems more likely that newly synthesized DNA will not be displaced over the length of a complete chromosome. Rather it will be synthesized semiconservatively and, as BIR is defined as a one-ended recombination event, a single HJ will be resolved for each episome end independently. Here, two HJ structures are shown similar to the classical DSB repair model (Szostak et al. 1983), however resolution of single HJs might occur independently and non-simultaneously. **D** One major anticipated chimeric *Nap1* knockin product. For other tract marker-combinations found in targeting products see (Lankenau et al. 2003)

the S-phase of the cell cycle together with replication initiated at multiple origins of replication (ORIs), the replication fork of the BIR reaction triggered by an end of the broken episome will be naturally merged with endogenous replication-forks born in nascent ORIs during the S-phase. If there were no strand displacement, topological problems will arise, and HJs will be resolved at some point as suggested by Lankenau et al. (2003) (Fig. 4C). Both versions lead to the anticipated knockin product (Figs. 3C and 4D), but because heteroduplex formation and subsequent mismatch repair can be a consequence of the reactions alternative tract marker combinations will be found as well (see SSA below).

Both models, SDSA and BIR, draw attention to the importance of the exonucleolytic processing of the DSB ends to create 3′-OH overhangs. This process particularly appears to have significant influence on the outcome of all targeted products and the corresponding recombination tracts (Lankenau 1995; Lankenau et al. 2003). Probably, both processes involve the proofreading-like feature of the MRN (Mre11/Rad50/Nbs1) protein complex (Hopfner 2005). Further, the universal evolutionary conservation of these proteins (i.e. SMC-like ATPases) in all kingdoms of life suggests a universal and ancient significance of these recombination reactions during evolution (Aravind et al. 1999). BIR and SDSA therefore seem to be fundamentally im-

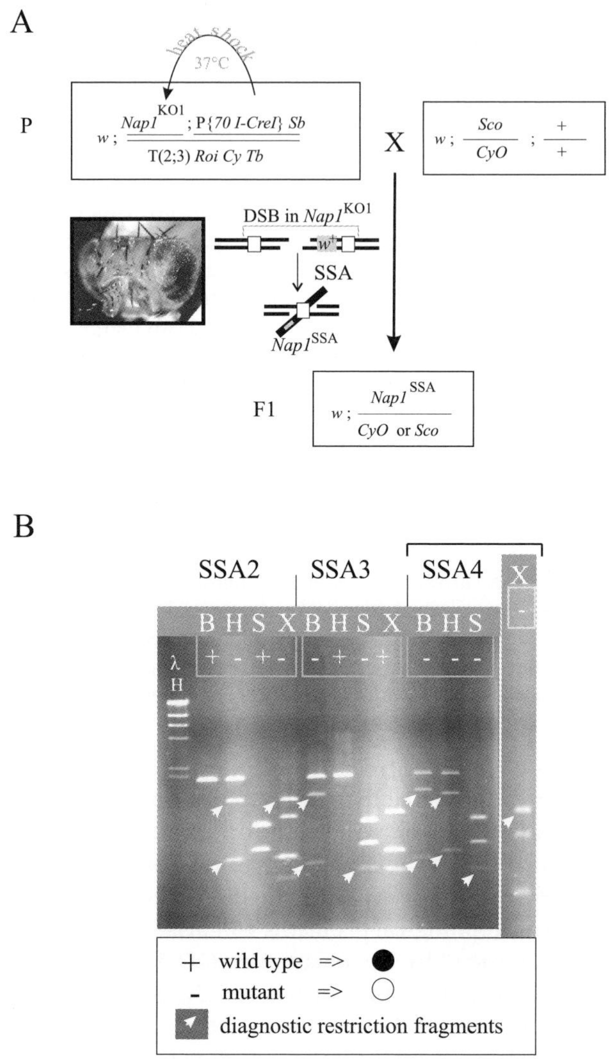

portant while HJ formation and resolution remains a significant topological consequence of variant pathways.

5.2 SSA-Recombination Tracts, Mismatch Repair and Knockout Targeting

Aiming at understanding the molecular function of *Nap1* in replication-related chromatin assembly it was inappropriate to use the targeted knockin mutations as published (Lankenau et al. 2003) because the locus under investigation consists of a chimeric, direct duplication of the *Nap1* gene rather

◀ **Fig. 5** Production of a single copy *Nap1* knockout mutant fly. **A** The chimeric knockin-allele *Nap1*^{KO1} on chromosome no. 2 was obtained as published (Lankenau et al. 2003). Its sequence structure and the *I-CreI* meganuclease cutting site are as indicated in Fig. 3C. The *Nap1*^{KO1} knockin allele was combined with a third chromosome obtained from Kent Golic. It contained a transgenic *I-CreI* meganuclease gene from *Chlamydomonas reinhardii* whose expression is driven by a heatshock hsp70 promoter. In the parental flies (P) both chromosomes were balanced over a compound translocation chromosome [T(2;3)] obtained from Bob Kreber (Univ. Wisconsin at Madison). Larvae of these flies were heat treated triggering expression of *I-CreI* endonuclease (*bent arrow*). Mosaic-eyed offspring indicating somatic activity of *I-CreI* were crossed to CyO-balancer flies. *I-CreI* produces a DSB between the two chimeric *Nap1* knockin copies. It induces preferentially SSA repair. Flap overhangs as intermediate structures of SSA are thought to be removed by proteins like Fen1. F1 flies which have lost the red-eyed phenotype were collected and either crossed to maintain the strain or used in single-fly PCR experiments for recombination tract analysis. Confirmed single copy knockout flies were termed *Nap1*$^{SSA1...n}$. **B** Examples of three diagnostic restriction digests of PCR products confirming *Nap1*^{SSA} knockout alleles. B, H, S, X designate the restriction digests with BamHI, HindIII, SalI and XhoI respectively. SSA2, 3, 4 are three examples of independent *Nap1*^{SSA} knockout events. *Arrowheads* designate diagnostic restriction fragments. HindIII digested phage lambda DNA was used as a size marker. Digests diagnostic for wild-type and in vitro mutagenized tract markers are indicated (+ and –). This corresponds to the *filled* and *outlined circles* in Figs. 3, 4 and 6 as indicated (=>)

than a single copy mutant (Fig. 4D). Further, vector-related DNA fragments not representing the native *Nap1* locus may interfere artificially with gene function in complex ways. For example, *white* locus DNA is located between the two target gene chimeras (Fig. 4D). As the *white* gene DNA fragment originated from the X-chromosome it contains dosage compensation signals enhancing gene expression in male flies but not in females (Birchler et al. 2006). This might affect gene expression of other neighboring native genes leading to uncontrolled physiological and phenotypic effects. Therefore, in a second series of experiments, the chimeric *Nap1* knockin flies (*Nap1*^{KO1}) (Lankenau et al. 2003) were used to produce a true single-copy knockout mutant (D.-H. Lankenau, unpublished; G. E. Stephens, H. Xiao, D.-H. Lankenau, C. Wu, and S. C. R. Elgin, unpublished). Rong and Golic's vector fragment of the duplicated knockout *Nap1*^{KO1} mutant (Lankenau et al. 2003) contains an *I-CreI* meganuclease[21] cutting site (star in Fig. 3C). By combining the *Nap1*^{KO1} chromosome with a chromosome containing a heat inducible *I-CreI* meganuclease transgene a DSB was induced upon heat treatment of flies between the two gene copies of *Nap1*^{KO1} (Fig. 5A). This DSB takes advantage of the cell's DNA repair machinery to subsequently mend the break by SSA repair. The *Nap1*^{KO1} mutant contained eight restriction sites (Fig. 3C and 4D) which were used as markers to study the recombination tracts of the result-

[21] *I-CreI* meganuclease is a chloroplast-encoded mobile intron homing endonuclease derived from the ciliate *Chlamydomonas reinhardii*.

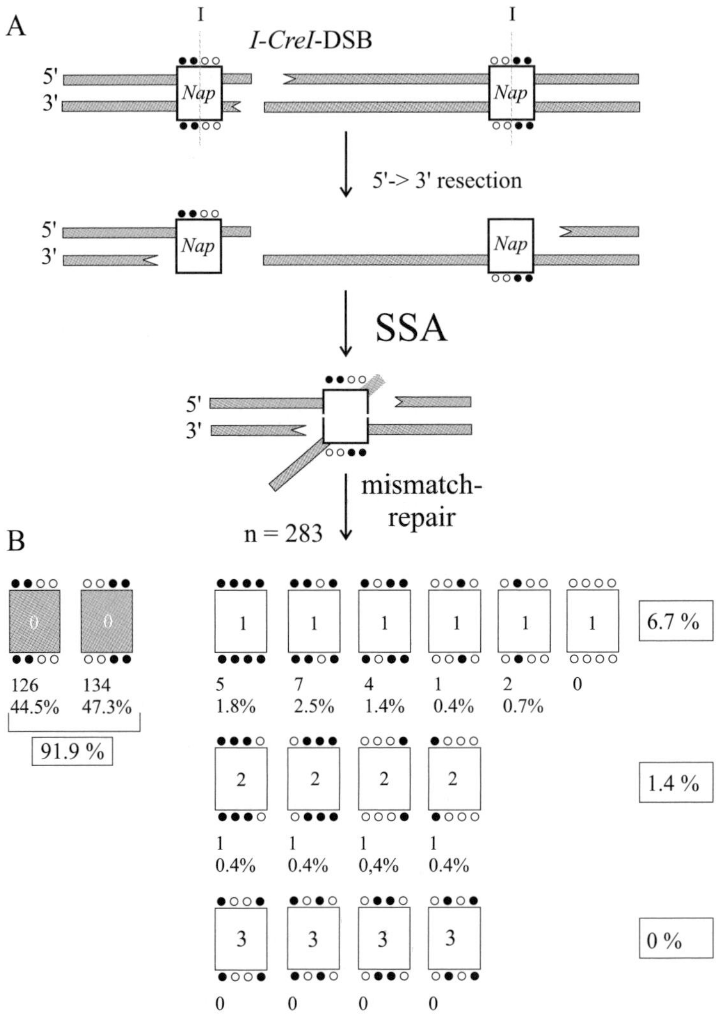

ing single copy knockout mutants termed *Nap1*^{SSA} (Fig. 5A). The production of *Nap1*^{SSA} knockout flies was highly efficient with about 30% of the F1 flies exhibiting loss of the *white*-marker phenotype. 283 *Nap1*^{SSA} flies each representing an independent germline SSA event were recovered and analyzed by PCR (Ming-Fei Lang, unpublished) (Fig. 5B). An important question was how the SSA repair machinery in germline cells handled the eight recombination tract markers. In theory the SSA reaction produces an annealing structure with single stranded flap overhangs (Fig. 6A) which will subsequently be removed by *Fen1* flap-endonuclease. Because each of the four markers on each of the chimeric *Nap1*^{KO1} duplicates will misbasepair with its counterpart to form a heteroduplex, the mispaired nucleotides will be recognized by the

◄ **Fig. 6** Recombination tract marker distribution in *Nap1*^{SSA} knockout flies. **A** The SSA reaction at the knockin allele *Nap1*^{KO1} (compare Figs. 3C, 4D and 5) was triggered after a DSB was induced between the chimeric tandem copy of *Nap1* using *I-CreI* endonuclease. The initial tract markers of each chimeric copy of *Nap1*^{KO1} is shown as *filled* and *outlined circles* for each DNA strand. After 5′ to 3′ strand resection the remaining two complementary strand of *Nap1* are thought to anneal. The SSA intermediate structure is a heteroduplex whose mispaired bases are thought to be repaired by mismatch repair enzymes. Independent germline events were analyzed by PCR and subsequent diagnostic restriction fragment analysis as shown in Fig. 5. **B** Recombination tracts of 283 independent SSA and mismatch repair events produced 91.9% of tract marker combinations either representing the left or right chimeric copy of *Nap1*^{KO1} where the left copy was found in 126 flies (44.5%) and the right copy occurred in 134 knockout flies (47.3%). Speculating that each of these frequent events is due to a mismatch repair complex which chooses one strand as a blueprint for repair of the complementary strand the less frequent tract marker distributions can be explained by independent docking steps (*boxed nos.* 1, 2, 3) of independent mismatch repair protein complexes. With increasing numbers of docking steps the frequency of tract marker combinations becomes increasingly infrequent. 6.7% of the tract marker combinations suggest a single template switch for the repair enzymes, 1.4% can be explained by two switches, and tract marker combinations requiring 3 docking steps were not found at all

germline mismatch repair machinery. If there is no strand bias for mismatch repair and one assumes random choice of choosing one strand as a template for repair at any position, there are 2^n theoretical combinations, where n represents the number of tract marker positions and 2 is the number of tandem-duplicates. Thus, the four tract markers of each chimeric *Nap1*^{KO1} duplicate should therefore end up in 16 different tract marker patterns if template choice during repair is random. Interestingly, the results were far from random. 92% (n = 260) of the *Nap1*^{SSA} knockout flies represented either the marker combination of the left or of the right copy of the *Nap1*^{KO1} duplicate. However, 23 (8.1%) alternative combinations of tract markers were also found. Figure 6B shows the results. The marker patterns of all independent *Nap1*^{SSA} mutants are grouped indicating their decreasing probability of occurrence. This dominant abundance of either the left chimeric *Nap1*^{KO1} copy (44.5%) or the right *Nap1*^{KO1} copy (47.3%) suggests that once a template strand has been chosen by mismatch repair enzymes this template will not be switched for a length of 1469 nucleotides which is the distance between the outer markers (Lankenau et al. 2003). The number within the boxes of Fig. 6B indicate possible changes of template strands by the mismatch repair proteins. The interpretation that the number of switched templates during a single SSA event reflects the real quantity of tract marker patterns fits well with the congruency of decreasing numbers of tract patterns (91.9%, 6.7%, 1.4%, 0%) versus a potential increase of template switches (0, 1, 2, 3). Four tract patterns could be explained only if three times a chosen template was switched. No such pattern was found in the analyzed sample size (n = 283). Four additional tract patterns could only be explained by two template events

(1.4%) and six patterns by only one switch of the template (6.7%). As the overall frequencies of scorable recombination events is very high this targeting system promises to be a suitable system in order to study SSA repair as well as mismatch repair simultaneously. The other good news from this data is that the knockout targeting products anticipated to alter gene function are produced with high efficiency, thus avoiding screening of large numbers.

6 Relative Usage of DSB Repair Pathways in the Germline

A genetic mutation fixed[22] in a population of any species originates as a rule in the DNA of germ cell chromosomes rather than in the genomes of somatic cells. Muller's experiment (Fig. 1A) established the gene as a physical entity through the correlation between the quantity of germline-induced lethal mutations and the amount of irradiation (see above, Sect. 3.2). The germline therefore is of particular relevance and represents a key cell lineage within the otherwise somatic gonads to focus analysis of DNA repair. An effort to address the specific role of DNA repair in mitotic germ cells in contrast to meiotic germ cells is experimentally not trivial, but first research steps have been initiated. First, for somatic mammalian cells only a small number of DNA repair genes (i.e. ~ 8 out of ~ 130 genes) were known to be inducible in mammalian cells upon their exposure to DNA-damaging agents (see this volume; Christmann et al., 2006). Induction of somatic mouse DNA repair genes[23] was monitored by Christmann et al. 6 hours after exposure of mouse fibroblast cells to UV-C radiation by isolation of total RNA and subsequent microarray analysis. The analysis revealed 24 genes with increased RNA quantities and in some cases the transcription factors p53 and c-Fos could be related to a changed transcript abundance (Christmann et al. 2006). This kind of study has not yet been carried out with regard to the soma or the germline of *Drosophila*. Thus, a similar experiment was initiated with *Drosophila* testes and ovaries (S. & D.-H. Lankenau, unpublished). After removal of accessory glands, roughly half of a dissected *Drosophila* gonad represents germline cells while the other half is somatic tissue (Fuller 1993; Hennig 1985). *Drosophila* spermatogenesis does not involve meiotic recombination preceding meiotic segregation, thus any expression data of DNA-repair genes would not interfere with meiotic expression of recombination genes. This promised a quite accurate comparative estimation of the relative transcript quantities in the germline versus the soma. Using Northern Blot analysis, poly A^+ RNA isolated from *Drosophila* testes and carcasses[24] was probed with

[22] but also **non-fixed mutations**, i.e. mutations only present in some individuals of a species (Dettai A, Volff JN 2006).

[23] In the presence and in the absence of p53 and c-Fos transcription factors.

[24] **carcass** is the total somatic tissue from flies whose gonads have been removed

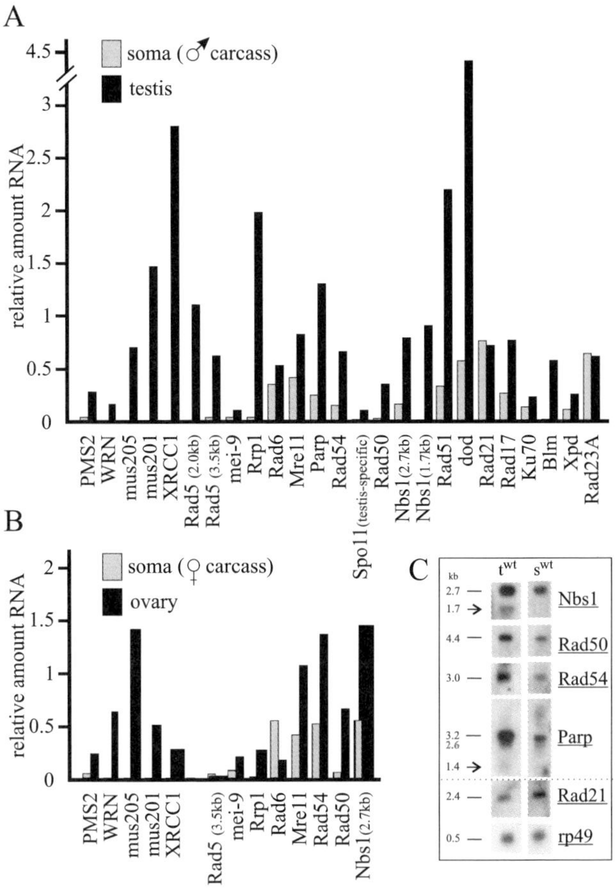

Fig. 7 Expression of polyA$^+$ transcripts of DNA repair genes in the *Drosophila* gonads vs. somatic tissues. **A** Expression of DSB repair-relevant genes from *Drosophila* testes (*filled bars*) and male carcasses (i.e. remaining somatic tissue) (*gray bars*). **B** Soma vs. germline expression profiles of *Drosophila* ovaries of a selection of DNA repair genes. **C** Examples of Northern filters hybridized to EST-derived dsDNA probes. rp49 and Rad21 were used as quantitative standards for equal loading. Quantitation was performed using a scanner and a commercial version of the Promega GeneStar software. For further data on the Northern hybridization experiments compare Fig. 10 in the Appendix, including expression profiles after γ-ray irradiation and a short recovery time (compare also Lankenau et al. 1999)

radioactively labeled EST[25]-derived double-stranded probes representing 23 different DNA-repair genes. The results are shown in Fig. 7. Interestingly, significantly elevated transcript levels were found in both, testes (Fig. 7A) and

[25] **EST** = enhanced sequence tags are small cDNA sequences derived from poly A$^+$ RNA preparations from specific tissues or organs. The applied ESTs were oligo-dT primed *Drosophila* cDNA clones from ResGen Invitrogen Incorp, Huntville, AL, USA.

ovaries (Fig. 7B) when compared to carcasses[23], while nearly no differences were found between γ-ray irradiated (5Gy) and non-irradiated gonads and carcasses (see Fig. 10). The same dominant abundance of transcripts, however, can be found for other housekeeping genes as well. Therefore, it comes as no surprise that also ovaries contained significantly elevated quantities of DNA-repair transcripts (S. & D.-H. Lankenau, unpublished). Future research should address, in analogy to what had been demonstrated for mouse irradiated fibroblast cells (Christmann et al. 2006), if particular transcription factors are correlated to irradiation-induced DNA-repair genes in the germline. To this end, DNA-repair proteins in germline cells need to be related to particular phases of the cell cycle, and the relative usage of DNA repair pathways at different developmental stages of germline development needs to be addressed. The *Drosophila* germline begins at early embryogenesis with the formation of the pole cells (Spradling 1993). Later during embryogenesis the dividing germ cells migrate through the midgut epithelium to form the primordial gonads, and associate with somatic cells within a germline stem cell niche at the apex of the gonad (Kai and Spradling 2004; Kai et al. 2005; Kiger et al. 2000, 2001; Ohlstein et al. 2004; Spradling et al. 2001; Xie and Spradling 2000). This differentiated behavior of germline cell lineages during development underscores the fact that each stage of germ cell differentiation might exhibit its own characteristic usage of different DNA repair pathways. And this indeed, most recently, has been found in a new experimental "*genetics of genetics*"-system: William R. Engels and colleagues set up a novel in vivo biochemistry system able to measure the differential usage of alternative pathways of DSB repair in the *Drosophila* germline (Preston et al. 2006). The group used a DNA-repair reporter construct termed Rr3 in which an *I-SceI* meganuclease-mediated DSB was produced at a specific location such that expression of the red fluorescent protein, DsRed, in the next generation could be used to measure the relative frequency at which either NHEJ, SSA, SDSA or DHJ repair occurred. Figure 8 shows the structure of the intelligent Rr3 reporter vector. In contrast to just vaguely determining the relative usage of competing DSB repair pathways at an undefined phase of the cell cycle or ill-defined stage of development, the study addressed quite precisely the major question if DSB repair pathways maintain their relative usage at different stages of germline development. To this end, the control of gene expression of the DSB-inducing meganuclease *I-SceI* was put under different promoters and allowed to address relative differences of DSB repair pathway usage at different stages of germ cell differentiation[26]. The β2-tubulin promoter is active specifically only in spermatogonia during the meiotic prophase (e.g. Lankenau et al. 1994) while the employed ubiquitin promoter triggered *I-SceI* production from zygote formation onward. Further, when *I-SceI* protein was only expressed in the maternal germline, it was deposited maternally in the

[26] Note: not usage within the cell-cycle.

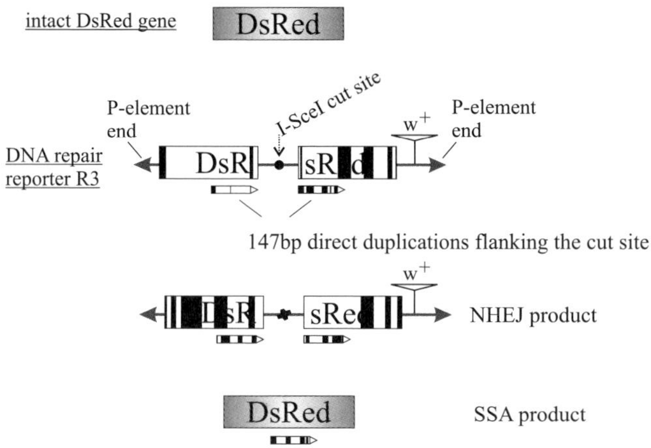

Fig. 8 The intelligent DNA repair reporter system developed by William Engels' lab (Preston et al. 2006). The Rr3 reporter construct contains a fluorescent protein gene DsRed2 (BD Biosciences/Clontech, Palo Alto, CA). DsRed is interrupted by the recognition sequence of the rare-cutting *I-SceI* meganuclease (*filled circle*). The *I-SceI* site is flanked by non-functional but partially overlapping DsRed gene fragments. The internal DsRed sequences form a 147 bp duplication of part of the DsRed sequence which can be part of a SSA reaction. The whole construct is carried within a P transposable element containing the standard *white* phenotypic marker gene. Reporters can be applied on homologous chromosomes at allelic positions but carrying intact or non-intact *I-SceI* cutting sites. When experiments are carried out in the male germline, no meiotic cross over will take place. Upon induction of a DSB at the *I-SceI* cutting site, repair occurs in alternative ways: SSA removes the cut site and results in a functional DsRed gene easily detected by fluorescence microscopy. NHEJ: does not lead to a functional DsRed gene but usually destroys the *I-SceI* cut site (*filled star*). This feature can be used as a discriminating marker for NHEJ. SDSA: removes the cutsite. DSB repair employing DHJs leads to recombination in the male germ line. The P element vector containing the intelligent reporter can be transformed into the germline and transposed into various positions in all *Drosophila* chromosomes, and different chromatin environments, opening a new linkage between epigenetics and in vivo DNA repair. The work by Engels demonstrated that *I-SceI* can also be driven by various promoters active at different developmental stages of germline development

oocyte to become increasingly reduced in quantity in post-zygotic development. This expression pattern of *I-SceI* produced DSBs in the early germline but presumably not in later stages. The results revealed that SSA dominated early in germ-cell development being quite infrequent just before meiosis. On the other hand, NHEJ became progressively more prevalent as germ-line cell development proceeded. BIR, DHJ-DSB repair and SDSA, jointly summarized under the term conversion repair, occurred only rarely in the early and late stages of germ-line development, but accounted for a substantial portion of the events during intermediate stages (Preston et al. 2006). While differences of the prevalence of SSA vs. NHEJ were confirmed also between mouse oocytes and preimplantation embryos (Fiorenza et al. 2001) there are many reasons which could account for these relative differences in DSB repair pathway usage.

Numerous questions now seem to be addressable, such as how germ line cells are selected during development, or does apoptosis remove cells with too much involvement of error-prone pathways? Do parthenogenetic species not engaging in sexual reproduction trim their germline DNA repair pathways to less error prone pathways? Are apoptotic germline cells more abundant in these species? A first step in addressing these questions has been made, and therefore again, this *Drosophila* in vivo system together with a still-to-be-developed complementing in vitro system has opened a new gateway for further research promising to link replicative DSB repair with developmental biology and other fields of importance.

7 Conclusions

The structure and organization of chromosomes and entire genomes, their dynamics and stability over evolutionary time scales is influenced and determined by multiple mechanisms. *Drosophila* genetics contributed significantly to our present understanding of the nature of the gene and how chromosomes work. No matter how difficult it may be to decipher the causes of long-range sequence patterns in chromosomes of a particular recent species, some molecular key mechanisms have become apparent which appear to explain most of the phenomenological characteristics of modern genome and chromosome organizations. Three main, and interlinked key processes appear to stand out: (1) DNA-repair with biochemical reactions keeping the chemical integrity of DNA polynucleotide chains intact thereby protecting genetic information. (2) RNA and DNA replication as fundamental units defining life are key motors structuring genomes at a large scale; and (3) the activity of transposable elements where class I elements transpose via RNA intermediates and class II elements transpose directly from DNA to DNA, the latter producing DSBs and making DSB repair a requirement for its own reason.

Especially, class I transposable elements such as HIV, *gypsy*, *micropia* or *Ty1* (with important contributions from *Drosophila*; Arkhipova I.R. et al. 1995) remind us of hypothetical early replicators on the primordial earth representing living molecular fossils from an RNA world whose present-day strategies of replication may have witnessed the transition to the DNA world not long after the beginnings of life. With respect to replicative DSB repair, in particular SDSA, where the broken ends independently undergo homology search, invasion and subsequent strand displacement it is interesting to note that the replication of retroviral transposons includes two essential steps of DNA strand transfers which restore the full redundant information of the two Long Terminal Repeats (LTRs) (Coffin et al. 1997). The involved mechanics occur universally in retroviruses or pararetroviruses (Varmus 1983), and it resembles intriguingly the independent template searches and template switches of each of the two ends of a DSB in SDSA, SSA and BIR although here megabases of sequence search are involved while retroviral strand transfers occur within ranges of kilobases (Fig. 2C) (Engels et al. 1994; Nassif et al. 1994; Paques and Haber 1999; Paques et al. 1998). Whether specialized proteins are universally involved in the process remains to be detailed. The lesson to be learned from this is that DSB repair does not necessarily require simultaneous involvement of both ends of a DSB triggering strand invasion restricted to identical template DNA-loci as depicted by the classical DSB repair model (Szostak et al. 1983). BIR as well as SDSA leave open the possibility of simultaneous vs. consecutive and independent strand invasions (Fig. 2C). Further, often DNA DSB repair has been classified into competing mechanistic pathways such as the "DSB-", "SSA-", "SDSA-", "BIR-" or "NHEJ-"repair pathways (Lankenau and Gloor 1998). However, there seem to be more subtle interconnections between these so-called pathways where differences mainly depend on the cell cycle phase-specific protein composition and the developmental stage of a particular cell as well as on real topological constraints. With regard to common and discriminating mechanical properties NHEJ, SDSA, BIR and SSA can now be depicted in a balalaika diagram (Fig. 9). SDSA appears to include most aspects of all DSB repair pathways (invasion, synthesis, annealing) except the resolution of HJs which may or may not play a role in BIR. In general however, it is the protein activities which decide as to which degree the ends of a DSB are processed and to which degree they get involved in search, invasion, annealing and synthesis activities while there does not appear to be a single rigid path towards a particular repair end-product. Chromatin (Swaminathan et al. 2005) and epigenetics[27], the cell cycle and germline specific functions seem to play important roles as well, while the invention of meiotic recombination transiently appears to be a separate major transition in the evolution of sex. The long history of *Drosophila* genetics together with microbial and vertebrate model systems has helped to arrive at a modern, more balanced view of DSB repair reactions promising to involve in the future other DNA repair aspects such as the here-described mismatch repair system. *Drosophila* also fulfills all properties of a key system for aspects of developmental biology especially with a focus on germline and stem cell development. A good starting platform to join DSB repair with germline development appear to be the most conserved and universal proteins with their related activities such as RecA/Rad51, Fen1, and the MRN complex (Aravind et al. 1999). Further, particular protein domains of the RecQ family (i.e. Bloom syndrome and Werner syndrome proteins) appear to be a link between the MRN complex and the DNA-PKcs proteinkinase to ancient proteins

[27] http://www.epigenome-noe.net/

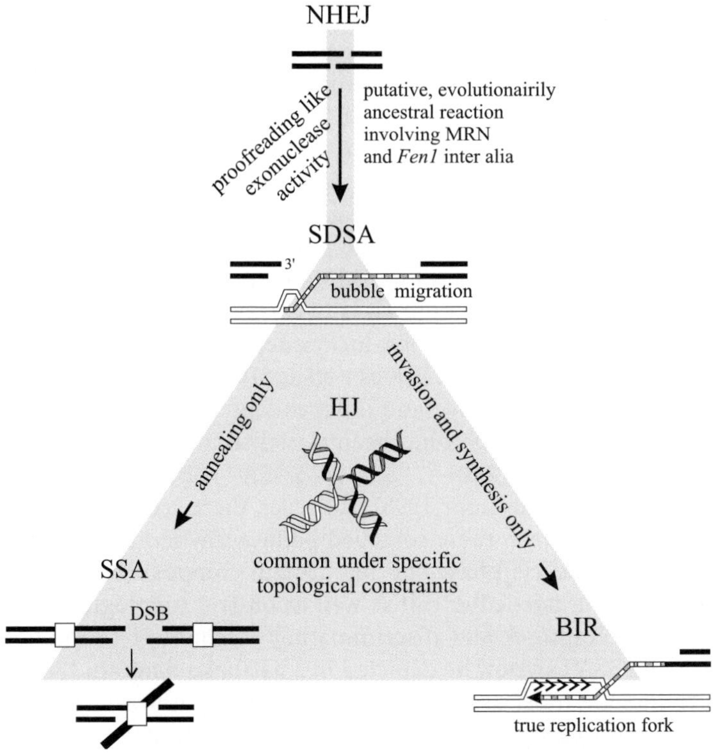

Fig. 9 Balalaika diagram linking the structural mechanisms of the major DSB repair pathways. Upon occurrence of a DSB NHEJ is the most straight forward repair process, and a minimum requirement to reinstall the continuity of chromosomal DNA. It is frequently termed "illegitimate recombination" as merely microhomologies give support to bringing the strands of the two ends together. This reaction is error prone altering original sequence information. Exonucleolytic activities initiate the complex processes of replicative recombination reactions. SDSA involves all of the major reactions, i.e. strand invasion, DNA synthesis and annealing. BIR includes the invasion and DNA replication steps only, while SSA encompasses only the reannealing of the processed ends. All other activities, such as removal of DNA flap structures through *Fen1* or the resolution of Holliday junctions are more specific activities which may or may not take place depending on the particular structural and topological constraints

such as RNAseD and the RNaseP ribozyme linking the RNA world with the DNA world. *Drosophila* will further continue to play a prime joint link between evolutionary biology and the many fields and subjects of scientific and biomedical interests.

Especially, gene targeting using designer zinc-finger endonucleases that trigger synthesis-dependent DSB repair at high frequencies at specific chromosomal loci promise to facilitate this development significantly (Porteus and Carroll 2005).

Note Added in Proof

While this manuscript was in press, two forthcoming publications report on an important break through of the described intelligent DNA repair reporter system and the relevance of the Drosophia germline stem cell niche: Stephen DiNardo and colleagues found that gonial stem cells are lost from the apical testicular gonial stem cell niche during aging correlated to the reduction of somatic hub cells. However, gonial stem cells can be regenerated by dedifferentiation of primary spermatogonia (Wallenfang et al. 2006). William Engels and colleagues now found that the relative usage of germline DSB repair pathways (Fig. 9) is dependent on the age of the organism. Homologous repair increased linearly from less than 14% in young individuals to more than 60% in old ones. Responsible for this may be the different age dependent proportion of early gonial stem cells versus regenerated gonial stem cells each employing a different set and/or quantity of pathway-specific DNA repair proteins (Preston et al. 2006).

Acknowledgements I thank Sabine Schreck and Susanne Lankenau for critically reading the manuscript; Susanne Lankenau, Ming-Fei Lang, Tobias Jursch, and Zhu Jiayun for help with the construction of the *Nap-1* knockout mutants and recombination tract analysis; the Institute of Zoology at Heidelberg for providing lab. space; Stephan Frings for help with software; the German Cancer Research Center (DKFZ) for consent to use their γ-radiation device and for the supply of fly-food; Frank Lyko for material help with the dDnmt2 experiment; the University of Heidelberg for providing access to the library and to the electronic journals. The work was supported in part by a joint grant of the Max-Planck Society, the State of Baden-Württemberg and the Chinese Academy of Sciences at Shanghai to Ming-Fei Lang and D.-H.L. I would also like to thank the Genetics Society of America for providing a free membership for 2006.

Appendix

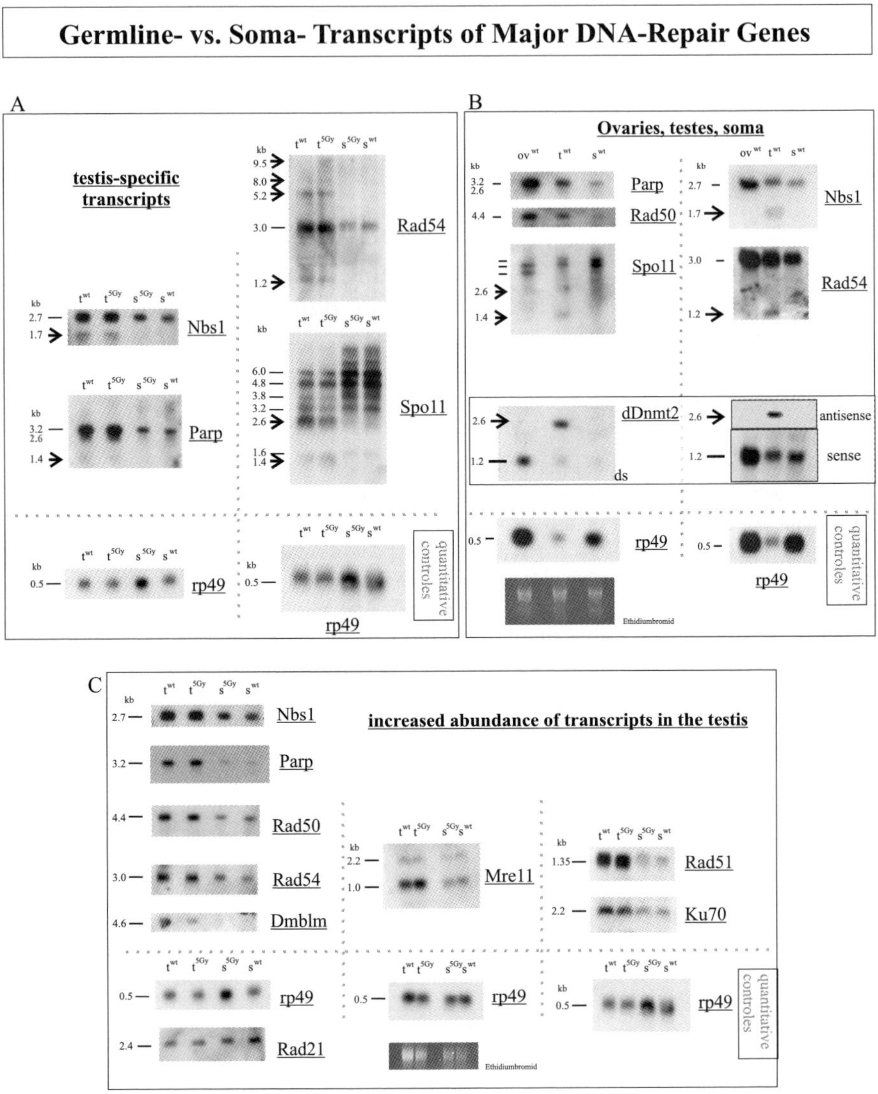
Germline- vs. Soma- Transcripts of Major DNA-Repair Genes
A
testis-specific transcripts
Nbs1
Parp
Rad54
Spo11
rp49
quantitative controles
B
Ovaries, testes, soma
Parp
Rad50
Spo11
Nbs1
Rad54
dDnmt2
antisense
sense
ds
rp49
Ethidiumbromid
C
increased abundance of transcripts in the testis
Nbs1
Parp
Rad50
Rad54
Dmblm
Mre11
Rad51
Ku70
rp49
Rad21
Ethidiumbromid
quantitative controles

◀ **Fig. 10** RNA expression of DNA-repair genes in the germline and soma of *Drosophila melanogaster*. Northern hybridization analysis using hybridization of ESt-derived double-stranded ^{32}P-labeled probes of 23 different DNA repair genes to poly A^+ RNA of dissected testes/ovaries vs. carcass (i.e. somatic tissues after gonad extirpation). Northern blots show significantly elevated transcript levels in germline tissues. Panels **A** and **B** show testis-specific transcripts expressed by some genes (Nbs1, PARP, Rad54, Spo11; *bold arrows*). **B** Hybridization of single-stranded RNA probes of the dDnmt2 gene revealed a testis specifically expressed antisense transcript (S. Lankenau and F. Lyko, unpublished). **C** Increased abundances of transcripts in *Drosophila* testes. Irradiation (5 Gy) was performed as published (Lankenau S, Bürkle A, Lankenau D-H 1999) with the difference that here living flies were bulk irradiated at 5 Gy. Subsequently, with a recovery time not exceeding 2h, gonads and carcasses were dissected and tissues frozen as published (Lankenau S, Corces VG, Lankenau DH 1994). In contrast to extended recovery times as published by Christmann in this volume (Christmann M, Fritz G, Kaina B 2006) no significant difference in expression levels were found. Future work will have to focus on increasing irradiation doses (flies can survive high doses) and varying recovery times

References

Adams MD, Celniker SE, Holt RA et al. (2000) The genome sequence of Drosophila melanogaster. Science 287:2185–2195

Adams MD, McVey M, Sekelsky JJ (2003) Drosophila BLM in double-strand break repair by synthesis-dependent strand annealing. Science 299:265–267

Aguilera A (2001) Double-strand break repair: are Rad51/RecA–DNA joints barriers to DNA replication? Trends Genet 17:318–321

Ahmad K, Golic KG (1998) The transmission of fragmented chromosomes in Drosophila melanogaster. Genetics 148:775–792

Allis CD, Underwood EM, Caulton JH, Mahowald AP (1979) Pole cells of Drosophila melanogaster in culture. Normal metabolism, ultrastructure, and functional capabilities. Dev Biol 69:451–465

Altenburg E (1928) The Limits of Radiation Frequency Effective in Producing Mutations. Am Nat 62:590–596

Aravind L, Walker DR, Koonin EV (1999) Conserved domains in DNA repair proteins and evolution of repair systems. Nucleic Acids Res 27:1223–1242

Arkhipova IR, Lyubomirskaya NV (1995) Drosophila Retrotransposons. Springer, Berlin Heidelberg New York

Arnold E, Jacobo-Molina A, Nanni RG, Williams RL, Lu X, Ding J, Clark AD, Zhang A, Ferris AL, Clark P, Hizi A, Hughes SH (1992) Structure of HIV-1 reverse transcriptase/DNA complex at 7A resolution showing active site locations. Nature 357:85–89

Barricelli NA (1956) A "chomosomic" recombination theory for multiplicity reactivation in bacteriophage. Acta Biotheor 11:107–120

Bartel DP, Szostak JW (1993) Isolation of new ribozymes from a large pool of random sequences. Science 261:1411–1418

Beall EL, Rio DC (1997) Drosophila P-element transposase is a novel site-specific endonuclease. Genes Dev 11:2137–2151

Bejerano-Sagie M, Oppenheimer-Shaanan Y, Berlatzky I, Rouvinski A, Meyerovich M, Ben-Yehuda S (2006) A checkpoint protein that scans the chromosome for damage at the start of sporulation in Bacillus subtilis. Cell 125:679–690

Biessmann H, Mason JM, Ferry K, d'Hulst M, Valgeirsdottir K, Traverse KL, Pardue M-L (1992a) Addition of Telomere-associated HeT DNA sequences "heals" broken chromosome ends in Drosophila. Cell 61:663–673
Biessmann H, Valgeirsdottir K, Lofsky A, Chin C, Ginther B, Levis RW, Pardue M-L (1992b) HeT-A, a transposable element specifically involved in "healing" broken chromosome ends. Mol Cell Biol 12(9):3910–3918
Birchler JA, Fernandez HR, Kavi HH (2006) Commonalities in compensation. Bioessays 28:565–568
Bosco G, Haber JE (1998) Chromosome break-induced DNA replication leads to nonreciprocal translocations and telomere capture. Genetics 150:1037–1047
Bottjer DJ (2005) The early evolution of animals. Sci Am 293:42–47
Boyd JB, Golino MD, Shaw KES, Osgood CJ, Green MM (1981) Third-chromosome mutagen sensitive mutants of Drosophila melanogaster. Genetics Genetics:607–623
Boyd JB, Mason JM, Yamamoto A, Brodberg RK, Banga SS, Sakaguchi K (1987) A genetic and molecular analysis of DNA repair in Drosophila. J Cell Sci Suppl 6:39–60
Bridges CB (1916) Non-disjunction as proof of the chromosome theory of heredity. Genetics 1:1–52
Briggs DEG, Erwin DH, Collier FJ (1994) The fossils of the Burgess Shale. Smithsonian Institution Press, Washington, DC
Brodsky MH, Weinert BT, Tsang G, Rong YS, McGinnis NM, Golic KG, Rio DC, Rubin GM (2004) Drosophila melanogaster MNK/Chk2 and p53 regulate multiple DNA repair and apoptotic pathways following DNA damage. Mol Cell Biol 24:1219–1231
Brosius J (2005) Disparity, adaptation, exaptation, bookkeeping, and the contingency at the genome level. Paleobiology 31:1–16
Capecchi MR (1989a) Altering the genome by homologous recombination. Science 244:1288–1292
Capecchi MR (1989b) The new mouse genetics: altering the genome by gene targeting. Trends Genet 5:70–76
Chen JY, Bottjer DJ, Oliveri P, Dornbos SQ, Gao F, Ruffins S, Chi H, Li CW, Davidson EH (2004) Small bilaterian fossils from 40 to 55 million years before the Cambrian. Science 305:218–222
Christmann M, Fritz G, Kaina B (2006) Induction of DNA Repair Genes in Mammalian Cells in Response to Genotoxic Stress. In: Lankenau DH (ed) Genome Integrity: Facets and Perspectives, vol 1. Springer, Berlin Heidelberg New York, p 16
Ciapponi L, Cenci G, Ducau J, Flores C, Johnson-Schlitz D, Gorski MM, Engels WR, Gatti M (2004) The Drosophila Mre11/Rad50 complex is required to prevent both telomeric fusion and chromosome breakage. Curr Biol 14:1360–1366
Coffin JM, Hughes SH, Varmus HE (1997) Retroviruses. Cold Spring Harb. Lab. Press, Cold Spring Harbor
Conway Morris S (1994) Why molecular biology needs palaeontology. Development 1994 (Supplement):1–13
Conway Morris S (1998) The crucible of creation: the Burgess Shale and the rise of animals. Oxford University Press, Oxford
Conway Morris S (2000) The Cambrian "explosion": slow-fuse or megatonnage? Proc Natl Acad Sci USA 97:4426–4429
Creighton HB, McClintock B (1931) A correlation of cytological and genetical crossing-over in Zea mays. Proc Natl Acad Sci USA 17:492–497
Dawkins R (1976) The selfish gene. Oxford University Press, Oxford

Dettai A, Volff JN (2006) Morphological Characters from the Genome: SINE Insertion Polymorphism and Phylogenies. In: Lankenau D-H (ed) Transposable Elements and Genome Dynamics, vol 2. Springer, Berlin Heidelberg New York
Dobzhansky TG (1937) Genetics and the Origin of Species. Columbia University Press, New York
Dognon A (1925) Coefficient de temperature de láction des rayons X sur lóeuf d'*Ascaris*. Compt Rend Soc Biol Par 92:1389–1392
Dover G, Brown S, Coen E, Dallas J, Strachan T, Trick M (1982) The Dynamics of Genome Evolution and Species Differentiation. In: Dover GA, Flavell RB (eds) Genome Evolution. Academic Press Inc., New York, p 343–372
Echalier G (1997) Drosophila Cells in Culture. Academic Press, San Diego
Egel R (2006) Chromosomal DNA Replication: on Replicases, Replisomes, and Bidirectional Replication Factories. In: Lankenau D-H (ed) Genome Integrity: Facets and Perspectives, vol 1. Springer, Berlin Heidelberg New York, p 1–24
Eigen M (1992) Steps Towards Life: a Perspective on Evolution. Oxford University Press, Oxford
Eigen M, Schuster P (1978) The Hypercycle/A Principle of Natural Self-Organisation. part C: The realistic Hypercycle. Naturwissenschaften 65:341–369
Engels WR (1989) P elements in Drosophila melanogaster. In: Berg DE, Howe MM (eds) Mobile DNA. American Society of Microbiology, Washingon, DC, p 437–484
Engels WR (2000) Reversal of fortune for Drosophila geneticists? Science 288:1973–1975
Engels WR, Johnson-Schlitz DM, Eggleston WB, Sved J (1990) High-Frequency P element Loss in Drosophila is homolog dependent. Cell 62:515–525
Engels WR, Preston CR, Johnson-Schlitz DM (1994) Long-range cis preference in DNA homology search over the length of a Drosophila chromosome. Science 263:1623–1625
Ferguson DO, Holloman WK (1996) Recombinational repair of gaps in DNA is asymmetric in Ustilago maydis and can be explained by a migrating D-loop model. Proc Natl Acad Sci USA 93:5419–5424
Finnegan DJ (1989) Eukaryotic transposable elements and genome evolution. TIG 5(4):103–107
Fiorenza MT, Bevilacqua A, Bevilacqua S, Mangia F (2001) Growing dictyate oocytes, but not early preimplantation embryos, of the mouse display high levels of DNA homologous recombination by single-strand annealing and lack DNA non-homologous end joining. Dev Biol 233:214–224
Fischer EP (1988) Thinking About Science: Max Delbrück and the Origins of Molecular Biology. WW Norton & Co Inc, New York
Fisher RA (1930) The Genetical Theory of Natural Selection. Oxford University Press, Oxford
Foe VE, Odell GM, Edgar BA (1993) Mitosis and Morphogenesis in the Drosophila Embryo: Point and Counterpoint. Cold Spring Harbor Laboratory Press, New York
Formosa T, Alberts BM (1986) DNA synthesis dependent on genetic recombination: characterization of a reaction catalyzed by purified bacteriophage T4 proteins. Cell 47:793–806
Friedberg EC (1997) Correcting the Blueprint of Life. Cold Spring Harbor Laboratory Press, Cold Spring Harbor, New York
Friedberg EC, Walker GC, Siede W (1995) DNA Repair and Mutagenesis. ASM Press, Washington, DC

Friedl A (2005) The Role of Chromatin Structure and Nuclear Architecture in the Cellular Response to DNA Double-Strand Breaks. In: Lankenau DH (ed) Genome Integrity: Facets and Perspectives, vol 1. Springer, Berlin Heidelberg New York, p 17

Fuller M (1993) Spermatogenesis. In: Martinez-Ariaz A, Bate M (eds) The Development of Drosophila melanogaster, vol 1. Cold Spring Harbor Press, Cold Spring Harbor, New York, p 71–147

Gager CS, Blakeslee AF (1927) Chromosome and Gene Mutations in *Datura* Following Exposure to Radium Rays. Proc Natl Acad Sci USA 13:75–79

Gaillard PH, Roche D, Almouzni G (1999) Nucleotide excision repair coupled to chromatin assembly. Methods Mol Biol 119:231–243

Gatz SA, Wiesmüller L (2005) Role of p53 in Double-Strand Break Repair. In: Lankenau DH (ed) Genome Integrity: Facets and Perspectives, vol 1. Springer, Berlin Heidelberg New York, p 24

Gloor GB, Lankenau D-H (1998) Gene conversion in mitotically dividing cells: a view from Drosophila. Trends Genet 14:43–46

Gloor GB, Nassif NA, Johnson-Schlitz DM, Preston CR, Engels WR (1991) Targeted Gene Replacement in Drosophila via P Element-Induced Gap Repair. Science 253:1110–1117

Goodspeed TH, Olson AR (1928) The production of variation in *Nicotiana* species by X-ray treatment of sex cells. Proc Natl Acad Sci USA 14:66–69

Gould SJ (1989) Wonderful life: the Burgess Shale and the nature of history. W.W. Norton, New York

Gould SJ (2002) The structure of evolutionary theory. Belknap Press of Harvard University Press, Cambridge, Mass

Haber JE (1995) In vivo biochemistry: physical monitoring of recombination induced by site-specific endonucleases. Bioessays 17:609–620

Haldane JBS (1932) The Causes of Evolution. Longmans Green, London

Harrison JWH, Garrett FC (1926) The Induction of Melanism in the *Lepidoptera* and its Subsequent Inheritance. Proc Royal Soc 99B:241–263

Hazen RM (2005) Genesis: The Scientific Quest for Life's Origins. Joseph Henry Press, Washington, DC

Hennig W (1985) Y chromosome function and spermatogenesis in Drosophila hydei. Adv Gen 23:179–234

Hiraizumi Y (1971) Spontaneous recombination in Drosophila melanogaster males. Proc Natl Acad Sci USA 68:268–270

Hiraizumi Y, Crow JF (1960) Heterozygous efects on variability, fertility, rate of development, and longevity of Drosophila chromosomes that are lethal when homozygous. Genetics 45:1071–1083

Holliday R (1964) A mechanism for gene conversion in fungi. Genet Res 5:282–304

Holliday R (1968) Genetic Recombination in Fungi. In: Peacock WJ, Brook R (eds) Replication and Recombination of Genetic Material. Australian Academy of Science, Canberra, p 157–174

Holthusen H (1920) Beiträge zur Biologie der Strahlenwirkung. Arch Ges Physiol 187:1–24

Hopfner K-P (2005) Structure and Function of Rad50/SMC Protein Complexes in Chromosome Biology. In: Lankenau D-H (ed) Genome Integrity: Facets and Perspectives, vol 1. Springer, Berlin Heidelberg New York, p 18

Hou X (2004) The Cambrian Fossils of Chengjiang, China: the Flowering of Early Animal Life. Blackwell, Malden, MA

Huxley JS (1942) Evolution: the Modern Synthesis. Allen and Unwin, London

Ivics Z, Kaufman CD, Zayed H, Miskey C, Walisko O, Izsvak Z (2004) The Sleeping Beauty transposable element: evolution, regulation and genetic applications. Curr Issues Mol Biol 6:43–55

Janssens F (1909) La théorie de la chiasmatypie. Nouvelle interpretation des cinèses de maturation. Cellule 22:387–411

Johnson RD, Jasin M (2000) Sister chromatid gene conversion is a prominent double-strand break repair pathway in mammalian cells. Embo J 19:3398–3407

Kai T, Spradling A (2004) Differentiating germ cells can revert into functional stem cells in Drosophila melanogaster ovaries. Nature 428:564–569

Kai T, Williams D, Spradling AC (2005) The expression profile of purified Drosophila germline stem cells. Dev Biol 283:486–502

Kaiser K, Sentry JW, Finnegan DJ (1995) Eukaryotic transposable elements as tools to study gene structure and function. In: Sherratt DJ (ed) Mobile Genetic Elements, vol 12. Oxford University Press, Oxford, p 69–100

Kidwell MG, Kidwell JF, Sved JA (1977) Hybrid dysgenesis in Drosophila melanogaster: a syndrome of aberant traits including mutation, sterility, and male recombination. Genetics 86:813–833

Kiger AA, Jones DL, Schulz C, Rogers MB, Fuller MT (2001) Stem cell self-renewal specified by JAK-STAT activation in response to a support cell cue. Science 294:2542–2545

Kiger AA, White-Cooper H, Fuller MT (2000) Somatic support cells restrict germline stem cell self-renewal and promote differentiation. Nature 407:750–754

Kohlstaedt LA, Wang J, Friedman JM, Rice PA, Steitz TA (1992) Crystal structure at 3.5 A resolution of HIV-1 reverse transcriptase complexed with an inhibitor. Science 256:1783–1790

Kraus E, Leung WY, Haber JE (2001) Break-induced replication: a review and an example in budding yeast. Proc Natl Acad Sci USA 98:8255–8262

Kusano K, Berres ME, Engels WR (1999) Evolution of the RECQ family of helicases: A Drosophila homolog, Dmblm, is similar to the human bloom syndrome gene. Genetics 151:1027–1039

Kusano K, Johnson-Schlitz DM, Engels WR (2001) Sterility of Drosophila with mutations in the Bloom syndrome gene—complementation by Ku70. Science 291:2600–2602

Kuzminov A (1996) Recombinational Repair of DNA Damage. Springer, Berlin Heidelberg New York

Lankenau D-H (1995) Genetics of genetics in Drosophila: P elements serving the study of homologous recombination, gene conversion and targeting. Chromosoma 103:659–668

Lankenau DH, Corces VG, Engels WR (1996) Comparison of targeted-gene replacement frequencies in Drosophila melanogaster at the forked and white loci. Mol Cell Biol 16:3535–3544

Lankenau D-H, Gloor GB (1998) In vivo gap repair in Drosophila: a one-way street with many destinations. BioEssays 20:317–327

Lankenau D-H, Peluso MV, Lankenau S (2000) The Su(Hw) Chromatin Insulator Protein Alters Double Strand-Break Repair Frequencies in the Drosophila Germ Line. Chromosoma 109:148–160

Lankenau S, Barnickel T, Marhold J, Lyko F, Mechler BM, Lankenau DH (2003) Knockout targeting of the Drosophila Nap1 gene and examination of DNA repair tracts in the recombination products. Genetics 163:611–623

Lankenau S, Bürkle A, Lankenau D-H (1999) Detection of poly(ADP-ribose) synthesis in Drosophila testes upon gamma-irradiation. Chromosoma 108:44–51

Lankenau S, Corces VG, Lankenau DH (1994) The Drosophila micropia retrotransposon encodes a testis-specific antisense RNA complementary to reverse transcriptase. Mol Cell Biol 14:1764–1775

Leach DRF (1996) Genetic Recombination. Blackwell Science, Oxford

Lohrmann R, Orgel LE (1976) Template-directed synthesis of high molecular weight polynucleotide analogues. Nature 261:342–344

Mattick JS (2004) The hidden genetic program of complex organisms. Sci Am 291:60–67

Maynard Smith J, Szathmary E (1997) The Major Transitions in Evolution. Oxford University Press, Oxford

Mayr E (1942) Systematics and the Origin of Species from the Viewpoint of a Zoologist. Columbia University Press, New York

McClintock B (1987) The Discovery and Characterization of Transposable Elements. Garland Publishing, New York

McVey M, Larocque JR, Adams MD, Sekelsky JJ (2004) Formation of deletions during double-strand break repair in Drosophila DmBlm mutants occurs after strand invasion. Proc Natl Acad Sci USA 101:15694–15699

Melnikova L, Biessmann H, Georgiev P (2005) The ku protein complex is involved in length regulation of Drosophila telomeres. Genetics 170:221–235

Melnikova L, Georgiev P (2002) Enhancer of terminal gene conversion, a new mutation in Drosophila melanogaster that induces telomere elongation by gene conversion. Genetics 162:1301–1312

Meselson M, Stahl FW (1958) The Replication of DNA in Escherichia Coli. Proc Natl Acad Sci USA 44:671–682

Meselson M, Weigle JJ (1961) Chromosome breakage accompanying genetic recombination in bacteriophage. Proc Natl Acad Sci USA 47:857–868

Miskey C, Izsvak Z, Kawakami K, Ivics Z (2005) DNA transposons in vertebrate functional genomics. Cell Mol Life Sci 62:629–641

Modrich P (1994) Mismatch repair, genetic stability, and cancer. Science 266:1959–1960

Mohr OL (1919) Mikroskopische Untersuchungen zu Experimenten über den Einfluß der Radiumstrahlen und der Kältewirkung auf die Chromatinreifung und das Heterochromosom bei *Decticus verruccivorus*. Arch mikr Anat Entw 92:300–368

Morgan TH, Sturtevant AH, Muller HJ, Bridges CB (1915) The Mechanism of Mendelian Heredity. H. Holt & Co., New York

Morozov V, Mushegian AR, Koonin EV, Bork P (1997) A putative nucleic acid-binding domain in Bloom's and Werner's syndrome helicases. Trends Biochem Sci 22:417–418

Muller HJ (1927a) Artificial transmutation of the gene. Science 66:84–87

Muller HJ (1927b) The Problem of Genetic Modification. Zeitschr ind Abst Vererb Sup-Bd 1:234–260

Muller HJ (1928a) The measurement of gene mutation rate in *Drosophila*, its high variability, and its dependence upon temperature. Genetics 13:279–357

Muller HJ (1928b) The Production of Mutations by X-Rays. Proc Natl Acad Sci USA 14:714–726

Muller HJ (1946) The Production of Mutations. In: Nobel Lectures, Physiology or Medicine 1942–1962. Elsevier Publishing Company, Amsterdam, 1964

Muller HJ (1966) The gene material as the initiator and organizing basis of life. American Naturalist 100:493–517

Nasheuer HP, Pospiech H, Syväoja J (2006) Towards an Anatomy of the Eukaryotic DNA Repliction Fork. In: Lankenau DH (ed) Genome Integrity: Facets and Perspectives, vol 1. Springer, Berlin Heidelberg New York

Nassif N, Penney J, Pal S, Engels WR, Gloor GB (1994) Efficient copying of non-homologous sequences from ectopic sites via P-element-induced gap repair. Mol Cell Biol 14:1613–1625

Nüsslein-Volhard C (2004) Das Werden des Lebens—Wie Gene die Entwicklung steuern. Verlag C.H. Beck oHG, München

Nüsslein-Vollhard C (1995) The Identification of Genes Controlling Development in Flies and Fishes. In, vol 1995. http://nobelprize.org/nobel_prizes/medicine/laureates/1995/nusslein-volhard-lecture.html

O'Brochta DA, Gomez SP, Handler AM (1991) P element excision in Drosophila melanogaster and related drosophilids. Mol Gen Genet 225:387–394

Ohlstein B, Kai T, Decotto E, Spradling A (2004) The stem cell niche: theme and variations. Curr Opin Cell Biol 16:693–699

Paques F, Haber JE (1999) Multiple pathways of recombination induced by double-strand breaks in Saccharomyces cerevisiae. Microbiol Mol Biol Rev 63:349–404

Paques F, Leung WY, Haber JE (1998) Expansions and contractions in a tandem repeat induced by double-strand break repair. Mol Cell Biol 18:2045–2054

Pardue ML, Danilevskaya ON, Lowenhaupt K, Slot F, Traverse KL (1996) Drosophila telomeres: new views on chromosome evolution. TIG 12(2):48–52

Patterson JT (1928) The effects of X-rays in producing mutations in the somatic cells of *Drosophila melanogaster*. Science 68:41–43

Pfeiffer P, Kuhfittig-Kulle S, Geoedecke W (2005) Mechanisms of Non-Homologous DNA End Joining: Aspects of In Vitro Assays. In: Lankenau DH (ed) Genome Integrity: Facets and Perspectives, vol 1. Springer, Berlin Heidelberg New York, p 24

Porteus MH, Carroll D (2005) Gene targeting using zinc finger nucleases. Nat Biotechnol 23:967–973

Preston CR, Flores CC, Engels WR (2006) Differential usage of alternative pathways of double-strand break repair in Drosophila. Genetics 172:1055–1068

Puchta H (2005) The repair of double-strand breaks in plants: mechanisms and consequences for genome evolution. J Exp Bot 56:1–14

Rensch B (1947) Neuere Probleme der Abstammungslehre (Die transspezifische Evolution). Ferdinand Enke Verlag, Stuttgart

Rio DC (2002) P Transposable Elements in Drosophila melanogaster. In: Craig NL, Craigie R, Gellert M, Lambowitz AM (eds) Mobile DNA II. ASM Press, Washington, DC, p 484–518

Rong YS (2002) Gene targeting by homologous recombination: a powerful addition to the genetic arsenal for Drosophila geneticists. Biochem Biophys Res Commun 297:1–5

Rong YS, Golic KG (2000) Gene targeting by homologous recombination in Drosophila. Science 288:2013–2018

Rong YS, Golic KG (2001) A targeted gene knockout in Drosophila. Genetics 157:1307–1312

Rong YS, Titen SW, Xie HB, Golic MM, Bastiani M, Bandyopadhyay P, Olivera BM, Brodsky MH, Rubin GM, Golic KG (2002) Targeted mutagenesis by homologous recombination in D. melanogaster. Genes & Dev 16:1568–1581

Rubin GM, Spradling AC (1982) Genetic Transformation of Drosophila with Transposable Element Vectors. In: Science 218:348–353

Rudolph C, Schürer KA, Kramer W (2006) Facing Stalled Replication Forks: The Intricacies of Doing the Right Thing. In: Lankenau DH (ed) Genome Integrity: Facets and Perspectives, vol 1. Springer, Berlin Heidelberg New York

Schrödinger E (1944) What is life? Cambridge University Press, London

Sentry JW, Kaiser K (1992) P element transposition and targeted manipulation of the Drosophila genome. Trends Genet 8:329–331
Simpson GG (1944) Tempo and Mode in Evolution. Columbia Univ. Press, New York
Spellman PT, Sherlock G, Zhang MQ, Iyer VR, Anders K, Eisen MB, Brown PO, Botstein D, Futcher B (1998) Comprehensive identification of cell cycle-regulated genes of the yeast Saccharomyces cerevisiae by microarray hybridization. Mol Biol Cell 9:3273–3297
Spradling A (1993) Developmental Genetics of Oogenesis. In: Martinez-Ariaz A, Bate M (eds) The Development of Drosophila melanogaster, vol 1. Cold Spring Harbor Press, Cold Spring Harbor, New York, p 1–70
Spradling A, Drummond-Barbosa D, Kai T (2001) Stem cells find their niche. Nature 414:98–104
Spradling AC, Rubin GM (1982) Transposition of cloned P elements into Drosophila germ line chromosomes. Science 218:341–347
Stadler LJ (1928a) Genetic Effects of X-Rays in Maize. Proc Natl Acad Sci USA 14:69–75
Stadler LJ (1928b) Mutations in barley induced by X-rays and radium. Science 68:186–187
Stahl FW (1956) The effects of the decay of incorporated radioactive phosphorus on the genome of bacteriophage T4. Virology 2:206–234
Stein E (1922) Über den Einfluß der Radiumbestrahlung auf *Antirrhinum*. Zeitschr ind Abst Vererb 29:1–15
Stern C (1931) Zytologisch-genetische Untersuchungen als Beweis für die Morgansche Theorie des Faktorenaustauschs. Biol Zentbl 51:547–587
Summons RE, Jahnke LL, Hope JM, Logan GA (1999) 2-Methylhopanoids as biomarkers for cyanobacterial oxygenic photosynthesis. Nature 400:554–557
Sved JA (1976) Hybrid dysgenesis in Drosophila melanogaster: a possible explanation in terms of spatial organization of chromosomes. Aust J Biol Sci 29:375–388
Swaminathan J, Baxter EM, Corces VG (2005) The role of histone H2Av variant replacement and histone H4 acetylation in the establishment of Drosophila heterochromatin. Genes Dev 19:65–76
Szostak JW, Orr-Weaver TL, Rothstein RJ, Stahl FW (1983) The double-strand-break repair model for recombination. Cell 33:25–35
Taft RJ, Mattick JS (2003) Increasing biological complexity is positively correlated with the relative genome-wide expansion of non-protein-coding sequences. Genome Biol 5:1.1–1.22
Temin HM, Mizutani S (1970) RNA-dependent DNA polymerase in virions of Rous sarcoma virus. Nature 226:1211–1213
Thomas KR, Capecchi MR (1987) Site-directed mutagenesis by gene targeting in mouse embryo-derived stem cells. Cell 51:503–512
Trinks H, Schröder W, Biebricher CK (2005) Ice and the origin of life. Orig Life Evol Biosph 35:429–445
Varmus HE (1983) Retroviruses. In: Shapiro JA (ed) Mobile Genetic Elements. Academic Press, New York, p 421–473
Walisko O, Izsvak Z, Szabo K, Kaufman CD, Herold S, Ivics Z (2006) Sleeping Beauty transposase modulates cell-cycle progression through interaction with Miz-1. Proc Natl Acad Sci USA 103:4062–4067
Watson JD, Baker TA, Bell SP, Gann A, Levine M, Losick R (2004) Molecular biology of the gene, 5th edn. Pearson/Benjamin Cummings, Cold Spring Harbor Laboratory Press, San Francisco

Watson JD, Crick FH (1953a) Genetical implications of the structure of deoxyribonucleic acid. Nature 171:964–967
Watson JD, Crick FH (1953b) Molecular structure of nucleic acids; a structure for deoxyribose nucleic acid. Nature 171:737–738
Whitehouse HL (1963) A theory of crossing-over by means of hybrid deoxyribonucleic acid. Nature 199:1034–1040
Whiting PW (1928) The production of mutations by X-rays in *Hadrobracon*. Science 58
Woese CR (2004) A new biology for a new century. Microbiol Mol Biol Rev 68:173–186
Wright S (1931) Evolution in Mendelian Populations. Genetics 16:97–159
Xie HB, Golic KG (2004) Gene deletions by ends-in targeting in Drosophila melanogaster. Genetics 168:1477–1489
Xie T, Spradling AC (2000) A niche maintaining germ line stem cells in the Drosophila ovary. Science 290:328–330
Preston CR, Flores C, Engels W (2006) Age-dependent usage of double-strand break repair pathways. Current Biology, in press
Wallenfang MR, Nayak R, DiNardo S (2006) Dynamics of the male germline stem cell population during aging of Drosophila melanogaster. Aging Cell 5, in press

Part C
Facets of Double-Strand Break Repair

Genome Dyn Stab (1)
D.-H. Lankenau: Genome Integrity
DOI 10.1007/7050_006/Published online: 20 October 2005

Structure and Function of Rad50/SMC Protein Complexes in Chromosome Biology

Karl-Peter Hopfner

Gene Center, University of Munich, Feodor-Lynen-Str. 25, 81377 Munich, Germany
hopfner@lmb.uni-muenchen.de

Abstract Sister-chromatid cohesion, chromosome condensation, as well as repair of DNA double-strand breaks, require multiprotein complexes that all contain a core ATPase of the Rad50/SMC protein family. Rad50/SMC proteins possess an ATP-binding cassette ATPase domain and a strikingly long coiled-coil dimerization/linker domain. In conjunction with associated subunits, Rad50/SMC proteins form large molecular rings or tethering structures that are well suited to link chromatids, DNA segments or DNA ends. This chapter will review recent structural and functional insights of how these remarkable molecular machines tether DNA using ATP binding and hydrolysis.

Abbreviations

ABC	ATP binding cassette
AT	ataxia telangiectasia
ATM	ataxia telangiectasia mutated
DSB	DNA double-strand break
dsDNA	double-strand DNA
ssDNA	single-strand DNA
HR	homologous recombination
NBS	Nijmegen breakage syndrome
NHEJ	non-homologous end joining
SMC	structural maintenance of chromosome

1 Rad50/SMC Protein Complexes in Chromosome Biology

The accurate inheritance of the genetic information from generation to generation is fundamental to life. In proliferating cells, this process not only involves duplication of each chromosome by DNA replication, compaction of all chromosomes before mitosis, and faithful segregation of the two copies of each chromosome (termed sister chromatids) into both daughter cells, but also the repair of DNA double-strand breaks (DSBs) by recombination or end-joining prior to cell division. The required chromosome associated processes – cohesion, condensation and DSB repair – require ATP hydrolysis by special ATPases, the Rad50/SMC (structural maintenance of chromosome) ATPase family. These ATPases are the core of multiprotein complexes that form large molecular rings or tether structures. In eukaryotes, six SMC proteins (SMC1-6) form three complexes in chromosome segregation (termed cohesin, condensin, SMC5/6 complex), each containing a pair of SMC proteins. The related Rad50 polypeptide associates with the nuclease Mre11 and cell cycle checkpoint protein NBS1 (mutated in Nijmegen breakage syndrome) into a DNA double-strand detection and repair factor (Fig. 1).

Cohesion of Sister chromatids between DNA replication and mitosis ensures the correct sorting of sister chromatids to daughter cells and requires ATP hydrolysis by the *cohesin* complex (see (Haering and Nasmyth 2003) and references therein). Cohesin consists of two SMC ATPases (SMC1 and SMC3), as well as two associated "non-SMC" subunits (SCC1 and SCC3). Cohesin is suggested to form a large proteinaceous ring that might directly embrace sister chromatids (Gruber et al. 2003; Haering et al. 2002). After binding to chromosomes in G1/S, cohesin is a stable, diffusible element on DNA until mitosis (Haering et al. 2004; Lengronne et al. 2004). In mitosis, cohesin is

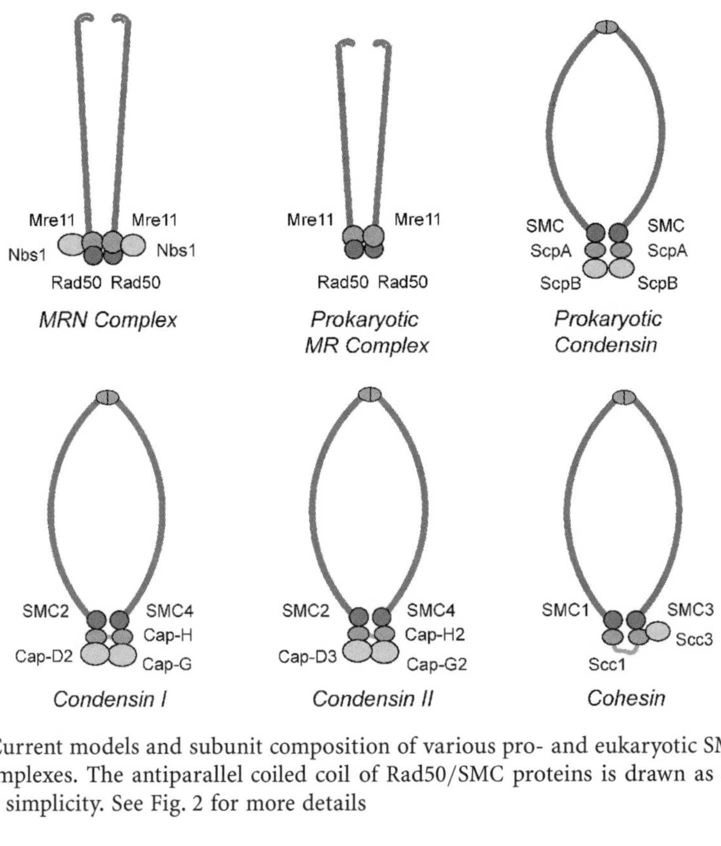

Fig. 1 Current models and subunit composition of various pro- and eukaryotic SMC protein complexes. The antiparallel coiled coil of Rad50/SMC proteins is drawn as a single line for simplicity. See Fig. 2 for more details

destroyed by proteolytic cleavage of SCC1 (see, e.g. Uhlmann 2003, and references therein). A cohesin variant consisting of SMC1β and SMC3 and is involved in meiotic chromosome dynamics and recombination (Revenkova et al. 2004).

Chromosome condensation at the onset of mitosis requires ATP hydrolysis by one or two condensin (condensin I and condensin II) complexes (see Hirano 2005, and references therein). 13S eukaryotic condensin I consists of SMC2 and SMC4, together with an 11S regulatory subcomplex, consisting of CAP-H, CAP-D2 and CAP-G. In vertebrates, condensin II shares the SMC proteins with condensin I, but has a different set of non-SMC proteins (CAP-H2, CAP-D3, CAP-G2) (Ono et al. 2003). In the presence of ATP, condensin can introduce positive supercoils into circular DNA and has been suggested to entrap loops of DNA using ATP hydrolysis (Bazett-Jones et al. 2002; Kimura and Hirano 1997; Strick et al. 2004).

The most recently discovered and less well understood family member of the SMC protein complexes is the SMC5/6 complex. It is conserved in all eukaryotes and consists of SMC5 and SMC6, as well as several associated subunits (Nse1, Nse2, Nse3, and Rad62). The SMC5/6 complex is required for the segregation of rDNA in budding yeast (Torres-Rosell et al. 2005), and is also involved in DNA repair as well as homologous recombination in meiosis (Lehmann et al. 1995; Verkade et al. 1999). In addition, the SMC5/6 complex possesses SUMO ligase activity (Andrews et al. 2005; Zhao and Blobel 2005). However, the precise mechanistic role of the SMC5/6 complexes in these cellular processes is not understood.

DSB repair and the DSB induced damage checkpoint by the Mre11/Rad50/Nbs1 (MRN) complex ensure that no part of the chromosome is lost before segregation (see details below). The MRN complex joins broken DNA ends to each other or to the undamaged sister chromatid and processes DNA ends for subsequent repair steps in recombination or end joining. In addition, the MRN complex is a key factor for the activation of the DNA damage checkpoint in response to DSBs.

Prokaryotes also contain a single condensin like complex and a Rad50/Mre11 (MR) complex (*E. coli* SbcC/SbcD). These complexes are biochemically and structurally related to the eukaryotic counterparts, suggesting that the principal architecture of SMC protein complexes is conserved in nature (Hirano and Hirano 2004; Soppa et al. 2002; Volkov et al. 2003). Finally, the nematode *Caenorhabditis elegans* contains a condensin-like complex that is involved in gene dosage compensation (for review, see Jessberger 2003, and references therein).

In the following, I will review our current understanding on the functional architecture and mechanism of these multisubunit factors in chromosome biology.

2 Structural Features and Mechanisms of Rad50/SMC Protein Complexes

2.1 The Rad50/SMC ATPase

Rad50/SMC proteins probably belong to the most bizarre polypeptides in nature (Fig. 2). They consist of a bipartite ATP-binding cassette (ABC) ATPase domain that contains a typically 900 amino acid long heptad repeat insertion. Electron micrographs of Rad50 and SMC proteins suggested that the N- and C-terminal parts of the bipartite ATPase domain of Rad50/SMC polypeptides associate into a functional ABC ATPase domain (head), while the heptad repeat region folds into an antiparallel 50 nm long coiled-coil domain that protrudes from the ABC ATPase domains (tail) (Anderson et al. 2001, 2002; Connelly and Leach 1996; de Jager et al. 2001; Haering et al. 2002; Hopfner et al. 2001, 2002). Both the ATPase and the coiled-coil domain are key elements in the functional architecture of Rad50/SMC protein complexes.

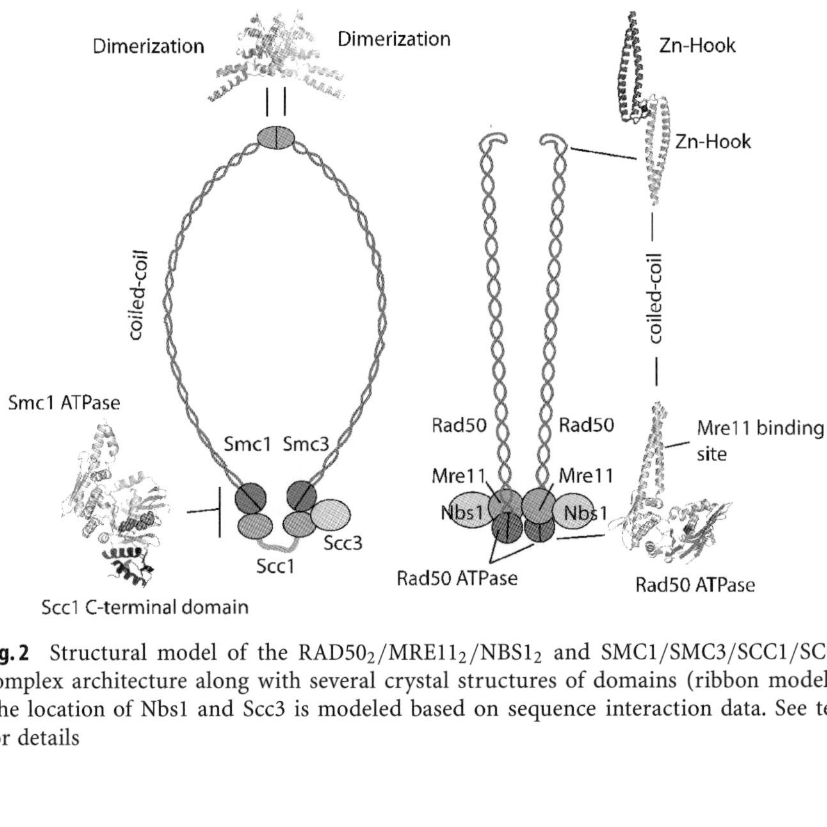

Fig. 2 Structural model of the $RAD50_2/MRE11_2/NBS1_2$ and SMC1/SMC3/SCC1/SCC3 complex architecture along with several crystal structures of domains (ribbon models). The location of Nbs1 and Scc3 is modeled based on sequence interaction data. See text for details

The ATPase domain of Rad50/SMC proteins is a member of the ATP-binding cassette (ABC) ATPase family. This family of ATPases is most prominent as the "engine" component of the ABC transporter, but is also found in some other DNA repair enzymes such as the bacterial nucleotide excision repair factor UvrA, as well as some translation associated proteins (see, e.g. Dassa and Bouige 2001; Hopfner and Tainer 2003; Schmitt and Tampe 2002, and references therein). ABC ATPases work as dimers. In the presence of ATP, two ABC domains exhibit a conformational change by sandwiching two ATP molecules in the ABC domain dimer interface (Fig. 2) (Hopfner et al. 2000; Lammens et al. 2004; Smith et al. 2002). The resulting ATP-driven engagement/disengagement cycles of two ABC ATPase domain dimers probably represent the "powerstrokes" of ABC enzymes (Chen et al. 2003; Liu et al. 1999). All SMC/Rad50 complexes analyzed in molecular detail contain a pair of SMC proteins or a homodimer of Rad50, raising the possibility that the powerstroke occurs between the two ABC domains of a single Rad50/SMC multisubunit complex.

The ATP dependent powerstroke of Rad50/SMC proteins is a key element for the function of these factors. ATPase deficient variants of Rad50 or SMC proteins behave as null, resulting in lethality in eukaryotes (SMC proteins) or vertebrates (Rad50). In general, the ATPase function of Rad50/SMC proteins is closely linked to association with or modulation of DNA, although the precise mechanistic role of the powerstroke of Rad50/SMC proteins is still unclear. For instance, ATP controls binding of DNA to the MRN complex (Connelly et al. 2003; de Jager et al. 2002; Paull and Gellert 1999; Trujillo and Sung 2001). ATP-induced structural switches in the Rad50 ABC domains also unwind DNA ends. This function facilitates DNA end processing by Mre11 (Paull and Gellert 1999) and is required for activation of the ATM kinase in the DNA damage response to double-strand breaks (Lee and Paull 2005). Although direct high affinity binding of DNA to SMC protein ATPase domains has not been observed yet, the powerstrokes of the SMC ATPase domains are presumably also linked directly or indirectly to association or structural modulation of DNA substrates. The ATPase activity of pro- and eukaryotic condensin complexes is highly stimulated by DNA (Hirano and Hirano 2002; Kimura and Hirano 1997). This DNA dependent stimulation seems to be a feature of the ATPase domain itself and requires a conserved arginine finger in the active site of SMC proteins that binds to the ATP α-phosphate (Lammens et al. 2004). Since ATP hydrolysis is required for DNA compaction by condensins, changes in DNA topology and ATP-induced conformational changes in the condensin ATPase domains are probably closely linked. In addition, the ATPase activity in both SMC1/3 ATPase domains is required for loading of cohesin onto DNA, and for formation of the fully assembled cohesin complex in yeast (Arumugam et al. 2003; Weitzer et al. 2003). In this respect, ATP-driven engagement of the two ATPase domains of cohesin could be an important mechanistic intermediate in the formation of ring structures around DNA.

2.2 The Coiled-Coil Domain

The 50 nm long coiled-coil domains of Rad50/SMC proteins protrude from the ATPase heads. The main function of the coiled-coil domain is presumably the assembly of large linker structures that form a structural element for the entrapment or tethering of DNA segments. On the other end of the coiled-coil domain, opposite from the ATPase domain, is an important dimerization motif in both Rad50 and SMC proteins that can join two coiled coils (Haering et al. 2002; Hopfner et al. 2002). In Rad50, this apex of the coiled-coil domain ends in a small "hook" loop that contains a conserved Cys-X-X-Cys motif. In the presence of metal, most likely zinc ions, two of these Rad50 hooks can form interlocked hook:zinc:hook bridges and join two Rad50 coiled-coil domains (de Jager et al. 2001; Hopfner et al. 2002). In that way, the hooks form a composite metal binding site, where each of the hooks contributes two cysteine residues to a tetrahedral metal coordinating site. Both coiled-coil arms protrude from the interlocked hooks at an angle of 140°. The interlock of two of these coiled-coil domains is not only found in crystallographic studies, but also observed in electron and scanning force micrographs of full MRN complexes (de Jager et al. 2001; Hopfner et al. 2002).

The apex of the SMC coiled-coil domains looks remarkably different to that of Rad50. In the crystal structure of the bacterial SMC protein, the coiled-coil apex ends in an approximately 150 amino acid containing domain that has the shape of half a donut (Haering et al. 2002). Two of these dimerization domains associate into a donut-shaped structure, whereby the two coiled coils protrude at a V-shaped angle. The molecular mechanistic reason for the structural difference of the dimerization domains of Rad50 and SMC proteins have not been explained yet. Perhaps, the much smaller hook:hook interface in Rad50 could lead to a more dynamic association during DNA damage detection, while the larger SMC dimerization domain could form a more stable association of the coiled-coil domains between replication and mitosis (Haering et al. 2004). In addition, the SMC dimerization domain features presumably also ensure that the correct pairs of SMC proteins associate, e.g. SMC1 and SMC3 in the case of cohesin.

Scanning force microscopy studies showed that Rad50 and SMC protein coiled-coil domains have specific regions of high flexibility (de Jager et al. 2004; van Noort et al. 2003). These regions co-localize with sequences of low coiled-coil propensities and could represent specifically designed hinge regions to allow for conformational flexibility of the coiled-coil domain. Besides serving as structural element, the coiled-coil domains of Rad50/SMC proteins also have additional functions. In Rad50, the coiled-coil contains the binding site for the associated Mre11 subunit (Hopfner et al. 2001). More precisely, the Mre11 binding site is situated at the coiled-coil stretch that is directly adjacent to the Rad50 ATPase domain. Because of this close interaction, Mre11 and the Rad50 ATPase domain presumably directly cooperate in the molecular recognition and processing of DNA ends.

In SMC1, the coiled-coil contains a phosphorylation site for the DNA damage checkpoint kinase ATM (ataxia telangiectasia mutated). This ATM dependent phosphorylation is important for the cellular checkpoint in response to DSBs (Kim et al. 2002; Yazdi et al. 2002). The mechanistic role of this phosphorylation has not been revealed. In principle, the phosphorylation of SMC1 could attract other protein factors or localize SMC1 (or cohesin) to double-strand breaks via phosphopeptide mediated protein-protein interactions. Alternatively, it could modulate the conformation or biophysical properties of the coiled coil.

2.3 Multisubunit Architecture

The functional mechanism of ABC ATPases – ATP-driven engagement/disengagement of two ABC domains – is reflected by the molecular architecture of Rad50/SMC protein complexes. SMC protein complexes consist of a homo- or heterodimer of the ATPase subunits, i.e. SMC1:SMC3 (cohesin), SMC2:SMC4 (condensin), SMC5:SMC6, or a bacterial SMC homodimer. The two SMC proteins are held together via their dimerization domains at the apex of the coiled coil (Chiu et al. 2004; Haering et al. 2002). The ATPase domains of cohesion and bacterial condensin directly interact with the ATPase domains of Scc1 and ScpA, respectively. Probably the best understood SMC protein complex is yeast cohesin. Cohesin forms a proteinaceous ring out of a heterotrimeric SMC1:SMC3:Scc1 complex. Scc1 binds to both the ATPase domains of SMC1 and SMC3 (which additionally interact via their dimerization domains). Hereby, the C-terminal winged-helix domain of Scc1 binds to the ATPase domain of SMC1 on the opposite side of the root of the coiled coil. The interaction of Scc1 with SMC1 and 3 – and the resulting tripartite proteinaceous ring – is very stable on chromosomes (Hearing et al. 2004). DNA loading into the ring requires ATP hydrolysis by both SMC proteins, and current models suggest that the engagment/disengagment of the two ATPase domains could provide the means for a transient opening/closing of the ring structure. However, it is still unclear how the ATP-dependent transient association of the two ABC domains, and the Scc1 mediated stable association of the two ABC domains, is functionally and mechanistically coupled.

Although the precise architecture of condensin-like proteins is less understood, the bacterial and eukaryotic condensin subunits ScpA and CAP-H share sequence homology with Scc1. Since both ScpA and the 11S regulatory subcomplex of eukaryotic cohesin bind to the ATPase domains of the respective SMC proteins, the principal architecture of condensin complexes could be similar to the architecture of cohesin. It is not clear, however, if ScpA or the 11S complex bridge both ABC domains in a similar manner to Scc1 bridges SMC1 and SMC3.

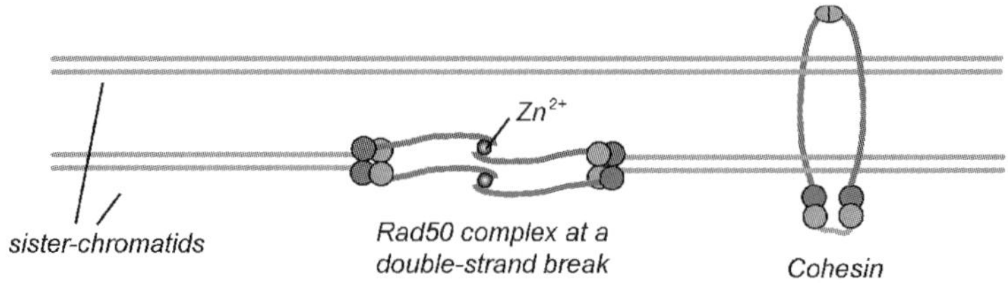

Fig. 3 Hypothetical model for the architectural function of MRN at DNA double-strand breaks. MRN could bridge DNA ends by forming zinc-hook mediated tether structures, while cohesin could embrace sister-chromatids

The MRN complex has a related but slightly different architecture. Two Rad50 molecules are presumably not linked by the hook structures, but rather by an Mre11 dimer that binds to the coiled coil near the ATPase domain. The two coiled-coil domains emerge as flexible tails from the globular head complex. The head consists of an Mre11 dimer plus the dimer of the Rad50 ATPase domains (Anderson et al. 2001; Chen et al. 2001; Connelly et al. 1998; de Jager et al. 2001; Hopfner et al. 2001). The head complex harbors the DNA binding and ATP-stimulated nuclease activity, while the two up to 60 nm long antiparallel coiled-coil with the terminal hook structures might form flexible "fishing rods" (Anderson et al. 2001; Chen et al. 2001; Connelly et al. 1998; de Jager et al. 2001; Hopfner et al. 2001; van Noort et al. 2003). In particular, the assembly of interlocked hook:zinc:hook bridges could join two or more MRN complexes, bound to each of the two DNA ends at a break (Fig. 3) (de Jager et al. 2001; Hopfner et al. 2002). Although the precise nature of the supermolecular architecture of these joined coiled-coils remains to be established the clustering of such heads at DNA ends could form a "molecular velcro"-like structure (a term coined by Roland Kanaar and coworkers) that could tether ends in both HR and NHEJ. Mutagenesis and biophysical data suggest that the third subunit of the MRN complex, NBS1, binds to the $Mre11_2$:$Rad50_2$ head, possibly forming a hexameric or even higher order $RAD50_2/MRE11_2/NBS1_2$ complex in eukaryotes (Lee et al. 2003). In this respect, Nbs1 (or the yeast ortholog Xrs2) could bind both DNA ends and further components of the DSB repair/checkpoint machinery (Falck et al. 2005; Trujillo et al. 2003).

3 The Mre11/Rad50/Nbs1 Complex: From Structure to Function

3.1 DNA Double Strand Break Repair and the MRN Complex

The MRN complex is a key component in the cellular response to DSBs. DSBs are among the most cytotoxic form of DNA damage and can lead to gross chromosomal aberrations, disruption of the genomic integrity and cancer (Carney et al. 1998; Game 1993; Luo et al. 1999; Stewart et al. 1999; Zdzienicka 1996). Under physiological conditions, the main cause of DSBs is problems in DNA replication, although DSBs can also arise from genotoxic agents or radiation (Costanzo et al. 2001; Gellert 2002; Haber 1998; Keeney 2001; Kuzminov 2001; Ward 1988). DSBs are repaired by two major pathways, homologous recombination (HR) and non-homologous end joining (NHEJ). In HR, the DSBs are resected into 3′ single-strand (ss) DNA tails. These tails pair with the homologous DNA segment of the sister chromatid and repair proceeds by DNA synthesis essentially without loss of genetic information (see, e.g. Haber 2000, and references therein). In NHEJ, the two broken ends are directly aligned and re-ligated. This process involves limited end processing and is potentially mutagenic (see details for HR and NHEJ in chapters by M Löbrich).

The MRN complex is required for genomic stability in yeast (Alani et al. 1990; Luo et al. 1999; Yamaguchi-Iwai et al. 1999), and is essential for viability in higher eukaryotes (Gorski et al. 2004; Luo et al. 1999; Yamaguchi-Iwai et al. 1999; Zhu et al. 2001). Hypomorphic mutations in NBS1 and MRE11 cause the genome instability syndromes Nijmegen breakage syndrome (NBS) and ataxia telangiectasia like disease (ATLD) (Carney et al. 1998; Matsuura et al. 1998; Stewart et al. 1999; Varon et al. 1998). A hypomorphic mutation in mouse RAD50, related to the separation of function mutations *Rad50S* in yeast, leads to partial embryonic lethality, hematopoietic failure and cancer susceptibility (Bender et al. 2002). Thus, even minor disturbances of the MRN complex function can have a large effect on genome stability (see also chapter by M. Digweed).

The MRN complex is implicated in both homologous recombination and NHEJ, as well as in several other aspects of the cellular responses to DSBs (Milne et al. 1996; Moore and Haber 1996; Tsukamoto et al. 1996; Udayakumar et al. 2003). In particular, the MRN complex is implicated in DSB detection (Mirzoeva and Petrini 2001), DNA end processing (Connelly and Leach 1996; Lewis et al. 2004; Mickelson and Wiberg 1981; Moreau et al. 1999; Trujillo et al. 1998), formation and processing of meiotic DSBs (Alani et al. 1990; Bressan et al. 1999; Tauchi et al. 2002; Yamaguchi-Iwai et al. 1999), telomere maintenance (Boulton and Jackson 1998; Chamankhah et al. 2000; Kironmai and Muniyappa 1997; Lundblad 2002; Teng et al. 2000; Zhu et al. 2000) and DNA damage checkpoint activation (Carney et al. 1998; D'Amours and Jackson 2001; Zhao et al. 2000). This puzzling variety of functions is explained by roles of the MRN complex in both processing and architectural tethering of DNA ends.

3.2
Enzymatic and Structural Roles of Rad50 in DSB Repair

Biochemically, the MRN complex is an ATP stimulated nuclease that acts endo- and exonucleolytically on ssDNA and hairpins, and exonucleolytically on various types of dsDNA ends, including ends with covalently attached proteins (Connelly et al. 2003; Lobachev et al. 2002; Ohta et al. 1998; Trujillo and Sung 2001). Mre11 by itself only degrades ssDNA. ATP hydrolysis by Rad50 is suggested to open DNA ends and hairpin structures for processing by Mre11 (Hopfner et al. 2001). With the notable exception of gene products 46/47, the bacteriophage T4 homologs of Mre11/Rad50, all Mre11/Rad50 proteins assayed so far degrade dsDNA in 3′-> 5′ direction (Connelly et al. 1999; Hopfner et al. 2001; Paull and Gellert 1998; Trujillo et al. 1998). This proofreading like nuclease activity argues that MRN does not directly generate formation of 3′-tails in recombination, but more likely processes DNA secondary structures or otherwise misfolded or damaged DNA ends for subsequent repair steps (Connelly et al. 2003; Farah et al. 2002; Lobachev et al. 2002; Paull and Gellert 1999; Trujillo and Sung 2001). The resulting clean 3′ – OH could for instance be important for subsequent DNA synthesis steps in recombinational repair.

Although DNA processing by MRN could be important in some instances, the structural role of the RMX (the yeast Rad50/Mre11/Xrs2 equivalent of MRN) complex in DNA end linkage seems to be its key function in genome integrity. Elegant studies in yeast have recently shown that the disruption of the zinc hook mediated formation of higher order RMX complex structures has the same effect than a null mutation in Rad50, while nuclease deficient but otherwise intact RMX complexes are hypomorphic (Wiltzius et al. 2005). Hereby, the zinc hook can be functionally replaced by a different dimerization element, showing that the dimerization function, but not the precise structure of the zinc hook is required for MRN complex activity. Thus, the key function of the MRN complex could be an ATP-dependent crosslinker of DNA ends or sister-chromatids. In this process, MRN appears to cooperate with cohesin complexes that relocate to DSBs (Kaye et al. 2004; Lobachev et al. 2004; Schar et al. 2004). Possibly, MRN links DSBs together, while cohesin ensures cohesion of the broken chromatid to the undamaged sister (Fig. 3).

3.3 Checkpoint Function of the MRN Complex

Besides the evolutionary conserved architectural and enzymatic roles in DSB repair, the eukaryotic MRN complex is also involved in the DSB activated checkpoint response. The damage checkpoint activation after DSBs is triggered by the large PI3-like kinase ATM (ataxia telangiectasia mutated). ATM resides in the nucleus as inactive dimer (Bakkenist and Kastan 2003). On introduction of DSBs and presumably other stress events, ATM autophosphorylates itself at Ser1981, dissociates into monomers and becomes active. The active ATM kinase phosphorylates a variety of downstream targets that trigger different cellular responses, such as DNA repair, transcription, cell cycle arrest or apoptosis (see below).

The MRN complex emerges as an important cofactor for the ATM dependent activation of the DNA damage checkpoint in response to DSBs. Hereby, the MRN complex physically binds to ATM (Lee and Paull 2004). This direct interaction between ATM and the MRN complex has recently been found to involve the extreme C-terminus of the Nbs1 subunit. Hereby, the C-terminus of Nbs1 appears to interact with a central region of ATM that is predicted to consist of HEAT repeats (Falck et al. 2005).

Exciting recent results show that the MRN complex mediated stimulation of ATM is dependent on DNA ends and needs the ATP-driven unwinding of DNA ends by the MRN complex (Lee and Paull 2005). While the precise mechanistic basis for this stimulation needs to be established, ATP-driven conformational changes in the MRN complex could for instance lead to structural alterations in DNA ends or in ATM. In addition to this direct activation, the MRN complex participates in the ATM dependent damage checkpoint probably also as a substrate recruitment factor (Carson et al. 2003; D'Amours and Jackson 2001; Gatei et al. 2000; Goldberg et al. 2003; Lou et al. 2003; Nakada et al. 2003; Stewart et al. 1999, 2003; Usui et al. 2001). Phosphorylation of a subset of ATM targets requires the MRN complex, in particular CHK2 (Buscemi et al. 2001; Lee and Paull 2004), histone H2AX (Lee and Paull 2004), SMC1 (Horejsi et al. 2004; Yazdi et al. 2002) and FANCD2 (Nakanishi et al. 2002). Perhaps, NBS1/Xrs2 could act as a substrate recruitment platform, e.g. for phosphopeptide binding domains on ATM target proteins.

4 Conclusions and Outlook

Progress over recent years has significantly advanced our understanding of the architecture and structural role of Rad50 and SMC protein complexes in chromosome-associated processes. However, several key questions still need to be addressed in future experiments. For instance, our understanding of Rad50/SMC protein complexes suffers significantly from a mechanistic gap between results obtained by structural biology and results obtained by cell biology. In particular, little is yet known about the biochemical role of ATP hydrolysis driven conformational changes in the ATPase domain and the macroscopic cellular functions of Rad50/SMC proteins. Exemplary questions that need to be addressed are: how is DNA loaded into a cohesin ring by the means of ATP hydrolysis? How is DNA mechanistically compacted by condensin using related ATP hydrolysis driven changes in the ATPase domains? How are DNA ends bound and processed by the MRN complex using ATP hydrolysis? A simplified working model could postulate that a conserved powerstroke of the ABC domains might be "translated" into the overall function of the respective Rad50/SMC protein complexes by associated subunits. This idea, while appealing, is probably oversimplified. For instance, SMC proteins can show a DNA stimulated ATPase activity that seems to be a property of the isolated ATPase domain, while Rad50 does not. In contrast, Rad50 shows ATP stimulated DNA binding activity, again a property of the isolated ATPase domains, while SMC proteins do not. Progress in the understanding of the mechanistic role of the structural and biochemical features of Rad50/SMC proteins is clearly a requirement for further understanding of the bizarre and fascinating Rad50/SMC multiprotein complexes in chromosome biology.

Acknowledgements I apologize to all colleagues whose key contributions could not been cited due to space restrictions and focus. Work in the KPHs laboratory is supported by grants from the Deutsche Forschungsgemeinschaft, the EU framework 6 integrated project "DNA Repair" and a Human Frontiers of Science Program Grant.

References

Alani E, Padmore R, Kleckner N (1990) Analysis of wild-type and rad50 mutants of yeast suggests an intimate relationship between meiotic chromosome synapsis and recombination. Cell 61:419–436

Anderson DE, Losada A, Erickson HP, Hirano T (2002) Condensin and cohesin display different arm conformations with characteristic hinge angles. J Cell Biol 156:419–424

Anderson DE, Trujillo KM, Sung P, Erickson HP (2001) Structure of the Rad50 x Mre11 DNA repair complex from *Saccharomyces cerevisiae* by electron microscopy. J Biol Chem 276:37027–37033

Andrews EA, Palecek J, Sergeant J, Taylor E, Lehmann AR, Watts FZ (2005) Nse2, a component of the Smc5-6 complex, is a SUMO ligase required for the response to DNA damage. Mol Cell Biol 25:185–196

Arumugam P, Gruber S, Tanaka K, Haering CH, Mechtler K, Nasmyth K (2003) ATP hydrolysis is required for cohesin's association with chromosomes. Curr Biol 13:1941–1953

Bakkenist CJ, Kastan MB (2003) DNA damage activates ATM through intermolecular autophosphorylation and dimer dissociation. Nature 421:499–506

Bazett-Jones DP, Kimura K, Hirano T (2002) Efficient supercoiling of DNA by a single condensin complex as revealed by electron spectroscopic imaging. Mol Cell 9:1183–1190

Bender CF, Sikes ML, Sullivan R, Huye LE, Le Beau MM, Roth DB, Mirzoeva OK, Oltz EM, Petrini JH (2002) Cancer predisposition and hematopoietic failure in Rad50(S/S) mice. Genes Dev 16:2237–2251

Boulton SJ, Jackson SP (1998) Components of the Ku-dependent non-homologous end-joining pathway are involved in telomeric length maintenance and telomeric silencing. Embo J 17:1819–1828

Bressan DA, Baxter BK, Petrini JH (1999) The Mre11-Rad50-Xrs2 protein complex facilitates homologous recombination-based double-strand break repair in *Saccharomyces cerevisiae*. Mol Cell Biol 19:7681–7687

Buscemi G, Savio C, Zannini L, Micciche F, Masnada D, Nakanishi M, Tauchi H, Komatsu K, Mizutani S, Khanna K, Chen P, Concannon P, Chessa L, Delia D (2001) Chk2 activation dependence on Nbs1 after DNA damage. Mol Cell Biol 21:5214–5222

Carney JP, Maser RS, Olivares H, Davis EM, Le Beau M, Yates JR 3rd, Hays L, Morgan WF, Petrini JH (1998) The hMre11/hRad50 protein complex and Nijmegen breakage syndrome: linkage of double-strand break repair to the cellular DNA damage response. Cell 93:477–486

Carson CT, Schwartz RA, Stracker TH, Lilley CE, Lee DV, Weitzman MD (2003) The Mre11 complex is required for ATM activation and the G(2)/M checkpoint. Embo J 22:6610–6620

Chamankhah M, Fontanie T, Xiao W (2000) The *Saccharomyces cerevisiae* mre11(ts) allele confers a separation of DNA repair and telomere maintenance functions. Genetics 155:569–576

Chen J, Lu G, Lin J, Davidson AL, Quiocho FA (2003) A tweezers-like motion of the ATP-binding cassette dimer in an ABC transport cycle. Mol Cell 12:651–661

Chen L, Trujillo K, Ramos W, Sung P, Tomkinson AE (2001) Promotion of Dnl4-catalyzed DNA end-joining by the Rad50/Mre11/Xrs2 and Hdf1/Hdf2 complexes. Mol Cell 8:1105–1115

Chiu A, Revenkova E, Jessberger R (2004) DNA interaction and dimerization of eukaryotic SMC hinge domains. J Biol Chem 279:26233–26242

Connelly JC, Leach DR (1996) The sbcC and sbcD genes of *Escherichia coli* encode a nuclease involved in palindrome inviability and genetic recombination. Genes Cells 1:285–291

Connelly JC, Kirkham LA, Leach DR (1998) The SbcCD nuclease of *Escherichia coli* is a structural maintenance of chromosomes (SMC) family protein that cleaves hairpin DNA. Proc Natl Acad Sci USA 95:7969–7974

Connelly JC, de Leau ES, Leach DR (1999) DNA cleavage and degradation by the SbcCD protein complex from *Escherichia coli*. Nucleic Acids Res 27:1039–1046

Connelly JC, de Leau ES, Leach DR (2003) Nucleolytic processing of a protein-bound DNA end by the *E. coli* SbcCD (MR) complex. DNA Repair (Amst) 2:795–807

Costanzo V, Robertson K, Bibikova M, Kim E, Grieco D, Gottesman M, Carroll D, Gautier J (2001) Mre11 protein complex prevents double-strand break accumulation during chromosomal DNA replication. Mol Cell 8:137–147

D'Amours D, Jackson SP (2001) The yeast Xrs2 complex functions in S phase checkpoint regulation. Genes Dev 15:2238–2249

Dassa E, Bouige P (2001) The ABC of ABCS: a phylogenetic and functional classification of ABC systems in living organisms. Res Microbiol 152:211–229

de Jager M, van Noort J, van Gent DC, Dekker C, Kanaar R, Wyman C (2001) Human Rad50/Mre11 is a flexible complex that can tether DNA ends. Mol Cell 8:1129–1135

de Jager M, Wyman C, van Gent DC, Kanaar R (2002) DNA end-binding specificity of human Rad50/Mre11 is influenced by ATP. Nucleic Acids Res 30:4425–4431

de Jager M, Trujillo KM, Sung P, Hopfner KP, Carney JP, Tainer JA, Connelly JC, Leach DR, Kanaar R, Wyman C (2004) Differential arrangements of conserved building blocks among homologs of the Rad50/Mre11 DNA repair protein complex. J Mol Biol 339:937–949

Falck J, Coates J, Jackson SP (2005) Conserved modes of recruitment of ATM, ATR and DNA-PKcs to sites of DNA damage. Nature 434:605–611

Farah JA, Hartsuiker E, Mizuno K, Ohta K, Smith GR (2002) A 160-bp palindrome is a Rad50Rad32-dependent mitotic recombination hotspot in *Schizosaccharomyces pombe*. Genetics 161:461–468

Game JC (1993) DNA double-strand breaks and the RAD50-RAD57 genes in *Saccharomyces*. Semin Cancer Biol 4:73–83

Gatei M, Young D, Cerosaletti KM, Desai-Mehta A, Spring K, Kozlov S, Lavin MF, Gatti RA, Concannon P, Khanna K (2000) ATM-dependent phosphorylation of nibrin in response to radiation exposure. Nat Genet 25:115–119

Gellert M (2002) V(D)J recombination: RAG proteins, repair factors, and regulation. Annu Rev Biochem 71:101–132

Goldberg M, Stucki M, Falck J, D'Amours D, Rahman D, Pappin D, Bartek J, Jackson SP (2003) MDC1 is required for the intra-S-phase DNA damage checkpoint. Nature 421:952–956

Gorski MM, Romeijn RJ, Eeken JC, De Jong AW, Van Veen BL, Szuhai K, Mullenders LH, Ferro W, Pastink A (2004) Disruption of *Drosophila* Rad50 causes pupal lethality, the accumulation of DNA double-strand breaks and the induction of apoptosis in third instar larvae. DNA Repair (Amst) 3:603–615

Gruber S, Haering CH, Nasmyth K (2003) Chromosomal cohesin forms a ring. Cell 112:765–777

Haber JE (1998) Mating-type gene switching in *Saccharomyces cerevisiae*. Annu Rev Genet 32:561–599

Haber JE (2000) Partners and pathways repairing a double-strand break. Trends Genet 16:259–264

Haering CH, Nasmyth K (2003) Building and breaking bridges between sister chromatids. Bioessays 25:1178–1191

Haering CH, Lowe J, Hochwagen A, Nasmyth K (2002) Molecular architecture of SMC proteins and the yeast cohesin complex. Mol Cell 9:773–788

Haering CH, Schoffnegger D, Nishino T, Helmhart W, Nasmyth K, Lowe J (2004) Structure and stability of cohesin's Smc1-kleisin interaction. Mol Cell 15:951–964

Hirano T (2005) Condensins: organizing and segregating the genome. Curr Biol 15:R265–275

Hirano M, Hirano T (2002) Hinge-mediated dimerization of SMC protein is essential for its dynamic interaction with DNA. Embo J 21:5733–5744

Hirano M, Hirano T (2004) Positive and negative regulation of SMC-DNA interactions by ATP and accessory proteins. Embo J 23:2664–2673

Hopfner KP, Tainer JA (2003) Rad50/SMC proteins and ABC transporters: unifying concepts from high-resolution structures. Curr Opin Struct Biol 13:249–255

Hopfner KP, Karcher A, Shin DS, Craig L, Arthur LM, Carney JP, Tainer JA (2000) Structural biology of Rad50 ATPase: ATP-driven conformational control in DNA double-strand break repair and the ABC-ATPase superfamily. Cell 101:789–800

Hopfner KP, Karcher A, Craig L, Woo TT, Carney JP, Tainer JA (2001) Structural biochemistry and interaction architecture of the DNA double-strand break repair Mre11 nuclease and Rad50-ATPase. Cell 105:473–485

Hopfner KP, Craig L, Moncalian G, Zinkel RA, Usui T, Owen BA, Karcher A, Henderson B, Bodmer JL, McMurray CT, Carney JP, Petrini JH, Tainer JA (2002) The Rad50 zinc-hook is a structure joining Mre11 complexes in DNA recombination and repair. Nature 418:562–566

Horejsi Z, Falck J, Bakkenist CJ, Kastan MB, Lukas J, Bartek J (2004) Distinct functional domains of Nbs1 modulate the timing and magnitude of ATM activation after low doses of ionizing radiation. Oncogene 23:3122–3127

Jessberger R (2003) SMC proteins at the crossroads of diverse chromosomal processes. IUBMB Life 55:643–652

Kaye JA, Melo JA, Cheung SK, Vaze MB, Haber JE, Toczyski DP (2004) DNA breaks promote genomic instability by impeding proper chromosome segregation. Curr Biol 14:2096–2106

Keeney S (2001) Mechanism and control of meiotic recombination initiation. Curr Top Dev Biol 52:1–53

Kim ST, Xu B, Kastan MB (2002) Involvement of the cohesin protein, Smc1, in Atm-dependent and independent responses to DNA damage. Genes Dev 16:560–570

Kimura K, Hirano T (1997) ATP-dependent positive supercoiling of DNA by 13S condensin: a biochemical implication for chromosome condensation. Cell 90:625–634

Kironmai KM, Muniyappa K (1997) Alteration of telomeric sequences and senescence caused by mutations in RAD50 of *Saccharomyces cerevisiae*. Genes Cells 2:443–455

Kuzminov A (2001) DNA replication meets genetic exchange: chromosomal damage and its repair by homologous recombination. Proc Natl Acad Sci USA 98:8461–8468

Lammens A, Schele A, Hopfner KP (2004) Structural biochemistry of ATP-driven dimerization and DNA stimulated activation of SMC ATPases. Curr Biol 14:1778–1782

Lee JH, Paull TT (2004) Direct activation of the ATM protein kinase by the Mre11/Rad50/Nbs1 complex. Science 304:93–96

Lee JH, Paull TT (2005) ATM activation by DNA double-strand breaks through the Mre11-Rad50-Nbs1 complex. Science 308:551–554

Lee JH, Ghirlando R, Bhaskara V, Hoffmeyer MR, Gu J, Paull TT (2003) Regulation of Mre11/Rad50 by Nbs1: effects on nucleotide-dependent DNA binding and association with ataxia-telangiectasia-like disorder mutant complexes. J Biol Chem 278:45171–45181

Lehmann AR, Walicka M, Griffiths DJ, Murray JM, Watts FZ, McCready S, Carr AM (1995) The rad18 gene of *Schizosaccharomyces Pombe* defines a new subgroup of the SMC superfamily involved in DNA repair. Mol Cell Biol 15:7067–7080

Lengronne A, Katou Y, Mori S, Yokobayashi S, Kelly GP, Itoh T, Watanabe Y, Shirahige K, Uhlmann F (2004) Cohesin relocation from sites of chromosomal loading to places of convergent transcription. Nature 430:573–578

Lewis LK, Storici F, Van Komen S, Calero S, Sung P, Resnick MA (2004) Role of the nuclease activity of *Saccharomyces cerevisiae* Mre11 in repair of DNA double-strand breaks in mitotic cells. Genetics 166:1701–1713

Liu PQ, Liu CE, Ames GF (1999) Modulation of ATPase activity by physical disengagement of the ATP-binding domains of an ABC transporter, the histidine permease. J Biol Chem 274:18310–18318

Lobachev KS, Gordenin DA, Resnick MA (2002) The Mre11 complex is required for repair of hairpin-capped double-strand breaks and prevention of chromosome rearrangements. Cell 108:183–193

Lobachev K, Vitriol E, Stemple J, Resnick MA, Bloom K (2004) Chromosome fragmentation after induction of a double-strand break is an active process prevented by the RMX repair complex. Curr Biol 14:2107–2112

Lou Z, Minter-Dykhouse K, Wu X, Chen J (2003) MDC1 is coupled to activated CHK2 in mammalian DNA damage response pathways. Nature 421:957–961

Lundblad V (2002) Telomere maintenance without telomerase. Oncogene 21:522–531

Luo G, Yao MS, Bender CF, Mills M, Bladl AR, Bradley A, Petrini JH (1999) Disruption of mRad50 causes embryonic stem cell lethality, abnormal embryonic development, and sensitivity to ionizing radiation. Proc Natl Acad Sci USA 96:7376–7381

Matsuura S, Tauchi H, Nakamura A, Kondo N, Sakamoto S, Endo S, Smeets D, Solder B, Belohradsky BH, Der Kaloustian VM, Oshimura M, Isomura M, Nakamura Y, Komatsu K (1998) Positional cloning of the gene for Nijmegen breakage syndrome. Nat Genet 19:179–181

Mickelson C, Wiberg JS (1981) Membrane-associated DNase activity controlled by genes 46 and 47 of bacteriophage T4D and elevated DNase activity associated with the T4 das mutation. J Virol 40:65–77

Milne GT, Jin S, Shannon KB, Weaver DT (1996) Mutations in two Ku homologs define a DNA end-joining repair pathway in *Saccharomyces Cerevisiae.* Mol Cell Biol 16:4189–4198

Mirzoeva OK, Petrini JH (2001) DNA damage-dependent nuclear dynamics of the Mre11 complex. Mol Cell Biol 21:281–288

Moore JK, Haber JE (1996) Cell cycle and genetic requirements of two pathways of non-homologous end-joining repair of double-strand breaks in *Saccharomyces cerevisiae.* Mol Cell Biol 16:2164–2173

Moreau S, Ferguson JR, Symington LS (1999) The nuclease activity of Mre11 is required for meiosis but not for mating type switching, end joining, or telomere maintenance. Mol Cell Biol 19:556–566

Nakada D, Matsumoto K, Sugimoto K (2003) ATM-related Tel1 associates with double-strand breaks through an Xrs2-dependent mechanism. Genes Dev 17:1957–1962

Nakanishi K, Taniguchi T, Ranganathan V, New HV, Moreau LA, Stotsky M, Mathew CG, Kastan MB, Weaver DT, D'Andrea AD (2002) Interaction of FANCD2 and NBS1 in the DNA damage response. Nat Cell Biol 4:913–920

Ohta K, Nicolas A, Furuse M, Nabetani A, Ogawa H, Shibata T (1998) Mutations in the MRE11, RAD50, XRS2, and MRE2 genes alter chromatin configuration at meiotic DNA double-stranded break sites in premeiotic and meiotic cells. Proc Natl Acad Sci USA 95:646–651

Ono T, Losada A, Hirano M, Myers MP, Neuwald AF, Hirano T (2003) Differential contributions of condensin I and condensin II to mitotic chromosome architecture in vertebrate cells. Cell 115:109–121

Paull TT, Gellert M (1998) The 3' to 5' exonuclease activity of Mre 11 facilitates repair of DNA double-strand breaks. Mol Cell 1:969–979

Paull TT, Gellert M (1999) Nbs1 potentiates ATP-driven DNA unwinding and endonuclease cleavage by the Mre11/Rad50 complex. Genes Dev 13:1276–1288

Revenkova E, Eijpe M, Heyting C, Hodges CA, Hunt PA, Liebe B, Scherthan H, Jessberger R (2004) Cohesin SMC1 beta is required for meiotic chromosome dynamics, sister chromatid cohesion and DNA recombination. Nat Cell Biol 6:555–562

Schar P, Fasi M, Jessberger R (2004) SMC1 coordinates DNA double-strand break repair pathways. Nucleic Acids Res 32:3921–3929

Schmitt L, Tampe R (2002) Structure and mechanism of ABC transporters. Curr Opin Struct Biol 12:754–760

Smith PC, Karpowich N, Millen L, Moody JE, Rosen J, Thomas PJ, Hunt JF (2002) ATP binding to the motor domain from an ABC transporter drives formation of a nucleotide sandwich dimer. Mol Cell 10:139–149

Soppa J, Kobayashi K, Noirot-Gros MF, Oesterhelt D, Ehrlich SD, Dervyn E, Ogasawara N, Moriya S (2002) Discovery of two novel families of proteins that are proposed to interact with prokaryotic SMC proteins, and characterization of the *Bacillus subtilis* family members ScpA, ScpB. Mol Microbiol 45:59–71

Stewart GS, Maser RS, Stankovic T, Bressan DA, Kaplan MI, Jaspers NG, Raams A, Byrd PJ, Petrini JH, Taylor AM (1999) The DNA double-strand break repair gene hMRE11 is mutated in individuals with an ataxia-telangiectasia-like disorder. Cell 99:577–587

Stewart GS, Wang B, Bignell CR, Taylor AM, Elledge SJ (2003) MDC1 is a mediator of the mammalian DNA damage checkpoint. Nature 421:961–966

Strick TR, Kawaguchi T, Hirano T (2004) Real-time detection of single-molecule DNA compaction by condensin I. Curr Biol 14:874–880

Tauchi H, Kobayashi J, Morishima K, van Gent DC, Shiraishi T, Verkaik NS, vanHeems D, Ito E, Nakamura A, Sonoda E, Takata M, Takeda S, Matsuura S, Komatsu K (2002) Nbs1 is essential for DNA repair by homologous recombination in higher vertebrate cells. Nature 420:93–98

Teng SC, Chang J, McCowan B, Zakian VA (2000) Telomerase-independent lengthening of yeast telomeres occurs by an abrupt Rad50p-dependent, Rif-inhibited recombinational process. Mol Cell 6:947–952

Torres-Rosell J, Machin F, Farmer S, Jarmuz A, Eydmann T, Dalgaard JZ, Aragon L (2005) SMC5 and SMC6 genes are required for the segregation of repetitive chromosome regions. Nat Cell Biol 7:412–419

Trujillo KM, Sung P (2001) DNA structure-specific nuclease activities in the *Saccharomyces cerevisiae* Rad50*Mre11 complex. J Biol Chem 276:35458–35464

Trujillo KM, Yuan SS, Lee EY, Sung P (1998) Nuclease activities in a complex of human recombination and DNA repair factors Rad50, Mre11, and p 95. J Biol Chem 273:21447–21450

Trujillo KM, Roh DH, Chen L, Van Komen S, Tomkinson A, Sung P (2003) Yeast xrs2 binds DNA and helps target rad50 and mre11 to DNA ends. J Biol Chem 278:48957–48964

Tsukamoto Y, Kato J, Ikeda H (1996) Effects of mutations of RAD50, RAD51, RAD52, and related genes on illegitimate recombination in *Saccharomyces cerevisiae*. Genetics 142:383–391

Udayakumar D, Bladen CL, Hudson FZ, Dynan WS (2003) Distinct pathways of nonhomologous end joining that are differentially regulated by DNA-dependent protein kinase mediated phosphorylation. J Biol Chem

Uhlmann F (2003) Chromosome cohesion and separation: from men and molecules. Curr Biol 13:R104–114

Usui T, Ogawa H, Petrini JH (2001) A DNA damage response pathway controlled by Tel1 and the Mre11 complex. Mol Cell 7:1255–1266

van Noort J, van Der Heijden T, de Jager M, Wyman C, Kanaar R, Dekker C (2003) The coiled-coil of the human Rad50 DNA repair protein contains specific segments of increased flexibility. Proc Natl Acad Sci USA 100:7581–7586

Varon R, Vissinga C, Platzer M, Cerosaletti KM, Chrzanowska KH, Saar K, Beckmann G, Seemanova E, Cooper PR, Nowak NJ, Stumm M, Weemaes CM, Gatti RA, Wilson RK, Digweed M, Rosenthal A, Sperling K, Concannon P, Reis A (1998) Nibrin, a novel DNA double-strand break repair protein, is mutated in Nijmegen breakage syndrome. Cell 93:467–476

Verkade HM, Bugg SJ, Lindsay HD, Carr AM, O'Connell MJ (1999) Rad18 is required for DNA repair and checkpoint responses in fission yeast. Mol Biol Cell 10:2905–2918

Volkov A, Mascarenhas J, Andrei-Selmer C, Ulrich HD, Graumann PL (2003) A prokaryotic condensin/cohesin-like complex can actively compact chromosomes from a single position on the nucleoid and binds to DNA as a ring-like structure. Mol Cell Biol 23:5638–5650

Ward JF (1988) DNA damage produced by ionizing radiation in mammalian cells: identities, mechanisms of formation, and reparability. Prog Nucleic Acid Res Mol Biol 35:95–125

Weitzer S, Lehane C, Uhlmann F (2003) A model for ATP hydrolysis-dependent binding of cohesin to DNA. Curr Biol 13:1930–1940

Wiltzius JJ, Hohl M, Fleming JC, Petrini JH (2005) The Rad50 hook domain is a critical determinant of Mre11 complex functions. Nat Struct Mol Biol 12:403–407

Yamaguchi-Iwai Y, Sonoda E, Sasaki MS, Morrison C, Haraguchi T, Hiraoka Y, Yamashita YM, Yagi T, Takata M, Price C, Kakazu N, Takeda S (1999) Mre11 is essential for the maintenance of chromosomal DNA in vertebrate cells. Embo J 18:6619–6629

Yazdi PT, Wang Y, Zhao S, Patel N, Lee EY, Qin J (2002) SMC1 is a downstream effector in the ATM/NBS1 branch of the human S-phase checkpoint. Genes Dev 16:571–582

Zdzienicka MZ (1996) Mammalian X ray sensitive mutants: a tool for the elucidation of the cellular response to ionizing radiation. Cancer Surv 28:281–293

Zhao S, Weng YC, Yuan SS, Lin YT, Hsu HC, Lin SC, Gerbino E, Song MH, Zdzienicka MZ, Gatti RA, Shay JW, Ziv Y, Shiloh Y, Lee EY (2000) Functional link between ataxia-telangiectasia and Nijmegen breakage syndrome gene products. Nature 405:473–477

Zhao X, Blobel G (2005) From the cover: A SUMO ligase is part of a nuclear multiprotein complex that affects DNA repair and chromosomal organization. Proc Natl Acad Sci USA 102:4777–4782

Zhao X, Blobel G (2005) A SUMO ligase is part of a nuclear multiprotein complex that affects DNA repair and chromosomal organization. Proc Natl Acad Sci USA

Zhu J, Petersen S, Tessarollo L, Nussenzweig A (2001) Targeted disruption of the Nijmegen breakage syndrome gene NBS1 leads to early embryonic lethality in mice. Curr Biol 11:105–109

Zhu XD, Kuster B, Mann M, Petrini JH, de Lange T (2000) Cell-cycle-regulated association of RAD50/MRE11/NBS1 with TRF2 and human telomeres. Nat Genet 25:347–352

Genome Dyn Stab (1)
D.-H. Lankenau: Genome Integrity
DOI 10.1007/7050_008/Published online: 2 December 2005

Mechanisms of Non-Homologous DNA End Joining: Aspects of In Vitro Assays

Petra Pfeiffer[1] (✉) · Steffi Kuhfittig-Kulle[2] · Wolfgang Goedecke[2]

[1]Institute of Genetics, University of Cologne, Zülpicher Str. 47, 50674 Cologne, Germany
petra.pfeiffer@uni-koeln.de

[2]Department of Genetics, Division of Biology and Geography,
University of Duisburg-Essen, Universitätsstr. 5, 45177 Essen, Germany

Abstract Double-strand breaks (DSB) in genomic DNA are a major threat to cell survival and chromosome integrity. In vertebrate cells, non-homologous DNA end joining (NHEJ) is the major pathway of DSB repair. Genetic studies in yeast, human and rodent cell lines displaying increased IR sensitivity and defects in DSB repair have provided insight in the genes involved in NHEJ. These genetic data have been confirmed and complemented by in vitro assays which have played a significant role in the elucidation and the dissection of the basic mechanisms underlying NHEJ. In vitro assays utilize model DNA substrates that carry defined DSB and thus provide information on the efficiency and fidelity of NHEJ in different cell systems. In contrast to investigations in living cells, in vitro assays facilitate the investigation of the functions of single proteins in the repair process itself so that their impact on the rejoining of different DNA end structures can be studied directly without interference by other cellular processes such as cell cycle and replication. In this chapter, we summarize the basic features of in vitro assays and give an overview over the different available cell-free systems which have facilitated the detailed analysis of NHEJ mechanisms in different vertebrate cells.

Keywords DSB repair · NHEJ (non-homologous DNA end joining) · Ligation · Cell-free extracts · In vitro assays

1 Introduction

Double-strand breaks (DSB) in genomic DNA are potentially lethal lesions because they can lead to the loss of essential genetic information. In contrast to base modifications, sugar damage, and single-strand breaks (SSB), DSB affect both DNA strands so that no complementary template is available for the restoration of the original sequence. Therefore, un-repaired DSB give rise to chromosome fragmentation and cell death, and mis-repaired DSB to mutations and chromosomal aberrations (CA) (van Gent et al. 2001) which can cause oncogenic transformation by activation of proto-oncogenes (e.g. insertions; translocations), and/or inactivation of tumor suppressor genes (e.g. deletions) (Khanna and Jackson 2001).

One important source of DSB is environmental mutagens such as ionizing radiation (IR) and clastogenic chemicals (e.g. bleomycin). However, DSB are also induced by a variety of endogenous factors produced in the normal cellular metabolism (e.g. reactive oxygen species: ROS) (Barnes 2002; Jackson 2002). An estimated ten DSB arise spontaneously in each S-phase because each SSB in a parental strand which is passed by the replication fork may be converted to a DSB in the corresponding sister chromatid (Haber 1999). Finally, DSB are intermediates in important processes such as meiotic recombination which is essential for germ-cell development (Dresser 2000), V(D)J recombination of immunoglobin (Ig) or T-cell receptor genes in B- and T-lymphocytes, respectively (Bassing et al. 2002), Ig heavy class switching (Honjo et al. 2002), and somatic hypermutation (Reynaud et al. 2003). All these examples demonstrate that DSB arise spontaneously in the cell at quite significant frequencies. Therefore, the efficient removal of DSB is, together with other DNA repair processes [base excision repair (BER); nucleotide excision repair (NER); mismatch repair (MMR)], of great importance for the maintenance of genome stability and thus for the prevention of oncogenic cell transformation. For this reason, DNA repair genes are often designated as "caretakers" in contrast to the classical "gatekeepers" which are involved in cell cycle control (Kinzler and Vogelstein 1997).

In mammalian cells, two main repair pathways are responsible for the removal of deleterious DSB: homologous recombination (HR) and non-homologous end joining (NHEJ) (Fig. 1). Both mechanisms complement each other, are strictly regulated and highly conserved in evolution from bacteria and yeast to mammals (Pfeiffer et al. 2004; Hefferin and Tomkinson 2005). Cells harbouring defects in NHEJ or HR, respectively, display increased sensitivity to IR (and other clastogenic agents) and chromosomal instability which indicates that both mechanisms are essential for DSB repair in mammalian cells and counteract genomic instability (Thacker and Zdzienicka 2004).

As indicated by the name, HR requires extended regions of sequence homology of several hundred base pairs and is able to restore the exact original sequence at the break by copying the missing sequence from the sister chromatid (Paques and Haber 1999). For the typical HR reaction, the Rad51 strand transfer protein and its five paralogues are essential with Rad51 forming a nucleoprotein filament with a DNA single-strand to invade a homologous double-strand (strand-invasion; d-loop formation) and to catalyse the subsequent strand exchange and Holliday junction resolution (Kawabata et al. 2005; Thacker 2005). Due to its dependence on sequence homology, HR is limited to late S- and G_2-phase of the mitotic cell cycle where the intact sister chromatid serves as a template for the repair of the broken chromosome by 2–3 orders of magnitude more often than the homologous or a heterologous chromosome (Johnson and Jasin 2000). The use of the sequence-identical sister chromatid in the mitotic cell cycle allows the restoration of the original

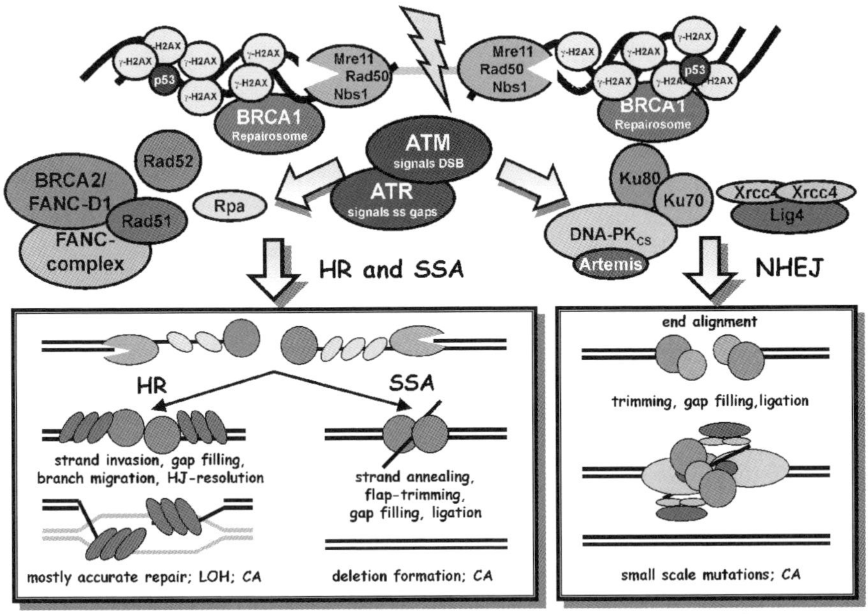

Fig. 1 Mechanisms of DSB repair and some of their most important proteins. The main features and proteins of HR, SSA and DNA-PK-dependent NHEJ are explained in the text. In addition, the schematic shows further important proteins that are involved in the regulation of general DSB-repair and the decision which of the three pathways is used. Interestingly, hypomorphic mutations in most of these genes cause certain human chromosome instability and cancer predisposition syndromes [MRE11: ataxia telangiectasia like disease (ATLD); NBS1: Nijmegen breakage syndrome; BRCA1 and 2: familial breast cancer; the 11 FANC genes: Fanconi Anemia (FA); ATM: ataxia telangiectasia (AT); ATR: ataxia telangiectasia related]. The phosphorylated variant form of the H2A core histone γH2AX occurs in chromatin within seconds after DSB induction and serves as a technical measure for the quantification of DSB induction and repair. Since these proteins are not discussed in detail in this chapter, the interested reader is referred to the following recent reviews (Fernandez-Capetillo et al. 2004; Kurz and Lees-Miller 2004; Shechter et al. 2004; Shivji and Venkitaraman 2004; Stracker et al. 2004; Surralles et al. 2004; Ting and Lee 2004; Wang and D'Andrea 2004; Jeggo and Lobrich 2005; Paull and Lee 2005)

sequence which is of great importance for the suppression of loss of heterozygosity (LOH) or translocations which would arise from usage of sequence homologies located on the homologous or a heterologous chromosome, respectively (Johnson and Jasin 2001). This is in contrast to the situation in meiosis where the sequence-similar homologous chromosome is used for HR to increase genetic diversity.

An error-prone variant of HR, the so-called single-strand annealing (SSA), often occurs between sequence repeats on the same or two heterologous chromosomes (ectopic recombination) (Pfeiffer et al. 2000, 2004). The error-proneness of this process is due to the fact that two homology regions interact directly with each other (without strand-exchange) leading to the loss of one repeat unit and the intervening sequence and thus giving rise to interstitial deletions or translocations. SSA is independent of Rad51 but requires the Rad52 strand-pairing protein and may function in mammalian cells as a back-up mechanism of accurate HR when no functional Rad51 nucleoprotein filament can be formed (Venkitaraman 2001).

In contrast to HR, NHEJ can principally dispense with sequence homology since it joins two DSB ends directly with each other. In mammalian cells, NHEJ occurs throughout the entire cell cycle and is used preferentially for the removal of IR-induced DSB and spontaneous DSB occurring outside of replication (Rothkamm et al. 2003). NHEJ is intrinsically error-prone because the original sequence is only restored if two complementary ends are precisely religated. If, however, two non-complementary ends (generated, e.g. by IR) are rejoined, they have to be converted first into a ligatable structure. The necessary enzymatic modifications (nucleolytic trimming, gap-filling) of the ends may generate base substitutions, insertions and deletions rendering this kind of DSB repair rather inaccurate.

In spite of this mutagenic potential, NHEJ is the primary mechanism of DSB repair in mammalian cells. The preference of NHEJ over HR in vertebrate cells may be explained, on the one hand, by the independence of NHEJ of sequence homology which facilitates DSB repair also during G_0 and G_1 and is responsible for the remarkable efficiency and velocity of the NHEJ reaction (most of IR-induced DSB are repaired within 2h by NHEJ). On the other hand, the mutagenic potential of HR may not be under-estimated either, because of the extremely high fraction ($\sim$ 45%) of repetitive sequences in the mammalian genome which can enhance genetic exchange between heterologous chromosomes by SSA and thus may cause deletions and translocations. This risk is minimized by a strict regulation (mediated by the breast cancer proteins BRCA1 and BRCA2) which couples HR to replication and thus limits the exchange of genetic material to the sister chromatids in the mitotic cell cycle (Shivji and Venkitaraman 2004; Ting and Lee 2004). Under these conditions, the risk of NHEJ-mediated DSB repair appears to be bearable since the probability that NHEJ-generated sequence alterations affect just a critical region of an essential expressed gene is estimated to be low due to the low proportion of coding sequences in the mammalian genome (1–3%) (Lieber et al. 2003; Pfeiffer et al. 2000, 2004).

The investigation of different Chinese hamster ovary (CHO) cell lines displaying extreme sensitivity to IR and defects in DSB repair led to the identification of the genes involved in NHEJ and V(D)J recombination (XRCC4 to XRCC7) (Jeggo 1998; Thacker and Zdzienicka 2003, 2004). The corresponding gene products are Xrcc4 which forms a protein complex with DNA ligase IV (Lig4) (Grawunder et al. 1997, 1998). Both proteins catalyse the final step of the NHEJ reaction and, in addition, perform essential functions in embryonic development (Barnes et al. 1998; Gao et al. 1998). XRCC5 and XRCC6 encode the DNA end-binding Ku70/80 heterodimer which interacts with the catalytic sub-unit of the DNA-dependent kinase DNA-PKcs (XRCC7) (Featherstone and Jackson 1999a,b). The loss of any of these three factors causes the murine SCID (severe combined immuno deficiency) phenotype (Jackson and Jeggo 1995; Jackson 2002). DNA-PK_{CS} not only participates in signal transduction after DNA damage and the recruitment of repair enzymes to the DSB site but also forms a complex with the Artemis nuclease (DNA-PK_{CS}:Art) whose inactivation causes the *human radiosensitive* (RS-SCID) phenotype (Moshous et al. 2001; Ma et al. 2002; Collis et al. 2005). Due to its endonucleolytic activity on single-stranded overhangs, Artemis plays a crucial role in the trimming of non-complementary DSB ends to convert them into ligatable structures which is necessary to facilitate the final step of DNA backbone ligation by Xrcc4/Lig4 (Jeggo and O'Neill 2002; Ma et al. 2005). Together, Ku70/80, DNA-PK_{CS}:Art, and Xrcc4/Lig4 are the three so-called core-factors of NHEJ and principally able to catalyse the rejoining of non-complementary DSB ends (Ma et al. 2004). Due to the nucleolytic trimming of the DSB ends by Artemis, the products of this reaction type contain deletions. In addition to the core factors, the DNA polymerases mu (Pol μ) and lambda (Pol λ) and terminal deoxynucleotidyl transferase (TdT) have been shown to be involved in the filling of small gaps (Mahajan et al. 1999, 2002; Lee et al. 2004). Finally, the Mre11/Rad50/Nbs1 (MRN) complex is, in addition to HR, also implicated in NHEJ (at least in yeast) (Zhang and Paull 2005). Although experimental evidence for a direct role in NHEJ of vertebrate cells is still lacking, it has been suggested that the nucleolytic activities of Mre11 are involved in the removal of unpaired bases and the long flexible hooks of Rad50 could function as bridges between either two sister chromatids (HR) or two DSB ends (NHEJ) (de Jager et al. 2001; Stracker et al. 2004).

Apart from the "classical" DNA-PK-dependent NHEJ pathway, there exists at least one more NHEJ pathway that is independent of DNA-PK and often generates small deletions at sites of microhomology (μhom; 3–7 bp) (Feldmann et al. 2000). Therefore, we designate this pathway as microhomology-mediated NHEJ (μhomNHEJ). Due to its feature of deletion formation at sites of homology, this mechanism is reminiscent of the Rad52-dependent SSA reaction. The latter, however, requires longer homology patches ($\geq$ 29 bp). μhomNHEJ takes over in cells in which the ~20-fold faster DNA-PK-dependent NHEJ reaction is inactivated due to mutations in XRCC4 to XRCC7, or LIG4, respectively (Cheong et al. 1999; DiBiase et al. 2000; Feldmann et al. 2000; Wang et al. 2001b, 2003). In spite of its similarity to the Rad52-dependent SSA reaction, the μhomNHEJ reaction, however, does not utilize gene products of the Rad52 epistasis group (Wang et al. 2001a). Instead, it appears that the Xrcc1/Lig3 complex and poly (ADP-Ribose) polymerase-1 (PARP-1) participate in μhomNHEJ (Göttlich et al. 1998; Wang et al. 2003, 2005; Audebert et al. 2004). Since these proteins are normally involved in BER and SSB repair, it is possible that other BER enzymes such as DNA polymerase beta and/or delta/epsilon (Pol β, Pol δ/ε) and the flap endonuclease 1 (Fen1) also play a role in µhomNHEJ. The unexpected involvement of BER factors in a "backup" pathway of DNA-PK-dependent NHEJ is interesting, since it demonstrates the existence of a tight network of different DNA repair pathways that can complement and substitute for each other.

2
Cell-Free Systems for the Analysis of NHEJ

Most of the data mentioned in the Introduction were obtained by genetic analysis of mutants derived from yeast, rodent and human cells displaying increased IR sensitivity and defects in DSB repair. However, in vitro assays are equally important to complement and confirm genetic data and thus have played a significant role in the elucidation of DSB repair, and especially the NHEJ pathway. In contrast to investigations in living cells, such biochemical approaches facilitate the characterisation of the functions of single proteins (e.g. by inhibition or stimulation) in the repair process itself so that their impact on the rejoining of different DNA end structures can be studied directly without the interference by other cellular processes like cell cycle and replication. The first NHEJ in vitro system was already available in 1988 in the form of cell-free extracts from eggs of the African clawed frog *Xenopus laevis* and has contributed significantly to the understanding of the basic mechanisms of NHEJ (Pfeiffer and Vielmetter 1988; Thode et al. 1990). Later, a variety of other cell-free systems derived from mammalian cells were developed which also opened the possibility of studying mutant cell lines with specific defects in NHEJ [e.g. (Daza et al. 1996; Feldmann et al. 2000); see also Table 1]. Progress in the development of such in vitro assays has recently culminated in the reconstitution of NHEJ by use of the purified core factors (Ku70/80; DNA-PK_{CS}:Art; Xrcc4/Lig4) and specific polymerases (Pol μ; Pol λ; TdT) (Ma et al. 2004).

In this chapter, we shall not discuss any data on NHEJ derived from in vivo studies but report only on NHEJ in vitro studies performed in cell-free extracts from frog eggs and mammalian cells. It is worth mentioning that there exists a large variety of different cell-free NHEJ systems which differ in their protocol of preparation, the types of substrates (see paragraph "Substrates") used and products formed, and the efficiency and fidelity of NHEJ. Since only few of these will be discussed in any detail here, the interested reader is referred to Labhard's review on the specific features of the various different cell-free systems established until 1999 (Labhart 1999b) and to Table 1 in which the majority of cell-free systems developed until 2005 are listed.

Table 1 List of different cell-free NHEJ systems developed in the period from 1988 until today. Abbreviations: CE: cell extract; NE: nuclear extract; WCE: whole cell extract; RE: restriction endonuclease; TFO: triplex forming oligo (see paragraph "Substrates")

Cells/extract type	Investigation	Substrates	Refs.
Xenopus egg WCE	NHEJ of complementary and non-complementary ends: overlap and fill-in pathway	RE-linearized plasmid-DNA	(Pfeiffer and Vielmetter 1988; Pfeiffer et al. 1994b; Pfeiffer et al. 1994a; Thode et al. 1990)
Xenopus oocyte WCE	NHEJ of complementary and non-complementary ends; deletion formation by μhomNHEJ	RE-linearized plasmid-DNA	(Lehman et al. 1994)
Human cells NE	Efficiency and fidelity of ligation; deletion formation by μhomNHEJ	RE-linearized plasmid DNA	(Fairman et al. 1992; North et al. 1990; Thacker et al. 1992)
Xenopus egg WCE	Requirements of the DNA substrates for NHEJ: minimal length of DNA duplex; maximal length of overhangs and ends	Hairpin oligos	(Beyert et al. 1994; Reichenberger et al. 1996; Reichenberger and Pfeiffer 1998)
Human cell NE	NHEJ of complementary and non-complementary ends: no ccc; only H:H and T:T multimers	RE-linealized plasmid DNA	(Derbyshire et al. 1994; Nicolas and Young 1994; Nicolas et al. 1995)
Human cell CE	NHEJ of complementary and non-complementary ends	RE-linealized plasmid DNA	(Boe et al. 1995)
HeLa WCE	NHEJ of radiation-induced DSB in genomic DNA	Nuclei from irradiated agarose-embedded human cells	(Ganguly and Iliakis 1995; Cheong et al. 1996)
Calf thymus WCE; protein fractions	NHEJ of complementary and non-complementary ends	RE-linealized plasmid DNA	(Johnson and Fairman 1996; Mason et al. 1996)
Human, mouse cells and *Xenopus* egg WCE	NHEJ of complementary and non-complementary ends	RE-linealized plasmid DNA	(Daza et al. 1996; Daza et al. 1997)
Xenopus egg WCE	Role of DNA-PK for NHEJ fidelity of free-radical-induced DSB	Linear plasmid DNA with 3′-phosphoglycolate ends	(Gu et al. 1996; Gu et al. 1998)

Table 1 (continued)

Cells/extract type	Investigation	Substrates	Refs.
HeLa WCE	NHEJ of bleomycin- and RE-induced DSB in genomic DNA	Nuclei from agarose-embedded human cells treated with bleomycin or RE	(Cheong and Iliakis 1997)
Xenopus egg WCE; protein fractions	Evidence for DNA-PK-independent NHEJ (μhomNHEJ)	RE-linearized plasmid DNA	(Göttlich et al. 1998)
Human cell CE	Role of Ku, DNA-PK, and Xrcc4/Lig4 for ligation	RE-linearized plasmid DNA	(Baumann and West 1998)
HeLa WCE	Comparison of NHEJ of radiation-induced DSB in genomic DNA and RE-induced DSB in naked plasmid DNA	Nuclei from irradiated agarose-embedded human cells; RE-linearized plasmid DNA	(Cheong et al. 1998)
Xenopus egg WCE	Role of Ku for NHEJ	RE-linearized plasmid DNA	(Labhart 1999a)
Mouse testis WCE	Ligation of complementary ends	RE-linearized plasmid DNA	(Sathees and Raman 1999)
Mitochondria	Ligation of complementary ends	RE-linearized plasmid DNA	(Lakshmipathy and Campbell 1999)
CHO wt & xrs6 mutant WCE	Role of Ku for NHEJ efficiency and fidelity	RE-linearized plasmid DNA	(Feldmann et al. 2000)
CHO wt & xrs6 mutant WCE	Role of Ku for NHEJ fidelity of a site-specific free-radical-induced DSB	Linear plasmid DNA with 3′-phosphoglycolate ends	(Chen et al. 2001)
Human 180BR WCE	Role of Lig4 for DNA-PK-dependent and -independent NHEJ	RE-linearized plasmid DNA	(Wang et al. 2001b)
Human HeLa CE	Inhibition of NHEJ by aphidicolin; potential role of pol α in NHEJ	RE-linearized plasmid DNA	(Pospiech et al. 2001)
Human Fanconi A anemia (FA) fibroblasts NE	Role of FANC-A, -C, -D for NHEJ	RE-linearized plasmid DNA	(Lundberg et al. 2001)

Table 1 (continued)

Cells/extract type	Investigation	Substrates	Refs.
Human cells CE; purified proteins	Association of Pol μ with Ku and Xrcc4/Lig4; role in NHEJ	oligo-nucleotides	(Mahajan et al. 2002)
CHO wt WCE	Efficiency and fidelity of NHEJ of a site-specific radiation-induced DSB	^{125}I-TFO-linearized plasmid DNA	(Odersky et al. 2002)
Human breast cancer cell lines WCE	Efficiency and fidelity of NHEJ	RE-linearized plasmid DNA	(Merel et al. 2002)
Human cells WCE	Role of BRCA1 for NHEJ	RE-linearized plasmid DNA	(Zhong et al. 2002)
Human HeLa cells CE; protein fractions	Role of phosphorylation by DNA-PKcs for two different NHEJ pathways	radio-labelled RE-linearized plasmid DNA	(Udayakumar et al. 2003)
CHO wt XR-1 mutant WCE; HeLa NE	Role of Xrcc4/Lig4 and Pol λ for NHEJ of a site-specific free-radical-induced DSB	RE-linearized plasmid DNA with 3′-phospho-glycolate ends	(Lee et al. 2003; Lee et al. 2004)
Human HeLa, MO59K and MO59J WCE	Role of Ku and DNA-PK_{CS} for DNA-PK-dependent and -independent NHEJ	RE-linearized plasmid DNA	(Wang et al. 2003; Perrault et al. 2004)
Human cell lines and primary cells of various origin CE	Minimisation of cell number required for extract preparation	radio-labelled RE-linearized plasmid DNA	(Diggle et al. 2003)
Human cell WCE	Dependence of NHEJ and substrate degradation on DNA and protein concentration	RE-linearized plasmid DNA	(Blanco et al. 2004)
Human HeLa, hamster AA8, EM9, mouse $PARP^{-/-}$ NE	Role of PARP-1 and Xrcc1/lig3 for NHEJ and μhomNHEJ	oligo-nucleotides	(Audebert et al. 2004)
Purified proteins: Ku, DNA-PK/Art.; Xrcc4/Lig4; Pol μ, λ, TdT	Reconstitution of various types of NHEJ	oligo-nucleotides	(Ma et al. 2004)
Human HeLa WCE and NE; protein fractionation	Role of Lig3 for DNA-PK-independent NHEJ	RE-linearized plasmid DNA	(Wang et al. 2005)

3 Basic Mechanisms of in Vitro NHEJ

The simplest NHEJ reaction is the accurate ligation of cohesive or blunt ends and would minimally require Ku70/80, DNA-K_{CS} and Xrcc4/Lig4 (Fig. 2A) (Ma et al. 2004). By contrast, the rejoining of non-complementary ends requires additional factors such as nucleases (e.g. Artemis) and polymerases which modify the ends in a way that they can be ligated in the final reaction step (Figs. 2B to E).

The type of NHEJ reaction is determined by the pair of DSB ends which will be rejoined, as well as by the proteins present in the cell-free system (e.g. Ku70/80; DNA-PK_{CS} etc.). In this context, it is important to define the terms "accurate" and "inaccurate" NHEJ. While it is obvious that "accurate" ligation of complementary RE ends restores the original restriction site, the definition of accurate NHEJ is not self-evident because the joining of non-complementary ends necessarily causes a change in the original sequence. Still, general rules for NHEJ were established because extracts from *Xenopus* eggs and NHEJ-proficient mammalian cells generate highly reproducible spectra of junctions by two main NHEJ pathways designated as "overlap"- and "fill-in"-mechanism (Figs. 2B and C) (Daza et al. 1996; Pfeiffer and Vielmetter 1988). While the overlap pathway joins anti-parallel single-stranded overhangs (5′/5′; 3′/3′), the fill-in pathway joins abutting DNA ends (bl/5′; bl/3′; 5′/3′). In the first case, the ends form, by pairing of single fortuitously complementary bases, short complete or partial overlaps which may contain mismatches at internal or external positions and which determine the patterns of subsequent gap-filling (probably by Pol μ and/or λ) and/or removal of unpaired bases (probably by Artemis) (Fig. 2B) (Pfeiffer et al. 1994a,b). In the second case, the sequences of participating 5′- and/or 3′-overhangs are

Fig. 2 Types of different NHEJ reactions as observed in cell-free systems from *Xenopus* eggs, rodent and human cells and their predicted protein requirements (Daza et al. 1996; Feldmann et al. 2000; Pfeiffer and Vielmetter 1988). The left column shows the different possible configurations of DSB ends as generated by restriction endocucleases. White letters on black ground indicate complementary base pairs. ligation (*black diamonds*) of complementary cohesive or blunt ends; accurate NHEJ of anti-parallel ends by formation of short overlaps (complete vs. partial) at sites of single base pairs [MM: mismatch at an internal (int.) or external (ext.) position of an overlap; *small arrowheads*: fill-in of 5′-overhangs; *black triangles*: fill-in of 3′-overhangs; *small arrows:* exonucleolytic trimming of unpaired bases]. (Pfeiffer et al. 1994a,b); accurate NHEJ of abutting ends by fill-in (Thode et al. 1990). inaccurate NHEJ by blunting of 5′ and/or 3′-overhangs. Blunting can occur on any terminus configuration either by fill-in of 5′-overhangs or exonucleolytic degradation (*small arrows*) of 5′- or 3′-overhangs. (Feldmann et al. 2000; Ma et al. 2004) inaccurate NHEJ by µhomNHEJ can occur on any terminus configuration (µhom-priming: microhomology priming; µhom-ligation: microhomology ligation; *oblique arrow:* site of endonuclease cleavage) (Göttlich et al. 1998) ▸

fully preserved by fill-in DNA synthesis in a process in which the ends are transiently held together by non-covalent interactions (alignment) while the 3′-hydroxyl group of the 5′-overhang or blunt end is used as a primer to direct

Terminus configuration	Mechanism	Proteins required
A) complementary ends GGGGATCCTC / CCCCTAGGAG 5′-cohesive ACCTGCAGCC / TGGACGTCGG 3′-cohesive CGCCC GGGGA / GCGGG CCCCT blunt	**accurate NHEJ** <u>**ligation**</u> nick ligation	**Ku70/80/DNA-PK$_{CS}$ Xrcc4/Lig4**
B) non-complementary <u>**antiparallel ends**</u> <u>**"complete overlaps"**</u> GGGGTACCGG / CCCCTAGGCC 5′/5′ (2 int. MM) TCGAGCTGCC / AGCACGTCGG 3′/3′ (ext. MM) <u>**"partial overlaps"**</u> GGG››TCGACCT / CCCCTAG‹‹GGA 5′/5′ (no MM) CCGGTAC›GCC / GGC‹ACGTCGG 3′/3′ (int. MM) TTGAGCT CGG / AAC‹CATGGCC 3′/3′ (2 ext. MM)	**accurate NHEJ** <u>**overlap of antiparallel ends**</u>**:** <u>"complete overlaps"</u> nick ligation <u>"partial overlaps"</u> gap-filling and/or nucleolytic trimming; nick ligation	**Ku70/80/DNA-PK$_{CS}$ Xrcc4/Lig4** **Ku70/80/DNA-PK$_{CS}$:Art. Xrcc4/Lig4 pol μ,λ**
C) non-complementary <u>**abutting ends**</u> CCCTCGACCT / GGG‹‹‹‹GGA bl/5′ (5′-fill-in) CCC››››GCC / GGGACGTCGG bl/3′ (3′-fill-in) GGG››››››››GCC / CCCCTAGACGTCGG 5′/3′ (5′/3′-fill-in)	**accurate NHEJ** <u>**fill-in of abutting ends**</u>**:** gap-filling; nick ligation	**Ku70/80/DNA-PK$_{CS}$:Art. Xrcc4/Lig4 pol μ,λ**
D) non-complementary ends (any; s. B & C) GGG›››› GCC / CCCCTAGACGTCGG	**inaccurate NHEJ** <u>**blunting**</u> nucleolytic trimming and/or gap-filling; nick ligation	**Ku70/80/DNA-PK$_{CS}$:Art. Xrcc4/Lig4 pol μ,λ** any of these factors can lack
E) non-complementary ends (any; s. B & C) CATGATCATGGG / GTACTAGTACCC GATCTGCGGGAGT / ACGTCTAGACGCCCTCA GATCTGC CATGATC›››GGGAGT / GTACTAGACGCCCTCA	**inaccurate NHEJ** <u>**μhomNHEJ**</u>**:** μhom-priming μhom-ligation	PARP-1 Xrcc1/Lig3 pol β,ε/δ ? Fen1 ? DNA helicase?

repair synthesis of the 3′-overhang (Fig. 2C) (Thode et al. 1990). Since overlap and fill-in are the main pathways in cells proficient in DNA-PK-dependent NHEJ and tend to preserve the sequences of 5′- as well as 3′-overhangs by fill-in DNA synthesis, both mechanisms are here defined as accurate NHEJ.

By contrast, accurate NHEJ is strongly reduced in extracts from mutant cells lacking crucial NHEJ core factors like Ku80, Xrcc4 or DNA-PK_{CS}. Instead, these cells generate by "blunting" or μhomNHEJ different types of junctions which contain deletions and are therefore designated here as "inaccurate" (Figs. 2D and E) (Feldmann et al. 2000). Blunting can occur on any terminus configuration either by fill-in of 5′-overhangs or exonucleolytic degradation of 5′- or 3′-overhangs. Although this mechanism is also found in NHEJ-proficient cells at very low frequencies, it apparently does not necessarily require all NHEJ core factors because it can occur in cells lacking either Ku70/80, DNA-PK_{CS}, or Xrcc4/Lig4 (see also Fig. 7) (Feldmann et al. 2000; Ma et al. 2004). Which additional factors are involved in the process of blunting is not clear at the moment.

Inaccurate NHEJ by μhomNHEJ can also join any terminus configuration but can occur in the absence of the NHEJ core factors (Göttlich et al. 1998; Feldmann et al. 2000). Microhomology (μhom) patches present in the duplex adjacent to the DSB are exposed on long single-strands by helicase unwinding and/or exonucleolytic degradation of the ends and used for annealing. The GATC patch of the trimmed left hand duplex in Fig. 2E is located directly at a 3′-end which can serve, upon annealing, for DNA fill-in synthesis (μhom-priming). The CTAG patch of the trimmed right hand duplex comes into direct adjunction with the recessed strand of the left hand duplex facilitating nick ligation (μhom-ligation). In the following step, unpaired flap ends can be removed by exonucleolytic digestion or by endonucleolytic cleavage (e.g. by Fen1). As indicated in the Introduction, it is likely that this process is independent of the NHEJ core factors but involves factors usually involved in BER and SSB repair (PARP-1; Xrcc1/Lig3) (Göttlich et al. 1998; Wang et al. 2005; Audebert et al. 2004).

4 The NHEJ in Vitro Assay

4.1 Substrates

Most in vitro studies of NHEJ make use of restriction endonucleases (RE) to introduce defined DSB in plasmids offered as substrates for NHEJ in cell-free extracts (Fig. 3; see also Table 1). The fact that RE induce no other lesions but DSB which are exactly defined with respect to their structure (depending on the enzyme used: single-stranded 5′- or 3′-overhangs or blunt ends;

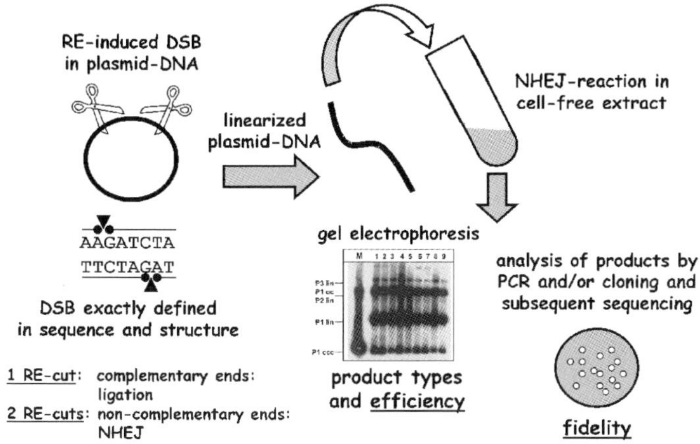

Fig. 3 Principle of the NHEJ in vitro assay. A plasmid linearized with one or two restriction endonuclease (RE) is incubated in a cell-free extract from *Xenopus* eggs or mammalian cells. The products resulting from the NHEJ reaction are separated by gel electrophoresis to identify the different types of products (circular monomers vs. linear multimers; see "Products of in vitro NHEJ" and Fig. 6) and to measure the efficiency of the reaction. PCR amplification of linear products and/or cloning of circular products in *E.coli* permits the analysis of the repair sites (junctions) and thus to assess the basic mechanisms and fidelity of the NHEJ reaction (see paragraph "Basic mechanisms of in vitro NHEJ")

always 3′-hydroxyl and 5′-phosphate) and position within a given DNA sequence has greatly facilitated the study of the mechanisms of NHEJ in the above-mentioned systems by comparing the original DSB ends and the resulting repair site (junction).

Plasmid substrates generated by cleavage with a single RE have complementary ends that permit the analysis of NHEJ by accurate ligation of cohesive 5′- or 3′-ends or blunt (bl) ends. Substrates generated by cleavage with two different RE contain non-complementary DNA ends (5′/5′; 3′/3′; bl/5′; bl/3′; 5′/3′) that permit the analysis of various types of NHEJ reactions which are more complex than "simple" ligation because the ends must be converted first into a ligatable structure by fill-in DNA synthesis and/or nucleolytic trimming of unpaired bases (for details see Fig. 2). In some cases, NHEJ restores a restriction site which then can be used for the easy analysis of the junctions and allows the quick and cheap evaluation of larger, statistically relevant numbers of junctions without the need for expensive sequence analysis. A list of RE-substrates generating junctions that can be studied by simple restriction analysis is shown in Fig. 4.

Other studies made use of plasmids containing a site-specific free radical-mediated DSB harboring 3′-phosphoglycolate ($-PO_4CH_2COO^-$) ends, a chemical modification often induced by IR (Bennett et al. 1996). In this procedure, an oligonucleotide is cleaved by treatment with bleomycin at a specific

substrate	terminus configuration	main junctions	RE-cut
1. Bam/Bam 5'-coh.	CGGG GATCCTCT GCCCCTAG CAGA	a) CGGGGATC | GATCCTCT 0/0: fill-in (blunting) b) CGGGGATC | CTCT 0/-4: ovlp; acc LIG	a) no b) Bam (0/-4)
2. Pst/Pst 3'-coh.	GACCTGCA GCCC CTGG ACGTCCCC	a) GACCTGCA | GCCC 0/-4: ovlp; acc LIG b) GACC | GCCC -4/-4: bl/bl (blunting)	a) Pst (0/-4) b) no
3. Sma/Sma blunt	TCGCCC GGGGAT AGCGGG CCCCTA	a) TCGCCC | GGGGAT 0/0: bl/bl: acc. LIG	a) Sma (0/0)
4. Bam/Asp 5'/5' NHEJ	CGGG GTACCGGA GCCCCTAG GCCT	a) CGGGGATC | GTACCGGA 0/0: fill-in (blunting) b) CGGGG(AT)C | CGGA 0/-4: ovlp; acc NHEJ	a) no b) Bam and / or Asp (ovlp)[1]
5. BstX/BstX 3'/3'	CCACTAAG GTGG GGTG AAACCACC	a) CCACT(AA)G | GTGG 0/-4: ovlp; acc NHEJ b) CCAC | GTGG -4/-4: bl/bl (blunting)	a) BstX1 (ovlp)[2] b) no
6. Ava/Hind 5'/bl	TCGC GACCTG AGCGGGCC GTCGAC	a) TCGCCCGG | GACCTG 0/0: fill-in; acc NHEJ	a) Ava (0/0)
7. Sma/Pst bl/3'	TCGCCC GCCC AGCGGG ACGTCGGG	a) TCGCCC | TGCAGCCC 0/0: fill-in; acc NHEJ b) TCGCCC | GCCC 0/-4: bl/bl (blunting) c) TCGCCC | 0/-8: µhomNHEJ	a) Pst (0/0; -1/0; -2/0) b) no c) no
8. Bam/Pst 5'/3'	CGGG GCCC GCCCCTAG ACGTCGGG	a) CGGGGATC | TGCAGCCC 0/0: fill-in; acc NHEJ b) CGGGGATC | CCC 0/0: fill/bl (blunting)	a) Pst (0/0) b) no
9. Sac/Sal 3'/5'	TTCGAGCT TCGACCTG AAGC GGAC	a) TTCGAGCT | TCGACCTG 0/0: fill-in; acc NHEJ b) TTCG | TCGACCTG -4/0: bl/fill (blunting) c) TTCGA | CCTG -3/-4: µhomNHEJ	a) no b) Sal (-4/0) detects inacc. NHEJ c) no
10. Ava/Kpn 5'/3'	TCGC CGGA AGCGGGCC CATGGCCT	a) TCGCCCGG | GTACCGGA 0/0: fill-in; acc NHEJ b) TCGCCCGG | CGGA 0/-4: fill/bl (blunting) c) TCGCCCGG | GGA 0/-5: fill/del	a) Ava (0/0) and Kpn (-1/0; 0/0) b) no c) Ava (0/-5; 0/-6)

Fig. 4 List of RE substrates with complementary ends (1–3) to measure NHEJ by accurate ligation and with various pairs of non-complementary ends (4–10) to measure NHEJ of antiparallel overhangs (4,5) and abutting ends (6–10). Corresponding structures and sequences of ends are shown in the column "terminus configuration" with the RE sequence highlighted in *light grey* and complementary base pairs and µhom patches marked in *white letters on black ground*. Main types of junctions resulting by accurate (acc) ligation or NHEJ [fill-in- or overlap (ovlp) mode; see Fig. 2 and text below], and µhomNHEJ are shown as top strand sequences with *vertical lines* indicating the junction breakpoint. Negative numerals indicate the number of bases lost from the left or right terminus, respectively. *Underlined* sequences mark restored RE-sites that can be used for RE-analysis (also see right column). Note that segregation in *E.coli* of the mismatches in the junctions of substrates # 4 and 5 generates two different sequences [[1]the T:T and A:A mismatches segregate in T:A and A:T (Asp718^S) and A:T and T:A (Bam^S); [2]the two A:A mismatches segregate in A:T and A:T (BstX^S) and T:A and T:A (BstX^S)] (Pfeiffer et al. 1994a)

site to generate a 3′-phosphoglycolate sugar damage in one strand. The damaged strand is gel purified and ligated to the end of an RE-linearized vector to generate a substrate mimicking a site-specific radiation-induced DSB. Another approach utilized the Auger-electron-emission of ^{125}Iodine coupled to a triplex-forming oligo (^{125}I-TFO) to induce site-specific radiation-induced DSB within a ~ 10 bp polypurine:polypyrimidine sequence of a given vector (Odersky et al. 2002). Both substrate types allow the investigation of the more complex NHEJ reactions dealing with radiation-damaged DNA (see Table 1).

The use of oligonucleotides in cell-free NHEJ assays has already early opened the possibility to determine the limitations of NHEJ with respect to the minimal length of the DNA duplex and the maximal length of the single-stranded overhangs to be joined (Beyert et al. 1994). For this, hairpin-shaped oligonucleotides carrying a (^{32}P)-phosphate-label at an internal phosphatase-

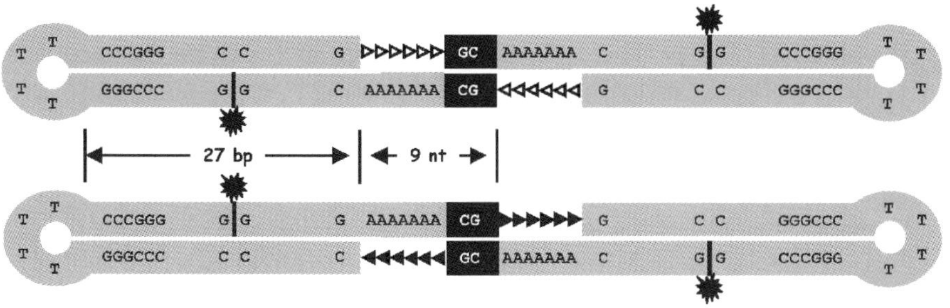

Fig. 5 General structure of hairpin-shaped oligonucleotides used as substrates in cell-free NHEJ-assays. 27 base pairs (bp) is the minimal length the duplex stems of both interacting hairpins must have for detectable NHEJ. At the same time, the length of the interacting single-stranded overhangs may not exceed 9 nucleotides (nt) (Beyert et al. 1994; Reichenberger et al. 1996). The positions of the internal (^{32}P)-phosphates used for detection are marked by asterisks. *White on black letters* in the single-stranded overhangs symbolize base matches used in the NHEJ reaction (see paragraph "Basic mechanisms of in vitro NHEJ"). *Open triangles* indicate fill-in of 5′-overhangs, *black triangles* fill-in of 3′-overhangs primed at the partner terminus. NHEJ of two hairpins generates dumbell-shaped products which, together with the substrates, are separated in sequencing gels and thus not only allow an exact analysis of the products formed but also to follow the fate of un-reacted substrates molecules

resistant position were constructed. The advantage of these substrates is that the length of the hairpin duplex as well as the length and sequence of the overhangs can be varied freely (Fig. 5). Use of these substrates in *Xenopus* egg extracts showed for the first time that NHEJ requires a minimal duplex length of 27 bp (corresponding to ∼ 2.5 double-helical turns) of both interacting DSB partners which corresponds well to the minimal duplex length required for Ku and DNA-PK$_{CS}$ (Yaneva et al. 1997). In addition, use of the hairpin substrates demonstrated that the length of the overhangs may not exceed 10nt (otherwise NHEJ will fail) and opened the possibility to follow the fate of the substrate DNA which, in the case of too long overhangs, was degraded by a "decanuclease" that releases single-stranded 10mers from overhanging and recessed 5′-ends (Beyert et al. 1994; Reichenberger et al. 1996). Interestingly, the shortened 5′-overhangs can subsequently serve as substrates for NHEJ. It is possible that this specific processing is conducted by the Artemis nuclease (Ma et al. 2002).

5 Products of in Vitro NHEJ

Early studies showed already that the extract-mediated NHEJ reaction can convert all types of the standard RE-cleaved plasmid substrates into monomeric open circular (oc) intermediates and covalently closed circles (ccc), and various linear multimers which can be separated by agarose gel electrophoresis (Fig. 6). The different types of products can be detected best by Southern blotting or by in situ gel hybridization. This is much more sensitive than the alternatively used direct staining with ethidium bromide and allows precise quantification of the NHEJ reaction in a phosphor imager (Pfeiffer et al. 2005).

Sequence analysis of the junctions created in the extract facilitates the investigation of the mechanisms and fidelity of the NHEJ reaction which is of particular interest for the comparison of the different NHEJ pathways in NHEJ-proficient cells and cells deficient in certain factors involved in the DNA-PK-dependent NHEJ machinery. Isolation of single NHEJ-events for sequence analysis of the junctions is achieved by two different strategies: firstly, transformation of total products in *E. coli* which results in preferential cloning of the junctions in circular products (with decreasing efficiency for ccc > oc ≫ lin.) and secondly, PCR amplification of gel-purified linear multimers (due to their palindromic structure, H:H and T:T multimers are resistant to PCR amplifica-

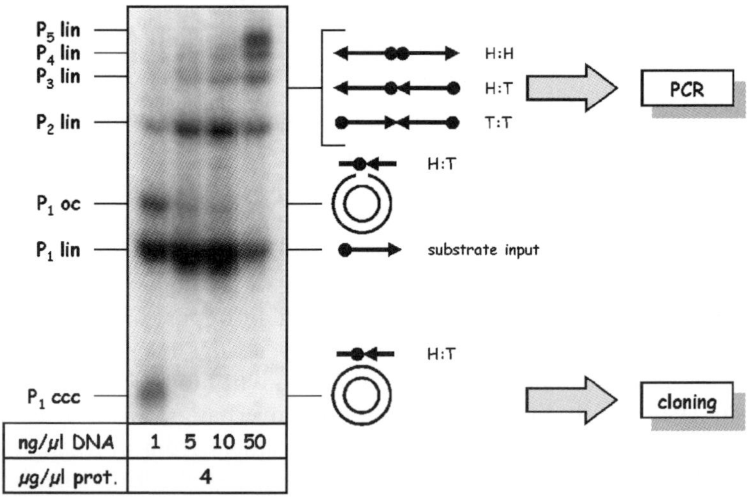

Fig. 6 Separation of a typical NHEJ reaction in an agarose gel (different concentrations of RE-cleaved plasmid 1–50 ng/µl tested with a fixed concentration of extracted protein of 4 µg/µl). Incubation of linear plasmid substrate (P1 lin) yields open circle (oc) and covalently closed circular (ccc) monomers (P1), linear (lin) dimers (P2) and higher multimers (P3–5). The corresponding orientations of head-to-tail (H:T), head-to-head (H:H), and tail-to-tail (T:T) junctions is shown on the *right side*. For further analysis, linear multimers are subjected to PCR and circular products to cloning in *E. coli*. The balance between circular monomers and linear multimers is dependent on the DNA/protein ratio used in the assay. The optimal balance allowing the formation of circular and linear products is achieved at a low DNA/protein ratio (1 ng/µl DNA and 4–8 µg/µl extract protein as final concentration in a standard 10 µl NHEJ assay) (Pfeiffer et al. 2005). Note that the same amount of DNA (2 ng) is loaded in each lane

tion so that only H:T multimers can be amplified) and subsequent sub-cloning in *E. coli* to produce single clones suitable for sequencing.

The advantage of those cell-free systems that generate both types of products is that both NHEJ pathways [the accurate DNA-PK-dependent and inaccurate DNA-PK-independent (blunting and μhomNHEJ)] are usually active under these conditions so that both mechanisms can be studied in parallel (Feldmann et al. 2000; Pfeiffer et al. 2005). Interestingly, the two mechanisms can be distinguished by the products they form: while the DNA-PK-dependent pathway forms both circular and multimeric products, the DNA-PK-independent pathway forms mainly linear multimers so that extracts made from cells deficient in the DNA-PK-dependent pathway can be easily identified by the absence of circular products (Fig. 8A) (Göttlich et al. 1998; Labhart 1999a; Feldmann et al. 2000; Wang et al. 2005).

	NHEJ of 3′/ 5′-fill-in substrate: (Sac/Sal)			delet.	RE-cut	CHO-K1	xrs6	XR-1	XR-C1
	ACGGAATTCGAGCT TCGACCTGCAGCCC TGCCTTAAGC GGACGTCGGG ACGGAATTCGAGCT TCGACCTGCAGCCC					Σ 49	Σ 40	Σ 21	Σ 31
1	ACGGAATTCGAGCT TCGACCTGCAGCCC*	0/ 0		0		39 80%			9 29%
2	ACGGAATTCGAGCT .CGACCTGCAGCCC	0/-1	-1/ 0	-1	Sac[S]	4 4%			2
3	ACGGAATTCG.... TCGACCTGCAGCCC	-4/ 0		-4	Sal[S]		21 53%	13 62%	6 19%
4	ACGGAATTCGAGC. ..GACCTGCAGCCC	-1/-2	-2/-1	-3		1			
5	ACGGAATTCGA... TCGACCTGCAGCCC	-3/ 0		-3					1
6	ACGGAATTCG.... .CGACCTGCAGCCC	-4/-1		-5			1	1	
7	ACGGAATTCGAG.. ...ACCTGCAGCCC	-2/-3	-3/-2	-5					1
8	ACGGAATTC..... TCGACCTGCAGCCC	-5/ 0		-5			1	1	
9	ACGGAATTCGAGC. .....CTGCAGCCC	-1/-5	-2/-4	-6		1			
10	ACGGAATTCG.... ..GACCTGCAGCCC	-4/-2		-6			1		
11	ACGGAATTCGAGCT ....GCAGCCC	0/-7	-2/-5	-7				1	
12	ACGGAATTCGA... ....CCTGCAGCCC	-3/-4	-7/ 0	-7		3 6%	8 20%	4 19%	10 32%
13	ACGGAATTCG.... ....CCTGCAGCCC	-4/-4		-8			1		
14	ACGGAATTCGA... ......TGCAGCCC	-3/-6		-9					1
15	ACGGAATTCG.... .....CTGCAGCCC	-4/-5		-9			1		
16	ACGGAATTCG.... ......TGCAGCCC	-4/-6		-10				1	
18	ACGGAATTC..... .....CTGCAGCCC	-5/-5	-6/-4	-10			1		1
19	ACGGA......... .CGACCTGCAGCCC	-9/-1		-10			1		
20	ACGGAATTCGAGC. CC	-1/-12	-4/-8	-12			1		
21	ACGGAATTCG.... CCC	-4/-11	-5/-10	-15			1		
	larger delet. n.d.					1	2		

Fig. 7 Fidelity of an NHEJ reaction in cell-free extracts from CHO-K1 wild-type cells and their NHEJ-deficient derivatives xrs6 (Ku80), XR-1 (XRCC4), and XR-C1 (DNA-PK_{CS}). The total number of junctions analyzed for each cell line is shown behind the *Σ*. Example of an RE-substrate with non-complementary abutting blunt/5′-ends (Sac/Sal). Note that in this particular case, the inaccurate junction is sensitive to cleavage with Sal (Sal^S) while the accurate junction is resistant (see Fig. 4: substrate #9). The *line highlighted* in *light grey* indicates the accurate junction preferentially formed by the DNA-PK-dependent NHEJ pathway (see also paragraph "Basic mechanisms of in vitro NHEJ" and Fig. 2). The *line highlighted in middle grey* indicates an inaccurate junction formed by "blunting" (Sal^S). The *line highlighted in dark grey* indicates an inaccurate junction formed by μhomNHEJ using a 4bp microhomology (TCGA). Note the increase in blunting (not a single junction of this type is found in the wild-type CHO-K1) and μhomNHEJ in the mutants compared to the wild-type. In summary, the Ku80-deficient xrs6 and the XRCC4-deficient XR-1 mutant show the most significant decrease in NHEJ fidelity (neither has formed a single accurate junction) whereas the DNA-PK_{CS}-deficient XR-C1 mutant exhibits a milder phenotype that is intermediate between the wild-type and the other two mutants

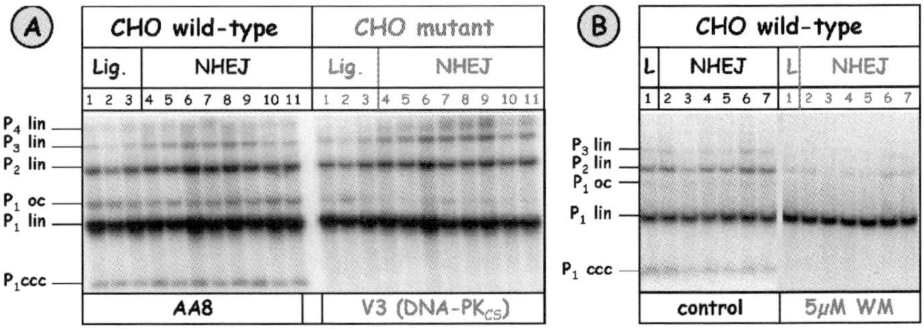

Fig. 8 Comparison of the NHEJ efficiencies of two Chinese hamster ovary (CHO) cell lines on different plasmid substrates. The V3 mutant cell line is defective in DNA-PK_{CS} and was derived from the wild-type AA8 parent (*left panel*) (Zdzienicka 1999). Numerals on top of each panel refer to the different substrates used (1–3 substrates with complementary ends to measure ligation: 1: Bam: 5′-cohesive; 2. Pst 3′-cohesive; 3: Sma: blunt; 4–11 substrates with non-complementary ends: 4: Bam/Asp: 5′/5′; 5: Bam/Sal: 5′/5′; 6: BstX/BstX: 3′/3′; 7: Kpn/Pst: 3′/3′; 8: Ava/Hind2: 5′/bl; 9: Sma/Pst: bl/3′; 10: Bam/Pst: 5′/3′; 11: Sac/Sal: 3′/5′; see also Fig. 4). Note that V3 is not able to form any ccc products (but only a small amount of oc products). The absence of ccc products is a hallmark of extracts prepared from cells defective in the DNA-PK-dependent NHEJ pathway (Feldmann et al. 2000; Göttlich et al. 1998; Labhart 1999b). Band designations are as in Fig. 6. Inhibition by Wortmannin (WM) of the DNA-PK-dependent NHEJ reaction in CHO-K1 wild-type cells. Numerals at the bottom of each panel refer to the different substrates used (1: Bam: 5′-cohesive; 2: Bam/Asp: 5′/5′; 3: BstX/BstX: 3′/3′; 4: Ava/Hind2: 5′/bl; 5: Sma/Pst: bl/3′; 6: Sac/Sal: 3′/5′; 7: Ava/Kpn; see also Fig. 4)

Remarkably, the loss of circle formation and maintenance of multimer formation – although at reduced efficiency – is also observed when the DNA-PK-dependent NHEJ-reaction is inhibited by the addition of Wortmannin, a competitive inhibitor of DNA-PK_{CS} (Fig. 8B). This indicates that the secondary DNA-PK-independent NHEJ pathway generates mainly linear multimers. This shift from circular to multimeric products appears to be actively enhanced by an as yet undescribed factor (Wang et al. 2005). Although linear multimers can, in principle, arise by ligation of the corresponding complementary ends (see above) it is worth mentioning that the multimers in NHEJ-deficient cells and after Wortmannin inhibition are mostly not formed by ligation but preferentially by "blunting" and/or μhomNHEJ (see paragraph "Basic mechanisms of in vitro NHEJ" and Fig. 7).

6
Concluding Remarks

Non-homologous DNA end joining in the form of the "accurate" DNA-PK-dependent and the "inaccurate" DNA-PK-independent (blunting and μhomNHEJ) pathways plays a major role in the repair of DSB in mammalian cells. Interestingly, NHEJ not only occurs in higher eukaryotes but also occurs in yeast and even – as discovered only recently – in bacteria where HR usually predominates because the small genomes hardly tolerate any changes due to the lack of redundancies and "junk DNA" (Wilson et al. 2003). This shows that NHEJ is an evolutionarily highly conserved mechanism which appears to be active mainly during G_0 and G_1 (and phases of prolonged mitotic exit during environmental stress in bacteria) where it guarantees the efficient rejoining of broken ends.

For the reconstitution of the main steps of NHEJ, in vitro assays with cell-free extracts and purified proteins have played an invaluable role. The picture emerging from these studies is a conserved series of reactions that are coordinated by protein-protein interactions. As proposed already in 1988, the key step in NHEJ is the physical alignment of the DNA ends which must be – due to the lack of sequence homology – mediated by protein-protein interactions between end binding factors (Pfeiffer and Vielmetter 1988; Thode et al. 1990). After binding to the DSB ends the Ku70/80 heterodimer (or possibly DNA-PK_{CS}) may serve as a platform for the assembly of proteins that mediate the alignment of the ends (Hefferin and Tomkinson 2005). Further biochemical studies are needed to elucidate the molecular mechanisms of end alignment and to dissect the multiple and redundant pathways of end processing by nucleases and polymerases to name only a few of the still open questions.

Acknowledgements Studies performed in the Pfeiffer laboratory were supported by grants of Wilhelm Sander Stiftung für Krebsforschung to P.P. (96.053.1-3 and 2002.108.1)

References

Audebert M, Salles B, Calsou P (2004) Involvement of poly(ADP-ribose) polymerase-1 and XRCC1/DNA ligase III in an alternative route for DNA double-strand breaks rejoining. J Biol Chem 279:55117–55126

Barnes DE (2002) DNA damage: air-breaks? Curr Biol 12:R262–R264

Barnes DE, Stamp G, Rosewell I, Denzel A, Lindahl T (1998) Targeted disruption of the gene encoding DNA ligase IV leads to lethality in embryonic mice. Curr Biol 8:1395–1398

Bassing CH, Swat W, Alt FW (2002) The mechanism and regulation of chromosomal VJ recombination. Cell 109 Suppl:S45–S55

Baumann P, West SC (1998) DNA end-joining catalyzed by human cell-free extracts. Proc Natl Acad Sci USA 95:14066–14070

Bennett RA, Gu XY, Povirk LF (1996) Construction of a vector containing a site-specific DNA double-strand break with 3′-phosphoglycolate termini and analysis of the products of end-joining in CV-1 cells. Int J Radiat Biol 70:623–636

Beyert N, Reichenberger S, Peters M, Hartung M, Göttlich B, Goedecke W, Vielmetter W, Pfeiffer P (1994) Nonhomologous DNA end joining of synthetic hairpin substrates in *Xenopus laevis* egg extracts. Nucleic Acids Res 22:1643–1650

Blanco MG, Boan F, Gomez-Marquez J (2004) A paradox in the in vitro end-joining assays. J Biol Chem 279:26797–26801

Boe SO, Sodroski J, Helland DE, Farnet CM (1995) DNA end-joining in extracts from human cells. Biochem Biophys Res Commun 215:987–993

Chen S, Inamdar KV, Pfeiffer P, Feldmann E, Hannah MF, Yu Y, Lee JW, Zhou T, Lees-Miller SP, Povirk LF (2001) Accurate in vitro end joining of a DNA double strand break with partially cohesive 3′-overhangs and 3′-phosphoglycolate termini: effect of Ku on repair fidelity. J Biol Chem 276:24323–24330

Cheong N, Iliakis G (1997) In vitro rejoining of double strand breaks induced in cellular DNA by bleomycin and restriction endonucleases. Int J Radiat Biol 71:365–375

Cheong N, Okayasu R, Shah S, Ganguly T, Mammen P, Iliakis G (1996) In vitro rejoining of double-strand breaks in cellular DNA by factors present in extracts of HeLa cells. Int J Radiat Biol 69:665–677

Cheong N, Perrault AR, Iliakis G (1998) In vitro rejoining of DNA double strand breaks: a comparison of genomic-DNA with plasmid-DNA-based assays. Int J Radiat Biol 73:481–493

Cheong N, Perrault AR, Wang H, Wachsberger P, Mammen P, Jackson I, Iliakis G (1999) DNA-PK-independent rejoining of DNA double-strand breaks in human cell extracts in vitro. Int J Radiat Biol 75:67–81

Collis SJ, DeWeese TL, Jeggo PA, Parker AR (2005) The life and death of DNA-PK. Oncogene 24:949–961

Daza P, Reichenberger S, Göttlich B, Hagmann M, Feldmann E, Pfeiffer P (1996) Mechanisms of nonhomologous DNA end-joining in frogs, mice and men. Biol Chem 377:775–786

Daza P, Schüßler H, McMillan TJ, Girod SC, Pfeiffer P (1997) Radiosensitivity and double-strand break rejoining in tumorigenic and non-tumorigenic human epithelial cell lines. Int J Radiat Biol 72:91–100

de Jager M, van Noort J, van Gent DC, Dekker C, Kanaar R, Wyman C (2001) Human Rad50/Mre11 is a flexible complex that can tether DNA ends. Mol Cell 8:1129–1135

Derbyshire MK, Epstein LH, Young CS, Munz PL, Fishel R (1994) Nonhomologous recombination in human cells. Mol Cell Biol 14:156–169

DiBiase SJ, Zeng ZC, Chen R, Hyslop T, Curran WJ Jr, Iliakis G (2000) DNA-dependent protein kinase stimulates an independently active, nonhomologous, end-joining apparatus. Cancer Res 60:1245–1253

Diggle CP, Bentley J, Kiltie AE (2003) Development of a rapid, small-scale DNA repair assay for use on clinical samples. Nucleic Acids Res 31:e83

Dresser ME (2000) Meiotic chromosome behavior in *Saccharomyces cerevisiae* and (mostly) mammals. Mutat Res 451:107–127

Fairman MP, Johnson AP, Thacker J (1992) Multiple components are involved in the efficient joining of double stranded DNA breaks in human cell extracts. Nucleic Acids Res 20:4145–4152

Featherstone C, Jackson SP (1999a) DNA-dependent protein kinase gets a break: its role in repairing DNA and maintaining genomic integrity. Br J Cancer 80 Suppl 1:14–19

Featherstone C, Jackson SP (1999b) Ku, a DNA repair protein with multiple cellular functions? Mutat Res 434:3–15

Feldmann E, Schmiemann V, Goedecke W, Reichenberger S, Pfeiffer P (2000) DNA double-strand break repair in cell-free extracts from Ku80-deficient cells: implications for Ku serving as an alignment factor in non-homologous DNA end joining. Nucleic Acids Res 28:2585–2596

Fernandez-Capetillo O, Lee A, Nussenzweig M, Nussenzweig A (2004) H2AX: the histone guardian of the genome. DNA Repair (Amst) 3:959–967

Ganguly T, Iliakis G (1995) A cell-free assay using cytoplasmic cell extracts to study rejoining of radiation-induced DNA double-strand breaks in human cell nuclei. Int J Radiat Biol 68:447–457

Gao Y, Sun Y, Frank KM, Dikkes P, Fujiwara Y, Seidl KJ, Sekiguchi JM, Rathbun GA, Swat W, Wang J, Bronson RT, Malynn BA, Bryans M, Zhu C, Chaudhuri J, Davidson L, Ferrini R, Stamato T, Orkin SH, Greenberg ME, Alt FW (1998) A critical role for DNA end-joining proteins in both lymphogenesis and neurogenesis. Cell 95:891–902

Göttlich B, Reichenberger S, Feldmann E, Pfeiffer P (1998) Rejoining of DNA double-strand breaks in vitro by single-strand annealing. Eur J Biochem 258:387–395

Grawunder U, Wilm M, Wu X, Kulesza P, Wilson TE, Mann M, Lieber MR (1997) Activity of DNA ligase IV stimulated by complex formation with XRCC4 protein in mammalian cells. Nature 388:492–495

Grawunder U, Zimmer D, Kulesza P, Lieber MR (1998) Requirement for an interaction of XRCC4 with DNA ligase IV for wild-type VJ recombination and DNA double-strand break repair in vivo. J Biol Chem 273:24708–24714

Gu XY, Bennett RA, Povirk LF (1996) End-joining of free radical-mediated DNA double-strand breaks in vitro is blocked by the kinase inhibitor wortmannin at a step preceding removal of damaged 3′ termini. J Biol Chem 271:19660–19663

Gu XY, Weinfeld MA, Povirk LF (1998) Implication of DNA-dependent protein kinase in an early, essential, local phosphorylation event during end-joining of DNA double-strand breaks in vitro. Biochemistry 37:9827–9835

Haber JE (1999) DNA recombination: the replication connection. Trends Biochem Sci 24:271–275

Hefferin ML, Tomkinson AE (2005) Mechanism of DNA double-strand break repair by non-homologous end joining. DNA Repair (Amst) 4:639–648

Honjo T, Kinoshita K, Muramatsu M (2002) Molecular mechanism of class switch recombination: linkage with somatic hypermutation. Annu Rev Immunol 20:165–196

Jackson SP (2002) Sensing and repairing DNA double-strand breaks. Carcinogenesis 23:687–696

Jackson SP, Jeggo PA (1995) DNA double-strand break repair and VJ recombination: involvement of DNA-PK. Trends Biochem Sci 20:412–415

Jeggo PA, O'Neill P (2002) The Greek Goddess, Artemis, reveals the secrets of her cleavage. DNA Repair (Amst) 1:771–777

Jeggo PA (1998) Identification of genes involved in repair of DNA double-strand breaks in mammalian cells. Radiat Res 150:S80–S91

Jeggo PA, Lobrich M (2005) Artemis links ATM to double strand break rejoining. Cell Cycle 4:359–362

Johnson AP, Fairman MP (1996) The identification and characterization of mammalian proteins involved in the rejoining of DNA double-strand breaks in vitro. Mutat Res 364:103–116

Johnson RD, Jasin M (2000) Sister chromatid gene conversion is a prominent double-strand break repair pathway in mammalian cells. EMBO J 19:3398–3407

Johnson RD, Jasin M (2001) Double-strand-break-induced homologous recombination in mammalian cells. Biochem Soc Trans 29:196–201

Kawabata M, Kawabata T, Nishibori M (2005) Role of recA/RAD51 family proteins in mammals. Acta Med Okayama 59:1–9

Khanna KK, Jackson SP (2001) DNA double-strand breaks: signaling, repair and the cancer connection. Nat Genet 27:247–254

Kinzler KW, Vogelstein B (1997) Cancer-susceptibility genes. Gatekeepers and caretakers. Nature 386:761, 763

Kurz EU, Lees-Miller SP (2004) DNA damage-induced activation of ATM and ATM-dependent signaling pathways. DNA Repair (Amst) 3:889–900

Labhart P (1999a) Ku-dependent nonhomologous DNA end joining in Xenopus egg extracts. Mol Cell Biol 19:2585–2593

Labhart P (1999b) Nonhomologous DNA end joining in cell-free systems. Eur J Biochem 265:849–861

Lakshmipathy U, Campbell C (1999) Double strand break rejoining by mammalian mitochondrial extracts. Nucleic Acids Res 27:1198–1204

Lee JW, Yannone SM, Chen DJ, Povirk LF (2003) Requirement for XRCC4 and DNA ligase IV in alignment-based gap filling for nonhomologous DNA end joining in vitro. Cancer Res 63:22–24

Lee JW, Blanco L, Zhou T, Garcia-Diaz M, Bebenek K, Kunkel TA, Wang Z, Povirk LF (2004) Implication of DNA polymerase lambda in alignment-based gap filling for nonhomologous DNA end joining in human nuclear extracts. J Biol Chem 279:805–811

Lehman CW, Trautman JK, Carroll D (1994) Illegitimate recombination in Xenopus: characterization of end-joined junctions. Nucleic Acids Res 22:434–442

Lieber MR, Ma Y, Pannicke U, Schwarz K (2003) Mechanism and regulation of human non-homologous DNA end-joining. Nat Rev Mol Cell Biol 4:712–720

Lundberg R, Mavinakere M, Campbell C (2001) Deficient DNA end joining activity in extracts from fanconi anemia fibroblasts. J Biol Chem 276:9543–9549

Ma Y, Pannicke U, Schwarz K, Lieber MR (2002) Hairpin opening and overhang processing by an Artemis/DNA-dependent protein kinase complex in nonhomologous end joining and VJ recombination. Cell 108:781–794

Ma Y, Lu H, Tippin B, Goodman MF, Shimazaki N, Koiwai O, Hsieh CL, Schwarz K, Lieber MR (2004) A biochemically defined system for mammalian nonhomologous DNA end joining. Mol Cell 16:701–713

Ma Y, Schwarz K, Lieber MR (2005) The Artemis: DNA-PKcs endonuclease cleaves DNA loops, flaps, and gaps. DNA Repair (Amst) 4:845–851

Mahajan KN, Gangi-Peterson L, Sorscher DH, Wang J, Gathy KN, Mahajan NP, Reeves WH, Mitchell BS (1999) Association of terminal deoxynucleotidyl transferase with Ku. Proc Natl Acad Sci USA 96:13926–13931

Mahajan KN, Nick McElhinny SA, Mitchell BS, Ramsden DA (2002) Association of DNA polymerase mu (pol mu) with Ku and ligase IV: role for pol mu in end-joining double-strand break repair. Mol Cell Biol 22:5194–5202

Mason RM, Thacker J, Fairman MP (1996) The joining of non-complementary DNA double-strand breaks by mammalian extracts. Nucleic Acids Res 24:4946–4953

Merel P, Prieur A, Pfeiffer P, Delattre O (2002) Absence of major defects in non-homologous DNA end joining in human breast cancer cell lines. Oncogene 21:5654–5659

Moshous D, Callebaut I, de Chasseval R, Corneo B, Cavazzana-Calvo M, Le Deist F, Tezcan I, Sanal O, Bertrand Y, Philippe N, Fischer A, de Villartay JP (2001) Artemis, a novel DNA double-strand break repair/VJ recombination protein, is mutated in human severe combined immune deficiency. Cell 105:177–186

Nicolas AL, Young CS (1994) Characterization of DNA end joining in a mammalian cell nuclear extract: junction formation is accompanied by nucleotide loss, which is limited and uniform but not site specific. Mol Cell Biol 14:170–180

Nicolas AL, Munz PL, Young CS (1995) A modified single-strand annealing model best explains the joining of DNA double-strand breaks mammalian cells and cell extracts. Nucleic Acids Res 23:1036–1043

North P, Ganesh A, Thacker J (1990) The rejoining of double-strand breaks in DNA by human cell extracts. Nucleic Acids Res 18:6205–6210

Odersky A, Panyutin IV, Panyutin IG, Schunck C, Feldmann E, Goedecke W, Neumann RD, Obe G, Pfeiffer P (2002) Repair of sequence-specific 125I-induced double-strand breaks by nonhomologous DNA end joining in mammalian cell-free extracts. J Biol Chem 277:11756–11764

Paques F, Haber JE (1999) Multiple pathways of recombination induced by double-strand breaks in *Saccharomyces cerevisiae.* Microbiol Mol Biol Rev 63:349–404

Paull TT, Lee JH (2005) The Mre11/Rad50/Nbs1 complex and its role as a DNA double-strand break sensor for ATM. Cell Cycle 4:737–740

Perrault R, Wang H, Wang M, Rosidi B, Iliakis G (2004) Backup pathways of NHEJ are suppressed by DNA-PK. J Cell Biochem 92:781–794

Pfeiffer P, Vielmetter W (1988) Joining of nonhomologous DNA double strand breaks in vitro. Nucleic Acids Res 16:907–924

Pfeiffer P, Thode S, Hancke J, Keohavong P, Thilly WG (1994a) Resolution and conservation of mismatches in DNA end joining. Mutagenesis 9:527–535

Pfeiffer P, Thode S, Hancke J, Vielmetter W (1994b) Mechanisms of overlap formation in nonhomologous DNA end joining. Mol Cell Biol 14:888–895

Pfeiffer P, Goedecke W, Obe G (2000) Mechanisms of DNA double-strand break repair and their potential to induce chromosomal aberrations. Mutagenesis 15:289–302

Pfeiffer P, Goedecke W, Kuhfittig-Kulle S, Obe G (2004) Pathways of DNA double-strand break repair and their impact on the prevention and formation of chromosomal aberrations. Cytogenet Genome Res 104:7–13

Pfeiffer P, Feldmann E, Odersky A, Kuhfittig-Kulle S, Goedecke W (2005) Analysis of DNA double-strand break repair by nonhomologous end joining in cell-free extracts from mammalian cells. Methods Mol Biol 291:351–371

Pospiech H, Rytkonen AK, Syvaoja JE (2001) The role of DNA polymerase activity in human non-homologous end joining. Nucleic Acids Res 29:3277–3288

Reichenberger S, Pfeiffer P (1998) Cloning, purification and characterization of DNA polymerase beta from Xenopus laevis – studies on its potential role in DNA-end joining. Eur J Biochem 251:81–90

Reichenberger S, Brüll N, Feldmann E, Göttlich B, Vielmetter W, Pfeiffer P (1996) A novel nuclease activity from Xenopus laevis releases short oligomers from 5′-ends of double- and single-stranded DNA. Genes Cells 1:355–367

Reynaud CA, Aoufouchi S, Faili A, Weill JC (2003) What role for AID: mutator, or assembler of the immunoglobulin mutasome? Nat Immunol 4:631–638

Rothkamm K, Kruger I, Thompson LH, Löbrich M (2003) Pathways of DNA double-strand break repair during the mammalian cell cycle. Mol Cell Biol 23:5706–5715

Sathees CR, Raman MJ (1999) Mouse testicular extracts process DNA double-strand breaks efficiently by DNA end-to-end joining. Mutat Res 433:1–13

Shechter D, Costanzo V, Gautier J (2004) Regulation of DNA replication by ATR: signaling in response to DNA intermediates. DNA Repair (Amst) 3:901–908

Shivji MK, Venkitaraman AR (2004) DNA recombination, chromosomal stability and carcinogenesis: insights into the role of BRCA2. DNA Repair (Amst) 3:835–843

Stracker TH, Theunissen JW, Morales M, Petrini JH (2004) The Mre11 complex and the metabolism of chromosome breaks: the importance of communicating and holding things together. DNA Repair (Amst) 3:845–854

Surralles J, Jackson SP, Jasin M, Kastan MB, West SC, Joenje H (2004) Molecular cross-talk among chromosome fragility syndromes. Genes Dev 18:1359–1370

Thacker J (2005) The RAD51 gene family, genetic instability and cancer. Cancer Lett 219:125–135

Thacker J, Zdzienicka MZ (2003) The mammalian XRCC genes: their roles in DNA repair and genetic stability. DNA Repair (Amst) 2:655–672

Thacker J, Zdzienicka MZ (2004) The XRCC genes: expanding roles in DNA double-strand break repair. DNA Repair (Amst) 3:1081–1090

Thode S, Schäfer A, Pfeiffer P, Vielmetter W (1990) A novel pathway of DNA end-to-end joining. Cell 60:921–928

Thacker J, Chalk J, Ganesh A, North P (1992) A mechanism for deletion formation in DNA by human cell extracts: the involvement of short sequence repeats. Nucleic Acids Res 20:6183–6188

Ting NS, Lee WH (2004) The DNA double-strand break response pathway: becoming more BRCAish than ever. DNA Repair (Amst) 3:935–944

Udayakumar D, Bladen CL, Hudson FZ, Dynan WS (2003) Distinct pathways of non-homologous end joining that are differentially regulated by DNA-dependent protein kinase-mediated phosphorylation. J Biol Chem 278:41631–41635

van Gent DC, Hoeijmakers JH, Kanaar R (2001) Chromosomal stability and the DNA double-stranded break connection. Nat Rev Genet 2:196–206

Venkitaraman AR (2001) Chromosome stability, DNA recombination and the BRCA2 tumour suppressor. Curr Opin Cell Biol 13:338–343

Wang H, Zeng ZC, Bui TA, Sonoda E, Takata M, Takeda S, Iliakis G (2001a) Efficient rejoining of radiation-induced DNA double-strand breaks in vertebrate cells deficient in genes of the RAD52 epistasis group. Oncogene 20:2212–2224

Wang H, Zeng ZC, Perrault AR, Cheng X, Qin W, Iliakis G (2001b) Genetic evidence for the involvement of DNA ligase IV in the DNA-PK-dependent pathway of non-homologous end joining in mammalian cells. Nucleic Acids Res 29:1653–1660

Wang H, Perrault AR, Takeda Y, Qin W, Wang H, Iliakis G (2003) Biochemical evidence for Ku-independent backup pathways of NHEJ. Nucleic Acids Res 31:5377–5388

Wang H, Rosidi B, Perrault R, Wang M, Zhang L, Windhofer F, Iliakis G (2005) DNA ligase III as a candidate component of backup pathways of nonhomologous end joining. Cancer Res 65:4020–4030

Wang X, D'Andrea AD (2004) The interplay of Fanconi anemia proteins in the DNA damage response. DNA Repair (Amst) 3:1063–1069

Wilson TE, Topper LM, Palmbos PL (2003) Non-homologous end-joining: bacteria join the chromosome breakdance. Trends Biochem Sci 28:62–66

Yaneva M, Kowalewski T, Lieber MR (1997) Interaction of DNA-dependent protein kinase with DNA and with Ku: biochemical and atomic-force microscopy studies. EMBO J 16:5098–5112

Zdzienicka MZ (1999) Mammalian X-ray-sensitive mutants which are defective in non-homologous (illegitimate) DNA double-strand break repair. Biochimie 81:107–116

Zhang X, Paull TT (2005) The Mre11/Rad50/Xrs2 complex and non-homologous end-joining of incompatible ends in *S. cerevisiae*. DNA Repair (Amst) 4:1281–1294

Zhong Q, Boyer TG, Chen PL, Lee WH (2002) Deficient nonhomologous end-joining activity in cell-free extracts from Brca1-null fibroblasts. Cancer Res 62:3966–3970

Genome Dyn Stab (1)
D.-H. Lankenau: Genome Integrity
DOI 10.1007/7050_009/Published online: 30 November 2005

Role of p53 in Double-Strand Break Repair

Susanne Andrea Gatz[1] · Lisa Wiesmüller[2] (✉)

[1]Universitätsklinik für Kinder und Jugendmedizin, Eythstr. 24, 89075 Ulm, Germany
susanne.gatz@uni-ulm.de

[2]Universitätsfrauenklinik, Prittwitzstr. 43, 89075 Ulm, Germany
lisa.wiesmueller@uni-ulm.de

Abstract In addition to p53's classical tumour suppressor functions in cell cycle control and apoptosis induction, p53 exhibits direct regulatory activities in double-strand break repair. In this review, we summarize data from biochemical, cell based and in vivo approaches demonstrating that wild-type p53 restrains excessive and mutagenic DNA exchange events. Thus, p53 appears to exhibit fidelity control of homologous recombination, and recent evidence also suggests that p53 counteracts error-prone non-homologous end-joining. Additionally, we discuss a novel role for p53: stimulation of recombination via pathways involving topoisomerase I and/or sequence-specific DNA binding. In summary, p53 either functions as a global suppressor of mutagenic genome rearrangements or contributes to recombinative repair, when it is recruited to specific repair complexes by molecular interactions with topoisomerase I and/or other binding partners. The p53 protein therefore combines two genetically separable regulatory functions in recombinative DNA repair that may have implications for a loss-of-function versus gain-of-function phenotype of p53 mutants in genetic destabilization.

1
Introduction

Genome integrity at the level of the whole organism is guaranteed by the coordination of the network of pathways including DNA repair, cell cycle control and apoptosis on the single cell level. Over recent years, p53 has been established as one of the key players in this network. The most detrimental DNA damage and strongest stimulus for p53 activation is a double-strand break (DSB) within DNA. Upon activation, the DNA binding protein p53 accomplishes many of its functions by transcriptional transactivation of target proteins. However, several DNA repair related activities of p53 are exerted independently of its transcriptional activities. This review will focus on p53's direct role in DSB repair, which over recent years has clearly been separated from its functions in cell cycle control and apoptosis induction.

1.1 p53 and its Multiple Functions

Human p53 is a protein consisting of 393 amino acids (aa) subdivided into three major domains (see Fig. 1): (i) the N-terminus comprising the transcriptional transactivation (aa 1–50) and the proline-rich domain for the induction of apoptosis (aa 63–97) (May and May 1999), (ii) the core domain (aa 102–292), which is the sequence-specific DNA binding domain and also harbours the 3′-5′-exonuclease activity (Mummenbrauer et al. 1996; May and May 1999) and (iii) the C-terminus (aa 300–393), which contains the p53 oligomerization domain, three nuclear import signal sequences and one nuclear export signal sequence (May and May 1999; Stommel et al. 1999) and a basic, non-sequence-specific DNA/RNA binding domain (Oberosler et al. 1993; Bakalkin et al. 1994; Brain and Jenkins 1994). The basic C-terminus (aa 363–393) regulates DNA binding via the core domain, and the mechanism may involve linear diffusion of p53 along DNA in vivo (McKinney et al. 2004).

p53 is the most frequently mutated gene found in human tumours, with about 50% of malignancies harbouring a mutated *p53* locus, and with > 97% of the mutations being located within the core domain (Olivier et al. 2002). Alternative mechanisms underlying loss of p53 function – given an intact

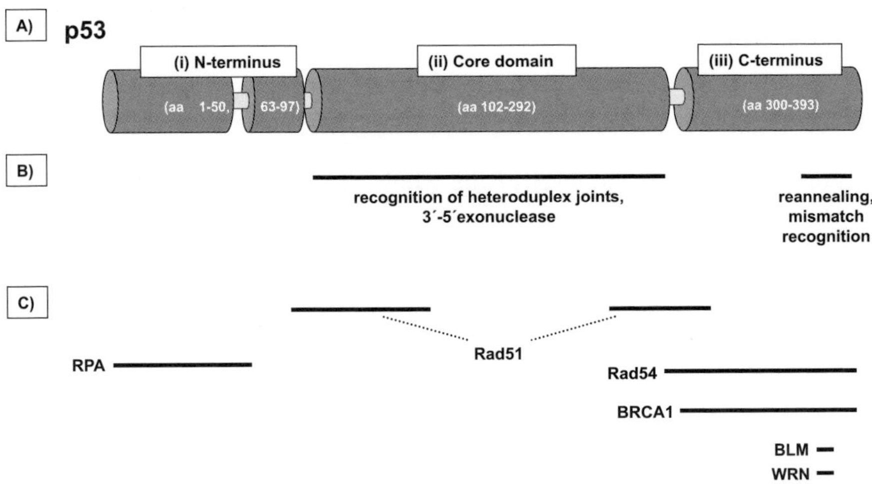

Fig. 1 Molecular model of p53 in DSB repair. **A** Illustrated is the 393 aa long p53 protein with its three major domains: (i) N-terminus (ii), core domain and (iii) C-terminus. **B** Biochemical activities of p53 with relevance for DSB repair are indicated. **C** p53 interacts physically with several DSB repair related proteins: Rad51, Rad54, BRCA1, RPA, BLM and WRN. *Black lines* symbolise the p53 protein regions carrying repair related activities and contact sites for DSB repair factors, respectively (for references see Table 1)

p53 gene – are nuclear exclusion, inactivation via protein complex formation and accelerated degradation (Vogelstein et al. 2000; Vousden and Prives 2005 and references therein). p53 activity is normally kept under very tight control, which is mainly achieved by its rapid turnover leading to extremely low expression levels in healthy, unstimulated cells. A number of ubiquitin ligases that target p53 protein for degradation through the proteasome have been identified, including Mdm2 (or Hdm2 in humans), Pirh2 and Cop1 (Corcoran et al. 2004). Indeed, as shown for many tumours that retain wild-type p53, inappropriately increasing Mdm2 activity counteracts p53 tumour suppression (Bond et al. 2005). Germline mutations in *p53* occur in 80% of families with the Li-Fraumeni cancer susceptibility syndrome (LFS). In support of the idea that p53 is a major factor in the maintenance of genome stability, structural chromosome aberrations are a hallmark of cells from LFS patients and from mice nullizygous for p53 (Harvey et al. 1993; Boyle et al. 1998).

p53 activates transcription through binding to specific DNA consensus sequences. This leads to transcriptional transactivation of at least 150 genes, among them the very well characterized target gene *Mdm2* which forms part of an autoregulatory feedback loop (Vogelstein et al. 2000). p53's cell cycle control function mainly relies on transcriptional transactivation of two genes, $p21^{\mathrm{WAF1/CIP1}}$ and 14-3-3σ, which execute cell cycle arrest at the G1/S transition and sustained G2 arrest, respectively. p53 dependent induction of apoptosis involves transcriptional transactivation of different proapoptotic target genes like *bax* and *PUMA*. In addition, there is a direct cytoplasmic role for p53 in apoptosis induction. This most recently discovered function relies on direct protein interactions with members of the BCL-2 protein family (Perfettini et al. 2004, and references therein).

However, soon after the establishment of these key roles of p53 in growth control and apoptosis induction, evidence accumulated indicating that p53 also plays a direct role in DNA repair, which can be separated from these well described functions (for review, see Albrechtsen et al. 1999; Christmann et al. 2003; Bertrand et al. 2004; Sengupta and Harris 2005). Thus, p53 is directly involved in the regulation of DSB repair and of base excision repair (BER), and is also active in nucleotide excision repair (NER). A major contribution of p53 in NER stems from transcriptional activation of $p48^{\mathrm{DDB2}}$ (Hwang et al. 1999). The respective gene product complexes with $\mathrm{p127}^{\mathrm{DDB2}}$, which together form part of one of the DNA lesion recognition complexes during NER. Additionally, p53 physically interacts with the TFIIH helicase subunits XPD and XPB (Wang et al. 2001) and plays a role in regulating chromatin accessibility during NER (Rubbi and Milner 2003). p53's activities in BER are possibly based on direct protein-protein interactions with polymerase β and AP endonuclease 1 (APE-1/Ref-1) (Offer et al. 1999; Zhou et al. 2001).

1.2
DSB Repair Pathways

Each replicating cell spontaneously acquires about 10 DSBs per day, due to errors in replication, recombination and mitosis. DSBs can also be introduced by external stimuli such as ionising radiation (IR), radiomimetic drugs and oxidative stress. These DSBs must be efficiently repaired to guarantee genome stability. There are three main pathways which are involved in the repair of DSBs (for review see Valerie and Povirk 2003): non-homologous end-joining (NHEJ), conservative homologous recombination (HR) and non-conservative single-strand annealing (SSA). NHEJ is the predominant pathway in mammalian cells and is active during the whole cell cycle (Rothkamm et al. 2003). NHEJ reseals DSBs in mitotically growing cells and is the repair mechanism that also underlies V(D)J rearrangements in developing lymphocytes and class switch recombination in B-lymphocytes. NHEJ does not rely on sequence homologies and may or may not be error-prone, i.e. may or may not be associated with the gain or loss of nucleotides at the breakage site. Important repair proteins which have been shown to be directly involved in the NHEJ pathway of DNA repair are the DNA end binding protein complex called DNA dependent protein kinase (DNA-PK), consisting of its catalytic subunit DNA-PK_{CS} and the regulatory heterodimer Ku70/80, the ligating compounds XRCC4 and DNA ligase IV and the nuclease Artemis. In contrast to NHEJ, homology based repair – as already suggested by the name – relies on the exchange of sequence homologies and can be subdivided into two categories depending on whether the homologous gene sequences pair after strand exchange events (HR) or are fused by strand annealing activities (SSA). In mammalian cells a minimal length of approximately 200 bp of homology between the two recombining DNA fragments is required to enable efficient HR (Rubnitz and Subramani 1984). HR underlies genetic mixing during meiosis. In mitotically growing cells, HR is mainly confined to the S- and G2 phases of the cell cycle when sister chromatids are available (Rothkamm et al. 2003). DSBs and other DNA lesions that remain unrepaired until being encountered by a replication fork are the targets of this repair mechanism (Saintigny et al. 2001). Genetic rearrangements are a possible outcome of HR, if the template is the second allele on the homologous chromosome. However, since in mitotically growing mammalian cells the preferred template for HR is the sister chromatid, HR in most cases is non-mutagenic.

HR starts with DSB detection and ssDNA resection. The MRN complex (containing Mre11, Rad50 and Nbs1) plays a crucial role in tethering the two DNA substrates and, via its Mre11 nuclease component, possibly also in end resection (Assenmacher and Hopfner 2004). So far, it has not unequivocally been determined whether the corresponding Mre11 function also applies to NHEJ. In HR the resulting 3′ssDNA tail is coated by replication protein A (RPA), followed by the assembly of pro-recombinogenic Rad51 (presynaptic complex). Rad52 stimulates filament formation of Rad51 on ssDNA, but is not essential for HR (Stark et al. 2004). The central step in HR is the DNA strand exchange reaction catalysed by Rad51, which is facilitated by the Swi/Snf2 protein family member Rad54 and its paralog Rad54B (Krejci et al. 2003; West 2003). Newly generated heteroduplex structures mark the synaptic phase of the reaction. Heteroduplex DNA extension and branch migration are indicative of the postsynaptic phase. DNA ligases join the newly synthesized fragments. Resolution of Holliday junctions is achieved by resolvases, and the complex consisting of Rad51C and XRCC3 is a good candidate according to a recent report (Liu et al. 2004). The breast cancer related gene products BRCA1 and BRCA2 are thought to play a role as molecular scaffolds for the assembly of the HR repair complex and facilitate Rad51 filament assembly, respectively (for review, see Jasin 2002). The RecQ helicases BLM and WRN, mutated in Bloom's and Werner's syndrome, respectively, have been found to unwind DNA. WRN additionally exhibits exonucleolytic activity. These enzymes may play a role in resolving aberrantly paired structures, particularly at stalled forks, to stabilise the fork during replication. This could explain the inhibitory effect on mitotic recombination by these enzymes and the aberrant increase in the frequency of sister chromatid exchanges (SCEs) in cells of Bloom's syndrome patients (Bachrati and Hickson 2003).

SSA is the second major homology based repair pathway and involves DNA repeats. Approximately half of the human genome consists of repetitive sequences. Annealing of complementary sequences within these genomic regions after resection of a DSB results in deletion of sequences between repeats. Therefore, this pathway is always mutagenic. SSA is mediated by Rad52. Additional factors of this repair pathway are RPA and the endonuclease Ercc1/XPF. Recent data from Stark et al. (2004) show that the two homology based repair pathways HR and SSA are competitive. Thus, Rad51 function is critically important to limit the mutagenic potential of homology based repair. Disruption of Rad51 can lead to a > 90-fold shift in pathway usage towards SSA. Depending on the DSB repair pathway involved, the repair process may end up with small deletions and insertions in the genome which may be responsible for loss of or disturbed gene function. Alternatively, it may result in allelic losses (loss of heterozygosity, LOH) and chromosomal translocations that may lead to the development of malignancies as was particularly well documented for different types of leukaemia (Rabbits et al. 1991).

2
What do we Know about the Effect of Activation of p53 on DNA Damage?

The p53 response is triggered most rapidly by DSBs, which can be introduced either directly, e.g. via IR, or indirectly after the conversion of DNA adducts or single-strand breaks by repair or replication. DNA damage is thought to prevent proteolysis and/or to enhance transcriptional transactivation of p53-responsive genes through posttranslational modifications by the DSB or DNA replication fork blockage sensing phosphatidylinositol 3-kinase (PI3-kinase) related kinases (PIKK) ATM and ATR, acetyltransferases such as CREB binding protein (CBP)/p300 and poly(ADP-ribose) polymerase 1 (PARP-1) (Vaziri et al. 1997; Wang et al. 1998; Saito et al. 2003; Meek 2004). 17 Acetylation and phosphorylation sites have been described (Saito et al. 2003). Consequently, p53 expressing cells may contain a large variety of p53 sub-populations with respect to the modification state. In vitro studies established a latent and an activated state of p53 in sequence-specific DNA binding on post-translational modification of the carboxyl-terminal negative regulatory domain (Hupp and Lane 1995). However, there is also evidence indicating that damage induced phosphorylation is dispensable for transcriptional transactivation (Fuchs et al. 1995; Jackson et al. 2004; Thompson et al. 2004). Similarly, acetylation of p53 does not affect the affinity for its target sequences in the physiological chromatin context (Espinosa and Emmerson 2001). However, phosphorylation of particular sites within the p53 molecule (e.g. serine 15) may very well be involved in DNA repair regulation (Okorokov et al. 2003; Sengupta et al. 2003; Restle et al. 2005). Using chromatin immunoprecipitation, ATM was found to localize to DSBs together with p53, indicating modification of p53 directly at the break (Perkins et al. 2002). Moreover, it was recently demonstrated that 100% of the p53 molecules colocalising with Rad51 are phosphorylated on serine 15 by ATM or ATR (Restle et al. 2005). Colocalization of p53 phosphorylated on serine 15 (p53pSer15) with other DSB repair proteins such as Mre11 and BLM was similarly observed, suggesting a repair related function particularly for this subpopulation of post-translationally modified p53 (Sengupta et al. 2003; Restle et al. 2005). Notably, when DNA replication is blocked, p53 becomes phosphorylated on serine 15, however, remains transcriptionally inactive (Gottifredi et al. 2001), suggesting that p53pSer15 is involved in tumour suppressor functions unrelated to transcriptional activities. On the contrary, from the results of a recent study PARP-1 mediated p53 modification is unlikely to exert a major influence on DSB repair regulation (Süsse et al. 2004).

3
Clues for Direct Influence of p53 on DSB Repair

Over the past few years a large body of biochemical evidence has emerged indicating that p53 might be directly involved in DNA replication or repair, more specifically in recombinative DNA repair: P53 binds ssDNA and dsDNA (Oberosler et al. 1993; Bakalkin et al. 1994), possesses a $3' \rightarrow 5'$ exonuclease activity (Mummenbrauer et al. 1996) and promotes reannealing and strand transfer of short DNA stretches (Oberosler et al. 1993; Bakalkin et al. 1994; Brain and Jenkins 1994; Jean et al. 1997). Further, p53 binds to mismatches, three-stranded heteroduplex joints and four-stranded Holliday junctions (Lee et al. 1995, 1997; Dudenhöffer et al. 1998). Importantly, exonucleolytic DNA degradation also shows a preference for three-stranded heteroduplex joint structures, particularly when comprising mismatches (Skalski et al. 2000; Süsse et al. 2000). An intact core domain is required for junction specific binding of heteroduplex joints in addition to sequence specific DNA binding, and also harbours the exonuclease activity (Mummenbrauer et al. 1996; Dudenhöffer et al. 1999; Janus et al. 1999; Janz et al. 2002). Sequence-specific DNA binding, binding of Holliday junctions and the $3' \rightarrow 5'$ exonuclease activity are differently regulated by phosphorylation of the p53 C-terminus suggesting that p53 may execute these biochemical functions in a mutually exclusive manner (Janus et al. 1999; Subramanian and Griffith 2005).

In addition to p53's biochemical activities, numerous reports on physical and functional interactions between p53 and DSB repair proteins further strengthened the proposal of direct participation of p53 in this repair pathway (see Fig. 1).

3.1
Molecular Interactions with DSB Repair Proteins

Wild-type p53 forms complexes with Rad51 as indicated by results from immunoprecipitation experiments using mammalian cell extracts, pull-down assays with bacterially expressed proteins and DNA-protein interaction studies applying electrophoretic mobility shift assays (Stürzbecher et al. 1996; Buchhop et al. 1997; Süsse et al. 2000; Linke et al. 2003; Sengupta et al. 2003). Rad51 contact sites within p53 encompass two segments of the central domain between aa 94–160 and 264–315, and within Rad51 between aa 125 and aa 220 (see Table 1). Core domain mutants p53(135Y), p53(249S) and p53(273H) showed reduced binding, mutant Rad51(186P) was completely inactive in complex formation. As was suspected from the interaction of p53 with the highly conserved region within Rad51, which in the *E. coli* homologue RecA is required for homo-oligomerization, p53 was shown to influence the formation of Rad51 polymers (Linke et al. 2003).

Furthermore, p53 interacts with RPA via its N-terminus, whereas mutation at aa 48 and 49 or 53 and 54 abolishes binding (Romanova et al. 2004, and references therein). Other DSB repair proteins reported to interact with p53 are Rad54 (Linke et al. 2003), BRCA1 (Ouchi et al. 1998; Zhang et al. 1998) and BRCA2 (Marmorstein et al. 1998) (see Table 1). p53 also forms complexes with BLM and WRN and inhibits unwinding activities on Holliday junctions in vitro (Blander et al. 1999; Spillare et al. 1999; Wang et al. 2001; Yang et al. 2002). Residues 373–383 are critical for interference with BLM and WRN helicase activity, but the core domain mutants p53(248W)

Table 1 Functional domains of the human p53 protein in the fidelity control of HR

P53 domain	HR related activity	Single amino acid exchanges causing impaired function	References
N-terminus aa 1–63	RPA binding	(48H,49H), (53S, 54S)	Romanova et al. 2004
Core domain aa 102–292	HR suppression	175H, 237I, 248Q, 248W, 273H, 273P, 281G	Bertrand et al. 1997; Saintigny et al. 1999; Dudenhöffer et al. 1999; Akyüz et al. 2002; Linke et al. 2003; Romanova et al. 2004
	Recognition of three-stranded heteroduplex joints	248P, 273H, 273P	Dudenhöffer et al. 1999; Süsse et al. 2000; Janz et al. 2002
	3′-5′exonuclease	175H, 273H	Mummenbrauer et al. 1996; Janus et al. 1999; Süsse et al. 2000
	Inhibition of branch migration by Rad51	248P, 273P	Yoon et al. 2004
	Inhibition of Holliday junction binding and unwinding by BLM and WRN	248W, 273H	Yang et al. 2002
Core domain aa 94–160 and 264–315	Rad51 binding	135Y, 249S, 273H	Buchhop et al. 1997; Süsse et al. 2000; Linke et al. 2003
C-terminal domain aa 300–393	BRCA1 binding		Ouchi et al. 1998; Zhang et al. 1998
C-terminal domain aa 294–393	Rad54 interaction		Linke et al. 2003
C-terminal oligomerization domain aa 323–356	HR suppression		Dudenhöffer et al. 1999; Akyüz et al. 2002
	Tetramerization for stable heteroduplex joint complex formation	(341K, 344E, 348E, 355K)	Janz et al. 2002

Table 1 (continued)

P53 domain	HR related activity	Single amino acid exchanges causing impaired function	References
C-terminal regulatory domain aa 363–393	Recognition of mismatches, particularly in heteroduplex joints		Lee et al. 1995; Dudenhöffer et al. 1999
	Non sequence-specific ssDNA and dsDNA binding		Brain and Jenkins 1994; Bakalkin et al. 1994
	DNA strand annealing and strand transfer activity		Brain and Jenkins 1994; Bakalkin et al. 1994; Jean et al. 1997
aa 373–383	Interaction with BLM and WRN helicases, modulation of Holliday junction unwinding		Yang et al. 2002

or p53(273H) are defective as well. More recently, extensive use has been made of immunofluorescence microscopy and chromatin immunoprecipitation analysis to delineate the assembly of DSB repair complexes in more detail. For p53, or its serine 15 phosphorylated form, significant association was seen with MRN, Rad51, Rad54, BLM and MSH2 foci immediately after a DSB repair stimulus in S-phase nuclei (Zink et al. 2002; Linke et al. 2003; Sengupta et al. 2003; Restle et al. 2005). In contrast, colocalization signals were negligibly low for p53pSer15 and Rad52 which carries an auxiliary function in HR and a central Rad51 independent role in SSA. Taken together, the observed protein associations strongly suggested a role for p53 in Rad51 dependent HR. From the kinetics of the interaction of p53 with HR proteins it is likely to be involved in the earliest steps of HR (Restle et al. 2005).

4
p53 Functions in Homology Based DSB Repair

To measure DSB repair with respect to p53, different test systems were applied, which rely on the reactivation of viral minichromosomes or on the reconstitution of reporter genes within extrachromosomal plasmids or cellular chromosomes (Wiesmüller et al. 1996; Willers et al. 2002). The introduction of targeted DSBs by use of the site-specific I-Sce I nuclease enabled Akyüz and colleagues (2002) to exclude that p53 dependent effects on DSB repair where induced indirectly by activities in NER or BER.

4.1
Suppression of HR in Mitotically Growing Cells

Using these test systems in cultured cells, several groups unanimously found that p53 suppresses inter- and intramolecular HR on both extrachromosomal and intrachromosomal DNA substrates by at least one to two orders of magnitude following ectopic expression of wild-type p53. Vice versa, inactivation of wild-type p53 by mutation or complex formation increases HR by several orders of magnitude (Wiesmüller et al. 1996; Bertrand et al. 1997; Mekeel et al. 1997; Dudenhöffer et al. 1998, 1999; Saintigny et al. 1999; Willers et al. 2000; Akyüz et al. 2002; Sengupta et al. 2003). In contrast, employing a plasmid shuttle assay with episomally replicating substrates, Willers et al. (2001) found HR not to be suppressed by wild-type p53 and postulated the necessity of the chromatin context for p53's suppressor function in HR. Data from other groups, however, indicated that the extent of recombination regulation by p53 is influenced by the p53/DNA substrate ratio rather than by the chromatin context (Akyüz et al. 2002; Sengupta et al. 2003). Therefore, conflicting results obtained by Willers et al. (2001) were reconcilable in the light of the low p53 expression levels relative to transfected plasmid copy numbers.

Different p53 mutants have been analysed so far, including the mutants with the most prevalent mutations in human tumours in Western Europe (alterations at aa 175, 248, 273). These experiments revealed significant HR inhibitory defects for most of these hotspot mutants (see Table 1) (Saintigny et al. 1999; Dudenhöffer et al. 1999; Akyüz et al. 2002). At first sight, somewhat contradictory results were obtained with the DNA contact mutant p53(273H). However, when utilising the p53-negative cell line K562 and thereby excluding any possible complementation or dominant negative effects between endogenous and exogenous p53 variants, Akyüz et al. (2002) verified severe impairment of HR suppression for p53(273H) as well. Importantly, aa exchanges 135V (murine p53), 138V, 143A and (22Q, 23S), which are known to reduce or even abolish p53's transcriptional transactivation and cell cycle regulatory capacity, did not significantly affect HR inhibition (Willers et al. 2000; Akyüz et al. 2002; Boehden et al. 2003; Linke et al. 2003; Sengupta et al. 2003). These findings provided convincing evidence for p53's direct role in recombination, which is separable from its functions as a transcriptionally transactivating and checkpoint control protein. The analysis of truncated p53 molecules revealed that the C-terminal 30 aa mediate mismatch recognition within heteroduplex recombination intermediates which in turn stimulates junction DNA binding and HR suppression (Dudenhöffer et al. 1998, 1999). Thus, as was previously seen for sequence specific DNA binding of the core, the C-terminal end of p53 negatively regulates heteroduplex joint binding and HR suppression, an effect which can be neutralized by concurrent mismatch recognition, truncation and post-translational modification of serine 392 (Hupp and Lane 1995; Jayaraman and Prives 1995; Lee et al. 1995; Subramanian and Griffith 2005). Further truncation beyond the tetramerization domain renders p53 inactive with respect to HR control and heteroduplex joint binding (Dudenhöffer et al. 1999). Analysis of p53 mutant protein with four aa exchanges specifically affecting tetramerization confirmed that oligomerization is required for stable heteroduplex joint complex formation (Janz et al. 2002) (Table 1).

More recently, in vivo approaches such as the pink eyed unstable (p^{un}) mouse model, in which homology based deletion events give rise to black coat pigmentation, were used for the analysis of p53 functions in DSB repair. Schiestl and co-workers (Bishop et al. 2003) demonstrated an increased frequency of deletion events after loss of p53 in the p^{un} mouse model, which also enabled the suppressing role of p53 to the early stages in mouse development. Further, $p53^{-/-}$ mice showed increased LOH at the adenine phosphoribosyltransferase and hypoxanthine-guanine phosphoribosyltransferease loci involving interstitial deletions after X-ray treatment (Liang et al. 2002). Aside from being involved in homology based recombination in somatic cells in vivo, p53 mRNA expression in testes of mice is high and specific for spermatocytes in zygotene to pachytene, i.e. in the meiotic stages, at which homologous chromosomes synapse for genetic exchange (Rotter et al. 1993; Sjöblom and Lähdetie 1996). However, Gersten and Kemp (1997) did not observe elevated rates for meiotic HR in *p53* knockout mice. Therefore, p53 seems to specifically contribute to HR suppression mechanisms in mitotically growing cells. During meiosis p53 may rather serve to facilitate alternative DNA repair pathways or to eliminate defective meiotic spermatocytes by apoptosis induction (Schwartz et al. 1999; Odorisio et al. 1998).

4.2
Possible Mechanisms of HR Suppression

To identify candidate pathways of HR, which are affected by p53, recombination substrates were designed such that they allowed to focus either on mechanisms initiated by strand invasion, i.e. mostly gene conversion, or to assay both gene conversion and non-conservative events, i.e. mostly SSA (Saintigny et al. 1999; Akyüz et al. 2002). Wild-type p53 was shown to affect HR processes with both types of substrates indicating an interference of p53 with recombination involving Rad51 dependent strand invasion. Linke and colleagues (2003) tested this notion by use of a host cell reactivation assay together with expression plasmids for mutant Rad51(186P), which cannot bind p53, and for dominant negative Rad51 or Rad54. The results confirmed that p53's anti-recombinogenic function depends on the Rad51 pathway. In search for the biochemical mechanism, Yoon et al. (2004) examined the influence of p53 on Rad51 mediated strand exchange in vitro. Their data suggested that wild-type but not mutant p53 interferes with Rad51 activities at two stages. First, via direct protein-protein interactions p53 inhibits Rad51-ssDNA filament formation and, therefore, hinders Rad51 from initiating strand exchange. Second, p53 inhibits continued strand exchange, and, thus counteracts branch migration promoted by Rad51. However, most recently, Yun and colleagues (2004) noticed pronounced enhancement of non-conservative events rather than Rad51 dependent gene conversion after p53 knockdown by siRNA. However, the cell lines used in this study show either defects in the ATM/ATR signalling pathway components Mlh1 and Mre11 (HCT116, see Giannini et al. 2002), are questionable with respect to the p53 status (HT1080, see Paulson et al. 1998) or constitutively express E1B, a known inhibitor of p53 in transcriptional transactivation. Therefore, it will be interesting to see whether the postulated, discriminatory recombination suppression is equally seen in other cell types as well.

Carefully designed cell based studies in combination with biochemical experiments further enlightened the mechanisms underlying p53's function in suppressing HR and provided a clue to the biological meaning of this somewhat paradoxical activity directed against a fairly safe DNA repair activity. Thus, knowing that p53 is a well established DNA binding protein, the influence of the DNA substrate was investigated more systematically. The corresponding set of experiments demonstrated consensus sequence independent HR suppression by p53 as well as inhibition of strand exchange promoted by Rad51 on randomly chosen DNA substrates (Boehden et al. 2004; Yoon et al. 2004). More importantly, two other studies unveiled that p53 preferentially suppresses recombination between sequences with short homologies, i.e. with extensively divergent sequences (Gebow et al. 2000; Akyüz et al. 2002). A threshold of homology length for recombination in mammalian cells was previously described, so that p53 may contribute to the barrier for recombination between repetitive sequences in the genome (Rubnitz and Subramani 1984). This has led to a model in which p53 represents one of the molecules responsible for the identification and alignment of sufficiently homologous sequences to permit homology based repair. Consistently, p53 was shown to recognize preformed recombination intermediates with certain mispairings most strongly and to preferentially suppress inappropriate recombination between correspondingly mispaired DNA sequences (Dudenhöffer et al. 1998). During exonucleolytic excision of nucleotides from defined synthetic DNA substrates p53 exhibits a preference for mispaired as compared with correctly paired DNA, particularly within the heteroduplex part of preformed recombination intermediates (Huang 1998; Skalski et al. 2000; Süsse et al. 2000). Moreover, exonucleolytic DNA attack by p53 was found to rise during Rad51 mediated strand transfer in vitro (Süsse et al. 2000). Together with the observed stabilization of p53-DNA complexes after preassembly of Rad51-nucleoproteins, these data suggest that p53 proofreads the heteroduplex immediately after Rad51 mediated strand invasion or continued strand exchange. In contrast to the mismatch repair and HR regulatory protein MSH2, which recognizes G-T mismatches best, mismatch recognition by p53 has a preference for A-G and C-C mismatches (Dudenhöffer et al. 1998; Huang 1998; Skalski et al. 2000; Degtyareva et al. 2001). Strikingly, both MSH2 and p53 were described to interact with Holliday junctions, to associate with recombinative repair complexes containing factors like Rad51 and Rad50, to regulate BLM helicase, and to sense heterologies between the substrates during HR (Alani et al. 1997; Dudenhöffer et al. 1998; Elliott and Jasin 2001; Zink et al. 2002; Yang et al. 2004). Subramanian and Griffith (2002) further demonstrated that binding of p53 to Holliday junctions is stimulated by MSH2. Given these facts, it is far more than speculation that these two molecules – p53 and MSH2 – play complementary or even synergistic roles in the fidelity control of HR.

P53 has also been shown to functionally interact with RPA to mediate suppression of HR. The p53-RPA interaction is independent of p53's transcriptional transactivation function, as mutant p53(22Q, 23S) still binds to RPA. In contrast, p53 mutants with amino acid exchanges which specifically abolish RPA binding, p53(48H, 49H) and p53(53S, 54S), failed to suppress HR in recombination assays. However, binding of p53 to RPA seems to be necessary but not sufficient for HR suppression, as was derived from the analysis of hotspot p53 mutants which bind RPA but do not counteract HR (Romanova et al. 2004).

Importantly, recent studies demonstrated that p53 inhibits HR not only following the introduction of an artificial DSB, but counteracts hyperrecombination also in response to replication fork stalling (Janz and Wiesmüller 2002; Saintigny and Lopez 2002). This replication associated activity may explain earlier observations indicating that p53 deficient cells must proceed through S-phase under conditions that induce DNA breakage for genetic instability to occur (Paulson et al. 1998). Harris and co-workers found BLM to be necessary for efficient accumulation of p53 within Rad51 foci at stalled replication forks (Sengupta et al. 2003). BLM-p53 interactions most likely also promote ATM/ATR mediated phosphorylation, which in turn may facilitate stable complex formation with other DSB repair proteins like Rad51, Rad54 and RPA. Because p53 was reported to counteract Holliday junction unwinding by BLM and WRN in vitro and WRN exonuclease activity on forked duplex DNA, it was initially thought that p53 indirectly regulates recombination by modulating BLM and WRN surveillance activities (Yang et al. 2002; Sommers et al. 2005). However, the extreme C-terminus, which interacts with BLM and WRN (Yang et al. 2002), is dispensable for p53 dependent HR suppression (Dudenhöffer et al. 1999; Willers et al. 2000). Moreover, cell based studies revealed complementary roles for p53 and BLM in the suppression of HR (Sengupta et al. 2003). Nevertheless, it is of interest that p53 counteracts regression of replication forks and branch migration of the resulting Holliday junctions (Yoon et al. 2004). Both replication fork regression and reversal of regression are likely to be executed by BLM or WRN (Bachrati and Hickson 2003), and could be regulated by p53. Therefore, the possibility remains that p53 cooperates with BLM or WRN particularly at sites of aberrant DNA exchange processes after replication blockage to prevent hyper-recombination at replication forks.

4.3 Enhancement of HR and Possible Mechanisms

The SV40 virus based assay system developed by Wiesmüller and colleagues (1996) was critically important for the discovery of HR stimulatory activities of p53, which were described in two recent publications. First, cells infected with virus containing DNA elements with topoisomerase I binding sites, but not alternative sequences, showed 4- fold elevated interchromosomal exchange frequencies when p53 was expressed (Boehden et al. 2004). Similar increases were seen after topoisomerase I inhibitor (camptothecin) treatment, suggesting that p53 and topoisomerase I act on the same HR promoting pathway. This activity was not restricted to wild-type p53, but was similarly seen for mutant p53(138V) with impaired transcriptional transactivation but retained HR suppression activities, which indicated that HR stimulation is separable from classical tumour suppressor functions. Earlier on, p53 had already been linked to topoisomerase I (Albor et al. 1998; Gobert et al. 1999; Stephan et al. 2002 and references therein). Thus, wild-type and mutant p53 form stable complexes with DNA bound topoisomerase I during periods of genotoxic stress. Further, p53 enhances topoisomerase I activity as biochemically measured by relaxation of supercoiled DNA or by a novel repair reaction that involves the formation of a specific cleavage complex with two topoisomerase I molecules encompassing the lesion. After the introduction of certain types of DNA damage, such as caused by γ-irradiation, p53 may therefore maintain genomic stability by promoting homology based repair indirectly via this newly discovered topoisomerase I dependent lesion excision mechanism followed by recombinative gap repair (Stephan et al. 2002; Boehden et al. 2004). Because p53 mutants exhibit topoisomerase I mediated HR stimulation without being competent in HR proofreading, error-prone recombinative rearrangements are expected to increase in a mutant as compared to a negative p53 background.

In the second report, HR stimulation was found upon sequence-specific binding of p53 to the ribosomal gene cluster repeat (RGC repeat) (Boehden et al. 2005). Although at first sight somewhat surprising, these findings are compatible with the proposed role of p53 as a caretaker of genomic stability as outlined in the following. Sequence-specific DNA binding of p53 primarily leads to transcriptional activation accomplishing p53's functions in cell cycle control and apoptosis. However, some of the sequence-specific binding sites are found in replication origins like the RGC repeats. In Boehden et al. (2005) sequence-specific binding of wild-type p53 to the RGC repeat led to a 5-fold enhancement of HR between SV40 minichromosomes. Importantly, the consensus DNA binding defective mutant p53(138V) exhibited a significantly reduced stimulatory effect, although p53(138V) is active in sequence independent HR suppression and topoisomerase I mediated HR stimulation. This newly discovered p53 dependent HR increase correlated with a replicational delay of viral genomes containing RGC repeats compared with controls. Consistently, HR enhancement was not seen with replication incompetent plasmid substrates. Thus, whereas genome wide replication arrest by aphidicolin or hydroxyurea led to HR suppression by p53 (Janz and Wiesmüller 2002; Saintigny and Lopez 2002), sequence-specific binding of p53 blocked DNA replication and enhanced HR at these sites, however, only in the absence of additional DNA breakage. In analogy to the role of the Fob1 protein in yeast, the nucleolar fraction of p53 could play a role in rDNA copy number adjustment and sequence homogenization (Boehden et al. 2005).

These studies were the first to identify stimulatory effects of p53 on HR. Further analysis will be required to fully understand the potential HR stimulatory role of p53 sequence-specifically bound to non-transactivating DNA-elements. The pathway of topoisomerase I dependent HR stimulation has major implications for the current concepts of the proposed oncogenic functions of certain p53 mutants.

5
p53 in the Fidelity Control of NHEJ

Despite the fact that the mammalian NHEJ pathway, including its specialized subtypes involved in lymphocyte maturation, has been characterized in detail, only little is known about the significance of p53 within its regulation. Moreover, results were reported that appeared contradictory by that time. Thus, utilising a human thyroid carcinoma cell line ectopically expressing temperature sensitive p53(138V) and mouse fibroblasts expressing wild-type, p53(135V) or no p53, respectively, two groups observed an increase in IR induced rejoining of linearized plasmid DNA in the presence of wild-type and mutant p53 (Yang et al. 1997; Tang et al. 1999). Unlike HR suppression, this stimulatory effect on end joining was seen to rely on the C-terminal reannealing activity of p53 (Tang et al. 1999). However, Mallya and Sikpi (1999) reported that p53 is required for the inhibition of IR induced DSB rejoining in human lymphoblast lines. Moreover, inhibition of NHEJ by p53 was also demonstrated in five other studies, in which cell based reporter gene reconstitution and integration assays or in vitro end-joining experiments were performed (Bill et al. 1997; Lee et al. 1999; Akyüz et al. 2002; Okorokov et al. 2002; Dahm-Daphi et al. 2005). In an attempt to bring these contradictory results together, Akyüz et al. (2002) proposed that p53 may particularly strongly inhibit error-prone NHEJ as compared to error-free NHEJ, i.e. proofread NHEJ. The repair of IR induced DSBs is generally error-prone, and, indeed, the down-regulatory effect of p53 on mutagenic NHEJ did apply to the rejoining of IR induced DSBs as assessed by the Comet assay (Bristow et al. 1998), but not by constant-field electrophoresis (Dahm-Daphi et al. 2005). Interestingly, from their data obtained with I-Sce I cleavable repair plasmid substrates integrated into the genome of mouse *Ltk*-cells in combination with pifithrin-α, a chemical inhibitor of transactivation and apoptosis inducing functions of p53, Lin et al. (2003) postulated that functional wild-type p53 facilitates precise ligation rather than inhibits error-prone NHEJ. In contrast, using similar genomic substrates in mouse embryo fibroblasts, Dahm-Daphi et al. (2005) noticed that p53 inhibited microhomology mediated (error-prone) NHEJ as well as precise ligation of two I-Sce I cleaved sites. The authors proposed that one possibility to support the concept of enhancement of precise ligation events is that p53 promotes rejoining exclusively on a single DSB, i.e. without DNA sequence deletion between two compatible sites. Clearly, further work will be required to understand the specific role of p53 in distinct subpathways of this important but potentially detrimental DSB repair process.

6
Summary of the Direct Effects of p53 on DSB Repair

Taken together, inhibition of excess and mutagenic genetic rearrangements by p53 can be explained by p53's classical tumour suppressor functions in cell cycle control and apoptosis as well as by direct regulatory activities in HR and possibly also in NHEJ, although an involvement in the latter pathway is much less understood. p53 has also been proven to accomplish stimulation of HR, when topoisomerase I targeted DNA lesions or ribosomal gene cluster repeats are involved. Putative roles of the latter sequence dependent activities are the enhancement of topoisomerase I dependent repair of specific DNA lesions and the maintenance of rDNA sequence homogeneity, respectively.

HR suppression represents a biochemically quite well characterized DNA repair function of p53. Up to now, evidence has accumulated for three different mechanisms of p53 mediated HR suppression that may act in a coordinated manner: first, in analogy to the mismatch repair protein MSH2, p53 blocks continued strand exchange by the recombinase Rad51 by strongly binding to nascent HR intermediates carrying mismatches. Second, considering the 3′-5′exonuclease activity of p53, proofreading of mispaired heteroduplexes by p53 may dissolve error-prone recombination junctions. Third, p53 at repair foci interacts with BLM, which bypasses stalled replication forks. p53 may cooperate with BLM at these sites to prevent replication fork collapse and thereby to counteract hyper-recombination after replication blockage.

Numerous physical and functional interactions of p53 with DSB repair related proteins have been elucidated in recent years, and it is likely that far more regulatory pathways remain to be discovered. When only considering the functional interactions known so far, p53 already influences DSB repair at different hierarchical levels as summarized in Fig. 2. Thus, as a direct regulator of DSB repair p53 exhibits caretaker activities in a highly coordinated manner with other well-known DSB repair surveillance and checkpoint fac-

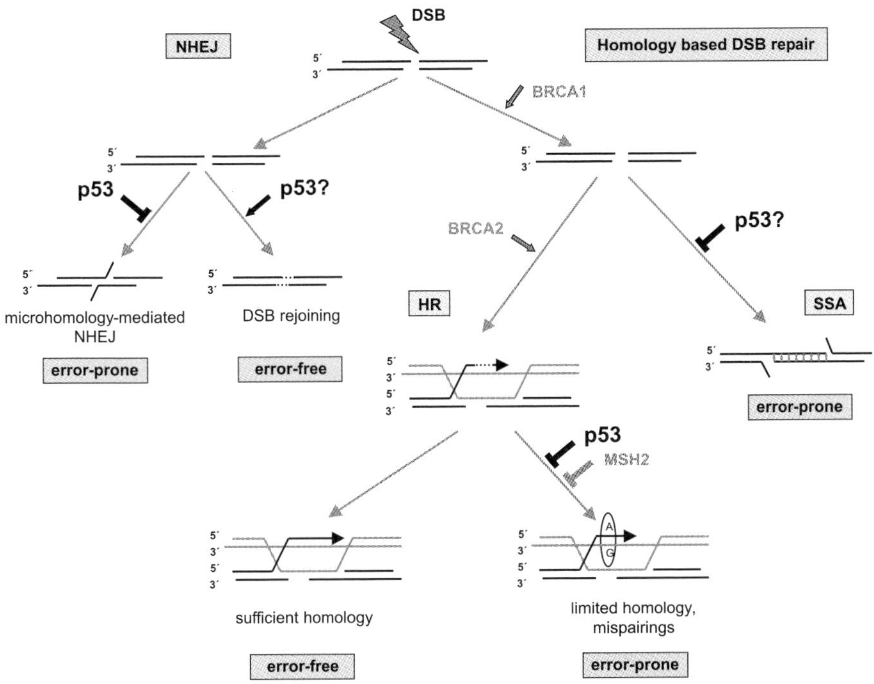

Fig. 2 p53 as a major factor of DSB repair surveillance. Schematic drawing of the different DSB repair pathways, NHEJ, SSA and HR, considering the stimulatory and inhibitory influence of central DSB repair surveillance factors which channel DSB repair into error-free pathways. The breast cancer related gene products BRCA1 and BRCA2 promote homology based repair. BRCA1 in this context has been shown to act at a step common to HR and SSA, whereas BRCA2 functions downstream, at the level of HR through enhancement of Rad51 filament formation (Stark et al. 2004). Similar to MSH2, p53 preferentially suppresses HR between sequences with short homologies, i.e. with extensively divergent sequences (Gebow et al. 2000; Akyüz et al. 2002). Consistently, p53 was shown to suppress inappropriate recombination between mispaired DNA sequences. However in contrast to MSH2, which recognizes G – T base mispairings best, mismatch recognition by p53 has a preference for A – G and C – C mismatches (Dudenhöffer et al. 1998). Thus, p53 may represent one of the molecules responsible for the identification of sufficiently homologous sequences to permit HR. Additionally, preferential suppression of non-conservative homology based DSB repair, i.e. mostly SSA, versus conservative HR has been postulated in one recent report (Yun et al. 2004). Finally, it will be a challenge of the future years to understand how p53 counteracts error-prone NHEJ and possibly also contributes to error-free DSB rejoining (Akyüz et al. 2002; Lin et al. 2003)

tors such as BRCA1, BRCA2 and MSH2, to protect cells from the detrimental outcome of error-prone repair.

References

Alani E, Lee S, Kane MF, Griffith J, Kolodner RD (1997) *Saccharomyces Cerevisiae* MSH2, a mispaired base recognition protein, also recognizes Holliday junctions in DNA. J Mol Biol 265:289–301

Albor A, Kaku S, Kulesz-Martin M (1998) Wild-type and mutant forms of p53 activate human topoisomerase I: a possible mechanism for gain of function in mutants. Cancer Res 5:2091–2094

Albrechtsen N, Dornreiter I, Grosse F, Kim E, Wiesmüller L, Deppert W (1999) Maintenance of genomic integrity by p53: complementary roles for activated and non-activated p53. Oncogene 18:7706–7717

Akyüz N, Boehden GS, Süsse S, Rimek A, Preuss U, Scheidtmann KH, Wiesmüller L (2002) DNA substrate dependence of p53 mediated regulation of double-strand break repair. Mol Cell Biol 22:6306–6317

Assenmacher N, Hopfner KP (2004) MRE11/RAD50/NBS1: complex activities. Chromosoma 113:157–166

Bachrati CZ, Hickson ID (2003) RecQ helicases: suppressors of tumorigenesis and premature aging. Biochem J 374:577–606

Bakalkin G, Yakovleva T, Selivanova G, Magnusson KP, Szekely L, Kiseleva E, Klein G, Terenius L, Wiman KG (1994) P53 binds single-stranded DNA ends and catalyzes DNA renaturation and strand transfer. Proc Natl Acad Sci USA 91:413–417

Bertrand P, Rouillard D, Boulet A, Levalois C, Soussi T, Lopez BS (1997) Increase of spontaneous intrachromosomal homologous recombination in mammalian cells expressing a mutant p53 protein. Oncogene 14:1117–1122

Bertrand P, Saintigny Y, Lopez BS (2004) P53's double life: transactivation-independent repression of homologous recombination. Trends Genet 20:235–243

Bill CA, Yu Y, Miselis NR, Little JB, Nickoloff JA (1997) A role for p53 in DNA end rejoining by human cell extracts. Mutat Res 385:21–29

Bishop AJ, Hollander MC, Kosaras B, Sidman RL, Fornace AJ Jr, Schiestl RH (2003) Atm-, p53-, Gadd45a-deficient mice show an increased frequency of homologous recombination at different stages during development. Cancer Res 63:5335–5343

Blander G, Kipnis J, Leal JFM, Yu CE, Schellenberg GD, Oren M (1999) Physical and functional interaction between p53 and Werner's syndrome protein. J Biol Chem 274:29463–29469

Boehden GS, Akyüz N, Roemer K, Wiesmüller L (2003) P53 mutated in the transactivation domain retains regulatory functions in homology-directed double-strand break repair. Oncogene 22:4111–4117

Boehden GS, Restle A, Marschalek R, Stocking C, Wiesmüller L (2004) Recombination at chromosomal sequences involved in leukaemogenic rearrangements is differentially regulated by p53. Carcinogenesis 25:1–9

Boehden GS, Baumann C, Siehler S, Wiesmüller L (2005) Wild-type p53 stimulates homologous recombination upon sequence-specific binding to ribosomal gene cluster repeat. Oncogene 24:4183–4192

Bond GL, Hu W, Levine AJ (2005) MDM2 is a central node in the p53 pathway: 12 years and counting. Curr Cancer Drug Targets 5:3–8

Brain R, Jenkins JR (1994) Human p53 directs DNA strand reassociation and is photolabelled by 8-azido ATP. Oncogene 9:1775–1780
Bristow RG, Hu Q, Jang A, Chung S, Peacock J, Benchimol S, Hill R (1998) Radioresistant MTp53-expressing rat embryo transformants exhibit increased DNA dsb rejoining during exposure to ionising radiation. Oncogene 16:1789–1802
Buchhop S, Gibson MK, Wang XY, Wagner P, Stürzbecher HW, Harris CC (1997) Interaction of p53 with the human Rad51 protein. Nucleic Acids Res 25:3868–3874
Bunz F, Dutriaux A, Lengauer C, Waldman T, Zhou S, Brown JP, Sedivy JM, Kinzler KW, Vogelstein B (1998) Requirement for p53 and p21 to sustain G2 arrest after DNA damage. Science 282:1497–1501
Christmann M, Tomicic MT, Roos WP, Kaina B (2003) Mechanisms of human DNA repair: an update. Toxicology 193:3–34
Corcoran CA, Huang Y, Sheikh MS (2004) The p53 paddy wagon: COP1, Pirh2 and MDM2 are found resisting apoptosis and growth arrest. Cancer Biol Ther 3:721–725
Dahm-Daphi J, Hubbe P, Horvath F, El-Awady RA, Bouffard KE, Powell SN, Willers H (2005) Nonhomologous end-joining of site-specific but not of radiation-induced DNA double-strand breaks is reduced in the presence of wild-type p53. Oncogene 24:1663–1672
Degtyareva N, Subramanian D, Griffith JD (2001) Analysis of the binding of p53 to DNAs containing mismatched and bulged bases. J Biol Chem 276:8778–8784
Dudenhöffer C, Kurth M, Janus F, Deppert W, Wiesmüller L (1998) Specific mismatch recognition in heteroduplex intermediates by p53 suggests a role in fidelity control of homologous recombination. Mol Cell Biol 18:5332–5342
Dudenhöffer C, Rohaly G, Will K, Deppert W, Wiesmüller L (1999) Dissociation of the recombination control and the sequence-specific transactivation function of p53. Oncogene 18:5773–5784
Elliott B, Jasin M (2001) Repair of double-strand breaks by homologous recombination in mismatch repair-defective mammalian cells. Mol Cell Biol 21:2671–2682
Espinosa JM, Emerson BM (2001) Transcriptional regulation by p53 through intrinsic DNA/chromatin binding and site-directed cofactor recruitment. Mol Cell 8:57–69
Fuchs B, O'Connor D, Fallis L, Scheidtmann KH, Lu X (1995) P53 phosphorylation mutants retain transcription activity. Oncogene 10:789–793
Gebow D, Miselis N, Liber HL (2000) Homologous and nonhomologous recombination resulting in deletion: effects of p53 status, microhomology and repetitive DNA length and orientation. Mol Cell Biol 20:4028–4035
Gersten KM, Kemp CJ (1997) Normal meiotic recombination in p53-deficient mice. Nature Genet 17:378–379
Giannini G, Ristori E, Cerignoli F, Rinaldi C, Zani M, Viel A, Ottini L, Crescenzi M, Martinotti S, Bignami M, Frati L, Screpanti I, Gulino A (2002) Human MRE11 is inactivated in mismatch repair-deficient cancers. EMBO Rep 3:248–254
Gobert C, Skladanowski A, Larsen AK (1999) The interaction between p53 and DNA topoisomerase I is regulated differently in cells with wild-type and mutant p53. Proc Natl Acad Sci USA 96:10355–10360
Gottifredi V, Shieh S, Taya Y, Prives C (2001) p53 accumulates but is functionally impaired when DNA synthesis is blocked. Proc Natl Acad Sci USA 98:1036–1041
Harvey M, Sands AT, Weiss RS, Hegi ME, Wiseman RW, Pantazis P, Giovanella BC, Tainsky MA, Bradley A, Donehower LA (1993) In vitro growth characteristics of embryo fibroblasts isolated from p53-deficient mice. Oncogene 8:2457–2467
Huang P (1998) Excision of mismatched nucleotides from DNA: a potential mechanism for enhancing DNA replication fidelity by the wild-type p53 protein. Oncogene 17:261–270

Hupp TR, Lane DP (1995) Two distinct signaling pathways activate the latent DNA binding function of p53 in a casein kinase II-independent manner. J Biol Chem 270:18165–18174

Hwang BJ, Ford JM, Hanawalt PC, Chu G (1999) Expression of the p48 xeroderma pigmentosum gene is p53-dependent and is involved in global genomic repair. Proc Natl Acad Sci USA 96:424–428

Jackson MW, Agarwal MK, Agarwal ML, Agarwal A, Stanhope-Baker P, Williams BRG, Stark GR (2004) Limited role of N-terminal residues in the activation of transcription by p53. Oncogene 23:4477–4487

Janus F, Albrechtsen N, Knippschild U, Wiesmüller L, Grosse F, Deppert W (1999) Different regulation of the core domain activities 3′ to 5′exonuclease and sequence-specific DNA binding. Mol Cell Biol 19:2155–2168

Janz C, Wiesmüller L (2002) Wild-type p53 inhibits replication-associated homologous recombination. Oncogene 21:5229–5933

Janz C, Süsse S, Wiesmüller L (2002) P53 and recombination intermediates: role of tetramerization at DNA junctions in complex formation and exonucleolytic degradation. Oncogene 21:2130–2140

Jasin M (2002) Homologous repair of DNA damage and tumorigenesis: the BRCA connection. Oncogene 21:8981–8993

Jayaraman J, Prives C (1995) Activation of p53 sequence-specific DNA binding by short single strands of DNA requires the p53 C-terminus. Cell 81:1021–1029

Jean D, Gendron D, Delbecchi L, Bourgaux P (1997) p53-mediated DNA renaturation can mimic strand exchange. Nucleic Acids Res 25:4004–4012

Krejci L, Chen L, Van Komen S, Sung P, Tomkinson A (2003) Mending the break: two DNA double-strand break repair machines in eukaryotes. Prog Nucl Acid Res Mol Biol 74:159–201

Lee S, Elenbaas B, Levine A, Griffith J (1995) P53 and its 14 kDa C-terminal domain recognize primary DNA damage in the form of insertion/deletion mismatches. Cell 81:1013–1020

Lee S, Cavallo L, Griffith J (1997) Human p53 binds Holliday junctions strongly and facilitates their cleavage. J Biol Chem 272:7532–7539

Lee H, Sun D, Larner JM, Wu FS (1999) The tumor suppressor p53 can reduce stable transfection in the presence of irradiation. J Biomed Sci 6:285–292

Liang L, Shao C, Deng L, Mendonca MS, Stambrook PJ, Tischfield JA (2002) Radiation-induced genetic instability in vivo depends on p53 status. Mutat Res 502:69–80

Lin Y, Waldman BC, Waldman AS (2003) Suppression of high-fidelity double-strand break repair in mammalian chromosomes by pifithrin-alpha, a chemical inhibitor of p53. DNA Repair 2:1–11

Linke SP, Sengupta S, Khabie N, Jeffries BA, Buchhop S, Miska S, Henning W, Pedeux R, Wang XW, Hofseth LJ, Yang Q, Garfield SH, Stürzbecher HW, Harris CC (2003) P53 interacts with hRad51 and hRad54, and directly modulates homologous recombination. Cancer Res 63:2596–2605

Liu Y, Masson JY, Shah R, O'Regan P, West SC (2004) RAD51C is required for Holliday junction processing in mammalian cells. Science 303:243–246

Mallya SM, Sikpi MO (1999) Requirement for p53 in ionising-radiation-inhibition of double-strand-break rejoining by human lymphoblasts. Mutat Res 23:119–132

Marmorstein LY, Ouchi T, Aaronson SA (1998) The BRCA2 gene product functionally interacts with p53 and Rad51. Proc Natl Acad Sci USA 95:13869–13874

May P, May E (1999) Twenty years of p53 research: structural and functional aspects of the p53 protein. Oncogene 18:7621–7636

McKinney K, Mattia M, Gottifredi V, Prives C (2004) P53 linear diffusion along DNA requires its C-terminus. Mol Cell 16:413–424

Meek DW (2004) The p53 response to DNA damage. DNA Repair 3:1049–1056

Mekeel KL, Tang W, Kachnik LA, Luo CM, DeFrank JS, Powell SN (1997) Inactivation of p53 results in high rates of homologous recombination. Oncogene 14:1847–1857

Mummenbrauer T, Janus F, Müller B, Wiesmüller L, Deppert W, Große F (1996) P53 exhibits 3′- to 5′-exonuclease activity. Cell 85:1089–1099

Oberosler P, Hloch P, Ramsperger U, Stahl H (1993) P53 catalysed annealing of complementary single-stranded nucleic acids. EMBO J 12:2389–2396

Odorisio T, Rodriguez TA, Evans EP, Clarke AR, Burgoyne PS (1998) The meiotic checkpoint monitoring synapsis eliminates spermatocytes via p53-independent apoptosis. Nat Genet 18:257–261

Offer H, Wolkowicz R, Matas D, Blumenstein S, Livneh Z, Rotter V (1999) Direct involvement of p53 in the base excision repair pathway of the DNA repair machinery. FEBS Lett 450:197–204

Okorokov AL, Warnock L, Milner J (2002) Effect of wild-type, S15D and R175H p53 proteins on DNA end-joining in vitro: potential mechanism of DNA double-strand break repair modulation. Carcinogenesis 23:549–557

Olivier M, Eeles R, Hollstein M, Khan MA, Harris CC, Hainaut P (2002) The IARC TP53 database: new online mutation analysis and recommendations to users. Hum Mutat 19:607–614

Ouchi T, Monteiro AN, August A, Aaronson SA, Hanafusa H (1998) BRCA1 regulates p53-dependent gene expression. Proc Natl Acad Sci USA 95:2302–2306

Paulson TG, Almasan A, Brody LL, Wahl GM (1998) Gene amplification in a p53-deficient cell line requires cell cycle progression under conditions that generate DNA breakage. Mol Cell Biol 18:3089–3100

Perfettini JL, Kroemer RT, Kroemer G (2004) Fatal liaisons of p53 with Bax and Bak. Nat Cell Biol 6:386–388

Perkins EJ, Nair A, Cowley DO, Van Dyke T, Chang Y, Ramsden DA (2002) Sensing of intermediates in V(D)J recombination by ATM. Genes Dev 16:159–164

Rabbits TH (1991) Translocations, master genes and differences between the origins of acute and chronic leukemias. Cell 67:641–644

Restle A, Janz C, Wiesmüller L (2005) Differences in the association of p53 phosphorylated on serin 15 and key enzymes of homologous recombination. Oncogene 24:4380–4387

Romanova LY, Willers H, Blagosklonny MV, Powell SN (2004) The interaction of p53 with replication protein A mediates suppression of homologous recombination. Oncogene 23:9025–9033

Rothkamm K, Krüger I, Thompson LT, Löbrich M (2003) Pathways of DNA double-strand break repair during mammalian cell cycle. Mol Cell Biol 23:5706–5715

Rotter V, Schwartz D, Almon E, Goldfinger N, Kapon A, Meshorer A, Donehower LA, Levine AJ (1993) Mice with reduced levels of p53 protein exhibit the testicular giant-cell degenerative syndrome. Proc Natl Acad Sci USA 90:9075–9079

Rubbi CP, Milner J (2003) p53 is a chromatin accessibility factor for nucleotide excision repair of DNA damage. EMBO J 22:975–986

Rubnitz J, Subramani S (1984) The minimum amount of homology required for homologous recombination in mammalian cells. Mol Cell Biol 4:2253–2258

Saintigny Y, Lopez BS (2002) Homologous recombination induced by replication inhibition, is stimulated by expression of mutant p53. Oncogene 21:488–492

Saintigny Y, Rouillard D, Chaput B, Soussi T, Lopez BS (1999) Mutant p53 proteins stimulate spontaneous and radiation-induced intrachromosomal homologous recombination independently of the alteration of the transactivation activity and of the G1 checkpoint. Oncogene 18:3553–3565

Saintigny Y, Delacôte F, Varès G, Petitot F, Lambert S, Averbeck D, Lopez BS (2001) Characterization of homologous recombination induced replication inhibition in mammalian cells. EMBO J 20:3861–3870

Saito S, Yamaguchi H, Higashimoto Y, Chao C, Xu Y, Fornace AJ Jr, Appella E, Anderson CW (2003) Phosphorylation site interdependence of human p53 post-translational modifications in response to stress. J Biol Chem 278:37536–37544

Schwartz D, Goldfinger N, Kam Z, Rotter V (1999) p53 controls low DNA damage-dependent premeiotic checkpoint and facilitates DNA repair during spermatogenesis. Cell Growth Differ 10:665–675

Sengupta S, Harris CC (2005) P53: traffic cop at the crossroads of DNA repair and recombination. Nat Rev Mol Cell Biol 6:44–55

Sengupta S, Linke SP, Pedeux R (2003) BLM helicase-dependent transport of p53 to sites of stalled DNA replication forks modulates homologous recombination. EMBO J 22:1210–1222

Sjöblom T, Lähdetie J (1996) Expression of p53 in normal and γ-irradiated rat testis suggests a role for p53 in meiotic recombination and repair. Oncogene 12:2499–2505

Skalski V, Lin ZY, Choi BY, Brown KR (2000) Substrate specificity of the p53-associated $3'$-$5'$ exonuclease. Oncogene 19:3321–3329

Sommers JA, Sharma S, Doherty KM, Karmakar P, Yang Q, Kenny MK, Harris CC, Brosh RM Jr (2005) p53 modulates RPA-dependent and RPA-independent WRN helicase activity. Cancer Res 65:1223–1233

Spillare EA, Robles AI, Wang XW, Shen JC, Yu CE, Schellenberg GD, Harris CC (1999) p53-mediated apoptosis is attenuated in Werner syndrome cells. Genes Dev 13:1355–1360

Stark JM, Pierce AJ, Oh J, Pastink A, Jasin M (2004) Genetic steps of mammalian homologous repair with distinct mutagenic consequences. Mol Cell Biol 24:9305–9316

Stephan H, Grosse F, Soe K (2002) Human topoisomerase I cleavage complexes are repaired by a p53-stimulated recombination-like reaction. Nucleic Acids Res 30:5087–5093

Stommel JM, Marchenko ND, Jimenez GS, Moll UM, Hope TJ, Wahl GM (1999) A leucine-rich nuclear export signal in the p53 tetramerization domain: regulation of subcellular localization and p53 activity by NES masking. EMBO J 18:1660–1672

Stürzbecher HW, Donzelmann B, Henning W, Knippschild U, Buchhop S (1996) p53 is linked directly to homologous recombination processes via Rad51/RecA protein interaction. EMBO J 15:1992–2002

Subramanian D, Griffith JD (2002) Interactions between p53, hMSH2-hMSH6 and HMG I(Y) on Holliday junctions and bulged bases. Nucleic Acids Res 30:2427–2434

Subramanian D, Griffith JD (2005) Modulation of p53 binding to Holliday junctions and 3-cytosine bulges by phosphorylation events. Biochemistry 44:2536–2544

Süsse S, Janz C, Janus F, Deppert W, Wiesmüller L (2000) Role of heteroduplex joints in the functional interactions between human Rad51 and wild-type p53. Oncogene 19:4500–4512

Süsse S, Scholz CJ, Bürkle A, Wiesmüller L (2004) Poly(ADP-ribose) polymerase (PARP-1) and p53 independently function in regulating double-strand break repair in primate cells. Nucleic Acids Res 32:669–680

Tang W, Willers H, Powell SN (1999) P53 directly enhances rejoining of DNA double-strand breaks with cohesive ends in gamma-irradiated mouse fibroblasts. Cancer Res 59:2562–2565

Thompson T, Tovar C, Yang H, Carvajal D, Vu BT, Xu Q, Wahl GM, Heimbrook DC, Vassilev LT (2004) Phosphorylation of p53 on key serines is dispensable for transcriptional activation and apoptosis. J Biol Chem 279:53015–53022

Valerie K, Povirk LF (2003) Regulation and mechanisms of mammalian double-strand break repair. Oncogene 22:5792–5812

Vaziri H, West MD, Allsopp RC, Davison TS, Wu YS, Arrowsmith CH, Poirier GG, Benchimol S (1997) ATM-dependent telomere loss in aging human diploid fibroblasts and DNA damage lead to the post-translational activation of p53 protein involving poly(ADP-ribose) polymerase. EMBO J 16:6018–6033

Vogelstein B, Lane D, Levine AJ (2000) Surfing the p53 network. Nature 408:307–310

Vousden KH, Prives C (2005) P53 and prognosis: new insights and further complexity. Cell 120:7–10

Wang FJ (1998) Analysis of downstream effectors of p53 on cell growth arrest and apoptosis induced by a temperature sensitive Val138 mutant. J Med Dent Sci 45:141–149

Wang X, Ohnishi K, Takahashi A, Ohnishi T (1998) Poly(ADP-ribosyl)ation is required for p53-dependent signal transduction induced by radiation. Oncogene 17:2819–2825

Wang XW, Tseng A, Ellis NA, Spillare EA, Linke SP, Robles AI, Seker H, Yang Q, Hu P, Beresten S, Bemmels NA, Garfield S, Harris CC (2001) Functional interaction of p53 and BLM DNA helicase in apoptosis. J Biol Chem 276:32948–32955

West S (2003) Molecular views of recombination proteins and their control. Nat Rev Mol Cell Biol 4:435–445

Wiesmüller L, Cammenga J, Deppert W (1996) In vivo assay of p53 function in homologous recombination between Simian Virus 40 chromosomes. J Virol 70:737–744

Willers H, McCarthy EE, Wu B, Wunsch H, Tang W, Taghian DG, Xia F, Powell SN (2000) Dissociation of p53-mediated suppression of homologous recombination from G1/S cell cycle checkpoint control. Oncogene 19:632–639

Willers H, McCarthy EE, Hubb P, Dahm-Daphi J, Powell SN (2001) Homologous recombination in extrachromosomal plasmid substrates is not suppressed by p53. Carcinogenesis 22:1757–1763

Willers H, Xia F, Powell SN (2002) Recombinational DNA repair in cancer and normal cells: the challenge of functional analysis. J Biomed Biotechnol 2:86–93

Yang X-J (2005) Multisite protein modification and intramolecular signaling. Oncogene 24:1653–1662

Yang T, Namba H, Takmura N, Nagayama Y, Fukata S, Ishikawa N, Kuma K, Ito K, Yamashita S (1997) P53 induced by ionizing radiation mediates DNA end-joining activity, but not apoptosis of thyroid cells. Oncogene 14:1511–1519

Yang Q, Zhang R, Wang XW, Linke SP, Sengupta S, Hickson ID, Pedrazzi G, Perrera C, Stagljar I, Littman SJ, Modrich P, Harris CC (2004) The mismatch DNA repair heterodimer, hMSH2/6, regulates BLM helicase. Oncogene 23:3749–3756

Yang Q, Zhang R, Wang XW, Spillare EA, Linke SP, Subramanian D, Griffith JD, Li JL, Hickson ID, Shen JC, Loeb LA, Mazur SJ, Appella E, Brosh RM Jr, Karamakar P, Bohr VA, Harris CC (2002) The processing of the Holliday junctions by BLM and WRN helicases is regulated by p53. J Biol Chem 277:31980–31987

Yoon D, Wang Y, Stapleford K, Wiesmüller L, Chen J (2004) P53 inhibits strand exchange and replication fork regression promoted by human Rad51. J Mol Biol 336:639–654

Yun S, Lie-A-Cheong C, Porter AC (2004) Discriminatory suppression of homologous recombination by p53. Nucleic Acids Res 32:6479–6489

Zhang H, Somasundram K, Peng Y, Tian H, Zhang H, Bi D, Weber BL, El-Deiry WS (1998) BRCA1 physically associates with p53 and stimulates its transcriptional activity. Oncogene 16:1713–1721

Zhou J, Ahn J, Wilson SH, Prives C (2001) A role for p53 in base excision repair. EMBO J 20:914–923

Zink D, Mayr C, Janz C, Wiesmüller L (2002) Association of p53 and MSH2 with recombinative repair complexes during S-phase. Oncogene 21:4788–4800

Genome Dyn Stab (1)
D.-H. Lankenau: Genome Integrity
DOI 10.1007/7050_001/Published online: 20 October 2005

The Role of Chromatin Structure and Nuclear Architecture in the Cellular Response to DNA Double-Strand Breaks

Anna A. Friedl

Radiobiological Institute, University of Munich, Schillerstr. 42, 80336 Munich, Germany
anna.friedl@lrz.uni-muenchen.de

Abstract The processes involved in the repair of DNA double-strand breaks (DSB) were thoroughly investigated at the genetic and biochemical level. The importance of chromatin structure and nuclear architecture in the outcome of repair has, however, only slowly emerged. Important recent developments in the field comprise the characterization of passive and active chromatin alterations following DSB induction, the function of chromatin modifying complexes and chromatin remodeling complexes in DSB processing, and the role of histone H2AX in the recruitment of these complexes. In addition, the development of new technologies, such as in vivo fluorescence labeling of chromatin and proteins, and the localized induction of breaks by use of endonucleases and microirradiation, has allowed insights to be gained into the mobility of damaged chromatin regions in the nucleus.

1 Introduction

Among the various types of DNA damage that threaten genetic integrity either spontaneously or after exposure to genotoxic agents, the DNA double-strand break (DSB) is considered the most deleterious. In recent years, numerous genetic and biochemical studies have enabled a detailed characterization of the molecular processes taking place during DSB repair. It is now known that there are several repair pathways that can roughly be grouped into homology-dependent and homology-independent branches. Various sub-pathways of homology-dependent repair were described (Paques and Haber 1999; Helleday 2003), which may, by and large, be regarded as variations of a common theme, since the proteins involved appear to overlap considerably. In contrast, there appear to exist at least two sub-pathways of homology-independent repair, which differ to a certain degree in the identity of the proteins participating (e.g. Löbrich, this book).

DSB repair takes place in the context of chromatin structure and nuclear architecture, but the potential influence of these parameters on repair efficiency and outcome is being recognized only slowly (Ehrenhofer-Murray 2004; Friedl 2004; Morrison and Shen 2005). This article intends to give an overview of recent developments in the field, with emphasis on work published since 2003.

2 Chromatin Alterations after Induction of Double-Strand Breaks

How is chromatin structure affected – both directly and indirectly – by DSB induction and repair? Finding an answer to this question is hampered by the fact that the three-dimensional organization of chromatin in the mammalian interphase nucleus is not yet understood in its entirety. There are several levels of organization in an hierarchical order (reviewed by Woodcock and Dimitrov 2001; Horn and Peterson 2002). Primary and secondary structures, which have been thoroughly investigated, comprise the wrapping of the DNA molecule around nucleosomes and formation of the so-called 30 nm fiber by interaction between nucleosomes. A nucleosome represents an octamer of core histone proteins, with two molecules each of H2A, H2B, H3, and H4. On the other end of hierarchical organization is the arrangement of chromosomes into rather well-defined territories and the arrangement of the territories within the nucleus. These levels of organization become more and more clear (Cremer and Cremer 2001). Less is known about the tertiary structure of chromatin, i.e. the way in which the 30 nm fiber is folded within the chromosomal territory. There are experimental data indicating that the fiber is arranged into loops with estimated loop sizes ranging from about 50 kbp to 200 kbp (reviewed by Heng et al. 2001), although alternative organization models cannot be excluded at the moment (e.g. Belmont et al. 1999; Dehghani et al. 2005). The loops are generally assumed to be fixed at their base points, thus allowing the formation of topologically independent functional chromatin domains (Labrador and Corces 2002). Indeed, in a variety of organisms some specific chromatin regions have been investigated and found to form loops via attachment to the nuclear matrix and/or interaction between so-called boundary elements (Ishii et al. 2002; Blanton et al. 2003; Byrd and Corces 2003; Yusufzai et al. 2004). Boundary elements (also called insulators) are specialized regulatory elements that are defined by their function: They can either block enhancer action when placed between enhancer and promoter, or they can act as barriers blocking the spread of chromatin states (reviewed by Capelson and Corces 2004). However, it remains unclear whether and to what extent these observations made on specific regions reflect the organization of the entire genome.

Consideration of chromatin loops as potential targets of radiation effects is not a new concept. Already in the 1980s transcriptional repression after irradiation was explained by loss of superhelical torsion in chromatin loops once breaks were introduced (Luchnik et al. 1988; Rodi and Sauerbier 1989). It is evident that loss of torsional conformation in closed circular molecules is a passive consequence of the presence of breaks, and the same may hold for chromatin loops in eukaryotes if fixed at their base points.

A more topical question is whether relaxation or other types of passive chromatin alteration can – even at low dose – affect the whole genome. This is an important question, since pan-nuclear chromatin alteration has been implicated in the activation of ATM (Bakkenist and Kastan 2003; Kitagawa et al. 2004). The protein kinase ATM is a key player in the cellular response to DSB (reviewed by Lavin et al. 2005). Once activated, ATM phosphorylates a large variety of substrates involved in cell cycle arrest, apoptosis and DNA repair, including p53, Nbs1 and Chk2. According to the model proposed by Kastan and co-workers, ATM, which in the unstressed cell is present as a catalytically inactive dimeric complex, is activated by rapid autophosphorylation of serine-1981 and subsequent monomerization. The major part of the cellular ATM pool becomes phosphorylated within minutes even if only a few DSB per cell are present. This was taken to suggest that direct contact between the DSB site and ATM is not required for autophosphorylation to occur. Rather, the authors propose that the ATM response is elicited by a global chromatin alteration, possibly associated with torsional relaxation (Bakkenist and Kastan 2004). The model is corroborated by the observation that ATM autophosphorylation can be induced by a variety of treatments which are known to cause chromatin alteration (e.g. chloroquine, inhibitors of histone deacetylases, and hypotonic conditions) but which do not elicit detectable direct DNA damage or even breakage (Bakkenist and Kastan 2003). In addition, a variety of genetic defects have been described which interfere with proper chromatin organization and which are associated with activation of ATM-dependent checkpoints (reviewed by Koundrioukoff et al. 2004). It cannot be excluded, however, that many of these defects cause DNA damage, e.g. as a consequence of replication stalling, in addition to the chromatin alterations.

While very interesting, the model proposed by Kastan and co-workers leaves many open questions with regard to the nature of the chromatin alteration, its generation (passive consequence of a loss of conformational structure or dependent on enzymatic activity), and the way ATM senses these alterations (directly or via adaptors or additional sensors). In addition, this model was recently challenged by the observation that ATM activation in vitro requires the presence of linear DNA (i.e. DNA ends) and of the MRN complex (Lee and Paull 2005). The MRN complex, comprising the proteins Mre11, Rad50 and Nbs1 (reviewed by Stracker et al. 2004; Hopfner, this book), has for some time been known to facilitate ATM activation (Carson et al. 2003; Uziel et al. 2003; Horejsi et al. 2004; Lee and Paull 2004). The crucial role played by DNA ends in this process has, however, not fully been recognized before. Clearly, it is difficult to reconcile the data of the Kastan group with those of Lee and Paull. While some discrepancies may be explained by the difference between the in vivo and in vitro situations (Abraham and Tibbetts 2005), further studies are required to clarify the role of chromatin structure in activation of the damage response pathway.

3
Chromatin Modification and Remodeling Associated with DSB Repair

A variety of chromatin modifications are known to constitute an active cellular response to the DNA damage. In general, DNA damage processing and repair require that the damaged site be accessible for the enzymes engaged in these processes. Initially proposed in the context of nucleotide excision repair (Green and Almouzni 2002), the "access-repair-restore" concept can also be applied to other repair pathways. The degree of chromatin compaction is to a large extent mediated by post-translational modifications, namely acetylation, methylation, phosphorylation, ubiquitylation, sumoylation, or ADP-ribosylation of histone proteins, mainly at their tails (Fischle et al. 2003). These modifications are brought upon by so-called chromatin modifying complexes. Histone acetyltransferases (HATs) are chromatin modifiying complexes which confer acetylation of specific residues in histone tails; this type of modification plays a critical role for a variety of biological processes. Generally speaking, hyper-acetylation of histones is associated with transcriptionally activated regions and thus chromatin de-compaction.

Several HATs have been identified in genetic studies as playing a role for DSB repair: The HAT activity of the human Tip60 protein is required for efficient DSB repair and apoptotic competence (Ikura et al. 2000), which may hint at a role in DSB-dependent signaling. Similarly, Esa1, a Tip60-related HAT in *Saccharomyces cerevisiae*, has been implicated in DSB repair (Bird et al. 2002). This function of Esa1 appears to be mediated by its ability to acetylate several lysine residues in the N-terminal tail of histone H4. Esa1 is a component of the NuA4 HAT complex, which is recruited to DSB sites (Bird et al. 2002; Downs et al. 2004). In addition, acetylation of several N-terminal lysines in yeast histone H3 was found to be required for DSB repair, especially recombination-dependent repair (Qin and Parthun 2002). Intriguingly, several histone deacetylases have also been implicated in DSB repair in yeast (Jazayeri et al. 2004), and mammalian histone deacetylase HDAC4 was found to accumulate in large number at damaged sites (Kao et al. 2003). In several instances, treatment of mammalian cells with histone deacetylase inhibitors was found to increase radiosensitivity (e.g. Camphausen et al. 2005), but whether this effect is due to chromatin decondensation remains to be established. Very recent work in *S. cerevisiae* demonstrates that ordered, dynamic changes in histone acetylation, conferred by histone acetyltransferases and histone deacetylases recruited to the break site, accompany recombinational repair (Tamburini and Tyler 2005). These authors proposed that the deacetylation step may signalize completion of repair.

Two examples of DSB-associated histone methylation were also described (Huyen et al. 2004; Sanders et al. 2004), as well as several examples for histone phosphorylation (Fernandez-Capetillo et al. 2004a; Harvey et al. 2005). Presumably the most prominent histone modification in the context of DSB repair is in mammalian cells the phosphorylation of histone H2A variant H2AX at serine-139 (Rogakou et al. 1998, 1999). In response to DSB, ATM is mainly responsible for H2AX phosphorylation (Burma et al. 2001); the contribution of other phosphatidylinositol 3-kinase (PI3KK)-like kinases such as ATR and DNA-PK is under debate (Paull et al. 2000; Park et al. 2003; Karlsson and Stenerlöw 2004; Stiff et al. 2004; Friesner et al. 2005). By virtue of antibodies specifically recognizing γ-H2AX, the phosphorylated form of H2AX, it was shown that H2AX phosphorylation in mammalian cells extends to an Mbp-sized chromatin region surrounding the break site. In immunofluorescence assays, chromatin regions thus decorated display a globular signal reminiscent of the signal generated by a variety of DNA repair and signaling factors (e.g Rad51, 53BP1, Mdc1, Mre11/Rad50/Nbs1, BRCA1, ATM, RPA32). In the first description of such protein accumulations at damaged DNA sites in mammalian cells the term "foci" was coined (Haaf et al. 1995), and the accumulations of both proteins and protein modifications in the vicinity of DNA double-strand breaks are now referred to as radiation-induced foci or repair foci. Since in microscopic analysis γ-H2AX foci were found to colocalize with the aforementioned protein accumulations, initial models postulated that phosphorylation of histone H2AX might facilitate recruitment of repair and signaling proteins to the damaged site, potentially by providing a binding platform. More recent models postulate that only the accumulation of these proteins into large foci, but not the initial recruitment to the break site, depend on γ-H2AX (reviewed by Fernandez-Capetillo et al. 2004b). Since many of the foci-forming proteins are substrates of ATM, one function of the foci may be to achieve signal amplification by bringing activated ATM and a large number of its target proteins into close vicinity of each other (Fig. 1).

Interestingly, phosphorylation of histone H2AX may also affect chromatin conformation (Fig. 1). This topic has mainly been investigated in the yeast *S. cerevisiae*. The major histone H2A species, called H2A, in this yeast is orthologous to mammalian H2AX. Like its mammalian counterpart, yeast H2A is phosphorylated (at serine-129) in response to DSB (Downs et al. 2000). Given that *S. cerevisiae* chromosomes are much smaller than mammalian

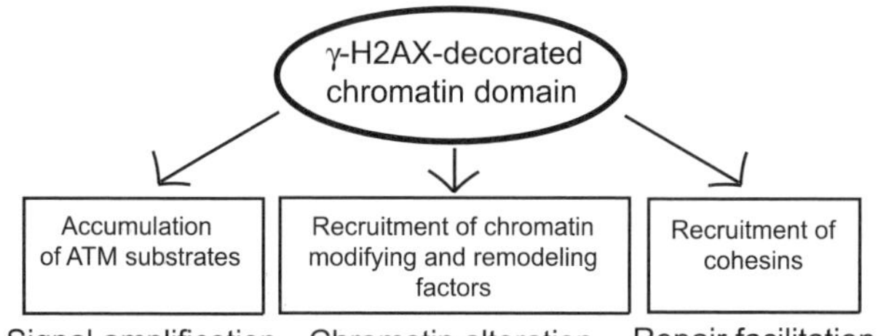

Fig. 1 Current models to explain why histone H2AX phosphorylation extends over an Mbp-sized chromatin region

chromosomes, it is not surprising that H2A phosphorylation extends over a smaller region (about 50 kbp) in this organism. First indications for a role in chromatin modification came from experiments with an H2A mutant (*hta1*-S129E) in which serine at position 129 was replaced by glutamic acid, thus mimicking constitutive phosphorylation (Downs et al. 2000). In this mutant, the overall chromatin compaction was decreased as compared to wild type cells, but it remained unclear whether this constituted a direct effect of histone phosphorylation or was due to the recruitment of some chromatin altering factors.

Recently, detailed chromatin immunoprecipitation analyses of the chromatin surrounding a site-specific, endonuclease-induced DSB in *S. cerevisiae*, revealed an H2A P-Ser129 dependent recruitment of both chromatin modifying complexes and chromatin remodeling complexes to the break site (Downs et al. 2004; Morrison et al. 2004; van Attikum et al. 2004). Recruitment of the chromatin modifying complex Nu4A, which contains the histone acetyltransferase Esa1, is mediated by an interaction of H2A P-Ser129 with Arp4 (Downs et al. 2004). While chromatin modifying complexes alter histone proteins, the chromatin remodeling complexes determine chromatin structure by changing the nucleosome positioning. Several groups demonstrated H2A P-Ser129 dependent recruitment of the INO80 chromatin remodeling complex to the vicinity of the break site in yeast (Downs et al. 2004; Morrison et al. 2004; van Attikum et al. 2004). The INO80 complex, which belongs to the SWI/SNF superfamily of chromatin remodeling complexes, has for some time been suspected of participating in break repair, because it contains homologs to the bacterial RuvB protein (Shen et al. 2000). RuvB is implicated in processing of Holliday junctions, but indications for similar functions of the homologs Rvb1 and Rvb2 are lacking (Yamada et al. 2004). Since formation of 3′ single-stranded overhangs is disturbed in strains that are unable to recruit the INO80 complex to the break site (van Attikum et al. 2004), one function of this complex may be to remove or slide nucleosomes lying close to the broken ends in order to facilitate end processing. One remaining difficulty with this model lies in the observation that H2A P-Ser129, while present in a region of about 50 kbp surrounding a DSB site, was not found to be enriched in the 1–2 kbp region directly flanking the break site (Shroff et al. 2004; Ünal et al. 2004). It is unclear at present whether the apparent non-enrichment of H2A P-Ser129 directly at the breaks sites may be due to experimental artifacts (Downs et al. 2004), and it is also not known whether for recruitment of the INO80 complex direct physical contact with H2A P-Ser129 (Morrison et al. 2004) is necessary.

Assuming that phosphorylated H2A (and by analogy its orthologs in other organisms) initiates a cascade of DSB- and repair-associated chromatin alterations, the question arises as to when and how the initiating signal is removed. In principle, removal of the signal could be accomplished by a phosphatase-dependent dephosphorylation reaction, leaving the histone protein in place. Such a model was proposed by Nazarov et al. (2003) for γ-H2AX. Alternatively, the phosphorylated H2A(X) protein could be replaced by an un-phosphorylated one. In *Drosophila melanogaster*, indications for this latter pathway have been found: here the phosphorylated H2AX-homolog, H2Av, is first acetylated by the Tip60 chromatin remodeling complex and then exchanged with unmodified H2Av (Kusch et al. 2004).

Another important question concerns the delimitation of the region which acquires a specific DSB-associated chromatin state. Generally, spreading of chromatin states is assumed to be limited either by specific insulator or boundary elements, or by encounter with antagonistic chromatin states of an adjacent domain (Fourel et al. 2004). This creates independently regulated functional domains, which may, as discussed above, correlate with structural features such as loops. It is reasonable to assume that damage/repair-associated chromatin alterations spread maximally to the boundaries of the functional domain which is affected. In microscopic analysis, mammalian γ-H2AX foci share similarities with so-called replication foci. These are chromosomal units, typically around 1 Mbp in size, that can be visualized by pulse-labeling and are defined by their temporal behavior during replication (Sadoni et al. 2004). It has been proposed that these domains may be formed by association of several chromatin loops in form of a rosette-like structure (Münkel et al. 1999). This model is substantiated by the observation of clustering boundary elements in *Drosophila* (for review, see Capelson and Corces 2004). In conclusion, transcription, replication, and repair may rely on the same spatially defined functional units.

The various DSB-associated histone modifications detected so far sustain the concept of repair-specific histone codes (Morrison and Shen 2005), where combinations of histone modifications coordinate damage response processes. It was proposed that the identity of histone modifications may even determine the choice of the repair pathway utilized in processing a certain lesion (Moore and Krebs 2004). Several researchers have begun to correlate specific histone modifications with specific DSB repair pathways (Bird et al. 2002; Downs et al. 2004; Tamburini and Tyler 2005), but further studies are necessary to obtain a conclusive picture.

4 Mobility of Broken Sites

If after DSB induction both ends started to move away from each other, the efficiency of correct repair would be compromised and the probability of mis-rejoining with ends originating from different break sites might increase. It is therefore important to investigate if some pre-existing or rapidly generated scaffold holds the ends within relative proximity to each other. Again, work in *S. cerevisae* has paved the path. In vivo fluorescence labeling of chromatin regions can be achieved by integrating lactose or tetracycline operator arrays into the DNA and expressing the respective binding proteins (lacI or TetR) fused to green fluorescent protein (GFP) or a spectral variant thereof (reviewed by Belmont 2001; see Fig. 2). This technology has been exploited in yeast to label regions flanking an endonuclease-generated DSB and to visualize under in vivo conditions the fate of the broken ends and their behavior

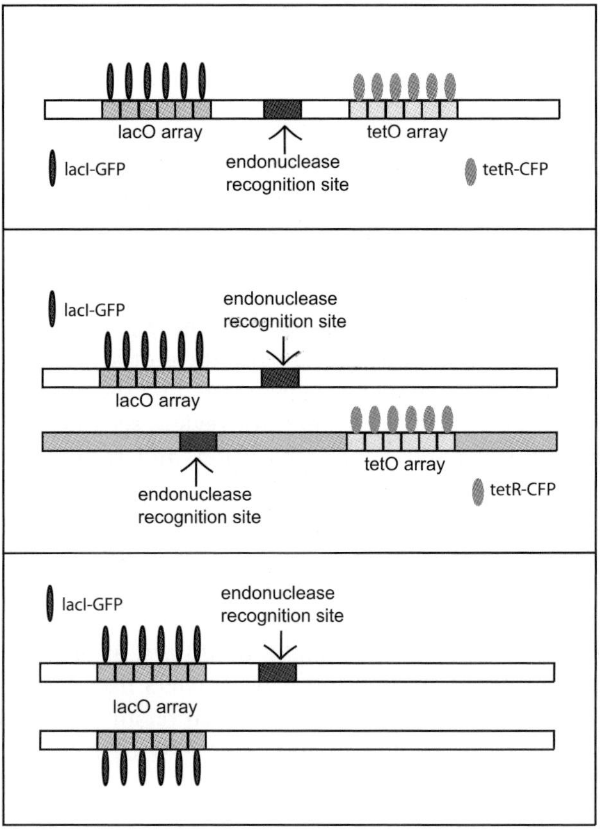

Fig. 2 Sketch of a method allowing in vivo fluorescence labeling of chromatin regions after inserting large arrays of lacO or tetO sequence elements into chromosomal DNA. Expression of lacI and tetR proteins fused to green fluorescent protein (GFP) or a spectral variant, such as CFP (cyan fluorescent protein) and binding of these proteins to their respective binding sites allows microscopic analysis of living cells. *Top panel*: Labeling both sides flanking an endonuclease recognition site with different fluorescent signals allows to investigate whether the ends separate after break induction. Such an approach was used by Lobachev et al. 2004. *Middle panel*: Labeling of two break sites on different chromosomes was used to demonstrate the existence of "repair garages" in yeast (Lisby et al. 2003). *Lower panel:* Labeling of a break site and a potential recombination donor site may be used to investigate genome-wide homology search

in relation to other nuclear components (Lisby et al. 2003; Kaye et al. 2004; Lobachev et al. 2004). If both sides flanking an endonuclease recognition site were labeled, both fluorescence signals coincided most of the time, and the frequency of clear separation of signals was the same with and without endonuclease induction (Kaye et al. 2004; Lobachev et al. 2004), indicating that the break ends are kept together. Loss of functional Rad50-Mre11-Xrs2 complex (the yeast homolog to the mammalian Mre11-Rad50-Nbs1 complex), mutation in the Rad50 hook structure (see Hopfner et al. this book), or loss of functional Rad52 protein lead to an increase in the number of cells with separated signal after DSB induction, suggesting that these factors are involved in holding the ends together (Kaye et al. 2004; Lobachev et al. 2004; Lisby and Rothstein 2004). It should be noted, however, that in *S. cerevisiae* apparently repair centers exist where multiple DSB associate with Rad52 proteins, giving rise to a single fluorescent focus if Rad52 proteins are tagged with GFP (Lisby et al. 2003). Thus, it cannot be excluded at the moment that these repair centers rather than specific association of both ends of a given break site were observed in the experiments described above.

Even when both ends are held together, one has to ask if the break site as a whole moves around in the nucleus in the course of repair. Among the various pathways for DSB repair, homologous recombination between sister chromatids bears the smallest risk for repair-associated sequence alterations, since both chromatids are identical (except for very rare sequence differences due to replication errors). The sister chromatid also comes in handy as a recombination partner because physical vicinity is guaranteed by so-called sister chromatid cohesion. Cohesion is established and maintained by cohesins, i.e. protein complexes which associate with chromosomes before the onset of S-phase. Presumably by their ring-like structure, they link the sister chromatids from the time of replication until mitosis, where the ring-like structured is opened by specific enzymes to allow chromosome segregation (reviewed by Haering and Nasmyth 2003; Riedel et al. 2004).

Sister chromatid cohesion is, however, not only required for proper chromosome segregation in mitosis, but also for DSB repair via sister chromatid recombination (SCR; Sjögren and Nasmyth 2001; Sonoda et al. 2001). Recent work demonstrates that after DSB induction additional cohesins are recruited to regions surrounding DSB in yeast and mammalian cells (Kim et al. 2002; Ünal et al. 2004; Ström et al. 2004; Ström and Sjögren 2005). This damage-specific cohesin recruitment is restricted to S/G2 phase and depends on functional Mre11/Rad50 complexes (Kim et al. 2002; Ünal et al. 2004) as does cohesin recruitment in the absence of exogenously induced damage (Kim et al. 2002; Warren et al. 2004). In yeast, the extension of cohesin recruitment (about 50 kb around a single DSB site) correlates with the extension of histone H2A phosphorylation (Shroff et al. 2004; Ström et al. 2004; Ünal et al. 2004), suggesting a functional link, and indeed cohesin recruitment was found to depend on H2A phosphorylation (Ünal et al. 2004). Similarly, it was found that γ-H2AX in mammalian cells is required to promote repair via SCR over single-strand annealing (Xie et al. 2004). Thus, cohesin recruitment may, at least in part, explain why H2A(X) phosphorylation occurs over such large chromosomal domains (Fig. 1). Ironically, while damage-induced cohesin recruitment facilitates DSB repair as long as cells do not undergo mitosis, it may contribute to disastrous mis-segregation of fragments in those cells which override checkpoint arrest and passage through mitosis in the presence of DSB (Kaye et al. 2004).

Interphase FISH experiments show that in the absence of functional cohesins, sister chromatids, although clearly separated, are still rather close to each other (Sonoda et al. 2001). Thus, the drastic SCR deficiency observed in the absence of cohesion may be taken to indicate that a high degree of nearness is required for this type of repair. It is, however, unclear to what extent this holds for other repair pathways. Repair by homologous recombination between homologous chromosomes occurs readily in some organisms, such as *Drosophila* (e.g. Rong and Golic 2003), but it is not regarded as a major repair pathway in mammalian cells. For example, in mouse embryonic stem cells DSB repair by inter-homolog recombination was found to occur about 1000× less frequently than repair by non-homologous end joining (Stark and Jasin 2003). One reason why inter-homolog interactions are not favored may be the absence of spatial association between homologous chromosomes in mammalian interphase cells (Leitch et al. 1994; Cremer and Cremer 2001), except for some minor effects due to identical radial positioning and satellite associations. Thus, the genome-wide search for homology and interactions between damaged site and its recombination partner would require major chromatin movements. Indeed, recently evidence for repositioning of heterochromatic regions in the course of repair events presumably requiring recombination was presented (Abdel-Halim et al. 2004, 2005; Monajembashi et al. 2005). In contrast, however, in micro-irradiation studies where damaged regions are arranged in geometric patterns (see below), no indications for large-scale movements were found (Hauptner et al. 2004). In *Drosophila* evidence has been obtained that constraints arising from the three-dimensional chromatin organization hamper the genome-wide homology search associated with recombinational repair (Lankenau et al. 2000).

The importance of proximity of damaged sites for the generation of exchange-type aberrations had already been recognized more than 60 years ago (reviewed by Savage 2000). In the case of mis-rejoining of free ends originating from two or more break sites, it is still not clear how close the later interaction partners have to be at the time of damage induction for sufficient closeness to occur during the repair process. It has long been suspected that the proximity effect, together with non-random positioning of chromosomes and chromosome regions within the nucleus (Tanabe et al. 2002), may lead to preferential interaction between certain loci (reviewed by Friedl 2004), but up to now, on the level of whole chromosomes, only small effects have been detected (e.g. Cornforth et al. 2002).

So far, the question of mobility of damaged chromatin after DSB induction has been addressed by two approaches which both rely on the use of radiation qualities with ionization densities sufficiently high to produce several breaks along the radiation track. In the first method, damaged chromatin is visualized via detection of γ-H2AX or one of the foci-forming proteins, and the localization of the foci along the track is analyzed. With standard radiation set-ups, this analysis suffers from the fact that the tracks run perpendicular to the microscopic plane, resulting in poor microscopic resolution along the track. To circumvent this problem, Jakob and co-workers (2003) introduced an elegant set-up involving irradiation at a small angle, where the tracks run almost parallel to the microscopic plane, allowing good microscopic resolution. These authors found globular foci along the tracks, again reminiscent of the replication foci described by Sadoni et al. (2004). Interestingly, even after irradiation with heavy ions which are expected to produce several hundred DSB when traversing a cell, the number of foci was limited and adjacent foci were clearly separated, strongly suggesting that each focus contains several break-sites (see also Taucher-Scholz, this book). Indeed, Aten et al. (2004), who used a similar irradiation set-up to study the time course of foci formation and localization, observed a clustering of small sub-foci into larger foci rather rapidly after irradiation, which is compatible with the assumption of repair centers in which several break-sites meet. This clustering occurred mainly in G1 phase cells, whereas in S/G2 phase cells sub-foci remained separated. It is important to stress that, as opposed to yeast cells, where only one or two putative repair centers can be found per nucleus (Lisby et al. 2003; Meister et al. 2003), in mammalian cells many more repair foci form. At low damage density, the number of γ-H2AX foci correlates nicely with the number of DSB expected (Rothkamm and Löbrich 2003), only at higher damage density, the number of foci falls short of the number of breaks. At the moment, the size of the catchment area of a single mammalian repair center is not clear, nor is its relationship to functional chromatin elements.

A second method to investigate the mobility of damaged chromatin is micro-irradiation in pre-determined geometric patterns and analysis whether these patterns are reflected in the pattern of repair foci forming after irradiation. After ion micro-irradiation, geometric foci pattern remain remarkably stable over up to 24 h post-irradiation (Fig. 3), thus excluding large-scale chromatin movements, while indications for small scale movements (up to 1–2 μm) are evident (Hauptner et al. 2004 and Friedl, Dollinger, Drexler, Hable, Hauptner, Greubel, unpublished data). Small scale movements were also observed by Aten et al. (2004) for non-clustered foci. Up to now, all studies on chromatin dynamics after DSB induction in mammalian cells involved fixation of cells. A major breakthrough in understanding the events taking place immediately after break induction is expected once in vivo flu-

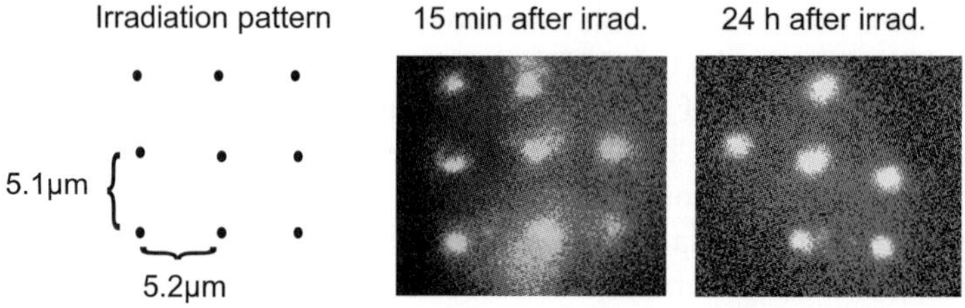

Fig. 3 Stability of 53BP1 foci pattern over 24 h post-irradiation. Hela cells were irradiated with 100 MeV oxygen ions in a geometric pattern (*left panel*) at the SNAKE ion microbeam facility (Hauptner et al. 2004) and incubated for 15 min (*middle panel*) or 24 h (*right panel*) before fixation and immunofluorescent detection of 53BP1 protein. Each picture depicts one nucleus

orescence labeling of break sites, similar to the approaches used in yeast (Fig. 2; Lisby et al. 2003; Kaye et al. 2004; Lobachev et al. 2004), is introduced. Combination with in vivo labeling of repair proteins with GFP or its variants (e.g. Essers et al. 2002; Lan et al. 2004; Mone et al. 2004), will allow similar insights into the spatiotemporal relationships among the factors involved in DSB processing as they are now available for yeast (Lisby et al. 2004).

5 Conclusions

For the last 2 or 3 years, we have seen an increasing consideration of the role played by chromatin structure and nuclear architecture in determining the cellular response to DSB and other types of DNA damage. Many important insights were obtained in model organisms, such as yeast, and they have still to be validated in mammalian cells. Further challenges in the field include the elucidation of the role played by chromatin in the initiation of cellular response mechanisms and the determination of temporal and causal relationships between individual histone modifications, chromatin alterations, and repair reactions. This will also shed light on the question of repair-type specific histone codes. Increasing use of live cell imaging techniques, coupled with localized damage infliction, are expected to contribute crucial insights on the way towards a comprehensive understanding of how cells deal with DNA damage.

Acknowledgements Work in the author's laboratory is supported by grants from Bundesamt für Strahlenschutz, Bundesministerium für Bildung und Forschung, and European Science Foundation. I thank all lab members and my collaborators for discussions and ideas. I apologize to all those whose work could not be cited for reasons of space and focus.

References

Abdel-Halim HI, Natarajan AT, Mullenders LH, Boei JJ (2005) Mitomycin C-induced pairing of heterochromatin reflects initiation of DNA repair and chromatid exchange formation. J Cell Sci 118:1757–1767

Abdel-Halim HI, Imam SA, Badr FM, Natarajan AT, Mullenders LH, Boei JJ (2004) Ionizing radiation-induced instant pairing of heterochromatin of homologous chromosomes in human cells. Cytogenet Genome Res 104:193–199

Abraham RT, Tibbetts RS (2005) Cell biology. Guiding ATM to broken DNA. Science 308:510–511

Aten JA, Stap J, Krawczyk PM, van Oven CH, Hoebe RA, Essers J, Kanaar R (2004) Dynamics of DNA double-strand breaks revealed by clustering of damaged chromosome domains. Science 303:92–95

Bakkenist CJ, Kastan MB (2003) DNA damage activates ATM through intermolecular autophosphorylation and dimer dissociation. Nature 421:499–506

Bakkenist CJ, Kastan MB (2004) Initiating cellular stress responses. Cell 118:9–17

Belmont AS (2001) Visualizing chromosome dynamics with GFP. Trends Cell Biol 11:250–257

Belmont AS, Dietzel S, Nye AC, Strukov YG, Tumbar T (1999) Large-scale chromatin structure and function. Curr Opin Cell Biol 11:307–311

Bird AW, Yu DY, Pray-Grant MG, Qiu Q, Harmon KE, Megee PC, Grant PA, Smith MM, Christman MF (2002) Acetylation of histone H4 by Esa1 is required for DNA double-strand break repair. Nature 419:411–415

Blanton J, Gaszner M, Schedl P (2003) Protein:protein interactions and the pairing of boundary elements in vivo. Genes Dev 17:664–675

Burma S, Chen BP, Murphy M, Kurimasa A, Chen DJ (2001) ATM phosphorylates histone H2AX in response to DNA double-strand breaks. J Biol Chem 276:42462–42467

Byrd K, Corces VG (2003) Visualization of chromatin domains created by the gypsy insulator of *Drosophila*. J Cell Biol 162:565–574

Camphausen K, Cerna D, Scott T, Sproull M, Burgan WE, Cerra MA, Fine H, Tofilon PJ (2005) Enhancement of in vitro and in vivo tumor cell radiosensitivity by valproic acid. Int J Cancer 114:380–386

Capelson M, Corces VG (2004) Boundary elements and nuclear organization. Biol Cell 96:617–629

Carson CT, Schwartz RA, Stracker TH, Lilley CE, Lee DV, Weitzman MD (2003) The Mre11 complex is required for ATM activation and the G2/M checkpoint. EMBO J 22:6610–6620

Cornforth MN, Greulich-Bode KM, Loucas BD, Arsuaga J, Vazquez M, Sachs RK, Bruckner M, Molls M, Hahnfeldt P, Hlatky L, Brenner DJ (2002) Chromosomes are predominantly located randomly with respect to each other in interphase human cells. J Cell Biol 159:237–244

Cremer T, Cremer C (2001) Chromosome territories, nuclear architecture and gene regulation in mammalian cells. Nat Rev Genet 2:292–301

Dehghani H, Dellaire G, Bazett-Jones DP (2005) Organization of chromatin in the interphase mammalian cell. Micron 36:95–108

Downs JA, Lowndes NF, Jackson SP (2000) A role for *Saccharomyces cerevisiae* histone H2A in DNA repair. Nature 408:1001–1004

Downs JA, Allard S, Jobin-Robitaille O, Javaheri A, Auger A, Bouchard N, Kron SJ, Jackson SP, Cote J (2004) Binding of chromatin-modifying activities to phosphorylated histone H2A at DNA damage sites. Mol Cell 16:979–990

Ehrenhofer-Murray AE (2004) Chromatin dynamics at DNA replication, transcription and repair. Eur J Biochem 271:2335–2349

Essers J, Houtsmuller AB, van Veelen L, Paulusma C, Nigg AL, Pastink A, Vermeulen W, Hoeijmakers JH, Kanaar R (2002) Nuclear dynamics of RAD52 group homologous recombination proteins in response to DNA damage. EMBO J 21:2030–2037

Fernandez-Capetillo O, Allis CD, Nussenzweig A (2004a) Phosphorylation of histone H2B at DNA double-strand breaks. J Exp Med 199:1671–1677

Fernandez-Capetillo O, Lee A, Nussenzweig M, Nussenzweig A (2004b) H2AX: the histone guardian of the genome. DNA Repair 3:959–967

Fischle W, Wang Y, Allis CD (2003) Histone and chromatin cross-talk. Curr Opin Cell Biol 15:172–183

Fourel G, Magdinier F, Gilson E (2004) Insulator dynamics and the setting of chromatin domains. Bioessays 26:523–532

Friedl AA (2004) Topological factors in radiation biology. In: Kiefer J (ed) Life Sciences and Radiation. Springer, Berlin Heidelberg New York, pp 69–77

Friesner JD, Liu B, Culligan K, Britt AB (2005) Ionizing radiation-dependent γ-H2AX focus formation requires ATM and ATR. Mol Biol Cell 16:2566–2576

Green CM, Almouzni G (2002) When repair meets chromatin. First in series on chromatin dynamics. EMBO Rep 3:28–33

Haaf T, Golub EI, Reddy G, Radding CM, Ward DC (1995) Nuclear foci of mammalian Rad51 recombination protein in somatic cells after DNA damage and its localization in synaptonemal complexes. Proc Natl Acad Sci USA 92:2298–2302

Haering CH, Nasmyth K (2003) Building and breaking bridges between sister chromatids. Bioessays 25:1178–1191

Harvey AC, Jackson SP, Downs JA (2005) *Saccharomyces cerevisiae* Histone H2A Ser122 facilitates DNA repair. Genetics, in press

Hauptner A, Dietzel S, Drexler GA, Reichart P, Krücken R, Cremer T, Friedl AA, Dollinger G (2004) Microirradiation of cells with energetic heavy ions. Radiat Environ Biophys 42:237–245

Helleday T (2003) Pathways for mitotic homologous recombination in mammalian cells. Mutat Res 532:103–115

Heng HH, Krawetz SA, Lu W, Bremer S, Liu G, Ye CJ (2001) Re-defining the chromatin loop domain. Cytogenet Cell Genet 93:155–161

Horejsi Z, Falck J, Bakkenist CJ, Kastan MB, Lukas J, Bartek J (2004) Distinct functional domains of Nbs1 modulate the timing and magnitude of ATM activation after low doses of ionizing radiation. Oncogene 23:3122–3127

Horn PJ, Peterson CL (2002) Molecular biology. Chromatin higher order folding–wrapping up transcription. Science 297:1824–1827

Huyen Y, Zgheib O, Ditullio RA Jr, Gorgoulis VG, Zacharatos P, Petty TJ, Sheston EA, Mellert HS, Stavridi ES, Halazonetis TD (2004) Methylated lysine 79 of histone H3 targets 53BP1 to DNA double-strand breaks. Nature 432:406–411

Ikura T, Ogryzko VV, Grigoriev M, Groisman R, Wang J, Horikoshi M, Scully R, Qin J, Nakatani Y (2000) Involvement of the TIP60 histone acetylase complex in DNA repair and apoptosis. Cell 102:463–473

Ishii K, Arib G, Lin C, Van Houwe G, Laemmli UK (2002) Chromatin boundaries in budding yeast: the nuclear pore connection. Cell 109:551–562

Jakob B, Scholz M, Taucher-Scholz G (2003) Biological imaging of heavy charged-particle tracks. Radiat Res 159:676–684

Jazayeri A, McAinsh AD, Jackson SP (2004) *Saccharomyces cerevisiae* Sin3p facilitates DNA double-strand break repair. Proc Natl Acad Sci USA 101:1644–1649

Karlsson KH, Stenerlow B (2004) Focus formation of DNA repair proteins in normal and repair-deficient cells irradiated with high-LET ions. Radiat Res 161:517–527
Kao GD, McKenna WG, Guenther MG, Muschel RJ, Lazar MA, Yen TJ (2003) Histone deacetylase 4 interacts with 53BP1 to mediate the DNA damage response. J Cell Biol 160:1017–1027
Kaye JA, Melo JA, Cheung SK, Vaze MB, Haber JE, Toczyski DP (2004) DNA breaks promote genomic instability by impeding proper chromosome segregation. Curr Biol 14:2096–2106
Kim JS, Krasieva TB, LaMorte V, Taylor AMR, Yokomori K (2002) Specific recruitment of human cohesin to laser-induced DNA damage. J Biol Chem 277:45149–45153
Kitagawa R, Bakkenist CJ, McKinnon PJ, Kastan MB (2004) Phosphorylation of SMC1 is a critical downstream event in the ATM-NBS1-BRCA1 pathway. Genes Dev 18:1423–1438
Koundrioukoff S, Polo S, Almouzni G (2004) Interplay between chromatin and cell cycle checkpoints in the context of ATR/ATM-dependent checkpoints. DNA Repair 3:969–978
Kusch T, Florens L, Macdonald WH, Swanson SK, Glaser RL, Yates JR 3rd, Abmayr SM, Washburn MP, Workman JL (2004) Acetylation by Tip60 is required for selective histone variant exchange at DNA lesions. Science 306:2084–2087
Labrador M, Corces VG (2002) Setting the boundaries of chromatin domains and nuclear organization. Cell 111:151–154
Lan L, Nakajima S, Oohata Y, Takao M, Okano S, Masutani M, Wilson SH, Yasui A (2004) In situ analysis of repair processes for oxidative DNA damage in mammalian cells. Proc Natl Acad Sci USA 101:13738–13743
Lankenau DH, Peluso MV, Lankenau S (2000) The Su(Hw) chromatin insulator protein alters double-strand break repair frequencies in the *Drosophila* germ line. Chromosoma 109:148–160
Lavin MF, Birrell G, Chen P, Kozlov S, Scott S, Gueven N (2005) ATM signaling and genomic stability in response to DNA damage. Mutat Res 569:123–132
Lee JH, Paull TT (2004) Direct activation of the ATM protein kinase by the Mre11/Rad50/Nbs1 complex. Science 304:93–96
Lee JH, Paull TT (2005) ATM activation by DNA double-strand breaks through the Mre11-Rad50-Nbs1 compex. Science 308:551–554
Leitch AR, Brown JK, Mosgoller W, Schwarzacher T, Heslop-Harrison JS (1994) The spatial localization of homologous chromosomes in human fibroblasts at mitosis. Hum Genet 93:275–280
Lisby M, Rothstein R (2004) DNA repair: keeping it together. Curr Biol 14:R994-R996
Lisby M, Mortensen UH, Rothstein R (2003) Colocalization of multiple DNA double-strand breaks at a single Rad52 repair centre. Nat Cell Biol 5:572–576
Lisby M, Barlow JH, Burgess RC, Rothstein R (2004) Choreography of the DNA damage response: spatiotemporal relationships among checkpoint and repair proteins. Cell 118:699–713
Lobachev K, Vitriol E, Stemple J, Resnick MA, Bloom K (2004) Chromosome fragmentation after induction of a double-strand break is an active process prevented by the RMX repair complex. Curr Biol 14:2107–2112
Lucknik AN, Hisamutdinov TA, Georgiev GP (1988) Inhibition of transcription in eukaryotic cells by X-irradiation: relation to loss of topological constraints on closed DNA loops. Nucleic Acids Res 16:5175–5184
Meister P, Poidevin M, Francesconi S, Tratner I, Zarzov P, Baldacci G (2003) Nuclear factories for signalling and repairing DNA double strand breaks in living fission yeast. Nucleic Acids Res 31:5064–5073

Monajembashi S, Rapp A, Schmitt E, Dittmar H, Greulich KO, Hausmann M (2005) Spatial association of homologous pericentric regions in human lymphocyte nuclei during repair. Biophys J 88:2309–2322

Mone MJ, Bernas T, Dinant C, Goedvree FA, Manders EM, Volker M, Houtsmuller AB, Hoeijmakers JH, Vermeulen W, van Driel R (2004) In vivo dynamics of chromatin-associated complex formation in mammalian nucleotide excision repair. Proc Natl Acad Sci USA 101:15933–15937

Moore JD, Krebs JE (2004) Histone modification and DNA double-strand break repair. Biochem Cell Biol 82:446–452

Morrison AJ, Highland J, Krogan NJ, Arbel-Eden A, Greenblatt JF, Haber JE, Shen X (2004) INO80 and gamma-H2AX interaction links ATP-dependent chromatin remodeling to DNA damage repair. Cell 119:767–775

Morrison AJ, Shen X (2005) DNA repair in the context of chromatin. Cell Cycle 4:568–571

Münkel C, Eils R, Dietzel S, Zink D, Mehring C, Wedemann G, Cremer T, Langowski J (1999) Compartmentalization of interphase chromosomes observed in simulation and experiment. J Mol Biol 285:1053–1065

Nazarov IB, Smirnova AN, Krutilina RI, Svetlova MP, Solovjeva LV, Nikiforov AA, Oei SL, Zalenskaya IA, Yau PM, Bradbury EM, Tomilin NV (2003) Dephosphorylation of histone gamma-H2AX during repair of DNA double-strand breaks in mammalian cells and its inhibition by calyculin A. Radiat Res 160:309–317

Paques F, Haber JE (1999) Multiple pathways of recombination induced by double-strand breaks in *Saccharomyces cerevisiae*. Microbiol Mol Biol Rev 3:349–404

Park EJ, Chan DW, Park JH, Oettinger MA, Kwon J (2003) DNA-PK is activated by nucleosomes and phosphorylates H2AX within the nucleosomes in an acetylation-dependent manner. Nucleic Acids Res 31:6819–6827

Paull TT, Rogakou EP, Yamazaki V, Kirchgessner CU, Gellert M, Bonner WM (2000) A critical role for histone H2AX in recruitment of repair factors to nuclear foci after DNA damage. Curr Biol 10:886–895

Qin S, Parthun MR (2002) Histone H3 and the histone acetyltransferase Hat1p contribute to DNA double-strand break repair. Mol Cell Biol 22:8353–8565

Riedel CG, Gregan J, Gruber S, Nasmyth K (2004) Is chromatin remodeling required to build sister-chromatid cohesion? Trends Biochem Sci 29:389–392

Rodi CP, Sauerbier W (1989) Structure of transcriptionally active chromatin: radiological evidence for requirement of torsionally constrained DNA. J Cell Physiol 141:346–352

Rogakou EP, Pilch DR, Orr AH, Ivanova VS, Bonner WM (1998) DNA double-stranded breaks induce histone H2AX phosphorylation on serine 139. J Biol Chem 273:5858–5868

Rogakou EP, Boon C, Redon C, Bonner WM (1999) Megabase chromatin domains involved in DNA double-strand breaks in vivo. J Cell Biol 146:905–916

Rong YS, Golic KG (2003) The homologous chromosome is an effective template for the repair of mitotic DNA double-strand breaks in *Drosophila*. Genetics 165:1831–1842

Rothkamm K, Löbrich M (2003) Evidence for a lack of DNA double-strand break repair in human cells exposed to very low x-ray doses. Proc Natl Acad Sci USA 100:5057–5062

Sadoni N, Cardoso MC, Stelzer EH, Leonhardt H, Zink D (2004) Stable chromosomal units determine the spatial and temporal organization of DNA replication. J Cell Sci 117:5353–5365

Sanders SL, Portoso M, Mata J, Bahler J, Allshire RC, Kouzarides T (2004) Methylation of histone H4 lysine 20 controls recruitment of Crb2 to sites of DNA damage. Cell 119:603–614

Savage JR (2000) Cancer. Proximity matters. Science 290:62–63

Shen X, Mizuguchi G, Hamiche A, Wu C (2000) A chromatin remodelling complex involved in transcription and DNA processing. Nature 406:541–544

Shroff R, Arbel-Eden A, Pilch D, Ira G, Bonner WM, Petrini JH, Haber JE, Lichten M (2004) Distribution and dynamics of chromatin modification induced by a defined DNA double-strand break. Curr Biol 14:1703–1711

Sjögren C, Nasmyth K (2001) Sister chromatid cohesion is required for postreplicative double-strand break repair in *Saccharomyces cerevisiae*. Curr Biol 11:991–995

Sonoda E, Matsusaka T, Morrison C, Vagnarelli P, Hoshi O, Ushiki T, Nojima K, Fukagawa T, Waizenegger IC, Peters JM, Earnshaw WC, Takeda S (2001) Scc1/Rad21/Mcd1 is required for sister chromatid cohesion and kinetochore function in vertebrate cells. Dev Cell 1:759–770

Stark JM, Jasin M (2003) Extensive loss of heterozygosity is suppressed during homologous repair of chromosomal breaks. Mol Cell Biol 23:733–743

Stiff T, O'Driscoll M, Rief N, Iwabuchi K, Lobrich M, Jeggo PA (2004) ATM and DNA-PK function redundantly to phosphorylate H2AX after exposure to ionizing radiation. Cancer Res 64:2390–2396

Stracker TH, Theunissen JW, Morales M, Petrini JH (2004) The Mre11 complex and the metabolism of chromosome breaks: the importance of communicating and holding things together. DNA Repair 3:845–854

Ström L, Sjögren C (2005) DNA damage-induced cohesion. Cell Cycle 4:536–539

Ström L, Lindroos HB, Shirahige K, Sjögren C (2004) Postreplicative recruitment of cohesin to double-strand breaks is required for DNA repair. Mol Cell 16:1003–1015

Tamburini BA, Tyler JK (2005) Localized histone acetylation and deacetylation triggered by the homologous recombination pathway of double-strand DNA repair. Mol Cell Biol 25:4903–4913

Tanabe H, Muller S, Neusser M, von Hase J, Calcagno E, Cremer M, Solovei I, Cremer C, Cremer T (2002) Evolutionary conservation of chromosome territory arrangements in cell nuclei from higher primates. Proc Natl Acad Sci USA 99:4424–4429

Ünal E, Arbel-Eden A, Sattler U, Shroff R, Lichten M, Haber JE, Koshland D (2004) DNA damage response pathway uses histone modification to assemble a double-strand break-specific cohesin domain. Mol Cell 16:991–1002

Uziel T, Lerenthal Y, Moyal L, Andegeko Y, Mittelman L, Shiloh Y (2003) Requirement of the MRN complex for ATM activation by DNA damage. EMBO J 22:5612–5621

van Attikum H, Fritsch O, Hohn B, Gasser SM (2004) Recruitment of the INO80 complex by H2A phosphorylation links ATP-dependent chromatin remodeling with DNA double-strand break repair. Cell 119:777–788

Warren CD, Eckley DM, Lee MS, Hanna JS, Hughes A, Peyser B, Jie C, Irizarry R, Spencer FA (2004) S-phase checkpoint genes safeguard high-fidelity sister chromatid cohesion. Mol Biol Cell 15:1724–1735

Woodcock CL, Dimitrov S (2001) Higher-order structure of chromatin and chromosomes. Curr Opin Genet Dev 11:130–135

Xie A, Puget N, Shim I, Odate S, Jarzyna I, Bassing CH, Alt FW, Scully R (2004) Control of sister chromatid recombination by histone H2AX. Mol Cell 16:1017–1025

Yamada K, Ariyoshi M, Morikawa K (2004) Three-dimensional structural views of branch migration and resolution in DNA homologous recombination. Curr Opin Struct Biol 14:130–137

Yusufzai TM, Tagami H, Nakatani Y, Felsenfeld G (2004) CTCF tethers an insulator to subnuclear sites, suggesting shared insulator mechanisms across species. Mol Cell 13:291–298

Part D
DNA Repair and Pleiotropic Interactions

Genome Dyn Stab (1)
D.-H. Lankenau: Genome Integrity
DOI 10.1007/7050_007/Published online: 18 November 2005

DNA Base Damage Recognition and Processing

Grigory L. Dianov[1] (✉) · Sarah L. Allinson[2]

[1]MRC Radiation and Genome Stability Unit, Harwell, Oxfordshire OX11 ORD, UK
g.dianov@har.mrc.ac.uk

[2]Department of Biological Sciences, Lancaster University, Lancaster LA1 4YQ, UK

Abstract In living cells DNA base lesions are formed continuously as a consequence of normal metabolism and are also generated by a number of external factors. Simple base damages are repaired by base excision repair that is initiated by a damage specific DNA glycosylase, which removes the damaged base creating an abasic site (apurinic/apyrimidinic, AP site). AP endonuclease cleaves the phosphodiester bond 5′ to the AP site and then DNA polymerase β adds the first nucleotide to the 3′-end of the incised AP site and at the same time removes the 5′-sugar phosphate residue. DNA ligase completes the repair by sealing the DNA ends. These processes are directed and co-ordinated by multiple protein-protein interactions. This review focuses primarily on mammalian base excision repair, and in particular addresses the enzymology of the repair process and co-ordination of repair reactions.

1 Introduction

DNA damage occurs relentlessly in all living cells and must be promptly repaired to prevent genome instability. Damage in mammalian cells can occur due to many factors, both exogenous and endogenous. Exogenous factors include, among others, DNA-reactive chemicals, UV light and ionizing radiation. Endogenous DNA damage is mainly due to the intrinsic instability of DNA molecule and the chemical reactivity of the cellular milieu, resulting in DNA base hydrolysis, oxidation and alkylation (Lindahl 1993). This instability of the primary structure of DNA may lead to a loss or corruption of the information held within the genome with potentially deleterious effects for the organism.

The aqueous environment of the cell is itself a source of endogenous DNA damage, with many of the chemical bonds of DNA being labile under physiological conditions. For example, it has been estimated that the instability of the *N*-glycosylic bond of purines to hydrolysis may result in the formation of abasic (AP) sites at a rate of approximately 10 000 per human genome per day (Lindahl and Nyberg 1972). The accumulation of unrepaired AP sites can block DNA synthesis and cause cell death and mutations (Guillet and Boiteux 2002; Loeb and Preston 1986).

In addition to the lability of the *N*-glycosylic bond, the exocyclic amino groups of the DNA bases are subject to hydrolytic deamination (Lindahl and Nyberg 1974). In the case of cytosine, this results in the formation of uracil which, if not repaired, can directly mispair with adenine, generating C to T transition mutations (Duncan and Miller 1980; Duncan and Weiss, 1982). Oxidative damage to DNA by reactive oxygen species, which arise as a result of normal oxidative metabolism, is also an important source of endogenous damage. Oxidation of DNA bases occurs predominantly through radical addition to the π-bonds of the purine or pyrimidine rings (Burrows and Muller 1998). The products of such reactions, e.g. 8-oxoguanine, formamidopyrimidines and ring saturated pyrimidines such as thymine glycol, are often mutagenic or cytotoxic (Wallace 1998).

The ubiquity of DNA damaging agents and the need to ensure the integrity of the genome has necessitated the evolution of a system of DNA repair mechanisms which are able to reverse the effects of DNA damage. The majority of the simpler, non-bulky base lesions discussed above are repaired via base excision repair (BER) pathways (Friedberg et al. 1995) that have been functionally conserved throughout evolution and are present in all forms of life from bacteria to higher mammals (Lindahl and Wood 1999). BER includes two major pathways that involve different subsets of enzymes and result in replacement of one (short-patch pathway) or more (long-patch pathway) nucleotides (Dianov et al. 1992; Frosina et al. 1996). This review will focus primarily on mammalian short-patch BER.

2
Base Excision Repair Proteins

2.1
DNA Glycosylases

DNA glycosylases initiate the BER process by catalysing hydrolysis of the C1′-*N*-glycosidic bond to the damaged base, removing the base and leaving behind an AP site. The first such enzyme to be identified was uracil-DNA glycosylase (UNG), isolated from *E. coli* in the mid 1970s (Lindahl 1974). In the three decades since this discovery many more glycosylases have been identified, covering the phylogenetic spectrum and a broad range of substrate specificities. The ubiquity and wide utility of the glycosylase family reflects the essential role of the BER process in protecting genomic DNA from being degraded through base hydrolysis, oxidation or alkylation. Eleven DNA glycosylases have so far been implicated in mammalian BER, and their various specificities are summarised in Table 1.

High-resolution structures of a number of DNA glycosylases have been obtained in recent years (Barrett et al. 1998; Bjoras et al. 2002; Bruner et al. 2000; Drohat et al. 2002; Fromme and Verdine 2002; Gilboa et al. 2002; Hollis et al. 2000; Labahn et al. 1996; Lau et al. 1998, 2000; Mol et al. 1995; Norman et al.

Table 1 Mammalian DNA glycosylases

Glycosylase	Bifunctional?	Substrates	Refs.
UNG2	No	U	(Krokan et al. 2001; Olsen et al. 1989; Slupphaug et al. 1993)
SMUG1	No	U, OHMeU	(Boorstein et al. 2001; Haushalter et al. 1999)
TDG	No	U:G, T:G, ethenoC	(Neddermann et al. 1996; Wiebauer and Jiricny 1990)
MBD4	No	U:G, T:G in CpG sites	(Hendrich et al. 1999; Petronzelli et al. 2000)
OGG1	Yes	8oxoG:C, FapyA, FapyG	(Radicella et al. 1997; Roldan-Arjona et al. 1997; Rosenquist et al. 1997)
MYH	No	A:8oxoG	(Slupska et al. 1996; Slupska et al. 1999)
NTH1	Yes	Oxidised and fragmented pyrimidines (e.g. Tg, DHU), Fapy	(Aspinwall et al. 1997; Ikeda et al. 1998; Sarker et al. 1998)
AAG (MPG)	No	3-meA, 7-meG, ethenoA, ethenoG, hypoxanthine	(Chakravarti et al. 1991; O'Connor and Laval 1991; Samson et al. 1991)
NEIL1	Yes	FapyA, FapyG, dihydrouracil, thymine glycol, 8-oxoG (low)	(Jaruga et al. 2004; Mokkapati et al. 2004; Rosenquist et al. 2003)
NEIL2	Yes	Cytosine oxidation products (5-OHU, DHU)	(Dou et al. 2003)
NEIL3	?	uncharacterised	

2001; Parikh et al. 1998, 2000; Savva et al. 1995; Thayer et al. 1995) and have provided an insight into how these enzymes overcome the significant challenge of recognizing small base modifications in the presence of a vast excess of unmodified bases. All of the DNA glycosylases characterized thus far use nucleotide flipping to facilitate access by the active site to the anomeric center of the target nucleotide, which would otherwise be inaccessibly buried in the DNA structure. Flipping is initiated through minor groove contacts and the local structure is subsequently stabilized by insertion of one or more amino acids into the helix at the orphan base. The flipped out base is inserted into the active site pocket of the DNA glycosylase which confers the specificity of the glycosylase through aromatic residues which stack against the base, polar residues which make hydrogen bonding contacts and hydrophobic residues that provide steric control of binding.

Although DNA glycosylases share very little sequence homology, they can be divided into four structural subfamilies: the UNG family, comprising UNG,

TDG, SMUG1 and MUG; the helix-hairpin-helix (HhH) family which contains such DNA glycosylases as MBD4, OGG1, MYH and NTH1; the helix 2 turn helix (H2TH) or Fpg/Nei family; and finally AAG, which at present is unique in structure. Glycosylases can additionally be broadly divided into two different mechanistic subclasses: the monofunctional or "pure" glycosylases and the bifunctional glycosylases, which have an associated AP lyase activity that cleaves the phosphodiester backbone at the abasic site. In the former case, the base is displaced by nucleophilic attack at the anomeric center by an activated water molecule whereas in the latter, an amino group on the enzyme acts as the nucleophile, thereby forming a transient Schiff base intermediate. These Schiff bases are primed to undergo either β- or $\beta\,\delta$ elimination reactions resulting in cleavage of the phosphodiester backbone, and generating either a 5′-phosphate and a 3′-$\alpha\beta$-unsaturated aldehyde, or 5′ and 3′ phosphates respectively. Alternatively, the Schiff base can undergo hydrolysis to leave the phosphodiester backbone intact. The degree to which the glycosylase and lyase activities of bifunctional glycosylases are coupled depends on both the particular enzyme and the reaction conditions. The lyase activity of OGG1 has been estimated to be an order of magnitude less than its glycosylase activity (Zharkov et al. 2000) and does not seem to play a role in the repair process (Allinson et al. 2001). In contrast, the $\beta\delta$ lyase activities of NEIL1 and NEIL2, H2TH family members with a catalytic *N*-terminal proline, seem to be tightly coupled to the associated glycosylase activity and repair initiated by these glycosylases may therefore proceed through an APE1-independent, PNK-dependent mechanism (Wiederhold et al. 2004).

Mouse knockout models have been constructed for AAG (Engelward et al. 1997; Hang et al. 1997), OGG1 (Klungland et al. 1999; Minowa et al. 2000), UNG2 (Nilsen et al. 2000), NTH1 (Elder and Dianov 2002; Ocampo et al. 2002; Takao et al. 2002), MBD4 (Millar et al. 2002) and MYH (Xie et al. 2004) and have proven invaluable in furthering our understanding of BER *in vivo*. Surprisingly, these mice are viable and display few overt phenotypic abnormalities, in stark contrast to knockouts of a number of other BER enzymes which are embryonic lethal. The reason for this may lie in the functional redundancy of many glycosylases, with alternative glycosylases being able to play a back-up role in the repair of certain lesions. This possibility is supported by the phenotype of the double knockout $Ogg1^{-/-}Myh^{-/-}$ mouse which shows a marked increase in both 8-oxoguanine levels (Russo et al. 2004) and tumorigenesis (Xie et al. 2004) compared to the single knockouts.

The redundancy in substrate specificity of certain glycosylases may also reflect differing roles played within the cell cycle. UNG2, NTH1 and MYH have all been demonstrated to interact with the replication-associated processivity factor PCNA (Otterlei et al. 1999; Oyama et al. 2004; Parker et al. 2001), while UNG2 and MYH have additionally been shown to associate with the replication factor RPA (Nagelhus et al. 1997; Otterlei et al. 1999; Parker et al.

2001). Since UNG2 and MYH co-localise with PCNA and RPA at replication foci (Tini et al. 2002) and MYH also interacts with the post-replicative mismatch repair protein MutSα (Gu et al. 2002) it is highly probable that these glycosylases are involved in a post-replicative repair mechanism, specifically removing damaged nucleotides misincorporated by polymerases. Interestingly all three of these PCNA-associated glycosylases are up-regulated in proliferating tissue (Boldogh et al. 2001; Luna et al. 2000; Nagelhus et al. 1997; Otterlei et al. 1999): lending further support to the idea that these enzymes may act in post-replicative repair.

It seems increasingly likely that post-translational modification of glycosylases plays an important role in both their function and regulation in vivo. SUMO-ylation of TDG stimulates enzyme turnover by relieving product inhibition (Hardeland et al. 2002). The molecular interactions of TDG are also subject to modulation by p300-mediated acetylation (Tini et al. 2002). It has recently been demonstrated that NEIL2 is also a target for acetylation by p300; however, in this case acetylation results in the total inactivation of the glycosylase (Bhakat et al. 2004). Phosphorylation of MYH (Gu and Lu 2001), OGG1 (Dantzer et al. 2002) and UNG2 (Muller-Weeks et al. 1998) have all been reported and in the case of UNG2 phosphorylation may play a role in mediating the ubiquitin dependent proteolytic degradation of the enzyme during late S-phase (Fischer et al. 2004).

2.2
AP Endonucleases

Removal of damaged bases from DNA by DNA glycosylases leads to the formation of AP sites. AP sites also arise spontaneously at a substantial rate (Lindahl and Karlström 1973; Lindahl and Nyberg 1972). In addition to this, DNA damaging chemicals such as free radicals and alkylating agents promote the release of bases, often by introducing base modifications that destabilize the *N*-glycosylic bond by generating a better leaving group moiety (Lindahl 1993; Lindahl and Andersson 1972). AP sites are a threat to cellular viability and genomic integrity as they can prevent normal DNA replication and are cytotoxic and mutagenic (Guillet and Boiteux, 2002; Loeb and Preston 1986). A critical step in the BER pathway involves the cleavage of baseless sites in DNA by an AP endonuclease. The major AP endonuclease in mammalian cells is APE1 also known as Ref1 or HAP1, an enzyme that is homologous to bacterial exonuclease III (Demple et al. 1991; Robson and Hickson 1991). APE1 is a multifunctional protein that is not only responsible for repair of AP sites, but also functions as a reduction-oxidation (redox) factor maintaining transcription factors in an active reduced state, and has been shown to be closely linked to apoptosis (reviewed in Evans et al. 2000; Wilson and Barsky 2001). The functional importance of APE1 is underscored by the findings that mice nullizygous for the APE1 gene are embryonic lethal (Xanthoudakis et al. 1996) and downregulation of

APE1 levels in human cells using RNAi led to accumulation of AP sites, stopped cell proliferation and triggered apoptosis (Fung and Demple 2005).

APE1 is the major human AP endonuclease that comprises more than 95% of cellular AP site incision activity (Chen et al. 1991) and it cleaves phosphodiester bonds hydrolytically in a Mg^{2+}-dependent manner, leaving a 3′-hydroxyl group and a 5′-deoxyribose phosphate (dRP) group flanking the nucleotide gap. The APE1 protein is modest in size; 317 amino acids in length and approximately 35 kDa. It contains two distinct domains; the *N*-terminal domain contains the nuclear localization sequence and is essential for redox activity, while the endonuclease activity resides in the C-terminal region (Xanthoudakis et al. 1992). APE1 cleaves abasic sites in double-stranded DNA far more efficiently ($\sim$ 20 fold) than in single-stranded substrates (Demple and Harrison 1994). However, the ring structure of an abasic site, the opposing base and the helical conformation prior to binding are not critical components for APE1 complex formation (Erzberger et al. 1998; Wilson et al. 1995). It has been postulated that since DNA is already prone to deformation by the presence of an abasic site APE1 may bind to DNA and induce a conformational change (Chaudhry and Weinfeld 1995). In support of this hypothesis, it has been shown that APE1-bound DNA is distorted (Mol et al. 2000). This bending of DNA could enable the abasic site to flip out of the DNA and move into position within the active site (Mol et al. 2000; Parikh et al. 1999).

In addition to AP endonuclease activity, APE1 has 3′-diesterase and phosphatase activity, albeit its phosphatase activity is approximately 100-fold lower than its AP endonuclease activity (Chen et al. 1991; Wilson and Barsky 2001) and most probably has a limited physiological importance (Wiederhold et al. 2004). It has been recently demonstrated that APE1 is the major activity in human cells responsible for removal of 3′-phosphoglycolate or 8-oxoguanine that occur at single strand breaks created by oxidative agents (Parsons et al. 2004, 2005a). Additionally, it has been shown that APE1 has a DNA exonuclease activity on mismatched or modified deoxyribonucleotides at the 3′-termini of nicked or gapped DNA molecules. This finding suggests that APE1 may increase the fidelity of base excision repair by correcting insertion errors made by DNA polymerase β, which lacks its own proofreading activity (Chou and Cheng 2002, 2003; Chou et al. 2000).

APE1 expression is ubiquitous; however, it exhibits a complex and heterogeneous staining pattern that differs among tissue types and even differs between neighboring cells (Rivkees and Kelley, 1994; Wilson et al. 1996). As expected, APE1 is found in the nucleus of cells which corresponds to its repair functions. In some cell types, however, APE1 is also found in the cytoplasm and mitochondria (Tell et al. 2001) which may correspond either to its role as a DNA repair protein in the mitochondria (Sawyer and Van Houten 1999) or, through its redox activity, in maintaining newly synthesized transcription factors in a reduced state during transportation to the nucleus (Evans et al. 2000; Kakolyris 1998).

APE1 is regulated at both the transcriptional and post-translational level. Oxidative agents induce APE1, characterized by a transient increase in APE1 protein and mRNA (Grosch et al. 1998; Grosch and Kaina 1999; Ramana et al. 1998). This reactive oxygen species-mediated increase in protein and mRNA has been shown to involve *de novo* synthesis since various agents that block protein synthesis also block the induction of APE1 (Ramana et al. 1998). Additionally, induction of APE1 is associated with an increase in AP endonuclease activity and an increase in resistance to H_2O_2, methyl methane sulfonate, bleomycin, γ-radiation, but not UV light (Grosch et al. 1998; Ramana et al. 1998). Such resistance is also transient and is consistent with the reduction in APE1 mRNA and protein levels (Ramana et al. 1998).

Mammals also possess a second protein (APE2) with sequence homology to exonuclease III, but this protein exhibits comparatively weak AP site-specific and 3′-nuclease activities (Hadi et al. 2002), despite having a highly conserved nuclease domain with APE1. APE1 and APE2 do, however, contain distinguishing amino and carboxyl-terminal ends, APE2 was found to contain a mitochondrial targeting sequence and was also found to localize in the mitochondria suggesting that it may play a role in mitochondrial base excision repair (Tsuchimoto et al. 2001).

2.3 Poly(ADP-ribose)polymerase

The enzyme poly(ADP-ribose) polymerase (PARP-1) has been strongly implicated in BER, although its precise role is still not yet fully understood. Poly(ADP)-ribose polymerase-1 (PARP-1) has a very high affinity for strand breaks and binds incised AP site before any other repair proteins (Parsons et al. 2005b). Binding strand breaks allosterically activates PARP-1's catalytic activity of modifying nuclear proteins with polymers of ADP-ribose that it synthesises from NAD^+ (see Ame et al. 2004; Herceg and Wang 2001; Smith 2001 for review). One of the major targets of poly(ADP-ribosyl)ation is PARP-1 itself and the resulting accumulation of negative charge on the enzyme reduces its affinity for DNA via electrostatic repulsion. As a result, single strand break repair exhibits a dependence on NAD^+ due to the activity of PARP-1 (Satoh and Lindahl 1992; Satoh et al. 1993). Similarly, the repair of damaged DNA bases via BER also exhibits a dependence on NAD^+ according to how rapidly the APE1 incision intermediate accumulates, which in itself depends on how rapidly the base is removed by a glycosylase and ultimately on the identity of the damaged base (Allinson et al. 2004). In vivo inhibition of PARP-1 catalysis by the use of either chemical inhibitors or expression of the catalytically inactive PARP-1 DNA binding domain results in inhibition of poly(ADP-ribosyl)ation and a concomitant retardation of strand break repair (Aoufouchi et al. 1999; Molinete et al. 1993; Perkins et al. 2001). Interestingly, PARP-1 null mice are hypersensitive to alkylating agents and ionising radiation (Ménessier-de Murcia et al. 1997; Wang et al. 1997), and immortalised cells from these animals manifest characteristic features of genomic instability, including delayed rejoining of strand breaks, mixed ploidy and elevated frequencies of SCE and micronuclei formation upon treatment with DNA-damaging agents (Ménessier-de Murcia et al. 1997; Trucco et al. 1998; Wang et al. 1997). Taken together, these results suggest that efficient strand break repair requires both PARP-1 and the poly(ADP-ribosyl)ation reaction that it catalyses.

An explanation for these observations lies in a possible co-ordinating role for PARP-1 in BER. PARP-1 has been shown to interact physically and functionally with the BER enzymes XRCC1 (Masson et al. 1998), DNA ligase III (Leppard et al. 2003), and Pol β (Caldecott et al. 1996; Dantzer et al. 2000). For XRCC1 and DNA ligase III, interaction with PARP-1 occurs preferentially with the auto-modified enzyme. Furthermore, it has been shown that relocation of XRCC1 and DNA ligase III to cellular DNA repair foci was notably decreased by PARP inhibitors and in PARP-1 deficient cells (El-Khamisy et al. 2003; Okano et al. 2003). These results suggest that poly(ADP-ribosyl)ation and PARP-1 dissociation from repair sites is required for active recruitment of repair enzymes to the repair sites in vivo and also indicate a more direct role for PARP-1 in recruitment of repair proteins to the sites of DNA damage.

The observations above all add up to a possible co-ordinating and regulatory role for PARP-1 in BER; however, there is evidence to suggest that the net effect of PARP-1 binding and automodification might actually result in an overall slowing of repair kinetics. Whole cell extract-mediated repair of plasmids containing a site-specifically located normal AP site showed that the kinetics of repair in PARP-1 deficient extracts were faster than in wild-type extracts (Allinson et al. 2003) and titration of PARP-1 into PARP-1-deficient extracts has been shown to inhibit, rather than stimulate repair (Allinson et al. 2003; Vodenicharov et al. 2000).

Given that activation of PARP-1 can lead to massive depletion of cellular NAD^+ levels and even trigger necrosis (Tentori et al. 2001), and that PARP-1 binding to BER intermediates might result in a decrease in the rate of individual repair reactions (Allinson et al. 2003), what is the purpose of PARP-1's involvement in the BER pathway? It was recently demonstrated that in vitro repair reactions carried out in the absence of PARP-1 result in increased nucleolytic degradation of the base excision repair intermediates and it has been suggested that PARP-1 functions to protect strand breaks and repair intermediates from degradation or recombination where the amount of damage outstrips the repair capacity of the cell (Parsons et al. 2005b). In support of this idea, it has been recently reported that the genomes of PARP-1 null mice show an increased incidence of deletion mutations upon DNA damage (Shibata et al. 2005). It is also possible that by increasing the local concentration of repair enzymes at sites of DNA damage through interaction with other BER proteins PARP-1, while protecting strand breaks may simultaneously be involved in targeting the machinery of BER to damage sites in vivo.

2.4 BER DNA Polymerases

Excision of a damaged base and subsequent processing of the resulting AP site by APE1 generally results in an intermediate containing a 3′-hydroxyl, suitable for priming strand resynthesis by a polymerase, and a 5′-deoxyribose phosphate (dRP) that must be removed prior to completion of repair by a DNA ligase. The majority of BER proceeds through the so-called "short-patch" subpathway in which a single nucleotide is removed and replaced. In this pathway, both DNA resynthesis (Dianov et al. 1992; Sobol et al. 1996; Wiebauer and Jiricny 1990) and removal of the blocking 5′-dRP residue (Matsumoto and Kim 1995) are carried out by Pol β, a member of the X family of polymerases.

Like other X family polymerases Pol β is a compact (39 kDa) single subunit protein. It contains two domains connected by a hinge region: a C-terminal 31 kDa polymerase domain (Kumar et al. 1990) and an *N*-terminal 8kDa dRP lyase domain (Matsumoto and Kim 1995), both of which activities are required for short-patch BER (Podlutsky et al. 2001a). The dRP lyase activity of Pol β removes the dRP residue through catalysis of a β-elimination reaction (Matsumoto and Kim 1995). The two activities of Pol β are likely to be tightly co-ordinated since co-complementation of Pol β null cell extracts with both dRP lyase and polymerase deficient mutants of Pol β is unable to restore single nucleotide patch repair to these extracts (Podlutsky et al. 2001a).

Pol β lacks any intrinsic proof-reading activity yet carries out DNA synthesis of moderately high fidelity compared to that of the so-called error-prone polymerases. The error frequency of short patch base excision repair by mammalian cell extracts has been calculated to be of the order of 5–10×10^{-4}, with one nucleotide deletions predominating (Bennett et al. 2001; Sanderson and Mosbaugh 1998; Zhang and Dianov 2005). These mutation frequencies well reflect the fidelity of in vitro filling of a single nucleotide gap by Pol β (Chagovetz et al. 1997).

Pol β is known to interact with a number of BER factors, including XRCC1 (Caldecott et al. 1996; Kubota et al. 1996), DNA ligase I (Dimitriadis et al. 1998; Prasad et al. 1996), PCNA (Kedar et al. 2002) and PARP-1 (Dantzer et al. 2000) and also forms a ternary complex with APE1 and an uncleaved AP site, which is thought to help co-ordinate the cleavage and polymerase steps (Bennett et al. 1997). The XRCC1-Pol β interaction plays an important role for coordinating the efficiency of the BER process; a point mutation in the XRCC1 protein which disrupts its functional interaction with DNA polymerase β, affected the ligation efficiency of the mutant XRCC1-DNA ligase III heterodimer in reconstituted base excision repair reactions (Dianova et al. 2004). It was also demonstrated that the plasmid encoding the mutated XRCC1 protein has reduced ability to rescue the hydrogen peroxide sensitivity of XRCC1-deficient cells (Dianova et al. 2004).

Mice nullizygous for the Pol β gene are non-viable and die early in embryogenesis (Gu et al. 1994). Embryonic fibroblasts derived from these mice are hypersensitive to the methylating agent, MMS (Sobol et al. 1996). As might be expected, under normal circumstances Pol β is present in all tissues although, with the exception of testis, generally only at low levels (Hirose et al. 1989). It is expressed constitutively and without cell-cycle dependence (Zmudzka et al. 1988). However, it seems increasingly likely that expression can be up-regulated on DNA damage (Cabelof et al. 2002; Fornace et al. 1989). Levels of Pol β are most probably kept low in undamaged cells due its mis-insertion frequency compared to replicative polymerases. Overexpression of Pol β results in a mutator phenotype in cells (Canitrot et al. 1998) and has been associated with tumourigenesis (Bergoglio et al. 2001) and early onset cataractogenesis (Sobol et al. 2003) in mouse models.

As discussed above, Pol β employs a β-elimination reaction to remove the blocking 5′-dRP moiety in BER. However, if the dRP has been modified, e.g. by oxidation or reduction, such that it is no longer labile to β-elimination then an alternative mechanism for its removal must be employed. In this case the repair is re-routed through the PCNA-dependent long-patch pathway in which between 2 and 8 nucleotides are removed and replaced (Klungland and Lindahl 1997; Matsumoto et al. 1994). It should be noted that the pathway switch from short- to long-patch repair occurs after insertion of the first nucleotide by Pol β (Podlutsky et al. 2001b). Synthesis thereafter is performed by the replicative polymerase, Pol δ (Dianov et al. 2003; Frosina et al. 1996) which also carries out repair synthesis in the nucleotide excision repair and mismatch repair pathways (Wood et al. 2001) in addition to its well-documented role in synthesis of the leading strand during replication.

Pol λ is a new emerging member of X family DNA polymerases that may play a backup role in BER (Garcia-Diaz et al. 2002). Pol λ has an intrinsic dRP lyase activity (Garcia-Diaz et al. 2001) and is responsible for background repair activity in Pol β-deficient cell extracts (Braithwaite et al. 2005). However, it is not clear yet whether its major role is within BER, since Pol λ is also involved in processing of DNA ends during non-homologous end joining (DeRose et al. 2003; Lee et al. 2004; Ma et al. 2004).

2.5 DNA Ligases

BER reactions result in formation of a nicked repair intermediate which must be sealed by a DNA ligase for repair to be completed. Two DNA ligases have been implicated in BER: DNA ligase I and DNA ligase III (Tomkinson and Levin 1997). DNA ligase I (102 kDa) is conserved in all eukaryotes, with orthologs being found in organisms from yeast to mammals (Tomkinson and Mackey 1998) and plays an essential role in DNA replication (Waga and Stillman 1998). DNA ligase I associates with replication forks (Montecucco et al. 1998), where it is active in joining Okazaki fragments (Levin et al. 1997), and is up-regulated during cell proliferation (Soderhall 1976). In addition to its replicative role, DNA ligase I has also been implicated in several DNA repair pathways including both nucleotide excision repair (Aboussekhra et al. 1995) and base excision repair (Levin et al. 2000). Human cell-lines containing a partially inactive DNA ligase I exhibit hypersensitivity to such DNA-damaging agents as ionizing radiation and alkylating agents (Barnes et al. 1992; Prigent et al. 1994; Webster et al. 1992), suggesting a role for DNA ligase I in BER. However, a mouse model with an identical mutation was defective in DNA replication, but not in DNA repair (Harrison et al. 2002). Moreover, DNA ligase I was found to be essential for normal development, but not essential for DNA repair (Bentley et al. 1996, 2002). On the other hand, there is evidence to suggest the involvement of DNA ligase I in the long-patch pathway of BER. In addition to the observed deficiency in long-patch BER of the human DNA ligase I mutant, 46BR.1G1 (Levin et al. 2000), in vitro studies demonstrated in particular that the interaction between PCNA and DNA ligase I plays a key role in the long-patch pathway (Levin et al. 2000). Interaction with PCNA stimulates DNA ligase I activity through increased binding to nicked substrates (Tom et al. 2001). Further support for DNA ligase I being involved in long-patch repair derives from the stimulation of its catalytic efficiency by replication protein A (RPA) (Ranalli et al. 2002) which has also been shown to specifically stimulate long-patch BER in both mammalian cell extracts (Dianov et al. 1999) and a reconstituted system (DeMott et al. 1998).

DNA ligase III exists in the nucleus as two isoforms, denoted DNA ligase IIIα (103 kDa) and DNA ligase IIIβ (96 kDa), which arise through differential splicing of RNA transcripts (Mackey et al. 1997). DNA ligase IIIα is ubiquitously expressed, whereas DNA ligase IIIβ is found only in testes and is therefore believed to be involved in homologous recombination during meiosis (Tomkinson and Mackey 1998). The two isoforms share many features in common and differ only at their C-termini, with the C-terminal 77 amino acids of DNA ligase IIIα being replaced with 17 different amino acids in DNA ligase IIIβ (Tomkinson et al. 2001). Both contain a putative zinc finger motif homologous to that of PARP-1 (Caldecott et al. 1996; Taylor et al. 2000b; Wei et al. 1995) and this is believed to function as a nick sensor. However, this motif does not seem to be essential for activity on simple nicks (Taylor et al. 1998) and may instead be required for carrying out ligation at sites with complex secondary structure (Taylor et al. 2000b) or stabilising nicks at physiological salt concentrations (Mackey et al. 1999).

DNA ligase IIIα was originally purified in complex with XRCC1 (Caldecott et al. 1994, 1995). This interaction has been characterized in vitro and is mediated through the BRCT domain located at the C-terminus of DNA ligase IIIα (Nash et al. 1997). XRCC1 null cell lines have 4-6 fold reduced levels and activity of DNA ligase IIIα, indicating that the XRCC1 protein is important for maintaining cellular levels of DNA ligase IIIα (Caldecott et al. 1995). In support of this finding, it has been shown that the activity of DNA ligase III can be restored in mutant XRCC1 cells (EM9) by the XRCC1 gene product (Ljungquist et al. 1994). Since XRCC1 null cells are characteristically deficient in short-patch BER, it has been suggested that the XRCC1-DNA ligase IIIα complex plays an important role in this sub-pathway (Cappelli et al. 1997), although biochemical data have suggested that DNA ligase I could be also active in this pathway (Kubota et al. 1996; Prasad et al. 1996). This suggestion was supported by the recent finding that DNA ligase I was able to efficiently substitute for the XRCC1-DNA ligase IIIα complex during short- patch repair. In contrast, DNA ligase IIIα is not able to substitute for DNA ligase I during long-patch repair (Sleeth et al. 2004).

2.6
XRCC1

Originally identified as a result of its absence from a mutant Chinese hamster ovary cell line (EM9), which showed hypersensitivity to both alkylating agents and ionising radiation (Thompson et al. 1990), XRCC1 has been increasingly implicated as a key player in the BER process. Cells deficient in XRCC1 show many of the hallmarks of defective BER including hypersensitivity to ionising radiation and alkylating agents, delayed single strand break rejoining and increased rates of sister chromatid exchange (SCE) and induced mutations (reviewed in Thompson and West 2000). Extracts from such cells show a partial defect in BER efficiency compared with the wild type (Cappelli et al. 1997; Sleeth et al. 2004). The importance of the *XRCC1* gene for cellular functioning is underlined by the fact that its targeted knockout results in embryonic lethality (Tebbs et al. 1999). Although it has no known enzymatic function of its own, XRCC1 interacts with a number of enzymes involved in BER including DNA ligase IIIα (Caldecott et al. 1994; Nash et al. 1997), Pol β (Caldecott et al. 1996; Kubota et al. 1996), PARP-1 (Caldecott et al. 1996; Masson et al. 1998), APE1 (Vidal et al. 2001a), OGG1 (Marsin et al. 2003) and PCNA (Fan et al. 2004).

XRCC1 is a 69.5 kDa multi-domain protein with an *N*-terminal DNA binding domain and two BRCT motifs. The first BRCT motif, BRCT I, is the site of interaction with PARP-1 (Masson et al. 1998), while the C-terminal BRCT motif (BRCT II) is responsible for the interaction with DNA ligase IIIα (Nash et al. 1997). Mutations in the BRCT domains of XRCC1 seriously compromise single-strand break repair (SSBR) in vivo, but with differing cell cycle dependencies. The BRCT II domain seems to be required only for SSBR during the G_0/G_1 phase of the cell cycle with SSBR being unaffected in S-phase (Taylor et al. 2000a). Disruption of this domain affects the survival of quiescent cells only: cycling cells with a mutated BRCT II domain show similar repair efficiencies and survival to those with normal XRCC1 (Moore et al. 2000). Conversely, the BRCT I domain is required for SSBR in both G_1

and S/G_2 (Taylor et al. 2002) and disruption of this domain compromises cells' ability to reinitiate DNA synthesis following exposure to an alkylating agent (Kubota and Horiuchi 2003).

The nick-binding *N*-terminal domain of XRCC1 interacts with the palm-thumb domain of Pol β allowing it to form a tertiary complex with Pol β and nicked DNA, in which both proteins almost completely surround the nicked site (Gryk et al. 2002; Marintchev et al. 1999, 2000). XRCC1 binds to the concave side of the 90° bent nicked DNA (Marintchev et al. 1999), whilst Pol β binds to the DNA's convex side (Sawaya et al. 1997), sequestering the damage from the cellular milieu. The biological importance of the XRCC1-Pol β interaction was recently confirmed: a mutant of XRCC1 incapable of interaction with Pol β was unable to fully reverse the hydrogen peroxide hypersensitivity of an XRCC1 deficient cell line and was less efficient at performing the ligation step in reconstituted repair reactions (Dianova et al. 2004).

3 Co-ordination of Base Excision Repair

Multiple interactions between BER proteins have been demonstrated by co-immunoprecipitation, GST-pull downs and the yeast two-hybrid system (reviewed in Caldecott 2001), instigating the proposal of several BER models. The "passing the baton" model of BER suggests a mechanism whereby each protein passes the repair intermediate to the next protein in the pathway in a co-ordinated manner (Mol et al. 2000; Wilson and Kunkel 2000). From this hypothesis, a damaged DNA base would be passed from a DNA glycosylase to APE1 to Pol β and finally to XRCC1-DNA ligase III during the course of repair. This mechanism would involve transient interaction of the proteins during the repair process.

In support of this model it has been demonstrated that the glycosylase UNG2 remains associated with the AP site following removal of uracil (Parikh et al. 1998), protecting the AP site from degradation. The glycosylase is subsequently displaced by the next enzyme in the pathway, APE1, which achieves this displacement through having a greater affinity for AP sites: making extensive contacts with the DNA and inducing significant kinking of the helical axis (Abner et al. 2001). This is further supported by the observed stimulation of turnover of a number of glycosylases by APE1 (Hill et al. 2001; Parikh et al. 1998; Marenstein et al. 2003; Vidal et al. 2001; Waters et al. 1999; Yang et al. 2001), which in the case of bifunctional glycosylases has wider implications for the mechanism of repair. As discussed above, bifunctional glycosylases form a Schiff base with C1′ of the deoxyribose during the course of the excision reaction. This covalent complex can either hydrolyze, leaving an intact AP site, or can undergo β-elimination, cleaving the phosphodiester backbone 3′ to the AP site, leaving a 3′-$\alpha\beta$-unsaturated aldehyde residue and a 5′-phosphate (McCullough et al. 1999). The blocking unsaturated aldehyde residue can then be excised by APE1 (Demple and Harrison 1994), although the 3′-phosphodiesterase activity of APE1 is significantly (100 fold) weaker than its AP endonuclease activity (Wilson and Barsky 2001). The product of sequential AP lyase and 3′-phosphodiesterase processing of an AP site is a one-nucleotide gap with flanking 3′-OH and 5′-phosphate termini. Repair may then be completed simply by insertion of a single nucleotide followed by ligation and indeed, short-patch BER predominates in the repair of the oxidative lesions, 8-oxoguanine and thymine glycol (Dianov et al. 1998, 2000; Fortini et al. 1999). However, for OGG1 and NTH1 APE1-mediated stimulation of turnover seems to be selective for the glycosylase activity with the associated AP lyase activity remaining unstimulated (Hill et al. 2001; Marenstein et al. 2003; Vidal et al. 2001). Work carried out in our laboratory has found that repair by whole cell extracts of oxidative base damage site-specifically incorporated into closed circular substrate DNA proceeds via a Pol β lyase-dependent short-patch mechanism (Allinson et al. 2001) suggesting that the majority of short-patch BER proceeds through a single general mechanism regardless of the AP lyase activity of the initiating glycosylase. However, this may not be the case for repair initiated by the recently discovered Nei homologues, NEIL1 and NEIL2, which carry out a tightly coupled $\beta\delta$ elimination of the Schiff base following hydrolysis of the damaged base, resulting in a one nucleotide gap flanked by 5′ and 3′ phosphates. The blocking 3′-phosphate residue can be subsequently removed by polynucleotide kinase (PNK) to allow DNA synthesis to proceed. Accordingly, PNK-dependent, APE1-independent repair of oxidative lesions has been recently demonstrated in human cells (Wiederhold et al. 2004).

Following APE1-processing of the AP site, Pol β recognizes the nicked AP site and adds a single nucleotide by displacing the 5′-terminal sugar phosphate. The 5′-dRP residue must be removed for single-nucleotide patch repair to proceed to completion. In mammals Pol β performs this task through a dRP lyase activity (Matsumoto and Kim 1995) that is tightly coupled to its polymerase function (Podlutsky et al. 2001a). In practice, Pol β has been demonstrated to have an affinity for the complex between APE1 and an uncleaved AP site and is recruited to the damage site as a ternary complex (Bennett et al. 1997). Following backbone cleavage Pol β retains affinity for the DNA and is then able to both remove the dRP and introduce a single nucleotide in a concerted fashion. Since Pol β also has an affinity for both DNA ligase I (Dimitriadis et al. 1998; Prasad et al. 1996) and DNA ligase III (mediated by XRCC1) (Caldecott et al. 1994; Nash et al. 1997) it is therefore possible for the final step of repair, i.e. ligation of the nick, to be coupled to the rest of the repair process.

However, it has recently been demonstrated that PARP-1 binds to the incised AP site at the very early stages of BER (Allinson et al. 2003; Lavrik et al. 2001; Parsons et al. 2005b) and that inhibition of poly(ADP-ribosyl)ation by 3-aminobenzamide blocks PARP-1 dissociation and completely prevents further repair (Parsons et al. 2005b). These data suggest that PARP-1 binding is always an integral part of the BER process and that BER may be divided into two major stages: removal of the damaged base and incision of the resulting AP site as a first stage and subsequent processing of the incised AP site as a second stage; with the two stages clearly demarcated by PARP-1 binding (Fig. 1). PARP-1 in turn is able to interact with a number of BER enzymes including Pol β, Ligase III and XRCC1 which also interact with each other and were therefore suggested to exist as a pre-formed complex or "repairosome" in vivo (Caldecott et al. 1995; Kubota et al. 1996). In practice it seems that the co-ordination of BER probably lies somewhere between the passing the baton and the repairosome models, with multi-protein complexes being formed dynamically as and when required. PARP-1 probably plays a key role in this process by protecting nicked BER intermediates from degradation or recombination until the

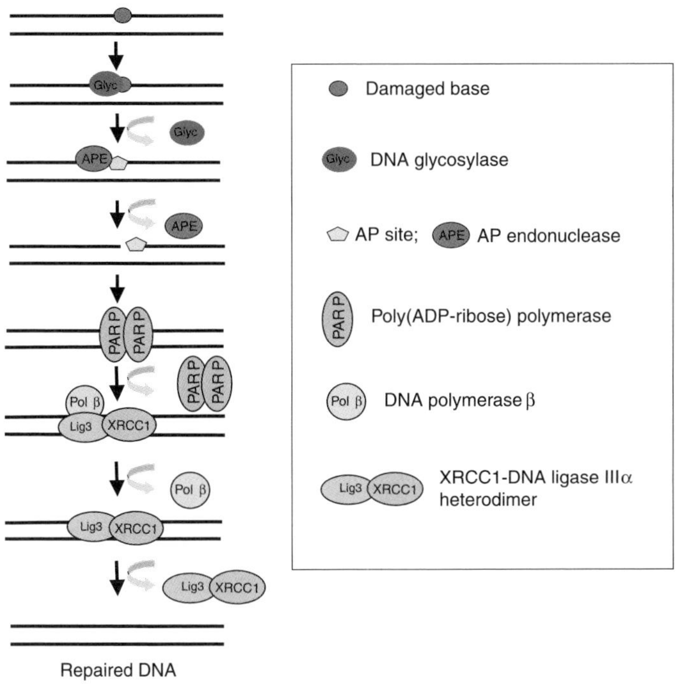

Fig. 1 Model for base excision repair. Short patch BER is initiated by a damage specific DNA glycosylase, which removes the damaged base creating an abasic site (AP site). AP endonuclease (APE) then cleaves the phosphodiester bond 5′ to the AP site containing 3′-OH and 5′-sugar phosphate ends. The strand break thus revealed is recognized by PARP-1 dimer. Following NAD^+-dependent poly(ADP-ribosyl)ation of PARP-1, automodified PARP-1 then dissociates from DNA. The release of automodified PARP-1 exposes the intermediate to DNA polymerase β (Pol β) and XRCC1-DNA ligase IIIα that perform the remainder of the repair process where Pol β adds one nucleotide to the 3′-end and simultaneously removes the 5′-sugar phosphate residue and DNA ligase complete the repair by sealing DNA ends

enzymes required for the subsequent steps of repair can be assembled at the repair site.

4
Conclusion

BER is a complex multi-step process involving a network of protein-protein interactions and modulated in part by post-translational processing of the proteins involved and by the cell cycle. Although the basic enzymology of the BER process is reasonably well understood, much remains to be uncovered about its co-ordination and regulation in vivo.

References

Abner CW, Lau AY, Ellenberger T, Bloom LB (2001) Base excision and DNA binding activities of human alkyladenine DNA glycosylase are sensitive to the base paired with a lesion. J Biol Chem 276:13379–13387

Aboussekhra A, Biggerstaff M, Shivji MK, Vilpo JA, Moncollin V, Podust VN, Protic M, Hubscher U, Egly JM, Wood RD (1995) Mammalian DNA nucleotide excision repair reconstituted with purified protein components. Cell 80:859–868

Allinson SL, Dianova II, Dianov GL (2003) Poly(ADP-ribose) polymerase in base excision repair: always engaged, but not essential for DNA damage processing. Acta Biochim Pol 50:169–179

Allinson SL, Dianova II, Dianov GL (2001) DNA polymerase β is the major dRP lyase involved in repair of oxidative base lesions in DNA by mammalian cell extracts. EMBO J 20:6919–6926

Allinson SL, Sleeth KM, Matthewman GE, Dianov GL (2004) Orchestration of base excision repair by controlling the rates of enzymatic activities. DNA Repair 3:23–31

Ame JC, Spenlehauer C, de Murcia G (2004) The PARP superfamily. Bioessays 26:882–893

Aoufouchi S, Yelamos J, Milstein C (1999) Inhibition of apoptosis of a PARP(–/–) cell line transfected with PARP DNA-binding domain mutants. J Mol Biol 290:943–949

Aspinwall R, Rothwell DG, Roldan-Arjona T, Anselmino C, Ward CJ, Cheadle JP, Sampson JR, Lindahl T, Harris PC, Hickson ID (1997) Cloning and characterization of a functional human homolog of *Escherichia coli* endonuclease III. Proc Natl Acad Sci USA 94:109–114

Barnes DE, Tomkinson AE, Lehmann AR, Webster HDB, Lindahl T (1992) Mutations in the DNA ligase I gene of an individual with immunodeficiencies and cellular hypersensitivity. Cell 69:495–504

Barrett TE, Savva R, Panayotou G, Barlow T, Brown T, Jiricny J, Pearl LH (1998) Crystal structure of a G:T/U mismatch-specific DNA glycosylase: mismatch recognition by complementary-strand interactions. Cell 92:117–129

Bennett RA, Wilson DM 3rd, Wong D, Demple B (1997) Interaction of human apurinic endonuclease and DNA polymerase β in the base excision repair pathway. Proc Natl Acad Sci USA 94:7166–7169

Bennett SE, Sung JS, Mosbaugh DW (2001) Fidelity of uracil-initiated base excision DNA repair in DNA polymerase β-proficient and -deficient mouse embryonic fibroblast cell extracts. J Biol Chem 276:42588–42600

Bentley DJ, Harrison C, Ketchen A-M, Redhead NJ, Samuel K, Waterfall M, Ansell JD, Melton DW (2002) DNA ligase I null mouse cells show normal DNA repair activity but altered DNA replication and reduced genome stability. J Cell Sci 115:1551–1561

Bentley DJ, Selfridge J, Millar JK, Samuel K, Hole N, Ansell JD, Melton DW (1996) DNA ligase I is required for fetal liver erythropoiesis but is not essential for mammalian cells. Nature Genetics 13:489–491

Bergoglio V, Canitrot Y, Hogarth L, Minto L, Howell SB, Cazaux C, Hoffmann JS (2001) Enhanced expression and activity of DNA polymerase beta in human ovarian tumor cells: impact on sensitivity towards antitumor agents. Oncogene 20:6181–6187

Bhakat KK, Hazra TK, Mitra S (2004) Acetylation of the human DNA glycosylase NEIL2 and inhibition of its activity. Nucleic Acids Res 32:3033–3039

Bjoras M, Seeberg E, Luna L, Pearl LH, Barrett TE (2002) Reciprocal flipping underlies substrate recognition and catalytic activation by the human 8-oxo-guanine DNA glycosylase. J Mol Biol 317:171–177

Boldogh I, Milligan D, Lee MS, Bassett H, Lloyd RS, McCullough AK (2001) hMYH cell cycle-dependent expression, subcellular localization and association with replication foci: evidence suggesting replication-coupled repair of adenine: 8-oxoguanine mispairs. Nucleic Acids Res 29:2802–2809

Boorstein RJ, Cummings A Jr, Marenstein DR, Chan MK, Ma Y, Neubert TA, Brown SM, Teebor GW (2001) Definitive Identification of Mammalian 5-Hydroxymethyluracil DNA N-Glycosylase Activity as SMUG1. J Biol Chem 276:41991–41997

Braithwaite EK, Prasad R, Shock DD, Hou EW, Beard WA, Wilson SH (2005) DNA Polymerase λ Mediates a back-up base excision repair activity in extracts of mouse embryonic fibroblasts. J Biol Chem 280:18469–18475

Bruner SD, Norman DPG, Verdine GL (2000) Structural basis for recognition and repair of the endogenous mutagen 8-oxoguanine in DNA. Nature 403:859–866

Burrows CJ, Muller JG (1998) Oxidative nucleobase modifications leading to strand scission. Chemical Reviews 98:1109–1151

Cabelof DC, Raffoul JJ, Yanamadala S, Guo Z, Heydari AR (2002) Induction of DNA polymerase β-dependent base excision repair in response to oxidative stress in vivo. Carcinogenesis 23:1419–1425

Caldecott KW, Aoufouchi S, Johnson P, Shall S (1996) XRCC1 polypeptide interacts with DNA polymerase β and possibly poly (ADP-ribose) polymerase, DNA ligase III is a novel molecular nick-sensor in vitro. Nucleic Acids Res 24:4387–4394

Caldecott KW, McKeown CK, Tucker JD, Ljungquist S, Thompson LH (1994) An interaction between the mammalian DNA repair protein XRCC1, DNA ligase III. Mol Cell Biol 14:68–76

Caldecott KW, Tucker JD, Stanker LH, Thompson LH (1995) Characterization of the XRCC1-DNA ligase III complex in vitro and its absence from mutant hamster cells. Nucleic Acids Res 23:4836–4843

Canitrot Y, Cazaux C, Frechet M, Bouayadi K, Lesca C, Salles B, Hoffmann JS (1998) Overexpression of DNA polymerase β in cell results in a mutator phenotype and a decreased sensitivity to anticancer drugs. Proc Natl Acad Sci USA 95:12586–12590

Cappelli E, Taylor R, Cevasco M, Abbondandolo A, Caldecott K, Frosina G (1997) Involvement of XRCC1 and DNA ligase III gene products in DNA base excision repair. J Biol Chem 272:23970–23975

Chagovetz AM, Sweasy JB, Preston BD (1997) Increased activity and fidelity of DNA polymerase β on single-nucleotide gapped DNA. J Biol Chem 272:27501–27504

Chakravarti D, Ibeanu GC, Tano K, Mitra S (1991) Cloning and expression in *Escherichia coli* of a human cDNA encoding the DNA repair protein N-methylpurine-DNA glycosylase. J Biol Chem 266:15710–15715
Chaudhry MA, Weinfeld M (1995) The action of *Escherichia coli* endonuclease III on multiply damaged sites in DNA. J Mol Biol 249:914–922
Chen DS, Herman VG, Demple B (1991) Two distinct human diesterases that hydrolyze 3′-blocking deoxyribose fragments from oxidised DNA. Nucleic Acids Res 19:5907–5914
Chou KM, Cheng YC (2002) An exonucleolytic activity of human apurinic/apyrimidinic endonuclease on 3′ mispaired DNA. Nature 415:655–659
Chou KM, Cheng YC (2003) The exonuclease activity of human apurinic/apyrimidinic endonuclease (APE1) Biochemical properties and inhibition by the natural dinucleotide Gp4G. J Biol Chem 278:18289–18296
Chou KM, Kukhanova M, Cheng YC (2000) A novel action of human apurinic/apyrimidinic endonuclease: excision of L-configuration deoxyribonucleoside analogs from the 3′ termini of DNA. J Biol Chem 275:31009–31015
Dantzer F, de la Rubia G, Murcia JMD, Hostomsky Z, de Murcia G, Schreiber V (2000) Base excision repair is impaired in mammalian cells lacking poly(ADP-ribose) polymerase-1. Biochemistry 39:7559–7569
Dantzer F, Luna L, Bjoras M, Seeberg E (2002) Human OGG1 undergoes serine phosphorylation and associates with the nuclear matrix and mitotic chromatin in vivo. Nucl Acids Res 30:2349–2357
DeMott M, Zigman S, Bambara RA (1998) Replication protein A stimulates long patch DNA base excision repair. J Biol Chem 273:27492–27498
Demple B, Harrison L (1994) Repair of oxidative damage to DNA: enzymology and biology. Annu Rev Biochem 63:915–948
Demple B, Herman T, Chen DS (1991) Cloning and expression of APE, the cDNA encoding the major human apurinic endonuclease: definition of a family of DNA repair enzymes. Proc Natl Acad Sci USA 88:11450–11454
DeRose EF, Kirby TW, Mueller GA, Bebenek K, Garcia-Diaz M, Blanco L, Kunkel TA, London RE (2003) Solution structure of the lyase domain of human DNA polymerase lambda. Biochemistry 42:9564–9574
Dianov G, Bischoff C, Piotrowski J, Bohr VA (1998) Repair pathways for processing of 8-oxoguanine in DNA by mammalian cell extracts. J Biol Chem 273:33811–33816
Dianov G, Price A, Lindahl T (1992) Generation of single-nucleotide repair patches following excision of uracil residues from DNA. Mol Cell Biol 12:1605–1612
Dianov GL, Jensen BR, Kenny MK, Bohr VA (1999) Replication protein A stimulates proliferating cell nuclear antigen-dependent repair of abasic sites in DNA by human cell extracts. Biochemistry 38:11021–11025
Dianov GL, Thybo T, Dianova II, Lipinski LJ, Bohr VA (2000) Single nucleotide patch base excision repair is the major pathway for removal of thymine glycol from DNA in human cell extracts. J Biol Chem 275:11809–11813
Dianov GL, Sleeth KM, Dianova II, Allinson SL (2003) Repair of abasic sites in DNA. Mutat Res 531:157–163
Dianova II, Sleeth KM, Allinson SL, Parsons JL, Breslin C, Caldecott KW, Dianov GL (2004) XRCC1-DNA polymerase β interaction is required for efficient base excision repair. Nucleic Acids Res 32:2550–2555
Dimitriadis EK, Prasad R, Vaske MK, Chen L, Tomkinson AE, Lewis MS, Wilson SH (1998) Thermodynamics of human DNA ligase I trimerization and association with DNA polymerase beta. J Biol Chem 273:20540–20550

Dou H, Mitra S, Hazra TK (2003) Repair of oxidized bases in DNA bubble structures by human DNA glycosylases NEIL1 and NEIL2. J Biol Chem 278:49679–49684

Drohat AC, Kwon K, Krosky DJ, Stivers JT (2002) 3-methyladenine DNA glycosylase is an unexpected helix-hairpin-helix superfamily member. Nature Struct Biol 9:659–664

Duncan BK, Miller J (1980) Mutagenic deamination of cytosine residues in DNA. Nature 287:560–561

Duncan BK, Weiss B (1982) Specific mutator effects of ung (uracil-DNA glycosylase) mutations in *Escherichia coli.* J Bacteriol 151:750–755

Elder RH, Dianov GL (2002) Repair of dihydrouracil supported by base excision repair in mNTH1 knock-out cell extracts. J Biol Chem 277:50487–50490

El-Khamisy SF, Masutani M, Suzuki H, Caldecott KW (2003) A requirement for PARP-1 for the assembly or stability of XRCC1 nuclear foci at sites of oxidative DNA damage. Nucleic Acids Res 31:5526–5533

Engelward BP, Weeda G, Wyatt MD, Broekhof JLM, deWit J, Donker I, Allan JM, Gold B, Hoeijmakers JHJ, Samson LD (1997) Base excision repair deficient mice lacking the Aag alkyladenine DNA glycosylase. Proceedings of the National Academy of Sciences of the United States of America 94:13087–13092

Erzberger JP, Barsky D, Scharer OD, Colvin ME, Wilson DM 3rd (1998) Elements in abasic site recognition by the major human and *Escherichia coli* apurinic/apyrimidinic endonucleases. Nucleic Acids Res 26:2771–2778

Evans AR, Limp-Foster M, Kelley MR (2000) Going APE over ref-1. Mutat Res 461:83–108

Fan J, Otterlei M, Wong HK, Tomkinson AE, Wilson DM 3rd (2004) XRCC1 co-localizes and physically interacts with PCNA. Nucleic Acids Res 32:2193–2201

Fischer JA, Muller-Weeks S, Caradonna S (2004) Proteolytic degradation of the nuclear isoform of uracil-DNA glycosylase occurs during the S-phase of the cell cycle. DNA Repair (Amst) 3:505–513

Fornace AJ Jr, Zmudzka B, Hollander MC, Wilson SH (1989) Induction of β-polymerase mRNA by DNA damaging agents in Chinese hamster ovary cells. Mol Cell Biol 9:851–853

Fortini P, Parlanti E, Sidorkina OM, Laval J, Dogliotti E (1999) The type of DNA glycosylase determines the base excision repair pathway in mammalian cells. J Biol Chem 274:15230–15236

Friedberg EC, Walker GC, Siede W (1995) DNA Repair and Mutagenesis (Washington DC, American Society of Microbiology Press)

Fromme JC, Verdine GL (2002) Structural insights into lesion recognition and repair by the bacterial 8-oxoguanine DNA glycosylase MutM. Nat Struct Biol 9:544–552

Frosina G, Fortini P, Rossi O, Carrozzino F, Raspaglio G, Cox LS, Lane DP, Abbondandolo A, Dogliotti E (1996) Two pathways for base excision repair in mammalian cells. J Biol Chem 271:9573–9578

Fung H, Demple B (2005) A vital role for ape1/ref1 protein in repairing spontaneous DNA damage in human cells. Mol Cell 17:463–470

Garcia-Diaz M, Bebenek K, Kunkel TA, Blanco L (2001) Identification of an intrinsic 5′-deoxyribose-5-phosphate lyase activity in human DNA polymerase λ: a possible role in base excision repair. J Biol Chem 276:34659–34663

Garcia-Diaz M, Bebenek K, Sabariegos R, Dominguez O, Rodriguez J, Kirchhoff T, Garcia-Palomero E, Picher AJ, Juarez R, Ruiz JF et al. (2002) DNA polymerase λ, a novel DNA repair enzyme in human cells. J Biol Chem 277:13184–13191

Gilboa R, Zharkov DO, Golan G, Fernandes AS, Gerchman SE, Matz E, Kycia JH, Grollman AP, Shoham G (2002) Structure of formamidopyrimidine-DNA glycosylase covalently complexed to DNA. J Biol Chem 277:19811–19816

Grosch S, Fritz G, Kaina B (1998) Apurinic endonuclease (Ref-1) is induced in mammalian cells by oxidative stress and involved in clastogenic adaptation. Cancer Res 58:4410–4416

Grosch S, Kaina B (1999) Transcriptional activation of apurinic/apyrimidinic endonuclease (Ape, Ref-1) by oxidative stress requires CREB. Biochem Biophys Res Commun 261:859–863

Gryk MR, Marintchev A, Maciejewski MW, Robertson A, Wilson SH, Mullen GP (2002) Mapping of the interaction interface of DNA polymerase β with XRCC1. Structure (Camb) 10:1709–1720

Gu H, Marth JD, Orban PC, Mossmann H, Rajewsky K (1994) Deletion of a DNA-Polymerase-β Gene Segment in T-cells using cell-type-specific gene targeting. Science 265:103–106

Gu Y, Lu AL (2001) Differential DNA recognition and glycosylase activity of the native human MutY homolog (hMYH) and recombinant hMYH expressed in bacteria. Nucleic Acids Res 29:2666–2674

Gu Y, Parker A, Wilson TM, Bai H, Chang DY, Lu AL (2002) Human MutY homolog, a DNA glycosylase involved in base excision repair, physically and functionally interacts with mismatch repair proteins human MutS homolog 2/human MutS homolog 6. J Biol Chem 277:11135–11142

Guillet M, Boiteux S (2002) Endogenous DNA abasic sites cause cell death in the absence of APN1, APN2 and Rad1/Rad10 in *Saccharomyces cerevisiae*. EMBO J 21:2833–2841

Hadi MZ, Ginalski K, Nguyen LH, Wilson DM 3rd (2002) Determinants in nuclease specificity of Ape1 and Ape2, human homologues of *Escherichia coli* exonuclease III. J Mol Biol 316:853–866

Hang B, Singer B, Margison GP, Elder RH (1997) Targeted deletion of alkylpurine-DNA-N-glycosylase in mice eliminates repair of 1,N-6-ethenoadenine and hypoxanthine but not of 3,N-4-ethenocytosine or 8-oxoguanine. Proc Natl Acad Sci USA 94:12869–12874

Hardeland U, Steinacher R, Jiricny J, Schar P (2002) Modification of the human thymine-DNA glycosylase by ubiquitin-like proteins facilitates enzymatic turnover. EMBO J 21:1456–1464

Harrison C, Ketchen A-M, Redhead NJ, O'Sillivan MJ, Melton DW (2002) Replication failure, genome instability, and increased cancer susceptibility in mice with a point mutation in the DNA ligase I gene. Cancer Res 62:4065–4074

Haushalter KA, Stukenberg PT, Kirschner MW, Verdine GL (1999) Identification of a new uracil-DNA glycosylase family by expression cloning using synthetic inhibitors. Curr Biol 9:174–185

Hendrich B, Hardeland U, Ng HH, Jiricny J, Bird A (1999) The thymine glycosylase MBD4 can bind to the product of deamination at methylated CpG sites. Nature 401:301–304

Herceg Z, Wang ZQ (2001) Functions of poly(ADP-ribose) polymerase (PARP) in DNA repair, genomic integrity and cell death. Mutat Res 477:97–110

Hill JW, Hazra TK, Izumi T, Mitra S (2001) Stimulation of human 8-oxoguanine-DNA glycosylase by AP-endonuclease: potential coordination of the initial steps in base excision repair. Nucleic Acids Res 29:430–438

Hirose F, Hotta Y, Yamaguchi M, Matsukage A (1989) Difference in the expression level of DNA polymerase beta among mouse tissues: high expression in the pachytene spermatocyte. Exp Cell Res 181:169–180

Hollis T, Ichikawa Y, Ellenberger T (2000) DNA bending and a flip-out mechanism for base excision by the helix-hairpin-helix DNA glycosylase, *Escherichia coli* AlkA. EMBO J 19:758–766

Ikeda S, Biswas T, Roy R, Izumi T, Boldogh I, Kurosky A, Sarker AH, Seki S, Mitra S (1998) Purification and characterization of human NTH1, a homolog of *Escherichia coli* endonuclease III. Direct identification of Lys-212 as the active nucleophilic residue. J Biol Chem 273:21585–21593

Jaruga P, Birincioglu M, Rosenquist TA, Dizdaroglu M (2004) Mouse NEIL1 protein is specific for excision of 2,6-diamino-4-hydroxy-5-formamidopyrimidine and 4,6-diamino-5-formamidopyrimidine from oxidatively damaged DNA. Biochemistry 43:15909–15914

Kakolyris S, Kaklamanis L, Giatromanolaki A, Koujourakis M, Hickson ID, Barzilay G, Turley H, Leek RD, Kanavaros P, Georgoulias V, Gatter KC, Harris AL (1998) Expression and subcellular localisation of human AP endonuclease 1 (HAP1/Ref-1) protein: a basis for its role in human disease. Histopathology 33:561–569

Kedar PS, Kim SJ, Robertson A, Hou E, Prasad R, Horton JK, Wilson SH (2002) Direct interaction between mammalian DNA polymerase beta and proliferating cell nuclear antigen. J Biol Chem 277:31115–31123

Klungland A, Lindahl T (1997) Second pathway for completion of human DNA base excision-repair: reconstitution with purified proteins and requirement for DNase IV (FEN1). EMBO J 16:3341–3348

Klungland A, Rosewell I, Hollenbach S, Larsen E, Daly G, Epe B, Seeberg E, Lindahl T, Barnes DE (1999) Accumulation of premutagenic DNA lesions in mice defective in removal of oxidative base damage. Proc Nat Acad Sci USA 96:13300–13305

Krokan HE, Otterlei M, Nilsen H, Kavli B, Skorpen F, Andersen S, Skjelbred C, Akbari M, Aas PA, Slupphaug G (2001) Properties and functions of human uracil-DNA glycosylase from the UNG gene. Prog Nucleic Acid Res Mol Biol 68:365–386

Kubota Y, Horiuchi S (2003) Independent roles of XRCC1's two BRCT motifs in recovery from methylation damage. DNA Repair (Amst) 2:407–415

Kubota Y, Nash RA, Klungland A, Schar P, Barnes D, Lindahl T (1996) Reconstitution of DNA base excision-repair with purified human proteins: interaction between DNA polymerase β and the XRCC1 protein. EMBO J 15:6662–6670

Kumar A, Widen S, Williams K, Kedar P, Karpel R, Wilson S (1990) Studies of the domain structure of mammalian DNA polymerase β. Identification of a discrete template binding domain. J Biol Chem 265:2124–2131

Labahn J, Scharer OD, Long A, Ezaz-Nikpay K, Verdine GL, Ellenberger TE (1996) Structural basis for the excision repair of alkylation-damaged DNA. Cell 86:321–329

Lau AY, Scharer OD, Samson L, Verdine GL, Ellenberger E (1998) Crystal structure of a human alkylbase-DNA repair enzyme complexed to DNA: Mechanisms for nucleotide flipping and base excision. Cell 95:249–258

Lau AY, Wyattt MD, Glassner BJ, Samson LD, Ellenberger T (2000) Molecular basis for discriminating between normal and damaged bases by the human alkyladenine glycosylase, AAG. Proc Natl Acad Sci USA 97:13573–13578

Lavrik OI, Prasad R, Sobol RW, Horton JK, Ackerman EJ, Wilson SH (2001) Photoaffinity labeling of mouse fibroblast enzymes by a base excision repair intermediate. Evidence for the role of poly(ADP-ribose) polymerase-1 in DNA repair. J Biol Chem 276:25541–25548

Lee JW, Blanco L, Zhou T, Garcia-Diaz M, Bebenek K, Kunkel TA, Wang Z, Povirk LF (2004) Implication of DNA polymerase λ in alignment-based gap filling for nonhomologous DNA end joining in human nuclear extracts. J Biol Chem 279:805–811

Leppard JB, Dong Z, Mackey ZB, Tomkinson AE (2003) Physical and functional interaction between DNA ligase IIIα and poly(ADP-ribose) polymerase 1 in DNA single-strand break repair. Mol Cell Biol 23:5919–5927

Levin DS, Bai W, Yao N, O'Donnell M, Tomkinson AE (1997) An interaction between DNA ligase I and proliferating cell nuclear antigen: implication for Okazaki fragment synthesis and joining. Proc Natl Acad Sci USA 94:12863–12868

Levin DS, McKenna AE, Motycka TA, Matsumoto Y, Tomkinson AE (2000) Interaction between PCNA and DNA ligase I is critical for joining of Okazaki fragments and long-patch base-excision repair. Current Biol 10:919–922

Lindahl T (1974) An N-glycosidase from *Escherichia coli* that releases free uracil from DNA containing deaminated cytosine residues. Proc Natl Acad Sci USA 71:3649–3653

Lindahl T (1993) Instability and decay of the primary structure of DNA. Nature 362:709–715

Lindahl T, Andersson A (1972) Rate of chain breakage of apurinic sites in double-stranded DNA. Biochemistry 11:3618–3623

Lindahl T, Karlström O (1973) Heat-induced depyrimidination of DNA. Biochemistry 12:5151–5154

Lindahl T, Nyberg B (1972) Rate of depurination of native deoxyribonucleic acid. Biochemistry 11:3610–3618

Lindahl T, Nyberg B (1974) Heat-induced deamination of cytosine residues in DNA. Biochemistry 13:3405–3410

Lindahl T, Wood RD (1999) Quality control by DNA R=repair. Science 286:1897–1905

Ljungquist S, Kenne K, Olsson L, Sandstrom M (1994) Altered DNA ligase III activity in the CHO EM9 mutant. Mutation Res 314:177–186

Loeb LA, Preston BD (1986) Mutagenesis by apurinic/apyrimidinic sites. Ann Rev Genet 20:201–230

Luna L, Bjoras M, Hoff E, Rognes T, Seeberg E (2000) Cell-cycle regulation, intracellular sorting and induced overexpression of the human NTH1 DNA glycosylase involved in removal of formamidopyrimidine residues from DNA. Mutat Res 460:95–104

Ma Y, Lu H, Tippin B, Goodman MF, Shimazaki N, Koiwai O, Hsieh CL, Schwarz K, Lieber MR (2004) A biochemically defined system for mammalian nonhomologous DNA end joining. Mol Cell 16:701–713

Mackey ZB, Niedergang C, Murcia JM, Leppard J, Au K, Chen J, de Murcia G, Tomkinson AE (1999) DNA ligase III is recruited to DNA strand breaks by a zinc finger motif homologous to that of poly(ADP-ribose) polymerase. Identification of two functionally distinct DNA binding regions within DNA ligase III. J Biol Chem 274:21679–21687

Mackey ZB, Ramos W, Levin DS, Walter CA, McCarrey JR, Tomkinson AE (1997) An alternative splicing event which occurs in mouse pachytene spermatocytes generates a form of DNA ligase III with distinct biochemical properties that may function in meiotic recombination. Mol Cell Biol 17:989–998

Marintchev A, Mullen MA, Maciejewski MW, Pan B, Gryk MR, Mullen GP (1999) Solution structure of the single-strand break repair protein XRCC1 N-terminal domain. Nat Struct Biol 6:884–893

Marintchev A, Robertson A, Dimitriadis EK, Prasad R, Wilson SH, Mullen GP (2000) Domain specific interaction in the XRCC1-DNA polymerase β complex. Nucleic Acids Res 28:2049–2059

Marenstein DR, Chan MK, Altamirano A, Basu AK, Boorstein RJ, Cunningham RP, Teebor GW (2003) Substrate specificity of human endonuclease III (hNTH1) Effect of human APE1 on hNTH1 activity. J Biol Chem 278:9005–9012

Marsin S, Vidal AE, Sossou M, Menissier-de Murcia J, Le Page F, Boiteux S, de Murcia G, Radicella JP (2003) Role of XRCC1 in the coordination and stimulation of oxidative DNA damage repair initiated by the DNA glycosylase hOGG1. J Biol Chem 278:44068–44074

Masson M, Niedergang C, Schreiber V, Muller S, Menissier-de Murcia J, de Murcia G (1998) XRCC1 is specifically associated with poly(ADP-ribose) polymerase and negatively regulates its activity following DNA damage. Mol Cell Biol 18:3563–3571

Matsumoto Y, Kim K (1995) Excision of deoxyribose phosphate residues by DNA polymerase β during DNA repair. Science 269:699–702

Matsumoto Y, Kim K, Bogenhagen DF (1994) Proliferating cell nuclear antigen-dependent abasic site repair in Xenopus laevis oocytes: an alternative pathway of base excision DNA repair. Mol Cell Biol 14:6187–6197

McCullough AK, Dodson ML, Lloyd RS (1999) Initiation of base excision repair: glycosylase mechanism and structure. Annu Rev Biochem 68:255–285

Ménessier-de Murcia J, Niedergang C, Trucco C, Ricoul M, Dutrillaux B, Mark M, Oliver FJ, Masson M, Dierich A, LeMeur M et al. (1997) Requirement of poly(ADP-ribose) polymerase in recovery from DNA damage in mice and in cells. Proc Natl Acad Sci USA 94:7303–7307

Millar CB, Guy J, Sansom OJ, Selfridge J, MacDougall E, Hendrich B, Keightley PD, Bishop SM, Clarke AR, Bird A (2002) Enhanced CpG mutability and tumorigenesis in MBD4-deficient mice. Science 297:403–405

Minowa O, Arai T, Hirano M, Monden Y, Nakai S, Fukuda M, Itoh M, Takano H, Hippou Y, Aburatani H et al. (2000) Mmh/Ogg1 gene inactivation results in accumulation of 8-hydroxyguanine in mice. Proc Natl Acad Sci USA 97:4156–4161

Mokkapati SK, Wiederhold L, Hazra TK, Mitra S (2004) Stimulation of DNA glycosylase activity of OGG1 by NEIL1: functional collaboration between two human DNA glycosylases. Biochemistry 43:11596–11604

Mol CD, Arvai AS, Slupphaug G, Kavli B, Alseth I, Krokan HE, Tainer JA (1995) Crystal-structure and mutational analysis of human uracil-dna glycosylase – structural basis for specificity and catalysis. Cell 80:869–878

Mol CD, Izumi T, Mitra S, Tainer JA (2000) DNA-bound structures and mutants reveal abasic DNA binding by APE1 and DNA repair coordination. Nature 403:451–456

Molinete M, Vermeulen W, Burkle A, Menissier-de Murcia J, Kupper JH, Hoeijmakers JHJ, de Murcia G (1993) Overproduction of the poly(Adp-ribose) polymerase DNA-binding domain blocks alkylation-induced DNA-repair synthesis in mammalian-cells. EMBO J 12:2109–2117

Montecucco A, Rossi R, Levin DS, Gary R, Park MS, Motycka TA, Giarrocchi G, Villa A, Biamonti G, Tomkinson A (1998) DNA ligase I is recruited to sites of DNA replication by an interaction with proliferating cell nuclear antigen: identification of a common targeting mechanism for the assembly of replication factories. EMBO J 17:3786–3795

Moore DJ, Taylor RM, Clements P, Caldecott KW (2000) Mutation of a BRCT domain selectively disrupts DNA single-strand break repair in noncycling Chinese hamster ovary cells. Proc Natl Acad Sci USA 97:13649–13654

Muller-Weeks S, Mastran B, Caradonna S (1998) The nuclear isoform of the highly conserved human uracil-DNA glycosylase is an Mr 36000 phosphoprotein. J Biol Chem 273:21909–21917

Nagelhus TA, Haug T, Singh KK, Keshav KF, Skorpen F, Otterlei M, Bharati S, Lindmo T, Benichou S, Benarous R, Krokan HE (1997) A sequence in the N-terminal region of human uracil-DNA glycosylase with homology to XPA interacts with the C-terminal part of the 34-kDa subunit of replication protein A. J Biol Chem 272:6561–6566

Nash RA, Caldecott KW, Barnes DE, Lindahl T (1997) XRCC1 protein interacts with one of two distinct forms of DNA ligase III. Biochemistry 36:5207–5211

Neddermann P, Gallinari P, Lettieri T, Schmid D, Truong O, Hsuan JJ, Wiebauer K, Jiricny J (1996) Cloning and expression of human G/T mismatch-specific thymine-DNA glycosylase. J Biol Chem 271:12767–12774

Nilsen H, Rosewell I, Robins P, Skjelbred CF, Andersen S, Slupphaug G, Daly G, Krokan HE, Lindahl T, Barnes DE (2000) Uracil-DNA glycosylase (UNG)-deficient mice reveal a primary role of the enzyme during DNA replication. Molecular Cell 5:1059–1065

Norman DP, Bruner SD, Verdine GL (2001) Coupling of substrate recognition and catalysis by a human base-excision DNA repair protein. J Am Chem Soc 123:359–360

Ocampo MT, Chaung W, Marenstein DR, Chan MK, Altamirano A, Basu AK, Boorstein RJ, Cunningham RP, Teebor GW (2002) Targeted deletion of mNth1 reveals a novel DNA repair enzyme activity. Mol Cell Biol 22:6111–6121

O'Connor TR, Laval J (1991) Human cDNA expressing a functional DNA glycosylase excising 3-methyladenine and 7-methylguanine. Biochem Biophys Res Commun 176:1170–1177

Okano S, Lan L, Caldecott KW, Mori T, Yasui A (2003) Spatial and temporal cellular responses to single-strand breaks in human cells. Mol Cell Biol 23:3974–3981

Olsen LC, Aasland R, Wittwer CU, Krokan HE, Helland DE (1989) Molecular cloning of human uracil-DNA glycosylase, a highly conserved DNA repair enzyme. EMBO J 8:3121–3125

Otterlei M, Warbrick E, Nagelhus TA, Haug T, Slupphaug G, Akbari M, Aas PA, Steinsbekk K, Bakke O, Krokan HE (1999) Post-replicative base excision repair in replication foci. EMBO J 18:3834–8344

Oyama M, Wakasugi M, Hama T, Hashidume H, Iwakami Y, Imai R, Hoshino S, Morioka H, Ishigaki Y, Nikaido O, Matsunaga T (2004) Human NTH1 physically interacts with p53 and proliferating cell nuclear antigen. Biochem Biophys Res Commun 321:183–191

Parikh SS, Mol CD, Hosfield DJ, Tainer JA (1999) Envisioning the molecular choreography of DNA base excision repair. Curr Opin Struct Biol 9:37–47

Parikh SS, Mol CD, Slupphaug G, Bharati S, Krokan HE, Tainer JA (1998) Base excision repair initiation revealed by crystal structures and binding kinetics of human uracil-DNA glycosylase with DNA. EMBO J 17:5214–5226

Parikh SS, Walcher G, Jones GD, Slupphaug G, Krokan HE, Blackburn GM, Tainer JA (2000) Uracil-DNA glycosylase-DNA substrate and product structures: Conformational strain promotes catalytic efficiency by coupled stereoelectronic effects. Proc Natl Acad Sci USA 97:5083–5088

Parker A, Gu Y, Mahoney W, Lee SH, Singh KK, Lu AL (2001) Human homolog of the MutY repair protein (hMYH) physically interacts with proteins involved in long patch DNA base excision repair. J Biol Chem 276:5547–5555

Parsons JL, Dianova II, Dianov GL (2004) APE1 is the major 3′-phosphoglycolate activity in human cell extracts. Nucleic Acids Res 32:3531–3536

Parsons JL, Dianova II, Dianov GL (2005) APE1-dependent repair of DNA single-strand breaks containing 3′-end 8-oxoguanine. Nucleic Acids Res 33:2204–2209

Parsons JL, Dianova I, Allinson SL, Dianov G (2005) Poly(ADP-ribose) polymerase-1 protects excessive DNA strand breaks from deterioration during repair in human cell extracts. FEBS J 272:2012–2021

Perkins E, Sun D, Nguyen A, Tulac S, Francesco M, Tavana H, Nguyen H, Tugendreich S, Barthmaier P, Couto J et al. (2001) Novel inhibitors of poly(ADP-ribose) polymerase/PARP1 and PARP2 identified using a cell-based screen in yeast. Cancer Res 61:4175–4183

Petronzelli F, Riccio A, Markham GD, Seeholzer SH, Stoerker J, Genuardi M, Yeung AT, Matsumoto Y, Bellacosa A (2000) Biphasic kinetics of the human DNA repair protein MED1 (MBD4), a mismatch-specific DNA N-glycosylase. J Biol Chem 275:32422–32429

Podlutsky AJ, Dianova II, Wilson SH, Bohr VA, Dianov GL (2001) DNA synthesis and dRPase activities of polymerase beta are both essential for single-nucleotide patch base excision repair in mammalian cell extracts. Biochemistry 40:809–813

Podlutsky AJ, Dianova II, Podust VN, Bohr VA, Dianov GL (2001) Human DNA polymerase beta initiates DNA synthesis during long-patch repair of reduced AP sites in DNA. EMBO J 20:1477–1482

Prasad R, Singhal RK, Srivastava DK, Molina JT, Tomkinson AE, Wilson SH (1996) Specific interaction of DNA polymerase beta and DNA ligase I in a multiprotein base excision repair complex from bovine testis. J Biol Chem 271:16000–16007

Prigent C, Satoh MS, Daly G, Barnes DE, Lindahl T (1994) Aberrant DNA repair and DNA replication due to an inherited enzymatic defect in human DNA ligase I. Mol Cell Biol 14:310–317

Radicella JP, Dherin C, Desmaze C, Fox MS, Boiteux S (1997) Cloning and characterization of hOGG1, a human homolog of the OGG1 gene of *Saccharomyces cerevisiae*. Proc Natl Acad Sci USA 94:8010–8015

Ramana CV, Boldogh I, Izumi T, Mitra S (1998) Activation of apurinic/apyrimidinic endonuclease in human cells by reactive oxygen species and its correlation with their adaptive response to genotoxicity of free radicals. Proc Natl Acad Sci USA 95:5061–5066

Ranalli TA, DeMott MS, Bambara RA (2002) Mechanism underlying replication protein A stimulation of DNA ligase I. J Biol Chem 277:1719–1727

Rivkees SA, Kelley MR (1994) Expression of a multifunctional DNA-repair enzyme gene, apurinic/apyrimidinic endonuclease (Ape-Ref-1) in the suprachiasmatic, supraoptic and paraventricular nuclei. Brain Res 666:137–142

Robson CN, Hickson ID (1991) Isolation of cDNA clones encoding a human apurinic/apyrimidinic endonuclease that corrects DNA repair and mutagenesis defects in E. coli xth (exonuclease III) mutants. Nucleic Acids Res 19:5519–5523

Roldan-Arjona T, Wei YF, Carter KC, Klungland A, Anselmino C, Wang RP, Augustus M, Lindahl T (1997) Molecular cloning and functional expression of a human cDNA encoding the antimutator enzyme 8-hydroxyguanine-DNA glycosylase. Proc Natl Acad Sci USA 94:8016–8020

Rosenquist TA, Zaika E, Fernandes AS, Zharkov DO, Miller H, Grollman AP (2003) The novel DNA glycosylase, NEIL1, protects mammalian cells from radiation-mediated cell death. DNA Repair (Amst) 2:581–591

Rosenquist TA, Zharkov DO, Grollman AP (1997) Cloning and characterization of a mammalian 8-oxoguanine DNA glycosylase. Proc Natl Acad Sci USA 94:7429–7434

Russo MT, De Luca G, Degan P, Parlanti E, Dogliotti E, Barnes DE, Lindahl T, Yang H, Miller JH, Bignami M (2004) Accumulation of the oxidative base lesion 8-hydroxyguanine in DNA of tumor-prone mice defective in both the Myh and Ogg1 DNA glycosylases. Cancer Res 64:4411–4414

Samson L, Derfler B, Boosalis M, Call K (1991) Cloning and characterization of a 3-methyladenine DNA glycosylase cDNA from human-cells whose gene maps to chromosome-16. Proc Natl Acad Sci of the USA 88:9127–9131

Sanderson RJ, Mosbaugh DW (1998) Fidelity and mutational specificity of uracil-initiated base excision DNA repair synthesis in human glioblastoma cell extracts. J Biol Chem 273:24822–24831

Sarker AH, Ikeda S, Nakano H, Terato H, Ide H, Imai K, Akiyama K, Tsutsui K, Bo Z, Kubo K et al. (1998) Cloning and characterization of a mouse homologue (mNthl1) of *Escherichia coli* endonuclease III. J Mol Biol 282:761–774

Satoh MS, Lindahl T (1992) Role of poly(ADP-ribose) formation in DNA repair. Nature 356:356–358

Satoh MS, Poirier GG, Lindahl T (1993) NAD(+)-dependent repair of damaged DNA by human cell extracts. J Biol Chem 268:5480–5487

Savva R, McAuley-Hecht K, Brown T, Pearl L (1995) The structural basis of specific base-excision repair by uracil-DNA glycosylase. Nature 373:487–493

Sawaya MR, Prasad R, Wilson SH, Kraut J, Pelletier H (1997) Crystal structures of human DNA polymerase beta complexed with gapped and nicked DNA: evidence for an induced fit mechanism. Biochemistry 36:11205–11215

Sawyer DE, Van Houten B (1999) Repair of DNA damage in mitochondria. Mutat Res-DNA Repair 434:161–176

Shibata A, Kamada N, Masumura K, Nohmi T, Kobayashi S, Teraoka H, Nakagama H, Sugimura T, Suzuki H, Masutani M (2005) Parp-1 deficiency causes an increase of deletion mutations and insertions/rearrangements in vivo after treatment with an alkylating agent. Oncogene 24:1328–1337

Sleeth KM, Robson RL, Dianov GL (2004) Exchangeability of mammalian DNA ligases between base excision repair pathways. Biochemistry 43:12924–12930

Slupphaug G, Markussen FH, Olsen LC, Aasland R, Aarsaether N, Bakke O, Krokan HE, Helland DE (1993) Nuclear and mitochondrial forms of human uracil-DNA glycosylase are encoded by the same gene. Nucleic Acids Res 21:2579–2584

Slupska MM, Baikalov C, Luther WM, Chiang JH, Wei YF, Miller JH (1996) Cloning and sequencing a human homolog (hMYH) of the *Escherichia coli* mutY gene whose function is required for the repair of oxidative DNA damage. J Bacteriol 178:3885–3892

Slupska MM, Luther WM, Chiang JH, Yang HJ, Miller JH (1999) Functional expression of hMYH, a human homolog of the *Escherichia coli* MutY protein. J Bacteriol 181:6210–6213

Smith S (2001) The world according to PARP. Trends in Biochemical Sciences 26:174–179

Sobol RW, Foley JF, Nyska A, Davidson MG, Wilson SH (2003) Regulated over-expression of DNA polymerase β mediates early onset cataract in mice. DNA Repair (Amst) 2:609–622

Sobol RW, Horton JK, Kuhn R, Gu H, Singhal RK, Prasad R, Rajewsky K, Wilson SH (1996) Requirement of mammalian DNA polymerase-b in base-excision repair. Nature 379:183–186

Soderhall S (1976) DNA ligases during rat liver regeneration. Nature 260:640–642

Takao M, Kanno S, Shiromoto T, Hasegawa R, Ide H, Ikeda S, Sarker AH, Seki S, Xing JZ, Le XC et al. (2002) Novel nuclear and mitochondrial glycosylases revealed by disruption of the mouse Nth1 gene encoding an endonuclease III homolog for repair of thymine glycols. Embo J 21:3486–3493

Taylor RM, Moore DJ, Whitehouse J, Johnson P, Caldecott KW (2000) A cell cycle-specific requirement for the XRCC1 BRCT II domain during mammalian DNA strand break repair. Mol Cell Biol 20:735–740

Taylor RM, Thistlethwaite A, Caldecott KW (2002) Central role for the XRCC1 BRCT I domain in mammalian DNA single-strand break repair. Mol Cell Biol 22:2556–2563

Taylor RM, Whitehouse CJ, Caldecott KW (2000) The DNA ligase III zinc finger stimulates binding to DNA secondary structure and promotes end joining. Nucleic Acids Res 28:3558–3563

Taylor RM, Whitehouse J, Cappelli E, Frosina G, Caldecott KW (1998) Role of the DNA ligase III zinc finger in polynucleotide binding and ligation. Nucleic Acids Res 26:4804–4810

Tebbs RS, Flannery ML, Meneses JJ, Hartmann A, Tucker JD, Thompson LH, Cleaver JE, Pedersen RA (1999) Requirement for the Xrcc1 DNA base excision repair gene during early mouse development. Dev Biol 208:513–529

Tell G, Crivellato E, Pines A, Paron I, Pucillo C, Manzini G, Bandiera A, Kelley MR, Di Loreto C, Damante G (2001) Mitochondrial localization of APE/Ref-1 in thyroid cells. Mutat Res, 485:143–152

Tentori L, Balduzzi A, Portarena I, Levati L, Vernole P, Gold B, Bonmassar E, Graziani G (2001) Poly (ADP-ribose) polymerase inhibitor increases apoptosis and reduces necrosis induced by a DNA minor groove binding methyl sulfonate ester. Cell Death Differ 8:817–828

Thayer MM, Ahern H, Xing DX, Cunningham RP, Tainer JA (1995) Novel DNA-Binding Motifs in the DNA-repair enzyme endonuclease-III crystal-structure. EMBO J 14:4108–4120

Thompson LH, Brookman KW, Jones JJ, Allen SA, Carrano AV (1990) Molecular cloning of the human XRCC1 gene, which corrects defective DNA repair and sister chromatid exchange. Mol Cell Biol 10:6160–6171

Thompson LH, West MG (2000) XRCC1 keeps DNA from getting stranded. Mutat Res 459:1–18

Tini M, Benecke A, Um SJ, Torchia J, Evans RM, Chambon P (2002) Association of CBP/p300 acetylase and thymine DNA glycosylase links DNA repair and transcription. Mol Cell 9:265–277

Tom S, Henricksen LA, Park MS, Bambara RA (2001) DNA ligase I and proliferating cell nuclear antigen form a functional complex. J Biol Chem 276:24817–24825

Tomkinson AE, Chen L, Dong Z, Leppard JB, Levin DS, Mackey ZB, Motycka TA (2001) Completion of base excision repair by mammalian DNA ligases. Prog Nucleic Acid Res Mol Biol 68:151–164

Tomkinson AE, Levin DS (1997) Mammalian DNA ligases. Bioessays 19:893–901

Tomkinson AE, Mackey ZB (1998) Structure and function of mammalian DNA ligases. Mutat Res 407:1–9

Trucco C, de Oliver FJMG, de Menissier MJ (1998) DNA repair defect in poly(ADP-ribose) polymerase-deficient cell lines. Nucleic Acids Res 26:2644–2649

Tsuchimoto D, Sakai Y, Sakumi K, Nishioka K, Sasaki M, Fujiwara T, Nakabeppu Y (2001) Human APE2 protein is mostly localized in the nuclei and to some extent in the mitochondria, while nuclear APE2 is partly associated with proliferating cell nuclear antigen. Nucleic Acids Res 29:2349–23460

Vidal AE, Boiteux S, Hickson ID, Radicella JP (2001) XRCC1 coordinates the initial and late stages of DNA abasic site repair through protein-protein interactions. EMBO J 20:6530–6539

Vidal AE, Hickson ID, Boiteux S, Radicella JP (2001) Mechanism of stimulation of the DNA glycosylase activity of hOGG1 by the major human AP endonuclease: bypass of the AP lyase activity step. Nucleic Acids Res 29:1285–1292

Vodenicharov MD, Sallmann FR, Satoh MS, Poirier GG (2000) Base excision repair is efficient in cells lacking poly(ADP-ribose) polymerase 1. Nucleic Acids Res 28:3887–3896

Waga S, Stillman B (1998) DNA replication fork in eukaryotic cells. Annu Rev Biochem 67:721–751

Wallace SS (1998) Enzymatic processing of radiation-induced free radical damage in DNA. Radiat Res 150:S60–S79

Wang ZQ, Stingl L, Morrison C, Jantsch M, Los M, Schulze-Osthoff K, Wagner EF (1997) PARP is important for genomic stability but dispensable in apoptosis. Genes & Development 11:2347–2358

Waters TR, Gallinari P, Jiricny J, Swann PF (1999) Human thymine DNA glycosylase binds to apurinic sites in DNA but is displaced by human apurinic endonuclease 1. J Biol Chem 274:67–74

Webster ADB, Barnes DE, Arlett CF, Lehmann AR, Lindahl T (1992) Growth retardation and immunodeficiency in a patient with mutations in the DNA ligase I gene. Lancet 339:1508–1509

Wei YF, Robins P, Carter K, Caldecott K, Pappin DJ, Yu GL, Wang RP, Shell BK, Nash RA, Schar P et al. (1995) Molecular cloning and expression of human cDNAs encoding a novel DNA ligase IV and DNA ligase III, an enzyme active in DNA repair and recombination. Mol Cell Biol 15:3206–3216

Wiebauer K, Jiricny J (1990) Mismatch-specific thymine DNA glycosylase and DNA polymerase β mediate the correction of G-T mispairs in nuclear extracts from human cells. Proc Natl Acad Sci USA 87:5842–5845

Wiederhold L, Leppard JB, Kedar P, Karimi-Busheri F, Rasouli-Nia A, Weinfeld M, Tomkinson AE, Izumi T, Prasad R, Wilson SH et al. (2004) AP endonuclease-independent DNA base excision repair in human cells. Mol Cell 15:209–220

Wilson DM 3rd, Takeshita M, Grollman AP, Demple B (1995) Incision activity of human apurinic endonuclease (Ape) at abasic site analogs in DNA. J Biol Chem 270:16002–16007

Wilson DM, Barsky D (2001) The major human abasic endonuclease: formation, consequences and repair of abasic lesions in DNA. Mutat Res 485:283–307

Wilson TM, Rivkees SA, Deutsch WA, Kelley MR (1996) Differential expression of the apurinic/apyrimidinic endonuclease (APE/ref-1) multifunctional DNA base excision repair gene during fetal development and in adult rat brain and testis. Mutat Res 362:237–248

Wood RD, Mitchell M, Sgouros J, Lindahl T (2001) Human DNA repair genes. Science 291:1284–1289

Xanthoudakis S, Graham M, Wang F, Pan YE, Curran T (1992) Redox activation of Fos-Jun DNA binding activity is mediated by a DNA repair enzyme. EMBO J 11:3323–3335

Xanthoudakis S, Smeyne RJ, Wallace JD, Curran T (1996) The redox/DNA repair protein, Ref-1, is essential for early embryonic development in mice. Proc Natl Acad USA 93:8919–8923

Xie Y, Yang H, Cunanan C, Okamoto K, Shibata D, Pan J, Barnes DE, Lindahl T, McIlhatton M, Fishel R, Miller JH (2004) Deficiencies in mouse Myh and Ogg1 result in tumor predisposition and G to T mutations in codon 12 of the K-ras oncogene in lung tumors. Cancer Res 64:3096–3102

Yang HJ, Clendenin WM, Wong D, Demple B, Slupska MM, Chiang JH, Miller JH (2001) Enhanced activity of adenine-DNA glycosylase by apurinic/apyrimidinic endonuclease (APE1) in mammalian base excision repair of an A/GO mismatch. Nucleic Acids Res 29:743–752

Zhang QM, Dianov GL (2005) DNA repair fidelity of base excision repair pathways in human cell extracts. DNA Repair (Amst) 4:263–270

Zharkov DO, Rosenquist TA, Gerchman SE, Grollman AP (2000) Substrate specificity and reaction mechanism of murine 8-oxoguanine-DNA glycosylase. J Biol Chem 275:28607–28617

Zmudzka BZ, Fornace A, Collins J, Wilson SH (1988) Characterization of DNA polymerase β mRNA: cell-cycle and growth response in cultured human cells. Nucleic Acids Res 16:9587–9596

Genome Dyn Stab (1)
D.-H. Lankenau: Genome Integrity
DOI 10.1007/7050_002/Published online: 18 November 2005

Inter- and Supramolecular Interactions of Protein Kinase CK2 and their Relevance for Genome Integrity

Birgitte B. Olsen[1] · Karsten Niefind[2] · Olaf-Georg Issinger[1] (✉)

[1]Institute of Biochemistry and Molecular Biology, University of Southern Denmark, Campusvej 55, 5230 Odense M, Denmark
ogi@bmb.sdu.dk

[2]Institute of Biochemistry, University of Cologne, Zülpicher Strasse 47, 50674 Cologne, Germany

Abstract CK2 is a pleiotropic and ubiquitous tetrameric serine/threonine protein kinase phosphorylating a plethora of substrates. Beside its typical enzymatic activity, it has been shown that it can undergo stoichiometric (non-enzymatic) interactions with other proteins. This interaction is not only limited to the tetrameric holoenzyme; the individual subunits of CK2, i.e. the catalytic α-subunit and the regulatory β-subunit, can also interact by themselves with other proteins. This is an exciting new development in understanding the structural and physiological role of CK2. We have focused in this overview on this aspect of CK2 function, with emphasis on molecules involved in genome integrity, i.e. p53, MDM2, Chk2, TFIIIB, topoisomerases I/II, Wnt, BRCA1, and APC. Special attention is given to the structural basis within the CK2 molecule, allowing for the discussed interactions in cell lines. We present a model in which the CK2β subunit serves as a docking station for other proteins than only the CK2α subunit, especially here for molecules playing a key role in genome integrity. Our model supports the notion of CK2 and its subunits being possible partners in high molecular mass complexes with dedicated functions known as "osomes" (for details, see below).

1 Introduction

Protein-protein interactions become increasingly important in describing the complexity of cellular organization (Alberts 1998). Traditionally, organelles and especially the cytoplasm were considered as entities, where reaction partners could freely move around; yet this idea has lately been abandoned in favor of a more complex understanding of how enzymatic and structural interactions function. Macromolecular complexes coined as "nucleosome", "splicesosome", "ATP-synthasome", "repairosome", "apoptosome" etc. are only a few examples of the current development in this field.

Protein kinase CK2: besides being a "mere" protein phosphorylating protein kinase, CK2 also is an example for undergoing non-catalytic interactions with other proteins.

This development can best be followed when looking at earlier CK2 reviews which appeared in the beginning of the 1990s (Pinna 1990; Tuazon and Traugh 1991; Issinger 1993) where interactions of CK2 with other proteins were not even mentioned, in contrast to reviews that appeared at the end of the 1990s and within the current decade, where whole chapters or articles were dedicated to the increasing number of evidences of CK2 interactions (Allende and Allende 1998; Guerra and Issinger 1999; Olsten and Litchfield 2004; Bibby and Litchfield 2005). Important for this new aspect of CK2 function were numerous investigations involving the yeast-two-hybrid system which led to the discovery of a number of interaction partners, especially for the CK2β-subunit [for review, see Kusk et al. (1999)].

In the meantime, CK2 is accepted as a potential interaction partner for other proteins. However, since the CK2 holoenzyme, the catalytic α-subunit and the regulatory β-subunit have been shown to interact with other proteins, it is difficult to distinguish between stoichiometric interactions of the catalytically active CK2 entities and enzyme-substrate interactions, especially since stoichiometric interactions can also involve phosphorylation reactions. Given the fact that more than 300 substrates have been described for CK2 (Meggio and Pinna 2003), a general classification between pure enzyme-substrate complexes and stoichiometric complexes is not generally possible.

Table 1 Protein kinase CK2 interacting proteins involved in DNA repair

Interacting protein	$\alpha_2\beta_2/\alpha/\beta$	Method of detection				
		Co – Pu	IP	PK	Moa	Y2H
Dishevelled	$\alpha_2\beta_2$	X	X	X	–	–
β-catenin	$\alpha_2\beta_2$	–	X	X	–	–
DNA topoisomerase I	$\alpha_2\beta_2$	X	X	X	–	–
DNA topoisomerase II	$\alpha_2\beta_2/\beta$	X	X	X	–	–
TFIIIB	$\alpha_2\beta_2$	X	X	–	–	–
p53	$\alpha_2\beta_2/\beta$	X	X	–	–	–
$p21^{WAF1}$	β	–	–	–	X	–
FACT	$\alpha_2\beta_2$	X	X	X	–	–
Chk1	β	–	X	–	–	–
Chk2	β	–	X	–	–	–
APC	$\alpha_2\beta_2$	–	X	–	–	–
BRCA1	β	–	X	–	–	X

Co – Pu, co-purification; IP, immunoprecipitation; PK, kinase activity test; Moa, membrane overlay assay; Y2H, yeast two-hybrid system.

Even protein-protein interactions which have been shown to be specific for the regulatory CK2β suggest in many cases that association of the holoenzyme occurs via the CK2β subunit and the same holds true for CK2α.

Therefore, the listing of the described CK2 interactions partners (Table 1) can only be considered as a tentative approach in order to throw some light on the increasing amount of data concerning CK2 and its subunits as stoichiometric interacting partners.

The major focus of this review will be on proteins which have some function in genome integrity in a wider sense, and their established interaction with CK2 and its subunits.

2 Biochemical and Structural Background of Protein Kinase CK2 and its Subunits

2.1 CK2α

CK2α is encoded and expressed in all eukaryotic cells investigated so far (for recent reviews, see Ahmed et al. 2002; Pinna 2002; Litchfield 2003). It belongs to the extended superfamily of eukaryotic protein kinases (EPK) and among this to the family of CMGC kinases (Manning et al. 2002). This means its closest relatives are prominent EPKs such as the cyclin-dependent kinases (CDK), the mitogen-dependent kinases (MAPK) and glycogen-synthase-kinase 3 (GSK3) all of which are strictly controlled and only activated upon specific signals (Huse and Kuriyan 2002). Against the background of this relationship, it is particularly conspicuous that CK2α is catalytically active both alone and in complex with CK2β, and that this activity is not controlled by any of the other typical regulatory mechanisms known for EPKs such as phosphorylation or second-messenger binding.

The "constitutive" activity of CK2α has been corroborated structurally in recent years. So far, CK2α structures have been solved from two species, namely from maize (Niefind et al. 1998) and from humans, the latter alone (Ermakova et al. 2003) and in complex with CK2β (Niefind et al. 2001). All these structures showed CK2α in a conformation typical for active EPKs. In particular, the two key elements of EPK activation – the helix αC and the activation loop – whose structural plasticity is normally most important for kinase regulation (Huse and Kuriyan 2002), are fixed in their active conformations by close contact to the N-terminal segment not found in any other EPK so far.

This remarkable finding was further confirmed by several structures of maize CK2α in complex with various ATP-competitive inhibitors (De Moliner et al. 2003; Sarno et al. 2003). While many EPK inhibitors work by stabilization of inactive conformations of the enzymes (Noble et al. 2004), the so far known CK2α inhibitors do not induce large structural changes, in particular not in the regions of the helix αC and the activation segment. In other words: even in complex with specific inhibitors, CK2α does not adopt any inactive conformation as do other EPKs in comparable situations.

Taken together, all our current biochemical and structural knowledge about CK2α supports the view that the enzyme has been optimized during evolution in favor of activity. It does not need any interaction partners (not even CK2β) to be active. Regulatory mechanisms to control CK2 activity – if existing at all – can only work on the level of the CK2 holoenzyme.

2.2 CK2β

Sequence and structural comparisons have shown that CK2β – unlike CK2α – possesses nearly no relatives among other proteins. A zinc-stabilized, highly hydrophobic patch on its surface guarantees that CK2β exists only as a dimer (Chantalat et al. 1999) which can recruit two CK2α chains (or possibly other EPKs; see below) to form the CK2 holoenzyme. The effect of CK2β on the activity of CK2α is highly diverse and substrate-dependent. Depending on the particular case, CK2β can obviously support or disturb the substrate recruitment of CK2α by allosteric interactions, and it is certainly no on/off-switch, as in the case of the various cyclins and the corresponding Cdks.

The constitutively active nature of CK2α and the moderate impact of CK2β on this activity generate the notion that CK2β is no obligatory companion of CK2α. This impression is further supported by the observation that orthologous proteins in eukaryotes may exhibit different degrees of functional importance. As CK2α function is an absolute requirement for all eukaryotic cells, CK2β shares this vital function with CK2α only in mammals while in yeast, knocking out CK2β only leads to minor phenotypic changes (Glover 1998).

These and other observations have changed our picture of CK2β completely during the last decade. As indicated by the nomenclature, CK2β was classically regarded exclusively as a member of the CK2 holoenzyme complex. Yeast-two-hybrid studies, however, provided the first evidence that the protein might have other functions, e.g. serving as a binding partner for various EPKs such as A-Raf (Boldyreff and Issinger 1997) and Mos (Chen et al. 1997; Liebermann and Rudermann 2004).

In addition to these contacts, which are competitive to CK2α, numerous non-competitive interactions of CK2β have been described (for review, see Guerra and Issinger 1999). Prominent examples are crucial regulators of cell division such as p53 (Appel et al. 1995), $p21^{WAF1}$ (Götz et al. 1996), $p27^{KIP1}$ (Tapia et al. 2004), Chk1 (Guerra et al. 2003) and Chk2 (Bjørling-Poulsen

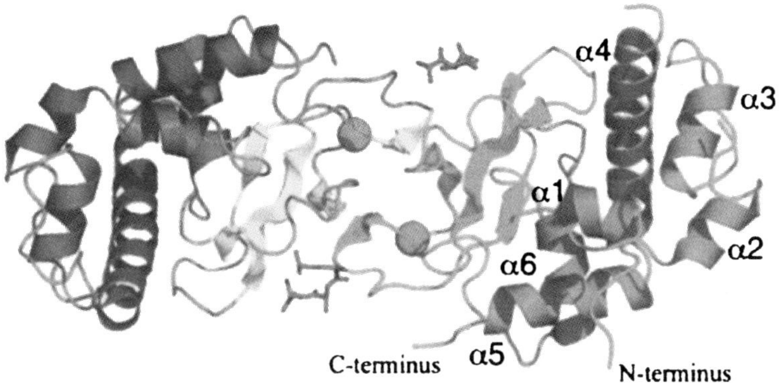

Fig. 1 Structure of CK2β^{1-178} crystallized in the presence of a peptide containing residues 46–65 of p21$^{\mathrm{WAF1}}$ (Bertrand et al. 2004, Protein Data Bank code 1RQF). The figure was reproduced from Bertrand et al. (2004) with kind permission of the International Union of Crystallography. It shows a dimer of two CK2β^{1-178} chains (*red/yellow and green*) each containing an α-helical N-terminal domain and a C-terminal dimerization domain with a central three-stranded antiparallel β-sheet and a binding site for a zinc ion (*green balls*). After fitting of the CK2β^{1-178} subunits pieces of residual electron density remained close to the dimerization interface. The authors modelled polyalanine peptides of eight residues length (*blue and purple*) into them and present evidence that these electron density zones are due to the p21$^{\mathrm{WAF1}}$ peptide in the crystallization solution

et al. 2005b). The general tendency of the CK2β dimer to interact with other proteins has been assigned to its non-compact shape and the pronounced negatively charged patches on its surface (Chantalat et al. 1999), where the "acidic loop" is the critical determinant for the ability of CK2 holoenzyme to form supra-molecular complexes under low-salt conditions (Niefind and Issinger 2005). On a structural level, only the non-competitive interaction of CK2β with p21$^{\mathrm{WAF1}}$ (Bertrand et al. 2004) has been partially characterized (Fig. 1).

Taken together, the speculation may be allowed that the primary role of CK2β is not controlling or even modulating the catalytic activity in a direct way but rather anchoring and networking CK2α and some other homologous protein kinases with substrates and other cellular components. It is possible that the increasing networking capability of CK2β during evolution made it essential for viability in higher eukaryotes (Buchou et al. 2003).

2.3
CK2 Holoenzyme

The last-mentioned conclusions are strongly supported by the structure of the human CK2 holoenzyme (Niefind et al. 2001) in which the CK2β dimer is found relatively far away from the active sites of the two associated CK2α

subunits. The most conspicuous feature of this structure is the small interface between CK2α and CK2β which has two consequences. First more than 92% of the surface of the CK2β dimer is still accessible for interaction partners, so that a major part of the interaction potential of CK2β can be transferred to the CK2 holoenzyme. In particular the negatively charged patches mentioned above are completely unaffected by CK2α. Second, the small size of the contact surface is consistent with a transient rather than a permanent complex which makes a dynamic equilibrium between the holoenzyme and its subunits conceivable. As a consequence, the aforementioned CK2α-competitive interactions of CK2β with other EPKs might be more important under physiological conditions than expected before. This is especially evident because on the side of the CK2α, the interface is formed by a part of its structure, which is highly conserved among EPKs, i.e. by the outer surface of the central β-sheet of the N-terminal domain.

In the architecture of the CK2 holoenzyme, the two catalytic subunits are far away from each other and bridged by the CK2β dimer. This overall organisation suggests that the binding of the two CK2α chains occurs independently from each other which led us speculate that CK2β might form mixed complexes containing two different EPK chains. In the case of CK2α and its human isoform CK2α' mixed holoenzymes with $\alpha\alpha'\beta_2$-stoichiometry have indeed been found (Chester et al. 1995).

3 Interactions with p53

3.1 p53-CK2$\alpha_2\beta_2$ and CK2β-subunit

Meek et al. (1990) were the first to show that p53 is phosphorylated by CK2 at the last serine of the C-terminal end of the molecule. This phosphorylation leads to an increase in sequence-specific DNA binding of p53, at least in vitro (Hupp et al. 1992).

Beside the CK2-catalyzed phosphorylation of p53, structural interactions between p53 and the regulatory CK2β subunit, but not with the catalytic CK2α-subunit have been described (Appel et al. 1995). Moreover, it was shown that the free CK2β subunit inhibits the DNA binding activity (Prowald et al. 1997) and the transactivation function of p53 (Schuster et al. 1999). In addition to these regulatory effects of free CK2β subunit on p53 activity, it was shown that the C-terminal fragment $p53^{264-393}$ stimulated CK2 holoenzyme activity using MDM2 as a substrate (Guerra et al. 1997). Such stimulation was not observed when the catalytic CK2α subunit alone was used to phosphorylate MDM2 in the presence of the p53 fragment, supporting the notion that the regulatory CK2β subunit within the tetrameric CK2 holoenzyme complex has an important function in mediating the stimulatory effect of the C-terminal p53 fragment.

In contrast, wild type full length p53 was shown to inhibit CK2 holoenzyme-catalyzed phosphorylation of MDM2 (Schuster et al. 2001). The essential binding site of CK2β was mapped to amino acids 330–339 of the p53 molecule (Götz et al. 1999). Taken together these observations suggest that apart from the enzyme-substrate interaction, there is also a stoichiometric type of interaction between CK2 and p53 which can influence the interaction of p53 with other proteins and substrates.

One of the central functions of p53 is to cause cell cycle arrest in response to DNA damage so that the damage can be repaired prior to the next round of replication. The molecular mechanism responsible for this cellular response involves the transcriptional induction of the p21^{WAF1} gene. p21^{WAF1}is a potent inhibitor of cyclin-dependent protein kinases and causes a pause in the G1 to S (and G2 to M) transition of the cell cycle. Götz et al. (1996) showed that p21^{WAF1} binds to the regulatory β-subunit of protein kinase CK2, causing a downregulation of kinase activity. Moreover, it was shown that p21^{WAF1} interaction with CK2β abrogates CK2-catalyzed phosphorylation of MDM2 (Guerra et al. 1997). Beside the well established interaction of p21^{WAF1} with the regulatory CK2β subunit, protein kinase CK2 can phosphorylate p21^{WAF1} (Romero-Oliva and Allende 2001), thereby making p21^{WAF1} a competitor for CK2-specific substrates. This competition is not related to the binding of p21^{WAF1} to CK2β, since competition is also seen with the catalytic CK2α subunit alone. Hence, p21^{WAF1} is another example, where a protein can serve as an interaction partner for CK2 and by the same token can be used as a substrate. The p21^{WAF1}/CK2β interaction is so far the only example for an interaction of a foreign protein with a CK2 subunit for which 3D-structural information exists. A small peptide derived from p21^{WAF1} (residues 46–65) was co-crystallized with a truncated form of CK2β (Bertrand et al. 2004). The resulting structure – though of limited quality – shows parts of the p21^{WAF1} peptide next to the dimerization interface of the CK2β dimer (Fig. 1). The p21^{WAF1}/CK2β contact region found in the structure was on the p21^{WAF1} side and is consistent with previous observations by Götz et al. (1996). On the CK2β side, it clarified contradictory reports about the location of the p21^{WAF1} binding region. It remains open how p21^{WAF1} manages to down-regulate the activity of CK2α within the CK2 holoenzyme.

3.2 p53-Chk2-CK2β-Interaction

Protein kinase Chk2 functions as a tumor suppressor protein by transducing DNA damage signals to cell cycle checkpoints. Among its established substrates are key components of cell homeostasis like Cdc25A, Cdc25C, BRCA1 and p53. In the case of p53, Chk2 catalyzes the phosphorylation of Thr18 and

Ser20. As a consequence, p53 is stabilized and improved in its ability to upregulate the expression of proteins involved in apoptosis, DNA repair and cell cycle control (for reviews, see Bartek et al. 2001; Bartek and Lukas 2003). An enigmatic point in this scenario has been the fact that p53 lacks the established consensus sequence for Chk2 phosphorylation, which is LXRXXS/T (O'Neill et al. 2002; Seo et al. 2003) arising the problem how then does p53 become a Chk2 substrate.

The first to address this question were Craig et al. (2003). These authors showed that p53 – within its DNA-binding domain – contains two additional docking sites for Chk2 binding and recognition, named Box-II (^{117}GTAKSVTCTYSPALNK^{132}) and Box-V (^{272}VRVCACPGRDRRTEEEN^{288}) Fig. 2. In the p53 sequence Box-II and Box-V are far away from Thr18 and Ser20 (located in Box-I (^{12}PPLSQETFSDLWKLL^{26})) and also on the 3D-structure level they are probably (no 3D-structure of the complete p53 molecule is known to date) located at considerable distance from these phosphorylation sites. This type of substrate recruitment via remote docking sites is well known from other protein kinases such as Cdk2 (Luciani et al. 2000), Pdk1, MAPK and GSK3 (Biondi and Nebreda 2003). It has been called an "allosteric effect" (Craig et al. 2003) because it resembles the activation of an enzyme by an allosteric effector and may have widespread significance for Chk2 function.

Based on results from Götz et al. (1999) and Schuster et al. (2001) who had shown previously interaction of p53 with CK2β (see previous chapter) Bjørling-Poulsen et al. (2005b) investigated the influence of CK2β on Chk2. These authors demonstrated that Chk2 specifically interacts with CK2β in vitro and in cultured cells, and that activation of Chk2 leads to a reduction of this interaction. Additionally, they showed that the presence of the CK2β-subunit significantly reduces the Chk2-catalysed phosphorylation of p53 in vitro. These findings support the notion that CK2β can act as a general modulator of remote docking sites in protein kinase-substrate interactions. Hence, Chk2 is therefore another example just like several other protein kinases which have been shown to interact with CK2β (e.g. A-Raf, Mos, Chk1, p90^{RSK}).

Thus, CK2β seems to have a dual function in inhibiting p53 activity. First, CK2β interacts with and directly inhibits binding of p53 to DNA (Prowald et al. 1997), and second the interaction of CK2β with p53 (and possibly also Chk2) inhibits Chk2-mediated phosphorylation and activation of p53. However, it still remains to be investigated, whether the interaction of p53 with CK2β is of physiological relevance.

As shown in Fig. 2, the interaction partners Chk2, p53 and CK2β are covered by electrostatic surfaces. The Box-II and Box-V regions of p53, which are known to function as docking sites for Chk2, are distinctly positive. A suitable counterpart on Chk2 is a stretch around Glu144, which is conspicuously negatively charged, and which is not too far away from the active site. If this notion is true, it is plausible to assume that the CK2β dimer might disturb

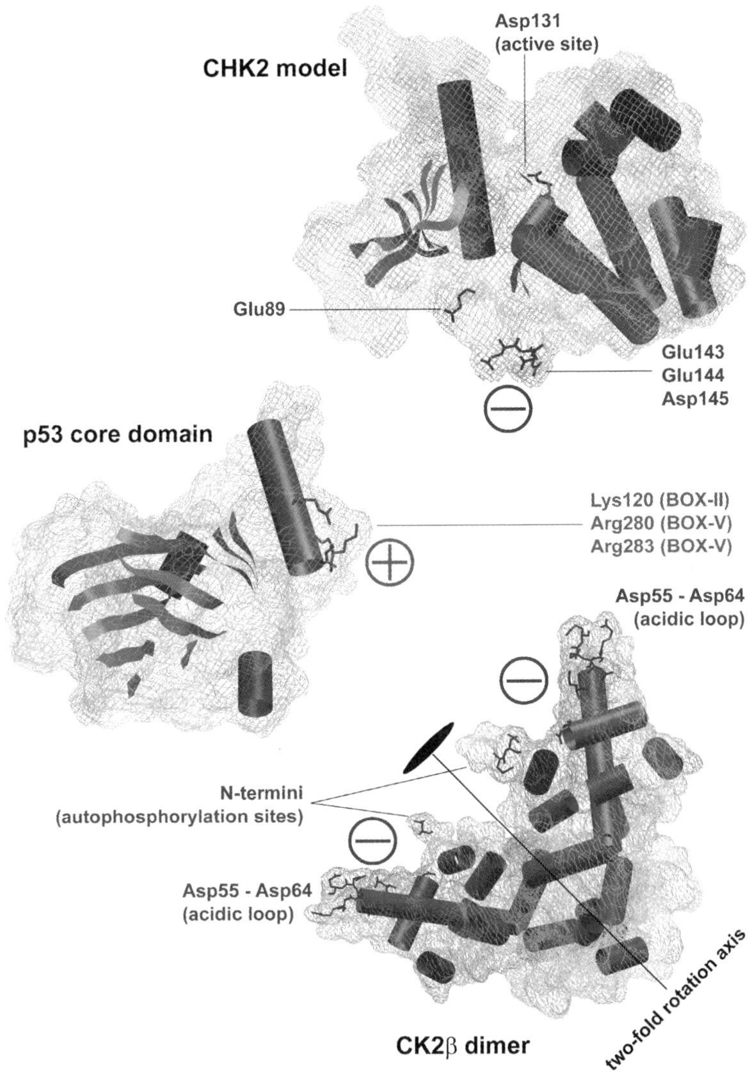

Fig. 2 Model of Chk2, p53 and CK2β. The interaction partners p53, Chk2 and CK2β are covered by electrostatic surfaces. The *red color* represents positively charged patches, and negatively charged zones are in *blue*. The figure was drawn with BRAGI (Schomburg and Reichelt 1988). The following PDB files were used: 1GZH for the p53 core domain (Derbyshire et al. 2002) and 1JWH for the CK2β dimer (Niefind et al. 2001). The Chk2 structure is a homology-based model obtained from the 3D-JIGSAW comparative modeling server (Bates et al. 2001)

the p53/Chk2 interaction with its distinct negative patches around the "acidic loops" and the two N-termini whose negative charges are typically enhanced by autophosphorylation.

3.3
CK2β – a General Provider/Modulator of Remote Docking Sites in Protein Kinase-Substrate Interactions?

The specificity of a protein kinase is determined considerably by the sequence of the protein substrate next to the phospho-acceptor site, which is recognized by the active site of the corresponding protein kinase. However, the exclusive role of this "consensus sequence" for protein kinase-substrate recognition has been questioned in recent years. It is well established now that interactions via "remote docking sites" play a considerable role in this context (Biondi and Nebreda 2003). They may be the basis for more complex signaling and regulatory networks in the cell. Furthermore, the manipulation of such functionally essential protein-protein interactions allows the development of a new generation of sophisticated drugs (Arkin and Wells 2004).

Remote docking sites are not necessarily located on the kinase molecule itself, but also on regulatory or anchoring interaction partners. In the case of Cdk2 for instance – one of the closest relatives of CK2α – such a remote site for substrate recognition has been found (Schulman et al. 1998) and structurally characterized (Lowe et al. 2002) on the associated cyclin A molecule. Similar to cyclin A, CK2β may provide positive and negative determinants for the recruitment of substrates for CK2α or alternative kinases (Mos, A-Raf, Chk1, Chk2) with which it can associate. Consistent with such a role for CK2β is the observation that it generally activates CK2α as long as short peptides become phosphorylated (Boldyreff et al. 1993). With macromolecular substrates, however, the effect of CK2β on CK2α strongly depends on the particular substrate proteins ranging from strongly activating in the case of casein (Grankowski et al. 1991) to strongly inhibitory in the case of calmodulin (Meggio et al. 1992). In other words, with macromolecular substrates such remote determinants come into play. The down-regulating effect of CK2β on the p53 phosphorylation by Chk2 (Bjørling-Poulsen et al. 2005b) is the most recent example of this CK2β property.

In summary, a role for CK2β as a general provider of docking sites is emerging. To be physiologically important, this requires a cellular pool of CK2β dimers not irreversibly associated with CK2α. The existence of such a pool has become very likely following the determination of the CK2 holoenzyme structure (Niefind et al. 2001) and other properties observed with free CK2 subunits (Niefind et al. 2001; Stigare et al. 1993; Goldberg 1999; Filhol et al. 2003).

3.4
Interactions of CK2 with the FACT Complex and its Consequences for the Phosphorylation of p53

A complicated, CK2-involving interplay between phosphorylation and non-substrate interaction has been discovered during the study of the cellular response to UV irradiation. A key component in this context is the tumor suppressor protein p53, a transcription factor of RNA polymerase II. At p53 several stress signals converge (Vogelstein et al. 2000). Normally p53 has a short lifetime in the cell and occurs at low concentrations. Under stress conditions, however, it is accumulated in the nucleus and activated by various phosphorylations and other covalent modifications. As a consequence, the expression of more than 150 genes is induced or repressed (Appella and Anderson 2001).

One of these p53 phosphorylations which occurs after UV stress-induced DNA damage (Kapoor and Lozano 1998; Lu et al. 1998) and which enhances the sequence-specific DNA-binding capability of p53 concerns a highly conserved serine residue close to the C-terminus (Ser392 in man, Ser389 in mouse). While already Meek et al. (1990) had shown that CK2 can catalyze the phosphorylation of p53 at this serine, only Keller et al. (2001) demonstrated its relevance for genome integrity: these authors isolated and characterized a p53/Ser392 kinase from cell cultures after UV stress. Surprisingly, the kinase was a high-molecular complex of CK2α, CK2α', CK2β and the so-called FACT heterodimer (after "**fa**cilitates **c**hromatin-mediated **t**ranscription") composed of the "structure-specific recognition protein-1" (SSRP1) and hSPT16, which is the human ortholog of yeast suppressor of Ty insertion mutations. The integration of CK2 into this complex has serious consequences for its substrate specificity, since as a member of the p53/Ser392 kinase complex, CK2 is no longer able to catalyze the phosphorylation of casein and other established substrates and strongly prefers p53 as a substrate (Keller et al. 2001).

In a subsequent paper, Keller and Lu (2002) mapped the interaction domains between the subunits of the p53/Ser392 kinase complex and presented a model reproduced in Fig. 3. Both CK2α and CK2β interact with the FACT components but – like in the CK2 holoenzyme – the CK2β dimer is still the core of the whole assembly.

Keller and Lu (2002) furthermore demonstrated that UV irradiation induces the association of the p53/Ser392 kinase complex, yet the detailed

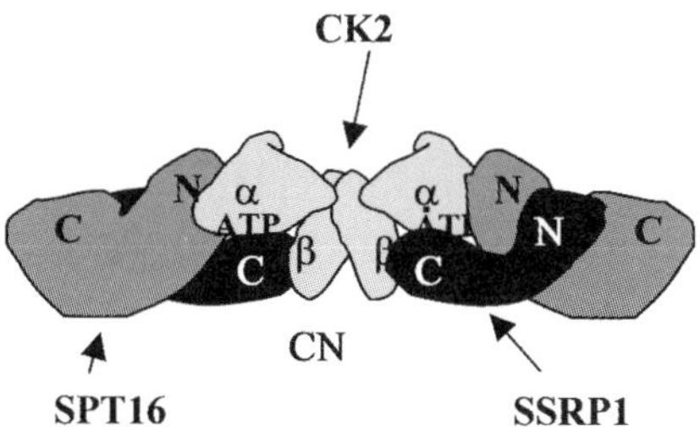

Fig. 3 Model of the CK2/FACT complex published by Keller and Lu (2002). The CK2 holoenzyme associates two FACT heterodimers each of which is composed of a SPT16 and a SSRP1 subunit. The figure is adopted from Keller and Lu (2002) with kind permission of the American Society for Biochemistry and Molecular Biology

mechanism of this process is so far unclear. One idea is that the phosphorylation of SSRP1 catalyzed by CK2 plays a role in this context. In fact SSRP1 is a CK2 substrate both in vitro (Keller et al. 2001) and in vivo (Li et al. 2005) whereby the in vivo phosphorylation occurs in response to UV irradiation. It is a plausible – though unproven – hypothesis that this SSRP1 phosphorylation might enhance the formation of the CK2/FACT complex suggesting that CK2 – via phosphorylation of its interaction partner – would catalyze its own integration into a functional complex.

3.5 p53-Chk2-HDM2-Interaction

HDM2 is a 491 amino acid long phosphoprotein that interacts through its N-terminal domain with an α-helix present in the N-terminal transactivation domain of p53. This binding to the N-terminal transactivation domain of p53 blocks its transcriptional activity directly. More importantly, HDM2 functions as the E3 ligase that ubiquitinates p53 for proteasome degradation (Honda et al. 1997).

HDM2, which is involved in the control of p53, undergoes dynamic conformational changes upon ligand binding (Schon et al. 2002), and it was also shown that the phosphorylation status of p53 modulates its affinity for HDM2 and other proteins such as p300. HDM2 binds to the Box-I domain located at the N-terminal part of the p53 molecule, where at least three different protein kinase phosphorylation sites, i.e. Ser15, Thr18 and Ser20, are located.

Protein kinase Chk2 interacts with the Box-II and Box-V domains of p53 and phosphorylates Thr18 and Ser20 in Box-I (see Sect. 3.2). Therefore one could expect an interference of p53 phosphorylation with HDM2 binding, and likewise the presence of HDM2 would be expected to affect protein kinases targeting amino acid residues in the Box-I domain, e.g. ATM, Chk1, Chk2, and CK1. Bjørling-Poulsen et al. (2005a) showed that Chk2 and HDM2 can physically interact, and that HDM2 can serve as a weak substrate for Chk2. In the presence of HDM2, Chk2-mediated phosphorylation of p53 is severely reduced, whereas HDM2 phosphorylation is increased by a factor of 2 in the presence of p53. Thus, the classical, well-established p53-MDM2 interaction becomes more complicated in the light of the many interaction partners that have been described for p53, and it may turn out that higher molecular mass complexes, involving p53, may be found at various stages of normal but also at pathological cellular states.

4 Transcription Factor TFIIIB

RNA polymerase III (Pol III) is responsible for about 10% of all nuclear transcription, including the synthesis of tRNA and 5S RNA. Transcription factor

TFIIIB is a complex composed of the TATA-binding protein (TBP) and two associated subunits, Bdp1 and Brf. It plays an essential role in Pol III transcription, by recruiting the polymerase and positioning it over the initiation site (Kassavetis et al. 1990).

From initial experiments in yeast, Ghavidel and Schultz (2001) concluded that CK2 regulates Pol III activity by phosphorylating the TBP subunit of TFIIIB. Whereas most stress signaling kinases are normally switched off in the absence of the appropriate stimulus, TBP-associated CK2 is most active in the absence of signaling from upstream components of the DNA damage pathway, and its regulation in response to the appropriate stress signal occurs by a negative regulatory mechanism. This system of signaling ensures constitutive activation of TFIIIB under low genotoxic stress.

It was speculated that one explanation for the repression of TBP-associated CK2 activity in response to DNA damage is due to dissociation of α/α' from the TBP-CK2 complex, implying that the regulatory CK2β-subunit remains bound (Ghavidel and Schulz 2001). Although all available evidences do not support such an event to occur, we cannot exclude this possibility to happen in yeast, because there are indications that the association between the CK2 subunits α and β in yeast and in plants are not as tight as for example in mammals. Maldonado and Allende (1999) used CK2 holoenzyme but also recombinant *Xenopus* CK2α and GST-β-subunits for in vitro phosphorylation of TBP to demonstrate that CK2β can stimulate CK2α activity, as described originally for the human recombinant CK2 subunits by Grankowski et al. (1991). Indeed, these authors could show that phosphorylation of yeast TBP by CK2 reduces its specific binding to DNA and that the CK2α-catalyzed reaction was stimulated by addition of CK2β. In yeast TBP, Ser128 is phosphorylated by CK2 (Ghavidel and Schulz 2001). Although a corresponding serine is found in human TBP, its surrounding residues do not match the CK2 consensus (Ghavidel et al. 1999). This may explain why human TBP is not a substrate for CK2.

Johnston et al. (2002) demonstrated that CK2 is also required for active mammalian Pol III transcription. Inhibiting human CK2 compromises the binding of TFIIIB to the assembly factor TFIIIC2, an interaction that is necessary to bring TFIIIB onto most Pol III templates.

Applying CK2 specific inhibitors, Johnston et al. (2002) showed that the CK2-TFIIIB interaction in mammalian cells is not dependent on the kinase activity of CK2. Moreover, they showed, that transcription by Pol III in human cell extracts can be ablated specifically by a peptide inhibitor of CK2, an effect reversed by recombinant CK2, suggesting that transcription by mammalian Pol III is highly dependent on CK2. In a recent paper on the role of CK2 phosphorylation (Hu et al. 2004), it was shown that CK2 inhibits Brf2-TFIIIB by specifically phosphorylating its Bdp1 component. Bdp1 is phosphorylated by CK2 during mitosis, and this is accompanied by Bdp1 dissociation from the U6 promoter and from chromatin in general and by transcription repression.

Whereas inhibition of CK2 in mitotic extracts restores Pol III transcription, inhibition of CK2 in active S phase extracts debilitates transcription. Thus, CK2 is directed to phosphorylate different targets within the basal Pol III transcription machinery at different times during the cell cycle with opposite transcriptional effects (Hu et al. 2004). Hence, the role of CK2 in mammalian cells may be different from what has been found in yeast (Ghavidel and Schulz 2001).

5
Topoisomerase I

DNA topoisomerase I (topo I) is an abundant nuclear enzyme mainly responsible for relaxation of DNA supercoils generated by transcription and replication (Wang 1996). Topo I catalyzes DNA relaxation by generating transient single-stranded breaks which then are religated. Numerous proteins have been reported to activate the relaxation catalyzed by topo I in vitro, e.g. HMG, H1 histone, p53, ARF protein, poly (ADP)-ribose) polymerase and protein kinase CK2.

Kaiserman et al. (1988) showed that CK2 co-purified over several purification steps with topo I. Furthermore, they demonstrated that phosphorylation of topo I by endogenous CK2 led to a several-fold increase in catalytic activity. Kordyiak et al. (1994) reported that CK2 stimulates topo I activity by more than one order of magnitude in the absence of ATP, although ATP resulted in additional stimulation. These authors concluded that modulation of the catalytic activity of topo I by direct physical association (protein-protein interactions) must therefore be considered, in addition to phosphorylation when assessing the physiological role of CK2 during regulation of DNA topo I activity in vivo. In a paper that appeared ca. 10 years later, dealing with activation of topo I by CK2, Kowalska-Loth et al. (2003) showed that CK2α but not CK2β activated topo I to a similar extent as the CK2 holoenzyme. Thus, two phenomena have to be considered in the regulation of topo I by CK2: (i) a physical interaction between both proteins (either CK2α or CK2 holoenzyme) and (ii) phosphorylation of topo I at sites accessible to CK2 (either CK2α or CK2 holoenzyme). So far, experiments involving a kinase-dead mutant have not been performed in order to further explore the contributions of physical interaction and phosphorylation of topo I activation.

6
Topoisomerase II

DNA topoisomerase II (topo II) is an enzyme that regulates the topological states of DNA by transient breakage and rejoining of double-stranded DNA

(Wang 1985). Topo II exists as a phosphoprotein in many different species. Whereas phosphorylation of topo II was observed in foremost proliferating cells, very low levels of topo II phosphorylation were found in quiescent cells.

Bojanowski et al. (1993) showed that DNA topo II associates with protein kinase CK2 in yeast to form a molecular complex that remains stable during immunoprecipitation. However, the isolated CK2α-subunit by itself was unable to form a stable complex with topo II, suggesting that the presence of the CK2β-subunit is required for the interaction. This observation clearly indicated that the stable topo II-CK2 association is not governed by a simple enzyme-substrate interaction and is independent of the phosphorylation reaction as such. In human cells, it was shown that phosphorylation of topoisomerase II by CK2 had no effect on the catalytic activity of topo II, supporting the notion that the catalytic activity of topo II is not regulated by phosphorylation, whether mediated by CK2 or by other kinases (Escargueil et al. 2000). Like the yeast enzyme, human topo II also forms stable molecular complexes with CK2 which are not affected by the phosphorylation state of the topoisomerase.

Studies involving the yeast topo II indicate that the central domain of CK2β (amino acids 51–110) is interacting with the C-terminal domain of topo II (Leroy et al. 1999).

The role of protein kinase CK2 interaction in a non-enzymatic way seems to be important for activation of topo II, however, the CK2 subunits involved in this interaction may play different roles, although a final answer regarding how CK2 and its subunits activate topo II by non-enzymatic interactions can only be derived after structure elucidation of binary complexes between topo II-CK2 and its subunits.

7
BRCA1

The known functions of BRCA1 include DNA repair, cell-cycle checkpoint control, protein ubiquitylation and chromatin remodeling and might underline its role in carcinogenesis (for review, see Narod and Foulkes 2004). As well as being involved in the repair of double-strand breaks, BRCA1 has also been implicated in nucleotide-excision repair. This involves two different mechanisms – transcription-coupled repair (TCR), in which the transcribed strand is preferentially repaired and global genome repair (GGR), which does not show strand bias. BRCA1 might have a role in both TCR and GGR. BRCA1 consists of 1863 amino acids, containing several important functional domains, i.e. RING finger, DNA binding and SQ sequences (clusters of serine threonine sequences), known as SQ-cluster domains (SCDs) that interact with a range of proteins such as ATM, Chk2 and Cdk2. Two regions at the carboxyl terminus – known as the BRCT domains – each 110 amino acids long comprise amino acids 1646–1859. BRCT domains are found in many proteins involved in DNA repair pathways, and bind to many proteins, including RNA polymerase II, p300, BACH1, histone deacetylases 1 and 2, p53, and RB, just to mention a few of them.

Through a yeast-two-hybrid screen the CK2β-subunit was identified to associate with a C-terminal region of BRCA1 (O'Brien et al. 1999). Furthermore it was shown that BRCA1 was phosphorylated by CK2 in vitro at Ser1572. The interesting aspect of these investigations is that the observed association was much weaker with the same fragment bearing a missense mutation (Met1775Arg) that has been identified in breast tumors. Since BRCA1 is a large protein with a molecular mass of ca. 220 kDa the extreme ends of the protein were among the first major target sites used to explore the functions of this protein. The C-terminus of BRCA1 has been implicated to play a role in growth control, in the cellular response to DNA damage, as well as in transcriptional regulation. Some of these processes are associated with changes in the phosphorylation status of BRCA1. In particular, increases in total phosphorylation have been noted both during cell cycle progression and following episodes of DNA damage (Scully et al. 1997; Thomas et al. 1997). Thus, regulation by phosphorylation may be an important determinant for BRCA1 activity. The CK2β interaction site is located at the extreme carboxy-terminus (amino acids 1712–1863), yet the phosphorylation site is located at Ser1571. Hence, one may postulate that the CK2β binding site may act as a docking site to place the holoenzyme into an active complex.

Based on the BRCT domain structure (Williams and Glover 2003), Met1775 is located in a hydrophobic cleft, which suggests that the CK2β-subunit indeed can interact but only via its CK2α interaction loop and not with the acidic loop (otherwise one would expect an increase in BRCA1/CK2β interaction when methionine is mutated to a basic arginine residue). The effect of a CK2β dimer by itself on BRCA1 function cannot be assessed at the moment. If one assumes that it is the CK2 holoenzyme (CK2$\alpha_2\beta_2$) that binds to the BRCT domain encompassing Met1775 via its CK2β subunits, a phosphorylation of Ser1572 can be excluded because of steric hindrance of the CK2α competitive interaction of CK2β with BRCT. The only means to achieve phosphorylation of Ser1572, when considering the Met1775 area as a docking site, would be to postulate a "hemi-holoenzyme" consisting of a CK2β dimer and one single CK2α molecule (CK2$\alpha\beta_2$).

Since several proteins (including protein kinases ATM, Chk2, Chk1, Plk1 and Cdk2) bind to the central region of BRCA1, especially to SQ-cluster domains (SCDs) located between amino acids 1280–1524 adjacent to the CK2 phosphorylation site at Ser1572 such an interaction may interfere with the binding and phosphorylation of the other BRCA1-specific kinases.

8 Wnt signaling

Wnts are highly conserved small secreted glycoproteins and so far 19 members of the Wnt family are known. Wnts act as ligands for the Frizzled receptor (Pruniers et al. 2004; Ilyas 2005).

The Wnt signaling pathway is crucial for both differentiation and proliferation of embryonic and adult tissues, and disruption of the pathway leads to cancer (Nelson and Nusse 2004). Three Wnt signaling pathways have been described. Here, we will focus on the canonical pathway leading to stabilization and transcriptional activity of β-catenin.

In both *Drosophila* and murine cells, activation of the Wnt pathway leads to phosphorylation of Dishevelled (Dvl), the most proximal cytoplasmic molecule known in the Wnt pathway and it is activated upon binding of Wnt to its receptor (Yanagawa et al. 1995; Lee et al. 1999). However, its precise function remains to be elucidated as does the mechanism behind Dvl activation. In *Drosophila* the kinase activity associated with Dvl was found to be CK2$\alpha_2\beta_2$, which also phosphorylates Dvl in vitro. Phosphopeptide mappings have indicated that CK2$\alpha_2\beta_2$ also phosphorylates Dvl in vivo (Willert et al. 1997).

These results from *Drosophila* show CK2$\alpha_2\beta_2$ associating with its substrate. The phosphorylation site in Dvl remains to be identified as does the physiological relevance of this phosphorylation. It is possible that phosphorylation leads to a conformational change, promoting the association between Dvl and CK2$\alpha_2\beta_2$, since CK2$\alpha_2\beta_2$ stronger associates with phosphorylated Dvl. Phosphorylation of Dvl is not sufficient for transduction of the Wnt signal; however, it is possible that the phosphorylation is a consequence of the Wnt signal (Willert et al. 1997).

In mouse mammary epithelial cells, CK2 is upregulated by Wnt and Dvl is found in a complex with β-catenin and CK2α. In this complex β-catenin is phosphorylated by CK2α. Inhibition of CK2 activity accelerates the degradation of β-catenin and Dvl proteins and causes cell cycle arrest in G2/M (Song et al. 2000). Hence, CK2 can be considered an important regulator of the Wnt signaling pathway in mammalian cells.

However, in these experiments Willert et al. (1997) only tested the presence of the catalytic CK2α, so whether Dvl interacts with only CK2α, CK2$\alpha_2\beta_2$ or CK2β is not known. It also cannot be distinguished whether Dvl and CK2α directly interact, or whether it is through the interaction with β-catenin.

β-catenin is a multifunctional protein with a structural role in cellular junctions. β-catenin binds cadherins and links these to the cytoskeleton (Aberle et al. 1994; Rimm et al. 1995). β-catenin also has a critical role in Wnt signaling, where it, upon activation of Wnt signaling, binds to Tcf-LEF-1 transcription factors; this complex translocates to the nucleus, binds to target DNA and activates gene expression (Huber et al. 1996; Molenaar et al. 1996). Hence, β-catenin can also act as a transcriptional co-activator.

Once formed, β-catenin is recruited to a multiprotein complex, where it is stabilized by GSK3β-mediated phosphorylation of APC and axin. Within the complex, GSK3β also phosphorylates the N-terminal part of β-catenin. However, this phosphorylation requires priming phosphorylation by CK1ε at Ser45, followed by sequential phosphorylation by GSK3β of Thr41, Ser37 and Ser33 (Hagen and Vidal-Puig 2002; van Noort and Clevers 2002).

Pharmacological and molecular inhibition of the catalytic activity of CK2 leads to accelerated proteasomal degradation of β-catenin and reduced transcriptional activity of Tcf (Song et al. 2000, 2003). Thr393 was shown to be a target for CK2 holoenzyme in vitro and substitution of Thr393Ala increased the turnover of β-catenin both in transfected mammalian cells and in *Xenopus* embryos (Song et al. 2003; Dominguez et al. 2004). Hence, phosphorylation of β-catenin by CK2 has a positively regulating role concerning stability; however, from these experiments, it cannot be ruled out that other CK2 phosphorylation sites may exist in β-catenin.

In immunoprecipitates from mammalian epithelial cell lysates CK2α was found together with β-catenin. Moreover, β-catenin is phosphorylated in vitro by CK2 holoenzyme. Mutation of Ser29, Thr102 and Thr112 inhibited the phosphorylation by CK2 in vitro. In vivo degradation of the triple mutant was delayed and mutant β-catenin was more stable than the wildtype β-catenin. In vitro experiments showed that CK2 and GSK3β phosphorylate wildtype β-catenin synergistically, and pre-phosphorylation by CK2 enhances the binding of GSK3β and β-catenin (Bek and Kemler 2002).

The stoichiometrical analyses from Song et al. (2003) indicated that two sites were phosphorylated by CK2, and identified Thr393 as the major phosphorylation site. Bek and Kemler (2002) do not show the results from the single mutants but claim a major reduction in phosphorylation. Therefore, it is tempting to speculate that not all three sites are phosphorylation sites but possibly only one of these, and actually the CK2 site at Ser29 flanks the reported GSK3β sites and indeed less phosphorylation was observed when Ser29 was mutated.

According to a model by Song et al. (2000), two cytoplasmic complexes in Wnt signaling exist in cells; one consists of β-catenin, Dvl, and CK2 having a positive regulatory role in enhancing β-catenin half-life. The other complex consists of β-catenin, GSK3β, axin, Frat, Dvl, APC leading to degradation of β-catenin.

9
Adenomatous Polyposis Protein (APC)

APC is a tumor suppressor protein with a molecular mass of ca. 312 kDa containing several functional domains. The N-terminal region contains an oligomerization domain (Su et al. 1993) with several repeats characteristic for coiled structures important for certain protein-protein interactions (Cohen and Parry 1994).

APC interacts with a large number of proteins and it has several functions within the cell; one is to regulate cellular levels of β-catenin and thereby act as a negative regulator of Wnt signaling (Fodde et al. 2001; Kaplan et al. 2001). The role of APC in β-catenin degradation is believed to be scaffolding, thereby facilitating the phosphorylation of β-catenin by GSK3β. APC also exports β-catenin from the nucleus (Henderson 2000; Rosin-Arbesfeld et al. 2000).

The C-terminal part of APC contains a stretch of approximately 200 amino acids with a high content of lysines and arginines, along with an abnormal high percentage of prolines. This domain has been known as the basic domain (Groden et al. 1991; Polakis 1997) and does not show significant homology to other known proteins; however, the existence of basic stretches interspersed with prolines is common for proteins associated with microtubules, e.g. microtubule-associated protein (MAP) and Tau (Lewis et al. 1988; Aizawa et al. 1990; Irminger-Finger et al. 1990); APC most likely binds to microtubules through this domain.

APC is often found inactivated by mutations in cancers, leading to C-terminal truncations (Nagase and Nakamura 1993). This activates the Wnt pathway, allowing β-catenin to accumulate and activate transcription of proliferative genes such as cyclin D1 and c-myc (He et al. 1998; Shtutman et al. 1999; Tetsu et al. 1999).

APC interacts directly with CK2$\alpha_2\beta_2$ as shown by immunoprecipitation. In vitro binding assays revealed the interaction to occur through the catalytic subunits of CK2, as no interaction was seen with the regulatory CK2β subunit. Hence, binding between APC and CK2 holoenzyme is mediated through the CK2α and CK2α' subunits.The interaction between CK2α and APC is cell cycle dependent and peaking in G_2/M. Interestingly, the C-terminal region of APC strongly suppresses the catalytic activity of CK2α and CK2$\alpha_2\beta_2$ (Homma et al. 2002).

APC shuttles between the nucleus and the cytoplasm in a tightly regulated process (Neufeld et al. 2000). The localization is predominantly cytoplasmic and the nuclear translocation is partly mediated by nuclear localization sequences (NLS), the second of these; NLS2 is flanked by CK2$\alpha_2\beta_2$ and PKA sites that positively and negatively regulate APC nuclear import, respectively as shown by mutagenesis studies (Zhang et al. 2000; Zhang et al. 2001). The MAP kinase p38 promotes APC nuclear localization; this has been shown to be due to p38 forming complexes with both PKA and CK2 and actually promoting CK2 activity and decreasing PKA activity, leading to promoted nuclear localization of APC (Hildesheim et al. 2005). The latter scenario is especially interesting, since it has been shown by Sayed et al. (2000) that p38 can bind to CK2 and thereby through an allosteric mechanism induce phosphorylation of p53 at Ser392.

10
More than an Enzyme: Protein Kinase CK2 as Essential Interaction Partner for Key Components of Genome Stabilization

Disruption of CK2 subunits in *S. cerevisiae* (Barz et al. 2003) leads to changes of the expression levels of roughly one quarter of cell growth/cell cycle genes supporting the notion that CK2 has an important function in the regulation of cell growth and cell survival (Ahmed et al. 2002).

Yamane and Kinsella (2005b) showed that down regulation of the catalytic CK2α and α' subunits by siRNA after induction of DNA double-strand breaks by ionizing radiation led to an increase in apoptosis. Since no enhancement in DNA breaks on chromosomes in CK2-reduced cells compared with controls was seen, it is likely that CK2 is not involved in initial double-strand break repair, but rather controls apoptosis downstream of IR-related DNA repair and its cell cycle effects. Similar results were obtained following DNA damage induced by 6-thioguanine (Yamane and Kinsella 2005a) supporting the notion of CK2 having anti-apoptotic properties.

Phosphorylation of proteins directly involved in DNA repair by CK2 has been reported earlier (Fritz and Kaina 1999; Loizou et al. 2004; Luo et al. 2004). Another report by Teitz et al. (1990) showed that CK2β confers partial UV resistance in xeroderma pigmentosum cells. Since these authors did not find a functional defect in the CK2β subunit in UV-sensitive xeroderma pigmentosum cell lines, one could also speculate that a higher expression level of CK2β in the transfected cell lines was responsible for the partial UV resistance, eventually caused by illegitimate interactions of the CK2β subunit with proteins involved in DNA repair. Indeed, an excess of CK2β molecules has been described to occur in various kidney tumors (Stalter et al. 1994) supporting the view of CK2β being a "wildcard" and as such disturbing physiological relevant pathways (Korn et al. 1999) and leading to a new aspect in CK2 function, beside its role as an enzyme.

Therefore, we have focused on CK2's interaction with proteins involved in genome integrity in this review.

Many reports in the literature support the notion that it is the tetrameric CK2 holoenzyme that is involved in the interaction with other proteins. Yet, there are reports that describe a specific interaction of either the CK2α or the CK2β subunit with various proteins. Results from yeast-two-hybrid screens indicate a specific interaction of either the one or the other CK2 subunit to be responsible for the interaction with a particular protein, yet it cannot be excluded that under physiological conditions it is the holoenzyme that interacts via the identified subunit with a particular protein. An example for such a scenario is FAF1 which was identified to interact specifically with CK2β in a yeast-two-hybrid screen (Kusk et al. 1999) yet, in immunoprecipitates from mammalian cell lysates FAF1 was found in association with the CK2 holoenzyme, presumably via the CK2β subunit (Guerra et al. 2001). On the other hand BRCA1 was identified in a yeast-two-hybrid screen to interact with CK2β and this specific binding was confirmed in immunoprecipates, devoid of the CK2α subunit (O'Brien et al. 1999).

More than 300 substrates have been described for protein kinase CK2 (Meggio and Pinna 2003) with no end in sight. This plethora of potential CK2 substrates has led to the notion of CK2 being a very unspecific protein kinase. However, given the fact that the CK2-specific consensus sequence is very simple, i.e. S/TXXD/E the likelihood to find such a sequence in an average size protein of ca. 50, 000 daltons is rather high. This is not a specific feature of CK2, since other protein kinases such as PKA or AKT, just to name two prominent protein kinases, also phosphorylate plenty of different substrates owing to the simple consensus sequence they target. Therefore, we tend to believe, that beside the consensus sequence, specific interaction sites on a particular protein, at least under physiological conditions, play a decisive role in the cellular function of a particular protein kinase either by acting as an enzyme or by being part of a "networking" higher molecular mass complex.

It will be one of the great challenges of the future years to come to characterize protein-protein interactions at the structural level and perhaps distinguish between true phosphorylation properties of a protein kinase and stoichiometric structural interactions involving higher molecular complexes.

References

Aberle H, Butz S, Stappert J, Weissig H, Kemler R, Hoschuetzky H (1994) Assembly of the cadherin-catenin complex in vitro with recombinant proteins. J Cell Sci 107:3655–3663

Ahmed K, Gerber DA, Cochet C (2002) Joining the cell survival squad: an emerging role for protein kinase CK2. Trends Cell Biol 12:226–230

Aizawa H, Emori Y, Murofushi H, Kawasaki H, Sakai H, Suzuki K (1990) Molecular cloning of a ubiquitously distributed microtubule-associated protein with Mr 190 000. J Biol Chem 265:13849–13855

Alberts B (1998) The cell as a collection of protein machines: preparing the next generation of molecular biologists. Cell 92:291–294

Allende CC, Allende JE (1998) Promiscuous subunit interactions: a possible mechanism for the regulation of protein kinase CK2. J Cell Biochem Suppl 30–31:129–136

Appel K, Wagner P, Boldyreff B, Issinger OG, Montenarh M (1995) Mapping of the interaction sites of the growth suppressor protein p53 with the regulatory beta-subunit of protein kinase CK2. Oncogene 11:1971–1978

Appella E, Anderson CW (2001) Post-translational modifications and activation of p53 by genotoxic stresses. Eur J Biochem 268:2764–2772

Arkin MR, Wells JA (2004) Small-molecule inhibitors of protein-protein interactions: progressing towards the dream. Nat Rev Drug Discov 3:301–317

Bartek J, Lukas J (2003) Chk1 and Chk2 kinases in checkpoint control and cancer. Cancer Cell 3:421–429

Bartek J, Falch J, Lukas J (2001) CHK2 kinase – a busy messenger. Nat Rev Mol Cell Biol 2:877–886

Barz T, Ackermann K, Dubois G, Eils R, Pyerin W (2003) Genome-wide expression screens indicate a global role for protein kinase CK2 in chromatin remodeling. J Cell Sci 116:1563–1577

Bates PA, Kelley LA, MacCallum RM, Sternberg MJ (2001) Enhancement of protein modeling by native intervention in applying the automatic programs 3D-JIGSAW and 3D-PSSM. Proteins 5:39–46

Bek S, Kemler R (2002) Protein kinase CKII regulates the interaction of beta-catenin with alpha-catenin and its protein stability. J Cell Sci 115:4743–4753

Bertrand L, Sayed MF, Pei XY, Parisini E, Dhanaraj V, Bolanos-Garcia VM, Allende JE, Blundell TL (2004) Structure of the regulatory subunit of CK2 in the presence of a p21WAF1 peptide demonstrates flexibility of the acidic loop. Acta Crystallogr D Biol Crystallogr 60:1698–1704

Bibby AC, Litchfield DW (2005) The multiple personalities of the regulatory subunit of protein kinase CK2: CK2 dependent and CK2 independent roles reveal a secret identity for CK2β. Int J Biol Sci 1:67–79

Biondi RM, Nebreda AR (2003) Signaling specificity of Ser/Thr protein kinases through docking-site-mediated interactions. Biochem J 372:1–13

Bjørling-Poulsen M, Meek D, Issinger OG (2005a) HDM2 negatively affects the Chk2-mediated phosphorylation of p53. FEBS Lett 579:2604–2608

Bjørling-Poulsen M, Siehler S, Wiesmüller L, Meek D, Niefind K, Issinger OG (2005b) The regulatory β-subunit of protein kinase CK2 negatively influences p53-mediated allosteric effects on Chk2 activation. Oncogene advance online publication, 23 May 2005 (DOI 10.1038/sj.onc.1208762)

Bojanowski K, Filhol O, Cochet C, Chambaz EM, Larsen AK (1993) DNA topoisomerase II and casein kinase II associates in a molecular complex that is catalytically active. J Biol Chem 268:22920–22926

Boldyreff B, Meggio F, Pinna LA, Issinger OG (1993) Reconstitution of normal and hyperactivated forms of casein kinase-2 by variable mutated beta-subunits. Biochemistry 32:12672–12677

Boldyreff B, Issinger OG (1997) A-Raf kinase is a new interacting partner of protein kinase CK2 beta subunit. FEBS Lett 403:197–199

Buchou T, Vernet M, Blond O, Jensen HH, Pointu H, Olsen BB, Cochet C, Issinger OG, Boldyreff B (2003) Disruption of the regulatory beta subunit of protein kinase CK2 in mice leads to a cell-autonomous defect and early embryonic lethality. Mol Cell Biol 23:908–915

Chantalat L, Leroy D, Filhol O, Nueda A, Benitez MJ, Chambaz EM, Cochet C, Dideberg O (1999) Crystal structure of the human protein kinase CK2 regulatory subunit reveals its zinc finger-mediated dimerization. EMBO J 18:2930–2940

Chen M, Li D, Krebs EG, Cooper JA (1997) The casein kinase II beta subunit binds to Mos and inhibits Mos activity. Mol Cell Biol 17:1904–1912

Chester N, Yu IJ, Marshak DR (1995) Identification and characterization of protein kinase CK2 isoforms in HeLa cells. J Biol Chem 270:7501–7514

Cohen C, Parry DA (1994) Alpha-helical coiled coils: more facts and better predictions. Science 263:488–489

Craig A, Scott M, Burch L, Smith G, Ball K, Hupp T (2003) Allosteric effects mediate Chk2 phosphorylation of the p53 transactivation domain. EMBO J 4:787–792

De Moliner E, Brown NR, Johnson LN (2003) Alternative binding modes of an inhibitor to two different kinases. Eur J Biochem 270:3174–3181

Derbyshire DJ, Basu BP, Serpell LC, Joo WS, Date T, Iwabuchi K, Doherty AJ (2002) Crystal structure of human 53BP1 BRCT domains bound to p53 tumour suppressor. EMBO J 21:3863–3872

Dominguez I, Mizuno J, Wu H, Song D, Symes, Seldin DC (2004) Protein kinase CK2 is required for dorsal axis formation in Xenopus embryos. Dev Biol 274:110–124

Ermakova I, Boldyreff B, Issinger OG, Niefind K (2003) Crystal structure of a C-terminal deletion mutant of human protein kinase CK2 catalytic subunit. J Mol Biol 330:925–934

Escargueil AE, Plisov SY, Filhol O, Cochet C, Larsen AK (2000) Mitotic phosphorylation of DNA topoisomerase II by protein kinase CK2 creates the MPM-2 phosphoepitope on Ser-1469. J Biol Chem 275:34710–34718

Filhol O, Nueda A, Gerber-Scokaert D, Benitez MJ, Souchier C, Saoudi Y, Cochet C (2003) Live-cell fluorescence imaging reveals the dynamics of protein kinase CK2 individual subunits. Mol Cell Biol 23:975–987

Fodde R, Kuipers J, Rosenberg C, Smits R, Kielman M, Gaspar C, van Es JH, Breukel C, Wiegant J, Ghavidel A, Schultz MC (2001) TATA binding protein-associated CK2 transduces DNA damage signals to the RNA polymerase III transcriptional machinery. Cell 106:575–584

Fritz G, Kaina B (1999) Phosphorylation of the DNA repair protein APE/REF-1 by CKII affects redox regulation of AP-1. Oncogene 18:1033–1040

Ghavidel A, Schultz MC (2001) TATA binding protein-associated CK2 transduces DNA damage signals to the RNA polymerase III transcriptional machinery. Cell 106:575–584

Ghavidel A, Hockman DJ, Schultz MC (1999) A review on progress toward elucidating the role of protein kinase CK2 in polymerase III transcription: regulation of the TATA binding protein. Mol Cell Biochem 191:143–148

Glover CV (1998) On the physiological role of casein kinase II in *Saccharomyces cerevisiae*. Progr Nucleic Acid Res Mol Biol 59:95–132

Goldberg Y (1999) Protein phosphatase 2A: who shall regulate the regulator? Biochem Pharmacol 57:321–328

Grankowski N, Boldyreff B, Issinger OG (1991) Isolation and characterization of recombinant human casein kinase II subunits alpha and beta from bacteria. Eur J Biochem 198:25–30

Groden J, Thliveris A, Samowitz W, Carlson M, Gelbert L, Albertsen H, Joslyn G, Stevens J, Spirio L, Robertson M, Sargeant L, Krapcho K, Wolff E, Burt R, Hughes JP, Warrington J, McPherson J, Wasmuth J, LePaslier D, Abderrahim H, Cohen D, Leppert M, White R (1991) Identification and characterization of the familial adenomatous polyposis coli gene. Cell 66:589–600

Guerra B, Issinger OG (1999) Protein kinase CK2 and its role in cellular proliferation, development and pathology. Electrophoresis 20:391–408

Guerra B, Götz C, Wagner P, Montenarh M, Issinger OG (1997) The carboxy terminus of p53 mimics the polylysine effect of protein kinase CK2-catalyzed MDM2 phosphorylation. Oncogene 14:2683–2688

Guerra B, Boldyreff B, Issinger OG (2001) FAS-associated factor 1 interacts with protein kinase CK2 in vivo upon apoptosis induction. Int J Oncol 19:1117–1126

Guerra B, Issinger OG, Wang JY (2003) Modulation of human checkpoint kinase Chk1 by the regulatory beta-subunit of protein kinase CK2. Oncogene 22:4933–4942

Götz C, Wagner P, Issinger OG, Montenarh M (1996) p21WAF1/CIP1 interacts with protein kinase CK2. Oncogene 13:391–398

Götz C, Scholtes P, Prowald A, Schuster N, Nastainczyk W, Montenarh M (1999) Protein kinase CK2 interacts with a multi-protein binding domain of p53. Mol Cell Biochem 191:111–120

Hagen T, Vidal-Puig A (2002) Characterization of the phosphorylation of β-catenin at the GSK-3 priming site Ser45. Biochem Biophys Res Commun 294:324–328

He TC, Sparks AB, Rago C, Hermeking H, Zawel L, da Costa LT, Morin PJ, Vogelstein B, Kinzler KW (1998) Identification of c-MYC as a target of the APC pathway. Science 281:1509–1512

Henderson BR (2000) Nuclear-cytoplasmic shuttling of APC regulates beta-catenin subcellular localization and turnover. Nat Cell Biol 2:653–660

Hildesheim J, Salvador JM, Hollander MC, Fornace AJ Jr (2005) Casein kinase 2- and protein kinase A-regulated adenomatous polyposis coli and β-catenin cellular localization is dependent on p38 MAPK. J Biol Chem 280:17221–17226

Homma MK, Li D, Krebs EG, Yuasa Y, Homma Y (2002) Association and regulation of casein kinase 2 activity by adenomatous polyposis coli protein. Proc Natl Acad Sci USA 99:5959–5964

Honda R, Tanaka H, Yasuda H (1997) Oncoprotein MDM2 is a ubiquitin ligase E3 for tumor suppressor p53. FEBS Lett 420:25–27

Hu P, Samudre K, Wu S, Sun Y, Hernandez N (2004) CK2 phosphorylation of Bdp1 executes cell cycle-specific RNA polymerase III transcription repression. Mol Cell 16:81–92

Huber O, Korn R, McLaughlin J, Ohsugi M, Herrmann BG, Kemler R (1996) Nuclear localization of beta-catenin by interaction with transcription factor LEF-1. Mech Dev 59:3–10

Hupp TR, Meek DW, Midgley CA, Lane DP (1992) Regulation of the specific DNA binding function of p53. Cell 71:875–886

Huse M, Kuriyan J (2002) The conformational plasticity of protein kinases. Cell 109:275–282

Ilyas M (2005) Wnt signaling and the mechanistic basis of tumor development. J Pathol 205:130–144

Irminger-Finger I, Laymon RA, Goldstein LSB (1990) Analysis of the primary sequence and microtubule-binding region of the *Drosophila* 205K MAP. J Cell Biol 111:2563–2572

Issinger OG (1993) Casein kinases: pleiotropic mediators of cellular regulation. Pharmacol Ther 59:1–30

Johnston IM, Allison SJ, Morton JP, Schramm L, Scott PH, White RJ (2002) CK2 forms a stable complex with TFIIIB and activates RNA polymerase III transcription in human cells. Mol Cell Biol 22:3757–3768

Kaiserman HB, Ingebritsen TS, Benbow RM (1988) Regulation of Xenopus laevis DNA topoisomerase I activity by phosphorylation in vitro. Biochemistry 27:3216–3222

Kaplan KB, Burds AA, Swedlow JR, Bekir SS, Sorger PK, Nathke IS (2001) A role for the adenomatous polyposis coli protein in chromosome segregation. Nat Cell Biol 3:429–432

Kapoor M, Lozano G (1998) Functional activation of p53 via phosphorylation following DNA damage by UV but not gamma radiation. Proc Natl Acad Sci USA 95:2834–2837

Kassavetis GA, Braun BR, Nguyen LH, Geiduschek EP (1990) *S. cerevisiae* TFIIIB is the transcription initiation factor proper of RNA polymerase III, while TFIIIA and TFIIIC are assembly factors. Cell 60:235–245

Keller DM, Lu H (2002) p53 serine 392 phosphorylation increases after UV through induction of the assembly of the CK2-hSPT16-SSRP1 complex. J Biol Chem 277:50206–50213

Keller DM, Zeng X, Wang Y, Zhang QH, Kapoor M, Shu H, Goodman R, Lozano G, Zhao Y, Lu H (2001) A DNA damage-induced p53 serine 392 kinase complex contains CK2, hSpt16, and SSRP1. Mol Cell 7:283–292

Kordiyak GJ, Jakes S, Ingebritsen TS, Benbow RM (1994) Casein kinase II stimulates *Xenopus laevis* DNA topoisomerase I by physical association. Biochemistry 33:13484–13491

Korn I, Gutkind S, Srinivasan N, Blundell TL, Allende CC, Allende JE (1999) Interactions of protein kinase CK2 subunits. Mol Cell Biochem 191:75–83

Kowalska-Loth B, Girstun A, Derlacz R, Staron K (2003) Activation of human topoisomerase I by protein kinase CK2. Mol Biol Rep 30:107–111

Kusk M, Ahmed R, Thomsen B, Bendixen C, Issinger OG, Boldyreff B (1999) Interactions of protein kinase CK2beta subunit within the holoenzyme and with other proteins. Mol Cell Biochem 191:51–58

Lee JS, Ishimoto A, Yanagawa S (1999) Characterization of mouse disheveled (Dvl) proteins in Wnt/Wingless signaling pathway. J Biol Chem 274:21464–21470

Leroy D, Alghisi GC, Roberts E, Filhol-Cochet O, Gasser SM (1999) Mutations in the C-terminal domain of topoisomerase II affect meiotic function and interaction with the casein kinase 2 beta subunit. Mol Cell Biochem 191:85–95

Lewis SA, Wang D, Cowan NJ (1988) Microtubule-associated protein MAP2 shares a microtubule binding motif with tau protein. Science 242:936–939

Li Y, Keller DM, Scott JD, Lu H (2005) CK2 phosphorylates SSRP1 and inhibits its DNA-binding activity. J Biol Chem 280:11869–11875

Lieberman SL, Ruderman JV (2004) CK2 beta, which inhibits Mos function, binds to a discrete domain in the N-terminus of Mos. Dev Biol 268:271–279

Litchfield DW (2003) Protein kinase CK2: structure, regulation and role in cellular decisions of life and death. Biochem J 369:1–15

Loizou JI, El-Khamisy SF, Zlatanou A, Moore DJ, Chan DW, Qin J, Sarno S, Meggio F, Pinna LA, Caldecott KW (2004) The protein kinase CK2 facilitates repair of chromosomal DNA single-strand breaks. Cell 117:17–28

Lowe ED, Tews I, Cheng KY, Brown NR, Gul S, Noble ME, Gamblin SJ, Johnson LN (2002) Specificity determinants of recruitment peptides bound to phospho-CDK2/cyclin A. Biochemistry 41:15625–15634

Lu H, Taya Y, Ikeda M, Levine AJ (1998) Ultraviolet radiation, but not gamma radiation or etoposide-induced DNA damage, results in the phosphorylation of the murine p53 protein at Serine-389. Proc Natl Acad Sci USA 95:6399–6402

Luciani MG, Hutchins JR, Zheleva D, Hupp TR (2000) The C-terminal regulatory domain of p53 contains a functional docking site for cyclin A. J Mol Biol 300:503–518

Luo H, Chan DW, Yang T, Rodriguez M, Chen PC, Leng M, Mu JJ, Chen D, Songyang Z, Wang Y, Qin J (2004) A new XRCC1-containing complex and its role in cellular survival of methyl methanesulfonate treatment. Mol Cell Biol 24:8356–8365

Maldonado E, Allende JE (1999) Phosphorylation of yeast TBP by protein kinase CK2 reduces its specific binding to DNA. FEBS Lett 443:256–260

Manning G, Whyte DB, Martinez R, Hunter T, Sudarsanam S (2002) The protein kinase complement of the human genome. Science 238:1912–1934

Meek DW, Simon S, Kikkawa U, Eckhart W (1990) The p53 tumour suppressor protein is phosphorylated at serine 389 by casein kinase II. EMBO J 9:3253–3260

Meggio F, Pinna LA (2003) One-thousand-and-one substrates of protein kinase CK2? FASEB J 17:349–368

Meggio F, Boldyreff B, Marin O, Marchiori F, Perich JW, Issinger OG, Pinna LA (1992) The effect of polylysine on casein-kinase-2 activity is influenced by both the structure of the protein/peptide substrates and the subunit composition of the enzyme. Eur J Biochem 205:939–945

Molenaar M, van de Wetering M, Oosterwegel M, Peterson, Maduro J, Godsave S, Korinek V, Roose J, Destree, Clevers H (1996) XTcf-3 transcription factor mediates beta-catenin-induced axis formation in *Xenopus* embryos. Cell 86:391–399

Nagase H, Nakamura Y (1993) Mutations of the APC (adenomatous polyposis coli) gene. Hum Mutat 2:425–434

Narod SA, Foulkes WD (2004) BRCA1 and BRCA2: 1994 and beyond. Nature Rev Cancer 4:665–676

Nelson WJ, Nusse R (2004) Convergence of Wnt, beta catenin, and cadherin pathways. Science 303:1483–1487

Neufeld KL, Nix DA, Bogerd H, Kang Y, Beckerle MC, Cullen BR, White RL (2000) Adenomatous polyposis coli protein contains two nuclear export signals and shuttles between the nucleus and cytoplasm. Proc Natl Acad Sci USA 97:12085–12090

Niefind K, Issinger OG (2005) Primary and secondary interactions between CK2α and CK2β lead to ring-like structures in the crystals of the CK2 holoenzyme. Mol Cell Biochem 274:3–14

Niefind K, Guerra B, Pinna LA, Issinger OG, Schomburg D (1998) Crystal structure of the catalytic subunit of protein kinase CK2 from Zea mays at 2.1 A resolution. EMBO J 17:2451–2462

Niefind K, Guerra B, Ermakowa I, Issinger OG (2001) Crystal structure of human protein kinase CK2: insights into basic properties of the CK2 holoenzyme. EMBO J 20:5320–5331

Noble ME, Endicott JA, Johnson LN (2004) Protein kinase inhibitors: insights into drug design from structure. Science 303:1800–1805

O'Brien KA, Lemke SJ, Cocke KS, Rao RN, Beckmann RP (1999) Casein kinase 2 binds to and phosphorylates BRCA1. Biochem Biophys Res Commun 260:658–664

O'Neill T, Giarratani L, Chen P, Iyer L, Lee CH, Bobiak M, Kanai F, Zhou BB, Chung JH, Rathburn (2002) Determination of substrate motifs for human Chk1 and hCds1/Chk2 by the oriented peptide library approach. J Biol Chem 277:16102–16115

Olsten MEK, Litchfield DW (2004) Order or chaos? An evaluation of the regulation of protein kinase CK2. Biochem Cell Biol 82:681–693

Pinna LA (1990) Casein kinase 2: an eminence grise in cellular regulation? Biochim Biophys Acta 1054:267–284

Pinna LA (2002) Protein kinase CK2: a challenge to canons. J Cell Sci 115:3873–3878

Polakis P (1997) The adenomatous polyposis coli (APC) tumor suppressor. Biochim Biophys Acta 1332:F127–142

Prowald A, Schuster N, Montenarh M (1997) Regulation of DNA binding of p53 by its interaction with protein kinase CK2. FEBS Lett 408:99–104

Pruniers C, Hocevar BA, Howe PH (2004) Wnt signaling: physiology and pathology. Growth Factors 22:141–150

Rimm DL, Koslov ER, Kebriaei P, Cianci CD, Morrow JS (1995) Alpha 1(E)-catenin is an actin-binding and -bundling protein mediating the attachment of F-actin to the membrane adhesion complex. Proc Natl Acad Sci USA 92:8813–8817

Romero-Oliva F, Allende JE (2001) Protein p21(WAF1/CIP1) is phosphorylated by protein kinase CK2 in vitro and interacts with the amino terminal end of the CK2 beta subunit. J Cell Biochem 81:445–452

Rosin-Arbesfeld R, Townsley F, Bienz M (2000) The APC tumour suppressor has a nuclear export function. Nature 406:1009–1012

Sarno S, de Moliner E, Ruzzene M, Pagano MA, Battistutta R, Bain J, Fabbro D, Schoepfer J, Elliott M, Furet P, Meggio F, Zanotti G, Pinna LA (2003) Biochemical and three-dimensional-structural study of the specific inhibition of protein kinase CK2 by [5-oxo-5,6-dihydroindolo-(1,2-a)quinazolin-7-yl]acetic acid (IQA). Biochem J 374:639–646

Sayed M, Kim SO, Salh BS, Issinger OG, Pelech SL (2000) Stress-induced activation of protein kinase CK2 by direct interaction with p38 mitogen-activated protein kinase. J Biol Chem 275:16569–16573

Schomburg D, Reichelt J (1988) BRAGI: a comprehensive protein modeling program system. J Mol Graphics 6:161–165

Schon O, Frielder A, Bycroft M, Freund SM, Fersht AR (2002) Molecular mechanisms of the interaction between MDM2 and p53. J Mol Biol 323:491–501

Schulman BA, Lindstrom DL, Harlow E (1998) Substrate recruitment to cyclin-dependent kinase 2 by a multipurpose docking site on cyclin A. Proc Natl Acad Sci USA 95:10453–10458

Schuster N, Prowald A, Schneider E, Scheidtmann KH, Montenarh M (1999) Regulation of p53 mediated transactivation by the beta-subunit of protein kinase CK2. FEBS Lett 447:160–166

Schuster N, Götz C, Faust M, Schneider E, Prowald A, Jungbluth A, Montenarh M (2001) Wild-type p53 inhibits protein kinase CK2 activity. J Cell Biochem 81:172–183

Scully R, Chen J, Ochs RL, Keegan K, Hoekstra M, Feunteun J, Livingston DM (1997) Dynamic changes of BRCA1 subnuclear location and phosphorylation state are initiated by DNA damage. Cell 90:425–435

Seo GJ, Kim SE, Lee YM, Lee JW, Lee RJ, Hahn MJ, Kim ST (2003) Determination of substrate specificity and putative substrates of Chk2 kinase. Biochem Biophys Res Commun 304:339–343

Shtutman M, Zhurinsky J, Simcha I, Albanese C, D'Amico M, Pestell R, Ben-Ze'ey A (1999) The cyclin D1 gene is a target of the beta-catenin/LEF-1 pathway. Proc Natl Acad Sci USA 96:5522–5527

Song D, Sussman DJ, Seldin DC (2000) Endogenous protein kinase CK2 participates in Wnt signaling in mammary epithelial cells. J Biol Chem 275:23790–23797

Song DH, Dominguez I, Mizuno J, Kaut M, Mohr SC, Seldin DC (2003) CK2 phosphorylation of the armadillo repeat region of β-catenin potentiates Wnt signaling. J Biol Chem 278:24018–24025

Stalter G, Siemer S, Becht E, Ziegler M, Remberger K, Issinger OG (1994) Asymmetric expression of protein kinase CK2 subunits in human kidney tumors. Biochem Biophys Res Commun 202:141–147

Stigare J, Buddelmeijer N, Pigon A, Egyhazi E (1993) A majority of casein kinase II alpha subunit is tightly bound to intranuclear components but not to the beta subunit. Mol Cell Biochem 129:77–85

Su LK, Johnson KA, Smith KJ, Hill DE, Vogelstein B, Kinzler KW (1993) Association between wild type and mutant APC gene products. Can Res 53:2728–2731

Tapia JC, Bolanos-Garcia VM, Sayed M, Allende CC, Allende JE (2004) Cell cycle regulatory protein p27KIP1 is a substrate and interacts with the protein kinase CK2. J Cell Biochem 91:865–879

Teitz T, Eli D, Penner M, Bakhanashvili M, Naiman T, Timme TL, Wood CM, Moses RE, Canaani D (1990) Expression of the cDNA for the beta subunit of human casein kinase II confers partial UV resistance on xeroderma pigmentosum cells. Mutat Res 236:85–97

Tetsu O, McCromick F (1999) Beta-catenin regulates expression of cyclin D1 in colon carcinoma cells. Nature 398:422–426

Thomas JE, Smith M, Tonkinson JL, Rubinfeld B, Polakis P (1997) Induction of phosphorylation on BRCA1 during the cell cycle and after DNA damage. Cell Growth Differ 8:801–809

Tuazon PT, Traugh JA (1991) Casein kinase I and II-multipotential serine protein kinases: structure, function and regulation. Adv Second Messenger Phosphoprotein Res 23:123–164

van Noort M, Clevers H (2002) TCF transcription factors, mediators of Wnt-signaling in development and cancer. Dev Biol 244:1–8

Vogelstein B, Lane D, Levine AJ (2000) Surfing the p53 network. Nature 408:307–310

Wang JC (1985) DNA topoisomerases. Annu Rev Biochem 54:665–697

Wang JC (1996) DNA topoisomerases. Annu Rev Biochem 65:635–692

Willert K, Brink M, Wodarz A, Varmus H, Nusse R (1997) Casein kinase 2 associates with and phosphorylates Dishevelled. EMBO J 16:3089–3096

Williams RS, Glover JNM (2003) Structural consequences of a cancer-causing BRCA1-BRCT missense mutation. J Biol Chem 278:2630–2635

Yamane K, Kinsella TJ (2005a) Casein kinase 2 regulates both apoptosis and the cell cycle following DNA damage induced by 6-thioguanine. Clin Cancer Res 11:2355–2363

Yamane K, Kinsella TJ (2005b) CK2 inhibits apoptosis and changes its cellular localization following radiation. Cancer Res 65:4362–4367

Yanagawa S, van Leeuwen F, Wodarz A, Klingensmith J, Nusse R (1995) The Dishevelled protein is modified by Wingless signaling in *Drosophila*. Genes Dev 9:1087–1097

Zhang F, White RL, Neufeld KL (2000) Phosphorylation near nuclear localization signal regulates nuclear import of adenomatous polyposis coli protein. Proc Natl Acad Sci USA 97:12577–12582

Zhang F, White RL, Neufeld KL (2001) Cell density and phosphorylation control the nuclear localization of adenomatous polyposis coli protein. Mol Cell Biol 21:8143–8156

Genome Dyn Stab (1)
D.-H. Lankenau: Genome Integrity
DOI 10.1007/7050_004/Published online: 20 October 2005

Role of Human Topoisomerase I in DNA Repair and Apoptosis

Kent Søe (✉) · Anja Rockstroh · Frank Grosse

Leibniz Institute for Age Research, Biochemistry, Beutenbergstrasse 11, 07745 Jena, Germany
kent@fli-leibniz.de

Abstract Human topoisomerase I (htopoI) is an enzyme that up to now was believed to function mainly in the removal of torsional stress generated during transcription and replication. In 1998, it was found that htopoI might play another important role in the cellular response to DNA damage – the so-called htopoI damage response. Since this initial discovery, many studies have suggested that the htopoI damage response is involved in DNA repair as well as in apoptosis. Here we discuss the earliest as well as the latest results in this field. Combining all of the published and as yet unpublished results, we suggest and discuss a model of how htopoI may function during DNA repair and apoptosis. Furthermore, numerous results show that the htopoI damage response is not a spontaneous event, but is strictly regulated by cellular signalling pathways. We discuss which pathways may be involved and how the htopoI damage response is activated. Although the htopoI damage response was discovered several years ago, research in this area is just beginning. The future will surely bring more clarity regarding the precise mechanism behind the involvement of htopoI in DNA repair and apoptosis, as well as its implications for a broader understanding of how the human organism ensures genomic stability.

1 Introduction

Genetic information is remarkably stable despite its constant challenges by UV light, smoke, combustion, gases, ionizing radiation and reactive oxygen species, to mention but a few. Not only exogenous sources but also endogenous events challenge genomic stability. Spontaneous depurinations, for example, have been estimated to take place 10 000 times/day/cell (Lindahl and Nyberg 1972). Nevertheless, the mutation rate of a human cell is kept as low as 1×10^{-9} mutations/bp/cell cycle. This low mutation frequency is due to the activity of several pathways that either repair damaged DNA or kill seriously damaged cells. Both pathways ensure a high degree of genomic stability of multicellular organisms. Thus, DNA repair and apoptosis pathways serve to avoid mutagenesis and oncogenesis. Four main DNA repair pathways are well characterized: nucleotide excision repair (NER), base excision repair (BER), mismatch repair (MMR) and recombination. If these pathways fail to repair DNA, a suicidal program of the cell, called apoptosis, is induced. Exciting new data show that the DNA relaxing enzyme topoisomerase I (topoI) plays an important role in both, DNA repair and apoptosis. This involvement of topoI in the cellular response to genotoxic stress is referred to as the "topoI damage response". This newly discovered pathway is the focus of the present review.

2 Eukaryotic DNA Topoisomerase I

2.1 Function

Eukaryotic topoI activity was first found in 1972, when Champoux and Dulbecco (Champoux and Dulbecco 1972) described its activity in secondary mouse embryo cell extracts, where it was able to relax positive as well as negative DNA supercoils without the use of energy. This seminal discovery was followed by the isolation and characterization of topoI from rat liver (Champoux and McConaughy 1976) and *Saccharomyces cerevisiae* (Durnford and Champoux 1978). After this pioneering work, a vast number of investigations subsequently characterized the mechanism and function of topoI in detail. TopoI was found to be involved in removing positive as well as negative supercoils that arise during transcription and replication. For example, the removal of tension during transcription was found to be necessary for obtaining full length transcripts (Tsao et al. 1989; Stewart et al. 1990; Merino et al. 1993). Despite its importance for transcription and replication, it is not essential in yeast, since knock-out of topoI only results in reduced growth (Uemura and Yanagida 1984). It is believed that topoisomerase II in such cases can substitute for the function of the missing topoI. However, in higher eukaryotes, topoI is essential. Both topoI-deficient *Drosophila* (Lee et al. 1993) and mice (Morham et al. 1996) display embryonic lethality at the 8-cell stage. Only the topoI present in the egg is enough to support the first three cell cycles before it becomes instable and is degraded. Using an inducible topoI gene, it was demonstrated that this enzyme is essential for the development of *Drosophila*, in particular for embryogenesis, larval and pupal growth, as well as oogenesis (Zhang et al. 2000). The requirement of topoI for developmental processes may also explain why it is essential for multicellular but not for unicellular organisms such as yeast.

2.2 Catalytic Mechanism

In order to relax supercoils, topoisomerases use a unique mechanism that includes a transient covalent intermediate. Depending on their catalytic mechanism, eukaryotic topoisomerases can be divided into two groups, type I and type II. Type I topoisomerases cleave one DNA strand, whereas type II cleave both. Eukaryotic type II topoisomerases form a covalent bond with the 5′-ends of both cleaved strands, generating two free 3′-OH groups (for example, human topoisomerase IIα and β). Type I topoisomerases are divided into two subgroups. Type IA enzymes form a covalent bond with the 5′-end and generate a free 3′-OH group (for example, human topoisomerase IIIα and β). Type IB enzymes become covalently attached to the 3′-end of DNA and generate a free 5′-OH group. Human topoI (htopoI) is a type IB topoisomerase.

In Fig. 1, a simplified overview of the catalytic cycle of htopoI is shown. The catalytic cycle can be divided into five steps: 1) DNA-binding, 2) DNA cleavage, 3) release of tension, 4) religation of DNA and 5) release of enzyme. The catalytic mechanism of htopoI has been thoroughly investigated by biochemical methods; these results were verified as the crystal structure of an *N*-terminally truncated htopoI was resolved (Redinbo et al. 1998; Stewart et al. 1998). Based on this structure, it is now possible to explain the catalytic cycle of htopoI in detail. As shown in Fig. 1, htopoI has the form of an open clamp. Upon binding, this clamp encloses the DNA within a 20 Å large binding cavity. Subsequently, a nucleophilic attack by an active site tyrosine (Tyr^{723}) breaks the DNA backbone and forms a covalent bond with the 3′-phosphate, generating a free 5′-hydroxyl group. This covalent intermediate of htopoI and DNA is referred to as "htopoI cleavage complex". The tension

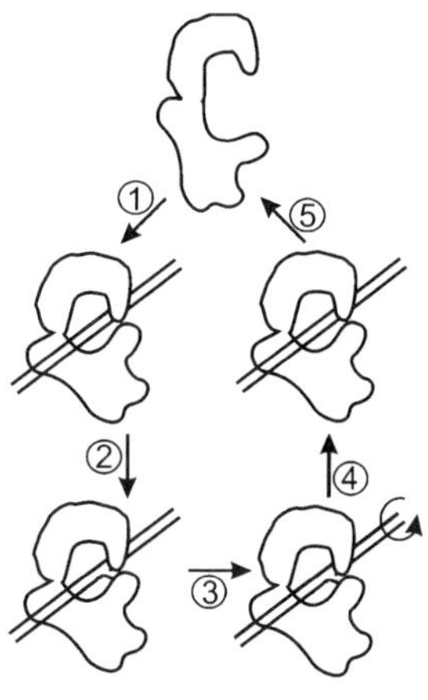

Fig. 1 The catalytic cycle of htopoI. Unbound htopoI has the shape of an open clamp. 1. Upon binding, this clamp closes tightly around the DNA. 2. An active site Tyr723 performs a nucleophilic attack on the DNA backbone and forms a covalent phospho-tyrosyl bond with the 3′-end and generates a nick with a free 5′ hydroxyl group. 3. The superhelical tension is released by controlled rotation of the scissile strand. 4. The 5′ hydroxyl group of the DNA performs a nucleophilic attack on the protein-DNA bond and the DNA is religated. 5. The htopoI clamp opens and the enzyme is released from the DNA. The entire catalytic cycle takes place without the use of external energy. The covalent intermediate (step 2) is referred to as a htopoI cleavage complex

is released through a controlled swivel of the incised strand around the intact strand (Koster et al. 2005). Subsequently, a rearrangement of the active site renders the protein-DNA bond unfavourable and through a nucleophilic attack of the 5′-OH on the protein-DNA bond the DNA is religated and the enzyme is released. The entire catalytic cycle does not require energy in the form of ATP hydrolysis.

The formation of a transient covalent attachment of the enzyme to the DNA backbone is an important characteristic of topoisomerases. The enzyme catalyzes its own release through religation. This makes it different from other endonucleases, which cannot reseal the cleaved bond.

2.3 Human Topoisomerase I as a Target of Chemotherapy

The ability of htopoI to induce transient protein-linked single strand (ss) breaks in DNA is exploited during chemotherapy. An alkaloid extracted from the bark, fruit, and wood of the tree *Camptotheca acuminata*, camptothecin (CPT), was identified as early as in 1966 to possess anti-tumour activities (Wall et al. 1966). In 1985 it was discovered that this compound is a specific inhibitor of htopoI (Hsiang et al. 1985). CPT reversibly inhibits the religation step of htopoI, thus stabilizing the htopoI cleavage complex. This stabilization results in transcription and replication mediated double strand breaks that eventually lead to apoptosis. CPT was initially used for chemotherapy, but soon abandoned due to severe and unforeseen side effects. The development of novel CPT analogues during the 1980s and 1990s led to new drugs with the same inhibitory effect as CPT, but with less severe side effects. Some of these compounds (topotecan and irinotecan) are now routinely used in chemotherapy of, for example, ovarian, lung, colon and cervical cancer.

3 The Human Topoisomerase I Damage Response

3.1 Pioneering Work in Vitro

In 1996 Lanza and co-workers (Lanza et al. 1996) reported that UV-irradiated DNA caused a decrease in htopoI relaxation activity in vitro. The reduced activity was due to partial inhibition of the religation step of htopoI, resulting in stabilization of htopoI cleavage complexes proximal to a UV photoproduct. The authors used UV-C irradiated plasmids to map the htopoI cleavage sites and their proximity to the UV photoproduct. This in vitro study strongly indicated that htopoI could “recognize” a UV photoproduct. Thereafter, numerous publications appeared in which several DNA lesions had been tested for their ability to be detected and subsequently stabilized as a htopoI cleavage complex in vitro. Almost exclusively, these studies made use of a hexadecameric preferential binding sequence of htopoI within a short oligodeoxynucleotide where various modified bases were incorporated at or across from the htopoI cleavage site. In this fashion, it was found that mismatches, abasic sites, 8-oxo-guanosine (8-oxo-G), arabinofuranosyl cytosine (ara-C), gemcitabine, nicks, methylated and alkylated bases placed immediately across from the cleavage site all result in stabilization of htopoI cleavage complexes (Pourquier et al. 1997, 1999, 2000, 2001, 2002; Christiansen and Westergaard 1999; Pommier et al. 2000). The preferential binding sequence was originally isolated from *Tetrahymena* rDNA and was found to bind htopoI with a K_d of 10^{-10} (Stevnsner et al. 1989). Since this sequence also directs cleavage to one specific position, these studies did not show that htopoI really recognized the lesions. It rather demonstrated that if htopoI is forced to cleave in the proximity of the induced lesions, a reduced ligation rate is observed. However, the use of this preferential binding sequence made it possible to crystallize htopoI cleavage complexes at such lesions and to identify how the observed inhibition was achieved. From these studies, it was found that in the case of ara-C, a lack of base pairing around the DNA lesion led to a slowing down of religation due to "breathing" of the 5′-end (Chrencik et al. 2003). Accordingly, in the case of 8-oxo-G, a rearrangement of the active site resulted in a similar decrease in religation (Lesher et al. 2002). These crystallographic studies supported previous publications demonstrating such a mode of inhibition biochemically (Pourquier et al. 1999, 2000). Based on these investigations, it is highly likely that lesions such as UV photoproducts and gemcitabine inhibit religation in a manner similar to ara-C.

3.2
Influence of DNA Lesions on HtopoI Activity in Cell Culture Systems

Modulation of htopoI activity has also been shown in vivo with, e.g. UV photoproducts, ara-C, gemcitabine, methylation, and alkylation (Subramanian et al. 1998; Pommier et al. 2000; Pourquier et al. 2000, 2001, 2002; Soe et al. 2004). These all result in stabilization of htopoI cleavage complexes, as measured by an increased number of covalent htopoI DNA complexes after treatment. This cellular accumulation of covalent htopoI cleavage complexes is also referred to as "the htopoI damage response". Considering the in vitro results described in Sect. 3.1, it is reasonable to conclude that htopoI recognizes DNA lesions, cleaves the DNA, and forms a stabilized cleavage complex due to the inhibition of religation. Unfortunately, this has never been proven until now. Therefore, it is not possible to judge whether htopoI cleaves directly at or distantly from the DNA damage site in vivo. Furthermore, it is not possible to conclude that htopoI spontaneously forms stable cleavage complexes on the DNA. Experimental data indicate that cleavage complex formation in response to damage is not a spontaneous event, but rather a regulated one and is possibly incorporated into a cellular signalling pathway (see following section).

3.3 Regulation of the Human Topoisomerase I Damage Response

3.3.1 Nucleotide Excision Repair

The first report to indicate that the htopoI damage response is not a spontaneous event was published in 1998 by Subramanian and co-workers (Subramanian et al. 1998). They found that the htopoI damage response to UV light-induced DNA lesions was dependent on an active NER pathway. Further details on this issue are given in Sect. 4.1.

3.3.2 The Tumour Suppressor Protein p53

Unexpectedly, the tumour suppressor protein p53 (for review, see, e.g. Vousden and Lu 2002) plays an important role in activating the htopoI damage response following UV-C irradiation (Mao et al. 2000a). The authors demonstrated that selected cell lines, which contain wild type (wt), mutant or deleted p53, showed different htopoI damage responses. The cells expressing wt p53 (MCF-7) displayed a damage response after treatment with 10 J/m^2 UV-C, whereas mutant (-/R175H) or p53-deficient (-/-) cells (SK-BR-3 and HL-60, respectively) showed no complex formation after the same treatment. Since these cell lines are not isogenic, direct comparison is limited. Thus, the authors reintroduced wt p53 by transfection into SK-BR-3 cells. This reactivated the htopoI damage response and thereby showed that wt p53 could restore the defect in this particular cell line. However, since these cells also express mutant p53 (R175H), which may have a dominant negative effect on wt p53 (Sigal and Rotter 2000), it is difficult to assess how the mutant form might have influenced this result. To this end, we have repeated the investigations of Mao and colleagues (Mao et al. 2000a), but using tetracycline regulated expression of wt p53 in a p53 deficient cell line. Our results confirm that p53 is one of the key factors for the induction of the htopoI damage response (unpublished data). We have tested several cell lines that express various forms of mutant p53. Some of them did show a htopoI damage response but others did not [(Soe et al. 2004) and unpublished data]. Similarly, transfecting various forms of mutant p53 in p53-deficient cell lines revealed a strong induction of htopoI cleavage complexes in response to UV-C with some mutants but not with others (unpublished data). Thus, for reasons as yet unknown, some but not all mutant forms of p53 support the htopoI damage response, as tested in isogenic cell systems.

It was reported that p53 is directly complexed with htopoI cleavage complexes (Mao et al. 2000a). Additionally, numerous publications have shown that p53 and htopoI interact directly, both in vitro and in vivo. Accordingly, after genotoxic stress, htopoI and wt p53 interact for a limited time, whereas some mutant forms of p53 constitutively associate with htopoI (Gobert et al. 1999). HtopoI interacts with p53 within amino acids (aa) 302–320 (Gobert et al. 1999; Soe and Grosse 2003) and p53 binds to aa 156–170 of htopoI (Mao et al. 2002; Soe et al. 2002). The binding of p53 to htopoI stimulates the relaxation of plasmid DNA (Gobert et al. 1996) by decreasing the DNA binding activity of htopoI (Soe and Grosse 2003). This results in a more rapid release of htopoI from DNA, i.e. a higher k_{off} value. It is therefore rather unlikely that the binding of p53 to htopoI could stabilize cleavage complex formation. In addition, the results obtained with HeLa cells are somewhat puzzling. HeLa cells show a very strong htopoI damage response (Subramanian et al. 1998; Soe et al. 2004), although these cells are transformed by human papilloma virus and express the E6 ubiquitin ligase that triggers the degradation of p53. In consequence, HeLa cells contain a very low level of p53 that, nevertheless, can be activated after genotoxic stress (Wesierska-Gadek et al. 2002). Our own data verify that cells lacking p53 can support a htopoI damage response, although in a strongly reduced manner and only at a high damage level. Therefore, p53 seems to be important but not essential for the htopoI induced damage response.

3.3.3 Posttranslational Modification of Human Topoisomerase I

As mentioned previously, there are situations, for example in absence of p53, in which the DNA of a cell is damaged but no htopoI cleavage complexes are detected. This indicates that htopoI has to be "activated" in order to form cleavage complexes on damaged DNA, but it is unclear how this is achieved. One likely mechanism would be a post-translational modification of htopoI. So far, it has been shown that htopoI becomes ubiquitinated, sumoylated, poly-ADP-ribosylated and phosphorylated under various conditions (Desai et al. 1997; Mao et al. 2000b; Horie et al. 2002; Kowalska-Loth et al. 2003; Cardellini et al. 2004; Malanga and Althaus 2004; Yu et al. 2004; Yung et al. 2004). Very little is known about the physiological effects of post-translational modifications of htopoI, and none have been reported to stabilize the htopoI cleavage complex following DNA damage.

3.3.3.1 Ubiquitination/Sumoylation

Polyubiquitination was shown to mark htopoI for degradation, in particular as a cellular response to CPT treatment in order to avoid cleavage complex formation (Desai et al. 1997, 2001). Thus, polyubiquitination seems a poor candidate for a post-translational modification that stabilizes htopoI cleavage complexes. Sumoylation of htopoI was found to occur on Lys^{117} and Lys^{153} (Horie et al. 2002). Furthermore, two reports showed that htopoI becomes sumoylated upon CPT treatment; the conclusions, however, differ significantly. Horie and co-workers report that sumoylation of htopoI enhances cleavage complex formation after CPT treatment (Horie et al. 2002) whereas Mao and co-workers propose sumoylation as a repair mechanism for htopoI cleavage complexes (Mao et al. 2000b). These contradictory opinions make it difficult to assess to what extent sumoylation is important for the htopoI damage response.

3.3.3.2 Poly-ADP-ribosylation

Bauer and co-workers (Bauer et al. 2001) reported that poly(ADP-ribose) polymerase-1 (PARP1) and htopoI interact in vitro, leading to the poly-ADP-ribosylation of htopoI. Binding of PARP1 to htopoI stimulates htopoI activity, whereas auto modified PARP1 does not bind the topoisomerase (Bauer et al. 2000, 2001). Physical contacts between these two enzymes increased the DNA cleavage activity of htopoI, whereas poly-ADP-ribosylation of htopoI completely abolished it (Bauer et al. 2001; Yung et al. 2004). Non-covalently bound poly-ADP-ribose was furthermore found to inhibit the cleavage of DNA by htopoI, as well as stimulating religation (Malanga and Althaus 2004). Both effects result in a reduced cleavage of DNA by htopoI. Thus, it seems that there is a consensus that poly-ADP-ribosylation of htopoI abolishes its enzymatic activity, most likely due to diminished DNA binding, which would imply reduced sensitivity to CPT in the poly-ADP-ribosylated state. This is supported by investigations showing that PARP1 deficiency results in an increased sensitivity to CPT in vivo (Chatterjee et al. 1989; Bowman et al. 2001). Consequently, PARP1 is involved in the regulation of htopoI activity after genotoxic stress, but rather as an inhibitor than stimulator. Thus, based on the present literature, it seems unlikely that poly-ADP-ribosylation is the missing activator that enhances and stabilizes htopoI cleavage complexes in response to DNA damage. On the other hand, investigations on the role of PARP1 in the htopoI response to treatments other than CPT are still lacking. Therefore, it is difficult to evaluate whether or not PARP may be an activator of the htopoI damage response.

3.3.3.3 Phosphorylation

Only very little is known about the phosphorylation of htopoI and the effects on its activity. So far (June 2005), only two kinases have been identified to phosphorylate htopoI. Casein kinase 2 (CK2) binds and influences htopoI activity even in the absence of ATP (Kowalska-Loth et al. 2003). Phosphorylation of htopoI has not yet been directly shown, but in a recent publication using a peptide screen to identify CK2 substrates, it was found to phosphorylate the N-terminal domain of htopoI (Cardellini et al. 2004). Support for an in vivo phosphorylation of htopoI by CK2 comes from investigations in *Xenopus laevis* (Kordiyak et al. 1994) and *Pisum sativum* (pea) (Tuteja et al. 2003). Both studies show that the respective topoI is phosphorylated by CK2 and, moreover, the activity of topoI is increased by this treatment. Phosphorylation of substrates by CK2 primarily activates transcription. Since CK2 does not seem to become activated by DNA damage, it is unlikely that phosphorylation of htopoI by CK2 serves to induce the htopoI damage response (for details, see Pinna 2002). Rather, CK2 may phosphorylate htopoI in order to activate transcription.

Just recently, another kinase has been found to phosphorylate and interact with htopoI, *c*-Abl (Yu et al. 2004). This kinase is activated upon cellular damage and is a component of the DNA-dependent protein kinase (DNA-PK) and the ataxia-telangiectasia mutated (ATM) pathway (Kharbanda et al. 1997; Shafman et al. 1997). c-Abl phosphorylates Tyr^{268} of htopoI and thereby stimulates htopoI activity in vitro as well as in vivo, in particular during genotoxic stress. Consequently, c-Abl expressing cells are particularly sensitive to CPT treatment, which seems to be due to an approximately 4-fold increase in the number of htopoI cleavage complexes (Yu et al. 2004). Interestingly, treatment of cells with ara-C activates both c-Abl (Kharbanda et al. 1995) and the htopoI damage response. At first sight, this may indicate that c-Abl phosphorylates htopoI after DNA damage and thereby induces htopoI cleavage complex formation. Despite this, the phosphorylation on Tyr^{268} on its own is not sufficient to stabilize htopoI cleavage complexes (Yu et al. 2004). Still, the phosphorylation of Tyr^{268} is an important step for cleavage complex formation, but other post-translational modifications or protein-protein interactions are probably also required.

4 The Function of the Human Topoisomerase I Damage Response

4.1 The Human Topoisomerase I Damage Response and DNA Repair

Already in the early 1990s it was believed that htopoI is involved in at least some aspects of DNA repair and particularly in NER. This repair pathway has been investigated in detail and the individual steps are well characterized. In short, NER starts with the recognition of the DNA lesion, most likely by the XPC protein, followed by local unwinding of the DNA double helix by the two helicases XPB and XPD. Subsequently, the damaged strand is incised and removed upstream and downstream of the lesion by the XPF and XPG endonucleases, possibly regulated by the XPA protein (Hoeijmakers 2001; Volker et al. 2001). About 25 nts are excised, depending on the type of lesion. It was speculated that htopoI might release tension generated during the unwinding of the DNA around the damaged site. Numerous investigations tried to answer this question, but never succeeded in demonstrating a convincing link between htopoI and NER (Jones et al. 1991; Stevnsner and Bohr 1993; Thielmann et al. 1993). Research activity on this question seemed to decline. It was only when Lanza and co-workers reported in 1996 (Lanza et al. 1996) that htopoI recognizes UV photoproducts in vitro that the original idea was resurrected. Subramanian and colleagues were the first to report a link between UV-B induced DNA lesions and the stabilization of htopoI cleavage complexes in vivo (Subramanian et al. 1998). The observation that the htopoI damage response to UV lesions was strongly impaired in XPA and XPD deficient cells gave further credibility to the view that this reaction may be a part of NER. In 2003, Mullers group showed that a reduced level of htopoI, by expression of anti-sense mRNA, resulted in increased sensitivity to UV and a decreased repair rate of UV photoproducts. The repair rate was measured by the number of PCNA foci detected by immunofluorescence as well as by the "host cell reactivation assay". Given the importance of htopoI for transcription, it should be noted, however, that a reduction in htopoI levels also influences the transcription rate. Since transcription coupled repair (TCR) is very effective in comparison with global genome repair, these results may simply mean that the decreased repair of UV photoproducts was caused by the reduced transcription rate. Since cells with defective XPA are compromised in the UV-induced htopoI damage response, a more direct role in NER is likely. How could htopoI participate in NER? Mao and Muller suggested two models: first, htopoI may be required to adjust the DNA topology at a lesion, and second, htopoI functions as a recruiting factor or sensor for DNA repair proteins. Mao and Muller favoured the latter model. If this were correct, it would imply that the htopoI damage response is a very early event in the repair process. This is in contrast to the findings that htopoI cleavage complex formation is absent in XPD-deficient cells, although XPD is involved in later processes (DNA unwinding) during NER. Additionally, it cannot explain why a htopoI damage response is down-regulated or lost in the absence of p53. This may rather indicate that the htopoI damage response is initiated after the cell has detected the damage. Thus, we suggest the following scenario (see Fig. 2). After cellular stress (for example UV light), the lesions are recognized by DNA repair pathways such as NER. These instigate a cellular signalling that may be mediated by p53. This signalling in turn "activates" htopoI and initiates a damage response. HtopoI may bind to sites of DNA damage and aid repair. When repair has been completed, htopoI activity is shut off again and the htopoI cleavage complexes disappear. This is in accordance with the finding that at low damage levels, the htopoI damage response is of a transient nature (Soe et al.

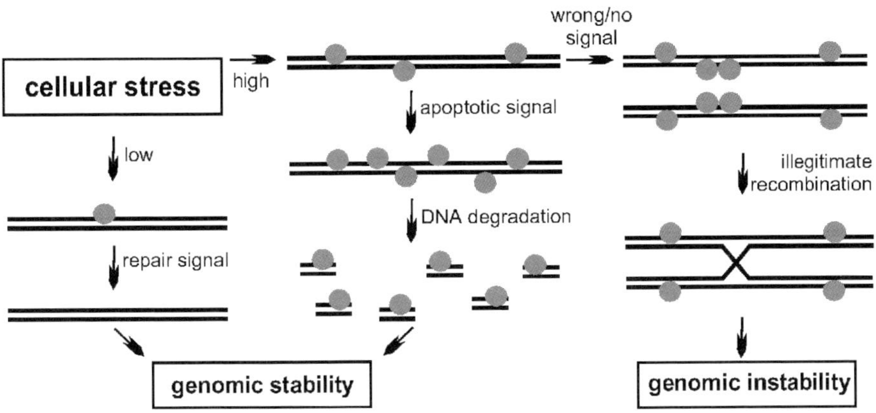

Fig. 2 Model for the purpose and regulation of the htopoI damage response. Please refer to text for further detail

2004). Alternatively, htopoI cleavage complexes may work as cellular sensors to signal how severely the cell is afflicted. If the DNA is only weakly damaged, a few htopoI cleavage complexes will form. These may participate in repair and contribute to cellular survival. However, if the damage level is high, more cleavage complexes accumulate due to an increased signal from, e.g. the NER machinery. Furthermore, a dramatic increase in htopoI cleavage complexes had been found when a genotoxic agent was applied at a high dosage; this was usually accompanied by extensive apoptosis (Soe et al. 2004). Furthermore, the htopoI damage response is strong and persistent in cells destined to die. Thus, there apparently exists a threshold for the number of htopoI cleavage complexes per cell. Below this threshold, htopoI may assist DNA repair processes, and above it, htopoI seems to play a role in the induction or execution of apoptosis.

4.2 The Human Topoisomerase I Damage Response and Apoptosis

In addition to its role in DNA repair, more and more results point to a direct contribution of the htopoI damage response in triggering and/or executing apoptosis. The first hint of its involvement in apoptosis was published in 2000 (Pourquier et al. 2000). Here, it was shown that ara-C incorporation into DNA resulted in a htopoI damage response. Since ara-C is a chemotherapeutic drug used for the treatment of acute leukaemias, but functions by a largely unknown mechanism, the authors wanted to find out whether htopoI contributed to the cell killing effect of this drug. To this end, they used a leukaemia mouse cell line, P388, and a daughter cell line that had previously been generated by selecting P388 cells resistant to 45 μM CPT. The resulting cell line, P388/CPT45, had a non-detectable level of topoI. A comparison between the sensitivity of

P388/CPT45 and P388 to ara-C showed a 7 to 10-fold reduction in sensitivity of P388/CPT45, indicating that the htopoI damage response is actively involved in cell killing. The same was shown for gemcitabine (Pourquier et al. 2002). However, since topoI is an essential enzyme in higher eukaryotic cells, P388/CPT45 may have compensatory mutations, which enable it to grow under such a high CPT concentration. Thus, the effect cannot unambiguously be ascribed to topoI. Recently, an association of a strong htopoI damage response and apoptosis was demonstrated (Soe et al. 2004). We found that a low dose of UV-C light resulted in a weak and transient damage response whereas a high dose was associated with a strong and persistent one. A direct link to apoptosis was shown by separately analysing living and dying cells within the same population. The result was that practically all the htopoI cleavage complexes detected in the heterogeneous cell population after UV irradiation were present in the dying fraction, demonstrating that the htopoI damage response at high damage levels is explicitly associated with apoptosis (Soe et al. 2004). This and further results of another group (Pourquier et al. 2000, 2002) strongly indicate that htopoI contributes to cell killing. Moreover, two recent papers show that after the generation of reactive oxygen species (ROS), the htopoI damage response is actively involved in the DNA laddering during apoptosis (Sordet et al. 2004a,b). This was investigated by the use of arsenic salts and staurosporine, which do not induce DNA damage on their own, but generate ROS by damaging mitochondria. Sordet and co-workers suggested that the htopoI damage response mainly takes place when the mitochondrial membrane potential breaks down during apoptosis and ROS are generated. These in turn generate oxidative DNA lesions that instigate htopoI cleavage complexes and thereby possibly trigger DNA laddering. This apoptotic involvement of htopoI can be overcome by over-expression of Bcl-2 (Sordet et al. 2004b) strongly supporting a direct and functional role of the htopoI damage response in apoptosis. In the case of staurosporine and arsenic salts the htopoI cleavage complexes were detected concomitantly with the occurrence of, e.g. PARP1 cleavage and DNA fragmentation. This was not the case with UV light. Furthermore, transient htopoI cleavage complexes form at UV dosages where no apoptosis is observed. Moreover, at higher UV doses, the htopoI cleavage complexes accumulate prior to DNA laddering and PARP1 cleavage (Soe et al. 2004). Thus, it seems that in the case of UV-induced damage, the htopoI cleavage complexes are formed at lower levels prior to a breakdown of the mitochondria and the resulting accumulation of oxidative damage. However, after UV irradiation and concomitantly with the cleavage of PARP1 and DNA laddering (indicating mitochondrial breakdown) a strong increase in htopoI cleavage complexes was observed. This strong increase is possibly due to a strong increase in ROS. To demonstrate this unambiguously, additional investigations are necessary.

Further support for an active function of topoI in cell killing comes from studies in *Saccharomyces cerevisiae*. Yeast cells also show topoI cleavage complex formation in response to various lesions (Nitiss et al. 2001; Daroui et al.

2004). Since topoI is not essential in yeast, the impact of topoI on cell killing can be measured directly. Nitiss and co-workers demonstrated that over-expression of topoI in yeast sensitized the cells for killing by UV, ionizing radiation and methyl methanesulphonate (MMS) about 5- to 10-fold. On the other hand, over-expression of a catalytically inactive mutant ($topoI^{Y723F}$) showed no effect (Nitiss et al. 2001). This strongly supports a direct role of eukaryotic topoI in cell killing. Reintroducing topoI into a topoI deficient yeast strain sensitized these cells for killing by H_2O_2, whereas expression of $topoI^{Y723F}$ had no effect (Daroui et al. 2004).

As aforementioned, it has not yet been shown in cells that htopoI indeed does cleave in the vicinity of a DNA lesion. Therefore, we considered if DNA lesions really are required to trigger the htopoI damage response. Recent results from our laboratory may indicate that this is not necessarily the case. To address this question, apoptosis was induced by colcemid (a microtubule inhibitor) or by receptor mediated apoptosis. Apoptosis induced by colcemid provoked a very strong htopoI damage response around the onset of apoptosis. This result indicates that a pre-existing DNA lesion is not required. Similarly, when apoptosis was induced through the tumour necrosis factor α (TNF-α) receptor, a very strong htopoI damage response was observed. Induction of apoptosis by TNF-α is exclusively dependent on the activation of caspase 8 (Afford and Randhawa 2000). Based on this result, we can conclude that the htopoI damage response does not require pre-existing DNA damage. It could still be argued that the generation of ROS during colcemid and TNF-α induced apoptosis is a cause for the htopoI damage response. This cannot be ruled out at present and, as discussed above, it is highly likely that ROS enhances the htopoI damage response, but it might not be the only reason for the induction of htopoI cleavage complexes. One supporting argument for this thesis is that the htopoI damage response can be turned on and off in the presence and absence of p53. Cells, which undergo p53-independent apoptosis after UV irradiation, do not show any htopoI damage response despite the fact that they also generate ROS (unpublished data). This suggests that activation of htopoI has to take place prior to the release of ROS in order to make a htopoI damage response possible. These new results clearly demonstrate that the research on function and regulation of the htopoI damage response is only at its beginning. Future experiments should bring more clarity on this complicated issue.

In summary, we suggest the following regulation and role of the htopoI damage response (see Fig. 2). If the DNA damage level is low, only few htopoI cleavage complexes form. According to the discussion above, these may be involved in DNA repair. When repair is completed the htopoI cleavage complexes are also repaired and disappear. However, if the level of DNA damage is very high, more htopoI cleavage complexes accumulate and trigger directly or indirectly an apoptotic signalling pathway. This causes the accumulation of further htopoI cleavage complexes. Such high levels of htopoI cleavage complexes are stable and either trigger or execute apoptosis. As suggested above, they may be an active part in fragmenting the genome during apoptosis. We have data showing that a pre-existing lesion is not necessary for triggering a htopoI damage response. Such cellular stress does not require DNA repair activity, and therefore the htopoI damage response to cellular stress cannot be involved in DNA repair. However, in the case of high cellular stress levels, apoptosis is induced and this is connected to a very strong htopoI damage response. Induction of cellular stress, which does not result in apoptosis, does not cause any htopoI damage response (unpublished data). Therefore, in the case of cellular stress, the htopoI damage response is only associated with a role in apoptosis. No matter if htopoI is active in DNA repair or apoptosis, both tasks insure the genomic stability of the organism as a whole (except single cell organisms). This also implies that if it is not regulated correctly, it may result in genomic instability as suggested in Fig. 2. This issue will be discussed in detail below.

4.3
Possible Role of a Deregulated Human Topoisomerase I Damage Response in Genomic Instability

In the previous sections, the possible function of a properly regulated htopoI damage response was discussed. However, indications in the literature, and our own results, suggest that the response under certain circumstances may trigger genomic instability. Studies in yeast have shown that topoI can support illegitimate recombination events that integrate viral DNA into the genome in vitro as well as in vivo (Wang and Rogler 1991; Zhu and Schiestl 1996). HtopoI was shown to perform several different recombination reactions in vitro (Christiansen and Westergaard 1994). These recombination reactions were relatively inefficient, since they only took place with a large excess of DNA oligomers that had to be ligated. However, at least in vitro, htopoI can induce a recombination reaction that is highly efficient. This reaction is supported by the formation of a so-called htopoI double cleavage complex and is shown in Fig. 3. Such a complex consists of two covalently attached htopoI cleavage complexes that are in direct protein-protein contact; the cleavage sites of these molecules are only about 13 nts apart (Soe et al. 2001). Due to the short distance between the two topoisomerases the double cleavage complex is relatively unstable. As a result, one htopoI molecule associated with a 13 nts long oligonucleotide is released. In vitro studies have shown that the remaining htopoI cleavage complex is very efficient in religating a foreign DNA strand that can hybridize fully or partly to the gap generated in front of this second cleavage complex. The ligation reaction results in the dissociation of the remaining htopoI molecule (see Fig. 3). Thus, the previous double cleavage complex has been repaired. The ligation reaction was called "topoisomerase I induced recombination-like repair" (TIRR)

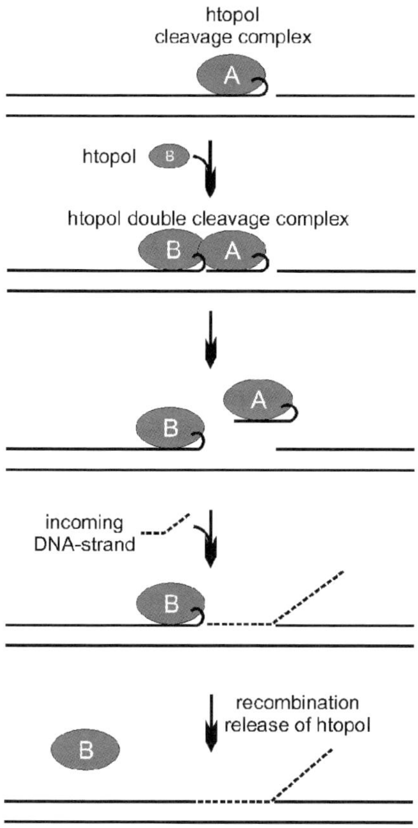

Fig. 3 The Topoisomerase I induced recombination-like repair, TIRR. Please refer to text for further detail

(Stephan et al. 2002). That htopoI double cleavage complexes may be generated on chromosomal DNA during a htopoI damage response in vivo was demonstrated by a cross-linking experiment (Mao et al. 2000a), which also indicated that p53 was physically associated with the htopoI double cleavage complex (as discussed above). This finding is very interesting, since in vitro studies have shown that p53 strongly stimulates TIRR (Soe et al. 2002; Stephan et al. 2002; Soe and Grosse 2003). It has been suggested that TIRR events may be responsible for genomic rearrangements after ionizing radiation, and that these in turn trigger apoptosis (Radford 2002). If, however, apoptosis cannot be induced possibly due to genetic defects, these illegitimate recombination events may cause genomic instability. In fact, there are in vivo data supporting this view from investigations on translocation sites in thyroid carcinomas of children exposed to ionizing radiation after the Chernobyl disaster (Klugbauer et al. 2001). Just recently, Lisa Wiesmüller's group found that TIRR most likely is responsible for one of the characteristic translocations (RARα) during acute promyelocytic leukaemia (APL) (Boehden et al. 2004). Such translocations may be mediated by TIRR-like events such as illustrated in Fig. 2 (right-hand side). Numerous stable htopoI cleavage complexes form during the htopoI damage response (a rough estimate is up to 100,000 htopoI cleavage complexes per cell during a strong htopoI damage response). If just a few of all these complexes are htopoI double cleavage complexes, numerous TIRR events within a cell that does not induce apoptosis properly can experience severe genomic instability. Further investigations will show when the htopoI damage response ensures genomic instability by either DNA repair or apoptosis and when it may result in genomic instability mediated by illegitimate recombination events.

5 Summary

HtopoI is an enzyme that previously was believed only to be involved in the removal of torsional stress generated during transcription and replication. However, in 1998 it was found that htopoI might play another important role in the cellular response to DNA damage – the so-called htopoI damage response. Since this initial discovery, many studies have suggested that the htopoI damage response is involved in DNA repair as well as in apoptosis. The choice of repair or apoptosis most likely depends on the damage level. The investigations within this research field are just at their beginnings. The future will surely bring more clarity on the precise mechanism behind the involvement of htopoI in DNA repair and apoptosis, as well as its implications for a broader understanding of how the human organism ensures genomic stability.

References

Afford S, Randhawa S (2000) Apoptosis. Mol Pathol 53:55–63

Bauer PI, Buki KG, Comstock JA, Kun E (2000) Activation of topoisomerase I by poly [ADP-ribose] polymerase. Int J Mol Med 5:533–540

Bauer PI, Chen HJ, Kenesi E, Kenessey I, Buki KG, Kirsten E, Hakam A, Hwang JI, Kun E (2001) Molecular interactions between poly(ADP-ribose) polymerase (PARP I) and topoisomerase I (Topo I): identification of topology of binding. FEBS Lett 506:239–242

Boehden GS, Restle A, Marschalek R, Stocking C, Wiesmuller L (2004) Recombination at chromosomal sequences involved in leukaemogenic rearrangements is differentially regulated by p53. Carcinogenesis 25:1305–1313

Bowman KJ, Newell DR, Calvert AH, Curtin NJ (2001) Differential effects of the poly (ADP-ribose) polymerase (PARP) inhibitor NU1025 on topoisomerase I and II inhibitor cytotoxicity in L1210 cells in vitro. Br J Cancer 84:106–112

Cardellini E, Felici F, Gianfranceschi GL (2004) Identification of peptides mimicking the ligands of proteins phosphorylated by protein kinase CK2. Peptides 25:191–197

Champoux JJ, Dulbecco R (1972) An activity from mammalian cells that untwists superhelical DNA - a possible swivel for DNA replication (polyoma-ethidium bromide-mouse-embryo cells-dye binding assay). Proc Natl Acad Sci USA 69:143–146
Champoux JJ, McConaughy BL (1976) Purification and characterization of the DNA untwisting enzyme from rat liver. Biochemistry 15:4638–4642
Chatterjee S, Cheng MF, Trivedi D, Petzold SJ, Berger NA (1989) Camptothecin hypersensitivity in poly(adenosine diphosphate-ribose) polymerase-deficient cell lines. Cancer Commun 1:389–394
Chrencik J, Burgin A, Pommier Y, Stewart L, Redinbo M (2003) Structural impact of the leukemia drug 1-beta-D-arabinofuranosylcytosine (Ara-C) on the covalent human topoisomerase I-DNA complex. J Biol Chem 278:12461–12466
Christiansen K, Westergaard O (1994) Characterization of intra- and intermolecular DNA ligation mediated by eukaryotic topoisomerase I. Role of bipartite DNA interaction in the ligation process. J Biol Chem 269:721–729
Christiansen K, Westergaard O (1999) Mapping of eukaryotic DNA topoisomerase I catalyzed cleavage without concomitant religation in the vicinity of DNA structural anomalies. Biochim Biophys Acta 1489:249–262
Daroui P, Desai SD, Li TK, Liu AA, Liu LF (2004) Hydrogen peroxide induces topoisomerase I-mediated DNA damage and cell death. J Biol Chem 279:14587–14594
Desai S, Liu L, Vazquez-Abad D, D'Arpa P (1997) Ubiquitin-dependent destruction of topoisomerase I is stimulated by the antitumor drug camptothecin. J Biol Chem 272:24159–24164
Desai SD, Li TK, Rodriguez-Bauman A, Rubin EH, Liu LF (2001) Ubiquitin/26S proteasome-mediated degradation of topoisomerase I as a resistance mechanism to camptothecin in tumor cells. Cancer Res 61:5926–5932
Durnford JM, Champoux JJ (1978) The DNA untwisting enzyme from *Saccharomyces cerevisiae*. Partial purification and characterization. J Biol Chem 253:1086–1089
Gobert C, Bracco L, Rossi F, Olivier M, Tazi J, Lavelle F, Larsen A, Riou J (1996) Modulation of DNA topoisomerase I activity by p53. Biochemistry 35:5778–5786
Gobert C, Skladanowski A, Larsen A (1999) The interaction between p53 and DNA topoisomerase I is regulated differently in cells with wild-type and mutant p53. Proc Natl Acad Sci USA 96:10355–10360
Hoeijmakers JH (2001) Genome maintenance mechanisms for preventing cancer. Nature 411:366–374
Horie K, Tomida A, Sugimoto Y, Yasugi T, Yoshikawa H, Taketani Y, Tsuruo T (2002) SUMO-1 conjugation to intact DNA topoisomerase I amplifies cleavable complex formation induced by camptothecin. Oncogene 21:7913–7922
Hsiang YH, Hertzberg R, Hecht S, Liu LF (1985) Camptothecin induces protein-linked DNA breaks via mammalian DNA topoisomerase I. J Biol Chem 260:14873–14878
Jones J, Stevnsner T, Mattern M, Bohr V (1991) Effect of specific enzyme inhibitors on replication, total genome DNA repair and on gene-specific DNA repair after UV irradiation in CHO cells. Mutat Res 255:155–162
Kharbanda S, Pandey P, Ren R, Mayer B, Zon L, Kufe D (1995) c-Abl activation regulates induction of the SEK1/stress-activated protein kinase pathway in the cellular response to 1-beta-D-arabinofuranosylcytosine. J Biol Chem 270:30278–30281
Kharbanda S, Pandey P, Jin S, Inoue S, Bharti A, Yuan ZM, Weichselbaum R, Weaver D, Kufe D (1997) Functional interaction between DNA-PK and c-Abl in response to DNA damage. Nature 386:732–735
Klugbauer S, Pfeiffer P, Gassenhuber H, Beimfohr C, Rabes H (2001) RET rearrangements in radiation-induced papillary thyroid carcinomas: high prevalence of topoisomerase I

sites at breakpoints and microhomology-mediated end joining in ELE1 and RET chimeric genes. Genomics 73:149–160

Kordiyak G, Jakes S, Ingebritsen T, Benbow R (1994) Casein kinase II stimulates *Xenopus laevis* DNA topoisomerase I by physical association. Biochemistry 33:13484–13491

Koster DA, Croquette V, Dekker C, Shuman S, Dekker NH (2005) Friction and torque govern the relaxation of DNA supercoils by eukaryotic topoisomerase IB. Nature 434:671–674

Kowalska-Loth B, Girstun A, Derlacz R, Staron K (2003) Activation of human topoisomerase I by protein kinase CK2. Mol Biol Rep 30:107–111

Lanza A, Tornaletti S, Rodolfo C, Scanavini M, Pedrini A (1996) Human DNA topoisomerase I-mediated cleavages stimulated by ultraviolet light-induced DNA damage. J Biol Chem 271:6978–6986

Lee MP, Brown SD, Chen A, Hsieh TS (1993) DNA topoisomerase I is essential in *Drosophila melanogaster*. Proc Natl Acad Sci USA 90:6656–6660

Lesher D, Pommier Y, Stewart L, Redinbo M (2002) 8-Oxoguanine rearranges the active site of human topoisomerase I. Proc Natl Acad Sci USA 99:12102–12107

Lindahl T, Nyberg B (1972) Rate of depurination of native deoxyribonucleic acid. Biochemistry 11:3610–3618

Malanga M, Althaus FR (2004) Poly(ADP-ribose) reactivates stalled DNA topoisomerase I and Induces DNA strand break resealing. J Biol Chem 279:5244–5248

Mao Y, Okada S, Chang L, Muller M (2000a) p53 dependence of topoisomerase I recruitment in vivo. Cancer Res 60:4538–4543

Mao Y, Sun M, Desai S, Liu L (2000b) SUMO-1 conjugation to topoisomerase I: A possible repair response to topoisomerase-mediated DNA damage. Proc Natl Acad Sci USA 97:4046–4051

Mao Y, Mehl I, Muller M (2002) Subnuclear distribution of topoisomerase I is linked to ongoing transcription and p53 status. Proc Natl Acad Sci USA 99:1235–1240

Merino A, Madden KR, Lane WS, Champoux JJ, Reinberg D (1993) DNA topoisomerase I is involved in both repression and activation of transcription. Nature 365:227–232

Morham SG, Kluckman KD, Voulomanos N, Smithies O (1996) Targeted disruption of the mouse topoisomerase I gene by camptothecin selection. Mol Cell Biol 16:6804–6809

Nitiss J, Nitiss K, Rose A, Waltman J (2001) Overexpression of type I topoisomerases sensitizes yeast cells to DNA damage. J Biol Chem 276:26708–26714

Pinna LA (2002) Protein kinase CK2: a challenge to canons. J Cell Sci 115:3873–3878

Pommier Y, Kohlhagen G, Pourquier P, Sayer J, Kroth H, Jerina D (2000) Benzo[a]pyrene diol epoxide adducts in DNA are potent suppressors of a normal topoisomerase I cleavage site and powerful inducers of other topoisomerase I cleavages. Proc Natl Acad Sci USA 97:2040–2045

Pourquier P, Ueng L, Kohlhagen G, Mazumder A, Gupta M, Kohn K, Pommier Y (1997) Effects of uracil incorporation, DNA mismatches, and abasic sites on cleavage and religation activities of mammalian topoisomerase I. J Biol Chem 272:7792–7796

Pourquier P, Ueng L, Fertala J, Wang D, Park H, Essigmann J, Bjornsti M, Pommier Y (1999) Induction of reversible complexes between eukaryotic DNA topoisomerase I and DNA-containing oxidative base damages. 7,8-dihydro-8-oxoguanine and 5-hydroxycytosine. J Biol Chem 274:8516–8523

Pourquier P, Takebayashi Y, Urasaki Y, Gioffre C, Kohlhagen G, Pommier Y (2000) Induction of topoisomerase I cleavage complexes by 1-beta -D-arabinofuranosylcytosine (ara-C) in vitro and in ara-C-treated cells. Proc Natl Acad Sci USA 97:1885–1890

Pourquier P, Waltman J, Urasaki Y, Loktionova N, Pegg A, Nitiss J, Pommier Y (2001) Topoisomerase I-mediated cytotoxicity of *N*-methyl-*N*′-nitro-*N*-nitrosoguanidine: trapping of topoisomerase I by the O6-methylguanine. Cancer Res 61:53–58

Pourquier P, Gioffre C, Kohlhagen G, Urasaki Y, Goldwasser F, Hertel L, Yu S, Pon R, Gmeiner W, Pommier Y (2002) Gemcitabine (2′,2′-difluoro-2′-deoxycytidine), an antimetabolite that poisons topoisomerase I. Clin Cancer Res 8:2499–2504

Radford I (2002) Model for the initiation of ionizing radiation-induced apoptosis in lymphoid cells by complex DNA double-strand breaks. Int J Radiat Biol 78:467–474

Redinbo M, Stewart L, Kuhn P, Champoux J, Hol W (1998) Crystal structures of human topoisomerase I in covalent and noncovalent complexes with DNA. Science 279:1504–1513

Shafman T, Khanna KK, Kedar P, Spring K, Kozlov S, Yen T, Hobson K, Gatei M, Zhang N, Watters D, Egerton M, Shiloh Y, Kharbanda S, Kufe D, Lavin MF (1997) Interaction between ATM protein and c-Abl in response to DNA damage. Nature 387:520–523

Sigal A, Rotter V (2000) Oncogenic mutations of the p53 tumor suppressor: the demons of the guardian of the genome. Cancer Res 60:6788–6793

Soe K, Dianov G, Nasheuer H, Bohr V, Grosse F, Stevnsner T (2001) A human topoisomerase I cleavage complex is recognized by an additional human topisomerase I molecule in vitro. Nucleic Acids Res 29:3195–3203

Soe K, Hartmann H, Schlott B, Stevnsner T, Grosse F (2002) The tumor suppressor protein p53 stimulates the formation of the human topoisomerase I double cleavage complex in vitro. Oncogene 21:6614–6623

Soe K, Grosse F (2003) p53 stimulates human topoisomerase I activity by modulating its DNA binding. Nucleic Acids Res 31:6585–6592

Soe K, Rockstroh A, Schache P, Grosse F (2004) The human topoisomerase I damage response plays a role in apoptosis. DNA Repair (Amst) 3:387–393

Sordet O, Khan QA, Plo I, Pourquier P, Urasaki Y, Yoshida A, Antony S, Kohlhagen G, Solary E, Saparbaev M, Laval J, Pommier Y (2004a) Apoptotic topoisomerase I-DNA complexes induced by staurosporine-mediated oxygen radicals. J Biol Chem 279:50499–50504

Sordet O, Liao Z, Liu H, Antony S, Stevens EV, Kohlhagen G, Fu H, Pommier Y (2004b) Topoisomerase I-DNA complexes contribute to arsenic trioxide-induced apoptosis. J Biol Chem 279:33968–33975

Stephan H, Grosse F, Soe K (2002) Human topoisomerase I cleavage complexes are repaired by a p53-stimulated recombination-like reaction in vitro. Nucleic Acids Res 30:5087–5093

Stevnsner T, Mortensen U, Westergaard O, Bonven B (1989) Interactions between eukaryotic DNA topoisomerase I and a specific binding sequence. J Biol Chem 264:10110–10113

Stevnsner T, Bohr VA (1993) Studies on the role of topoisomerases in general, gene- and strand-specific DNA repair. Carcinogenesis 14:1841–1850

Stewart A, Herrera R, Nordheim A (1990) Rapid induction of c-fos transcription reveals quantitative linkage of RNA polymerase II and DNA topoisomerase I enzyme activities. Cell 60:141–149

Stewart L, Redinbo M, Qiu X, Hol W, Champoux J (1998) A model for the mechanism of human topoisomerase I. Science 279:1534–1541

Subramanian D, Rosenstein B, Muller M (1998) Ultraviolet-induced DNA damage stimulates topoisomerase I-DNA complex formation in vivo: possible relationship with DNA repair. Cancer Res 58:976–984

Thielmann HW, Popanda O, Gersbach H, Gilberg F (1993) Various inhibitors of DNA topoisomerases diminish repair-specific DNA incision in UV-irradiated human fibroblasts. Carcinogenesis 14:2341–2351

Tsao YP, Wu HY, Liu LF (1989) Transcription-driven supercoiling of DNA: direct biochemical evidence from in vitro studies. Cell 56:111–118

Tuteja N, Reddy MK, Mudgil Y, Yadav BS, Chandok MR, Sopory SK (2003) Pea DNA topoisomerase I is phosphorylated and stimulated by casein kinase 2 and protein kinase C. Plant Physiol 132:2108–2115

Uemura T, Yanagida M (1984) Isolation of type I and II DNA topoisomerase mutants from fission yeast: single and double mutants show different phenotypes in cell growth and chromatin organization. Embo J 3:1737–1744

Volker M, Mone MJ, Karmakar P, van Hoffen A, Schul W, Vermeulen W, Hoeijmakers JH, van Driel R, van Zeeland AA, Mullenders LH (2001) Sequential assembly of the nucleotide excision repair factors in vivo. Mol Cell 8:213–224

Vousden K, Lu X (2002) Live or let die: the cell's response to p53. Nat Rev Cancer 2:594–604

Wall M, Wani M, Cook C, Palmer K, McPhail A, Sim G (1966) The isolation and structure of camptothecin, a novel alkaloid leukemia and tumor inhibitor from *Camptotheca acuminata*. J Am Chem Soc 88:3888–3890

Wang HP, Rogler CE (1991) Topoisomerase I-mediated integration of hepadnavirus DNA in vitro. J Virol 65:2381–2392

Wesierska-Gadek J, Schloffer D, Kotala V, Horky M (2002) Escape of p53 protein from E6-mediated degradation in HeLa cells after cisplatin therapy. Int J Cancer 101:128–136

Yu D, Khan E, Khaleque MA, Lee J, Laco G, Kohlhagen G, Kharbanda S, Cheng YC, Pommier Y, Bharti A (2004) Phosphorylation of DNA topoisomerase I by the c-Abl tyrosine kinase confers camptothecin sensitivity. J Biol Chem 279:51851–51861

Yung TM, Sato S, Satoh MS (2004) Poly(ADP-ribosyl)ation as a DNA damage-induced post-translational modification regulating poly(ADP-ribose) polymerase-1-topoisomerase I interaction. J Biol Chem 279:39686–39696

Zhang CX, Chen AD, Gettel NJ, Hsieh TS (2000) Essential functions of DNA topoisomerase I in *Drosophila melanogaster*. Dev Biol 222:27–40

Zhu J, Schiestl R (1996) Topoisomerase I involvement in illegitimate recombination in *Saccharomyces cerevisiae*. Mol Cell Biol 16:1805–1812

Genome Dyn Stab (1)
D.-H. Lankenau: Genome Integrity
DOI 10.1007/7050_013/Published online: 4 February 2006

Genomic Instability in Fanconi Anaemia and Nijmegen Breakage Syndrome

Martin Digweed

Institut für Humangenetik, Charité – Universitätsmedizin Berlin,
Campus Virchow-Klinikum, Augustenburger Platz 1, 13353 Berlin, Germany
martin.digweed@charite.de

Abstract An increased mutation rate in somatic cells is often manifested as spontaneous chromosomal instability. Patients suffering from the so-called chromosome instability syndromes have a greatly increased cancer risk as a major symptom. The complexity of the involvement of the underlying genes in the cellular response to DNA damage is well illustrated by the disorders Nijmegen Breakage Syndrome (NBS) and Fanconi anaemia (FA). These two Mendelian diseases show increased spontaneous chromosomal damage and hypersensitivity towards two of the most destructive environmental carcinogens–ionising radiation in the case of Nijmegen Breakage Syndrome and bifunctional alkylating agents in Fanconi anaemia. This review examines the similarities and differences between these two disorders of DNA repair.

1 Introduction

Nijmegen Breakage Syndrome is a rare autosomal recessive genetic disease with chromosomal instability, radiosensitivity, immunodeficiency, growth retardation, microcepahly and increased cancer risk, in particular for B cell lymphoma as major features. The gene mutated in NBS is *NBS1* and its protein product is nibrin (Varon et al. 1998). Fanconi anaemia is caused by the biallelic mutation of one of at least 11 different genes. FA is characterised by chromosomal instability, sensitivity towards agents that generate DNA interstrand crosslinks (ICLs), bone marrow failure and increased cancer risk, particularly for acute myelogenous leukemia (AML). Despite these differences in clinical picture, there is sufficient overlap that some patients have been erroneously diagnosed with FA although eventually disease-causing mutations were actually found in the *NBS1* gene (Nakanishi et al. 2002; Gennery et al. 2004; New et al. 2005).

The chromosomal instability observed in radiosensitive NBS cells is readily explained by the fact that ionising radiation (IR) produces DNA double strand breaks (DSBs), however, the chromosomal instability of FA cells is less readily explained. Unlike the DNA lesions produced by mono-functional alkylating agents, which can be efficiently repaired by the base excision repair pathway, ICLs affect both strands of the DNA. Error free repair of DNA cross-links must therefore necessarily involve an incision in both strands, a DSB. Indeed, cross-link repair in yeast and mammalian cells has been shown to require DSBs (Rothfuss and Grompe 2004). Interestingly, the induction of DSBs by ICLs is specifically limited to S-phase, only cells replicating their DNA show a cellular response to DNA-crosslinkers (Akkari et al. 2001), and it has been found that DSBs occur at the sites of cross-links when they are encountered at the replication fork by DNA polymerase (Dronkert and Kanaar 2001). These findings confirm the expectation, that just as in NBS, chromosomal instability in FA can be attributed to a DSB repair defect. The difference in carcinogen–sensitivity between FA and NBS reflects more the *time point* of DSB production rather than specificity for a particular primary lesion. The DSB-repair pathway disturbed in FA has the familial breast cancer genes, *BRCA1* and *BRCA2*, as further components, indeed *BRCA2* is the FA gene, *FANCD1*.

The repair pathways used in the cell to repair a DSB can be divided into two principle mechanisms: repair requiring homology between the rejoined sequences and repair which is independent of homology and thus usually termed non-homologous end joining non-homologous end joining (NHEJ; see also the chapter by Dr. Pfeiffer in this volume). Ligation of the free DNA ending at a DSB requires their prior preparation and this is a mutagenic process. Repair based on sequence homology is further classified based on the requirement for strand exchange between DNA molecules, instigated by the RecA family of proteins. Strand exchange is the basis for repair by homologous recombination and gene conversion, a non-reciprocal exchange of information from the intact DNA to the damaged DNA. Resolution of the exchange event may result in a reciprocal exchange with a cross-over event, a relatively rare event in human cells. In synthesis-dependent strand-annealing (SDSA), single stranded DNA from a DSB invades a homologous sequence to prime new DNA synthesis. The nascent DNA strand is released and can anneal to a single stranded DNA on the other side of the DSB to effect gap closure and again gene conversion without crossing over (Lankenau and Gloor 1998). Repair, which is dependent on DNA strand-exchange is termed homology-directed repair (HDR).

Another homology-based mechanism uses homologous sequences within a single DNA molecule to achieve repair. DNA ends are resected to regions of homologous sequence, single stranded DNA on either side of the original DSB then anneal to form a double-strand. DNA polymerase then replaces the missing sequence using the complementary strand as a template. This mechanism is termed single strand annealing (SSA). Due to its extremely high content of repetitive DNA, SSA may be of particular importance for the mammalian genome. Like NHEJ, SSA is a mutagenic process. Only HDR is potentially error free, although gene conversion results in loss of heterozygosity and so can unmask recessive mutations.

2
Fanconi Anaemia and Nijmegen Breakage Syndrome Genes

Currently twelve complementation groups are known in the genetically heterogeneous disease, Fanconi anaemia: A, B, C, D1, D2, E, F, G, I, J, L, M (Joenje et al. 1997; Levitus et al. 2004). The causative gene has now been identified for all of the groups, with the exception of FA-I, generally by the technique of functional cloning using the sensitivity of patient cells towards the crosslinker, Mitomycin C, to select complementing cDNAs (Strathdee et al. 1992) or by exploiting the fact that many of the FA proteins form a complex and using one (known) FA protein to fish for another (unknown) FA protein (Meetei et al. 2003, 2005). *FANCD2* was cloned by a positional cloning strategy (Timmers et al. 2001) as was *FANCA* (The Fanconi anaemia/breast cancer consortium 1996) and *FANCJ* (Levran et al. 2005; Levitus et al. 2005) Finally, *FANCD1* was found to be identical to *BRCA2* after mutation analysis in FA-D1 patients (Howlett et al. 2002). Table 1 gives a summary of the currently known complementation groups, genes and proteins in FA.

In contrast to FA, NBS is genetically much less heterogeneous: the major gene mutated in NBS was identified by positional cloning (Varon et al. 1998) after localisation to chromosome 8q21 by linkage analyses in NBS families (Saar et al. 1997). Simultaneously, the gene was identified as the functional human analogue of the yeast gene, *Xrs2* (Carney et al. 1998). The product of the *Xrs2* gene is found in a trimeric complex together with Rad50 and Mre11, which is apparently involved in both NHEJ and HDR. Similarly in mammalian cells, RAD50, MRE11 and the *NBS1* gene product, nibrin, form a complex (Carney et al. 1998).

Table 1 Fanconi anaemia – Groups, Genes, and Proteins

Group	Chromosome	Gene	Exons	Protein
FA-A	16q24.3	*FANCA*	43	163 kD
FA-B	Xp22.31	*FANCB*	10	95
FA-C	9q22.3	*FANCC*	14	65
FA-D1	13q12-13	*FANCD1/BRCA2*	27	384
FA-D2	3p25.3	*FANCD2*	44	155
FA-E	6p21-22	*FANCE*	10	59
FA-F	11p15	*FANCF*	1	42
FA-G	9p13	*FANCG/XRCC9*	14	68
FA-I				
FA-J	17q23	*FANCJ/BRIP1*	20	141
FA-L	2p16.1	*FANCL*	14	43
FA-M	14q21.2	*FANCM*	23	230

3 DNA Repair Pathways in Fanconi Anaemia and Nijmegen Breakage Syndrome

3.1 Fanconi Anaemia

The most characteristic chromosome aberrations observed in FA metaphases are translocations, manifested as triradial and quadriradial exchange figures (Schroeder and German 1974). These translocations, which are restricted to heterologous chromosomes, are an indication that some aspect of chromosome maintenance involving DNA recombination is probably defective in FA. A further indication that the recombination of DNA molecules is disturbed was given by the examination of nuclear foci formation in FA cells. Foci form as a response to DNA damage and represent aggregations of DNA-repair proteins at the sites of DNA lesions.

Foci usually contain many different DNA repair proteins, although their constitution varies with the type of DNA damage and also with cell cycle (Nelms et al. 1998; Wang et al. 2000; Taniguchi et al. 2002; Yuan et al. 2003). Many foci contain the human homologue of bacterial RecA, RAD51, an essential component of the HDR process. Examining FA cells for RAD51-foci after IR, we found that formation of these aggregates was diminished to about 30% of wild type cells (Digweed et al. 2002). Other researchers using either FA patient cells or rodent cells homologous to FA-A (Arwert et al. 1991) also found attenuation or delay in RAD51-foci formation (Pichierri et al. 2002; Larminat et al. 2004; Wang et al. 2004). We, and others, found foci formation in FA-D1 cells to be particularly strongly reduced, to less than 10% of wild type (Digweed et al. 2002; Godthelp et al. 2002), and this was later explained by the finding that the *FANCD1* gene is in fact *BRCA2* (Howlett et al. 2002), a well established interaction partner of RAD51 (Chen et al. 1998). FANCG also co-localizes with FANCD1/BRCA2 in nuclear foci (Hussain et al. 2003) and FANCD2 also interacts with FANCD1/BRCA2 (Hussain et al. 2004; Wang et al. 2004).

Since, in our hands, RAD51-foci formation was found, but was attenuated, in cells from all FA complementation groups, we interpreted this as a possible deficiency in recruitment of RAD51 to DSBs, rather than a direct deficiency in a core-component of HDR repair. This led to our suggestion that DSB repair by HDR might be generally diminished in FA cells and that the alternative use of inappropriate DNA repair pathways could lead to the chromosomal aberrations characteristic for FA (Digweed et al. 2002). We have now substantiated this hypothesis by using integrated plasmid reporters to follow the repair of a targeted DNA lesion.

Such reporter plasmids, pioneered particularly by Dr. Maria Jasin, exploit the rare cutter endonuclease, I-SceI (Richardson et al. 1999). There are apparently no recognition sequences for this enzyme in the mammalian genome, so that a unique double strand break is set in an integrated plasmid carrying an I-SceI site after expression of the enzyme. Repair of the DSB can then be followed using various assays, such as restoration of the expression of a fluorescent protein or of antibiotic resistance. In addition, amplification and analysis of the sequence containing the I-SceI site can allow determination of the exact repair pathway used.

In both *Fanca*$^{-/-}$ embryonic stem cells and *Fanca*$^{-/-}$ dermal fibroblasts, there was significantly less repair of a damaged integrated reporter plasmid via HDR than in control cells (Yang et al. 2005). The threefold reduction in HDR observed is comparable to that found in murine cells with hypomorphic mutations in *Brca1* (Moynahan et al. 2001a) or *Brca2* (Moynahan et al. 2001b) but much less than the 100-fold reduction found in rodent cells deficient in the *RAD51* paralogue, *XRCC2* (Johnson et al. 1999) or with mutations in *RAD51* itself (Stark et al. 2004). Not surprisingly, such a massive defect in HDR is incompatible with survival and indeed, *Xrcc2* and *RAD51* null mutant mice die before birth (Tsuzuki et al. 1996; Deans et al. 2000) even the above-mentioned hypomorphic *Brca1* and *Brca2* mutations lead to significant embryonic lethality and tumorigenesis (Cressman et al. 1999; Deng and Wang 2003; Donoho et al. 2003). Mice with the *Fanca* null mutation cited above are normally viable although they do have increased chromosomal aberrations, are ICL sensitive and male mice are infertile (Yang et al. 2005). Mutations in FA genes may therefore represent the strongest possible disturbances in the HDR repair pathway that are still compatible with survival.

Examination of HDR using integrated plasmid reporters in FA patient fibroblast cells has confirmed and extended these findings. In cells mutated in *FANCA*, *FANCG* and *FANCD2*, HDR is reduced to a level comparable to the murine *Fanca*$^{-/-}$ cells (Nakanishi et al. 2005). Since FANCA and FANCG are part of the nuclear core complex, which is essential for the monoubiquitination of FANCD2 on lysine 561 in response to DNA damage, the question arose whether this modification is perhaps even a critical step linking the FA proteins to HDR. FA-D2 patient cells were transduced with FANCD2 cDNA containing lysine or arginine at position 561. Only the cDNA with lysine at position 561 was able to stimulate normal levels of HDR (Nakanishi et al. 2005). The monoubiquitination of FANCD2 by the nuclear core-complex is truly an essential process for promotion of HDR in mammalian cells. A more or less severe impairment of HDR has also been reported in chicken DT40 cells with mutations in the homologues of *FANCC* and *FANCG* (Yamamoto et al. 2003; Niedzwiedz et al. 2004).

The HDR disturbance in FA cells parallels that found in cells with mutations in *FANCD1/BRCA2*. Interestingly, FA-A, FA-G and FA-D2 cells differed from FA-D1 cells in as much as repair by SSA was also reduced two-fold. In this respect, the FA proteins examined behave similarly to BRCA1 for which a similar reduction in both HDR and SSA has been found (Stark et al. 2004). Furthermore, direct interactions between BRCA1 and the FANCA and FANCD2 proteins have been reported (Garcia-Higuera et al. 2001; Folias et al. 2002) and BRCA1-interacting protein 1 (BRIP1) is in fact the FA gene, *FANCJ* (Levran et al. 2005). In contrast, *BRCA2/FANCD1* disruption leads to reduced HDR and *increased* SSA (Tutt et al. 2001). Mutation of the NHEJ protein, Ku70, leads in comparison, to increases in both HDR and SSA (Stark et al. 2004). Clearly, these findings suggest that the FA proteins control at least some aspects of both HDR and SSA via the monoubiquitination of FANCD2. BRCA2/FANCD1, which is not required for the monoubiquitination of FANCD2, on the other hand, may be involved only in HDR. Interestingly, chicken DT40 cells deficient in FANCJ/BRIP1 are proficient in HDR although they have typical sensitivity to DNA cross-linking agents (Bridge et al. 2005). Since FANCD2 monoubiquitination occurs normally in FA-J cells (Levitus et al. 2004), normal HDR in FANCJ/BRIP1-deficient cells is in-line with our own results (Nakanishi et al. 2005; Yang et al. 2005). It is tempting to speculate that FANCJ may play a role in an alternative DSB repair pathway such as SSA.

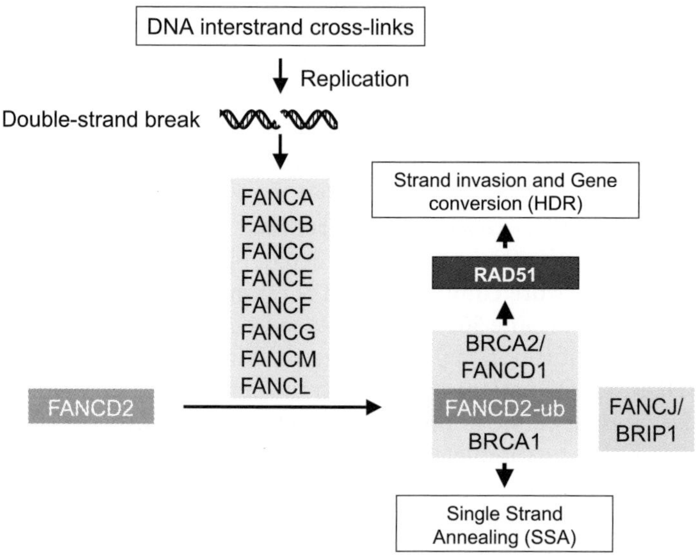

Fig. 1 The Fanconi anaemia DSB repair pathway. During replication, DNA interstrand cross-links lead to DSBs. The modification of FANCD2 by ubiquitination is dependent on the other FA proteins and is a cellular response to DSBs. FANCD2-ub co-operates with BRCA1 and with BRCA2, a further FA protein, in the efficient recruitment of RAD51 to the sites of DSBs for error-free repair by gene conversion. Monoubiquitination of FANCD2 is also important for DSB repair by error-prone SSA. FANCJ is not part of the core-complex for FANCD2 ubiquitination and functions downstream in the FA-pathway, its role suggested here in SSA is purely speculative

In FA-D1 cells, the compensatory increase in SSA, an error-prone process, is expected to increase the cellular mutation rate. In the FA-A, FA-G and FA-D2 cells, both HDR and SSA are reduced, however, a compensatory increase in NHEJ, like SSA also a mutagenic process, was not observed (Nakanishi et al. 2005). Figure 1 shows schematically the FA pathway for DSB repair. The role of the FANCJ/BRIP1 protein is unknown but is apparently situated downstream of FANCD2 monoubiquitination (Levitus et al. 2004) and, surprisingly, independent of its ability to interact with BRCA1, at least in chicken DT40 cells (Bridge et al. 2005).

3.2 Nijmegen Breakage Syndrome

In the yeast *S. cerevisiae*, Mre11 and Rad50 form a complex with Xrs2, the yeast functional analogue of nibrin. The Rad50 component has DNA-binding and ATPase activities whilst Mre11 is a 3′-5′ dsDNA exonuclease, an ssDNA endonuclease and a dsDNA endonuclease. In yeast, this MRX complex suppresses gross chromosomal rearrangements through its multiple functions in DNA repair by HDR, NHEJ and in the intra-S phase checkpoint (Smith et al. 2005). NHEJ was reduced 70-fold in *S. cerevisiae* deleted for *Mre11*, *Rad50* or *Xrs2*. Interestingly, the MRX complex was apparently more important for NHEJ occurring in S and G2 phases rather than in G1 (Moore and Haber 1996). In the fission yeast, Schizosaccharomyces pombe, Mre11 and Rad50 are not required for NHEJ (Manolis et al. 2001). This finding indicates that MRX/N-independent NHEJ exists and, indeed, in chicken DT40 cells disruption of Mre11 and the NHEJ protein, Ku70, has a strongly additive effect (Yamaguchi-Iwai et al. 1999). Ku-independent NHEJ may be mediated by deletion to microhomologous junctions of 8–10 bp a process found in yeast and which is dependent on Mre11 and Rad50 proteins (Ma et al. 2003). This repair mechanism is distinct from SSA which in *S. cerevisiae*, is not dependent upon *Rad50* or *Xrs2*, although mutation of these genes does retard the rate of both SSA and gene conversion (Ivanov et al. 1996).

The role of DSBs and DSB-processing in the rearrangement of immunoglobulin genes offers an apparent explanation for the immunodeficiency of NBS patients. V(D)J recombination, for example, is initiated by the RAG1 and RAG2 enzymes which introduce DSBs into Ig coding segments with DNA hairpin structures as an intermediate. In vitro, the MRE11/RAD50 complex has been shown to cleave DNA hairpins in an ATP-dependent fashion, an activity which is further greatly promoted by the addition of nibrin (Paull and Gellert 1999). Furthermore, immunofluorescence analysis of immature $CD4^+/CD8^+$ thymocytes has shown nibrin as a single fluorescent focus, together with the DSB-responsive histone, gamma-H2AX, at the sites of active V(D)J recombination in the *TCRα* gene (Chen et al. 2000). However, sequencing of the appropriate segments amplified from NBS lymphoblastoid

DNA has indicated normal rearrangements of Ig heavy chains (Petrini et al. 1994).

In confirmation of the above finding, when plasmid reporter substrates were used to examine V(D)J recombination frequencies and the quality of DNA signal-joins in NBS cells, both were found to be normal (Yeo et al. 2000). Surprisingly, however, when similar reporters were used to examine signal-joining in mutants of the yeast, *S. cerevisiae*, an absolute requirement for Rad50, Mre11 and Xrs2 was found (Clatworthy et al. 2005).

A further lymphocyte maturation event requiring the processing of DSBs is class switch recombination (CSR) leading to a switch from IgM to IgG and IgA expression. This is particularly pertinent to NBS since patients do indeed have a deficiency of serum IgG and IgA but essentially unaltered IgM levels (van Engelen et al. 2001). In further support of an involvement of nibrin in CSR, nibrin and the phosphorylated histone, gamma-H2AX, have been found localised together at the sites of class switch recombination in mouse B lymphocytes (Petersen et al. 2001). Examination of the switch recombination junctions in lymphocytes from AT, AT-LD and NBS patients has indicated less efficient and aberrant switching (Pan et al. 2002). In particular, there was an increased use of microhomology at the switch junctions in comparison to control cells (Lahdesmaki et al. 2004).

We have recently been able to prove a role for nibrin in CSR, a NHEJ-process, using a mouse model (Kracker et al. 2005). Since null mutation of the murine homolog of *NBS1*, *Nbn*, is embyonically lethal (Zhu et al. 2001; Dumon-Jones et al. 2003), we have generated mice with a conditional null mutation. After disruption of the second allele in cells from these animals, homozygous null mutant cells arise which are radiosensitive, have high levels of chromosome breakage and die with a half-life of approximately 10 days in culture (Demuth et al. 2004). When B lymphocytes from such animals were examined for IgG1 or IgG3 expression after stimulation to perform CSR, null mutant cells were significantly compromised in this function (Kracker et al. 2005).

Small changes in the abundance of DNA repair and recombination factors can have considerable physiological consequences, and this has been demonstrated for mutants with reduced Xrs2 protein expression and partial defects in formation of meiotic DSBs and maintenance of telomeres (Shima et al. 2005). With a similar expectation, Yabuki and colleagues examined the effect of ectopic over-expression of nibrin on Ig diversification in human B cells and chicken DT40 B cells. The experiments indicated that ectopic expression of nibrin accelerated hypermutation at Ig genes initiated by activation-induced deaminase (AID), both by hypermutation and by gene conversion (Yabuki et al. 2005).

Taken together these findings suggest that the MRN complex is involved in an, error-prone, NHEJ process utilised in vertebrates both for the generation of Ig diversity and for the repair of non-physiological DSBs. In yeast, at least, the MRX complex is also involved in homologous recombination (Bressan et al. 1999). Interestingly, the enzymatic domain of Mre11 is not required for repair of DSBs by HR whilst enzymatically intact Mre11 is required for meiosis suggesting a difference in DNA-end-processing between IR-DSBs and DSBs formed by the Spo11 protein (Moreau et al. 1999). Disrupting *Nbs1* in chicken DT40 cells led to a strong reduction in gene conversion rates in comparison to wild type cells as measured using I-*Sce*I plasmid reporters, however, this phenotype could not be fully rescued by transfecting wild type NBS1 cDNA (Tauchi et al. 2002).

The involvement of MRX/N in both NHEJ and HR suggests a common function in rejoining DSBs, possibly related to the various enzymatic activities of MRE11. In addition, however, mounting evidence suggests a structural role for the complex in tethering the DNA ends of a DSB together (Lobachev et al. 2004). The molecular architecture of the complex and its binding to DNA has been elucidated by electron microscopy and X-ray crystallography (Anderson et al. 2001; de Jager et al. 2001; Hopfner et al. 2001). Briefly, the RAD50 molecule forms an intramolecular coiled coil bringing together its N-terminal and C-terminal globular ATPase domains. These domains interact then with one molecule of an MRE11 dimer whilst two RAD50 molecules form a dimer by interaction in their coiled coil regions. The MRE11 dimer, associated with the RAD50 ATPase domains, binds to DNA ends, a function possibly promoted by a nibrin monomer. Thus, a bridge-like structure is formed that holds two DNA ends in close proximity either as part of a DSB repair process and/or for marking and signalling the presence of a DSB. The chapter by K. P. Hopfner in this book describes these structures in more detail. A purely structural role might explain why in some assays the enzyme domains of MRE11 are dispensable for MRX/N function (Bressan et al. 1999; Moreau et al. 1999; Lewis et al. 2002).

4 Cell Cycle Checkpoints in Fanconi Anaemia and Nijmegen Breakage Syndrome

4.1 Fanconi Anaemia

FA cells proliferate poorly in culture (Weksberg et al. 1979) and cell cycle disturbances can even be useful as a diagnostic parameter (Seyschab et al. 1995). However, there is little evidence, as yet, that the FA proteins are involved directly in cell cycle regulation. Disturbances in cell cycle progression often reflect the specific deficiency in HDR processes, which are expected to play a role in late S-phase and G2-phase. Thus, the interaction between FANCD2 and RAD51, a primary component of HR, is S-phase-specific (Taniguchi et al.

2002). However, direct roles for the FANCC protein in the G2/M checkpoint (Freie et al. 2004) and for FANCD2 in the intra-S-phase checkpoint (Pichierri and Rosselli 2004) have been suggested.

4.2 Nijmegen Breakage Syndrome

It has always been likely that the gene deficient in NBS plays a role not only in the active repair of DNA-DSBs, but also within the cellular signal cascades responding to changes in the status of the genome. In comparison to normal cells, NBS patient cells continue DNA synthesis despite irradiation, so-called radioresistant DNA synthesis (RDS). This is a clear indication that cell cycle regulation is non-functional in NBS, a feature shared with the autosomal recessive diseases Ataxia telangiectasia (AT) and Ataxia telangiectasia like disorder (AT-LD). AT-LD patients have mutations in the gene for MRE11.

Whilst a deficiency in the intra S-phase checkpoint in NBS is undisputed, the status of other checkpoints is controversial. Several studies have examined the G1/S-Phase checkpoint and while normal G1 arrest has been found in NBS fibroblast lines (Yamazaki et al. 1998) in lymphoblastoid NBS B-cells, G1 arrest was defective (Sullivan et al. 1997) or attenuated (Antoccia et al. 1999). In agreement with these latter findings, accumulation of p53 in response to IR is reported to be delayed with subsequently reduced activation of p21 and other p53-dependent genes (Jongmans et al. 1997).

The G2 checkpoint, which prevents cells with damaged DNA from entering mitosis, has been found to function normally in primary patient fibroblasts (Yamazaki et al. 1998) whilst in SV40 transformed fibroblasts there was a longer G2 accumulation than in control SV40-cells (Ito et al. 1999). In other virally transformed NBS cells, EBV immortalised B-lymphocytes, the G2/M checkpoint has been found to be intact (Xu et al. 2001). In one analysis of lymphoblastoid NBS cells; however, there was a remarkable defect in the G2/M checkpoint: cells still entered mitosis one hour after a 1.5Gy irradiation (Buscemi et al. 2001).

Analysis of NBS patient cells for nibrin function is problematic since the major *NBS1* mutation, 657del5, for which over 90% of patients are homoallelic, is in fact hypomorphic and leads to expression of a partially functional 70 kDa nibrin fragment (Maser et al. 2001). We have recently used a conditional null mutant mouse model to examine the G2/M checkpoint and find that it is clearly non-operative in the absence of nibrin (Demuth et al. 2004). The finding that at least some NBS patient cells have a normal G2/M checkpoint suggests that the 70 kDa nibrin fragment is sufficient to uphold this aspect of nibrin function.

Many components of the cell cycle checkpoints are activated after DNA damage events by the kinase, ATM (Ataxia telangiectasia mutated). These target-phosphorylations are generally deficient in NBS patient cells suggesting that nibrin is a promoting factor (Buscemi et al. 2001; Girard et al. 2002; Kitagawa et al. 2004). In elegant biochemical experiments, Lee and Paull have demonstrated that nibrin makes multiple contacts with ATM and facilitates the binding and phosphorylation of substrates such as p53 and CHK2 and, at least the phosphorylation of histone H2AX (Lee and Paull 2004). In cells from AT patients, ATM phosphorylation events can be alternatively carried out by the kinase ATR (ataxia-telangiectasia and RAD3-related). Interestingly, nibrin seems also to facilitate ATR-dependent phosphorylation of Chk1, c-jun and p53 (Stiff et al. 2005). Although AT and NBS have considerable clinical and cellular overlap, in terms of increased mutation rate and cancer risk, NBS patients are even more severely affected than AT patients, this may reflect the dual role of nibrin in both ATM and ATR signalling.

As shown schematically in Fig. 2, the relationship between ATM and nibrin is actually more complex, firstly, because nibrin is itself phosphorylated by ATM after DNA damage (Gatei et al. 2000). Secondly, the MRN complex has been shown to be required for activation of ATM itself (Bakkenist and Kastan 2003; Uziel et al. 2003). The complex senses DSBs and recruits inactive ATM-dimers to them; there, intermolecular autophosphorylation leads to monomerisation and activation of ATM (Lee and Paull 2005; You et al. 2005).

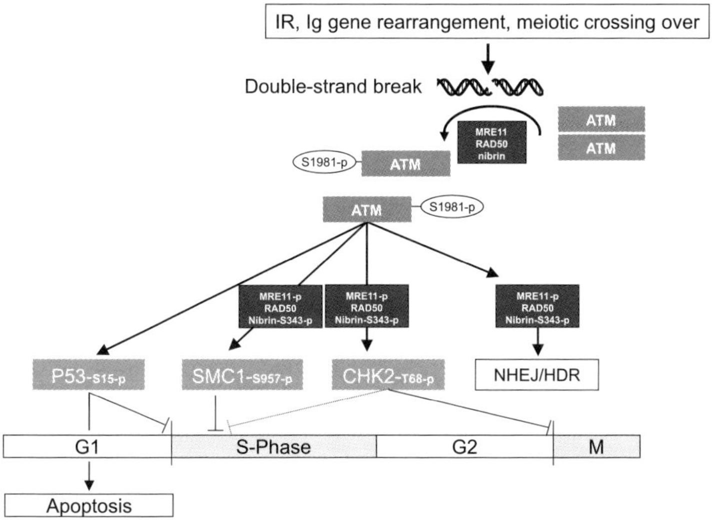

Fig. 2 The NBS DSB repair pathway. DSBs occur after IR exposure and also as a result of physiological processes such as rearrangement of immunoglobulin genes and meiotic crossing over. The trimeric complex of RAD50, MRE11 and NBS1 senses DSBs and binds to DNA before recruiting ATM which is then activated by monomerisation and autophosphorylation. The RAD50/MRE11/NBS1 complex then assists ATM in phosphorylation of CHK2 and SMC1, molecules that are involved in checkpoint regulation of the cell cycle. In addition, the complex is directly involved in DNA repair by NHEJ and/or HDR

Interestingly, the MR complex alone was able to recruit ATM to DNA, however, only the MRN complex was able to activate it. A complex containing not full-length nibrin but the 70 kDa fragment found in patient cells was less able to associate with ATM than the intact complex (Lee and Paull 2004).

In the in vitro experiments of Lee and Paul, nibrin enhanced the ATM-phosphorylation of CHK2 but was not required for phosphorylation of p53 (Lee and Paull 2004). We recently reported on the effects of targeted *Nbn* null mutation in neuronal precursors of mice. In these animals, there was a drastic reduction in cerebellum size and foliation. We were able to show that this phenotype was due to a p53-dependent inhibition of proliferation in neuronal stem cells and increased p53-dependent apoptosis of post-mitotic neurons (Frappart et al. 2005). Since these neurons were devoid of nibrin, it is clearly not required for the p53 response in vivo either.

5
Concluding Remarks

In conclusion, the human diseases Fanconi anaemia and Nijmegen Breakage Syndrome show the dramatic and life-threatening consequences of an inability to maintain genome integrity. In the case of FA, a multi-protein complex modulates a pathway leading to repair of DNA damage by a process utilizing molecular recombination. In fact, FA cells probably represent the greatest disruption of HDR compatible with survival. The central activation of one FA protein by a complex containing most of the other FA proteins explains very well the genetic heterogeneity of the disorder. In addition to HDR, the FA pathway may also control a less conservative DSB repair pathway based on the rejoining of homologous sequences, SSA. In the terms often used to describe genes involved in maintenance of genomic stability, the FA proteins are caretakers rather than gatekeepers.

NBS, in contrast, seems to represent a disorder in which cell cycle checkpoints and DNA repair are equally affected. Nibrin has a caretaker function in holding DNA ends together and preparing them for religation (NHEJ) or duplex invasion (HDR). The error-prone NHEJ repair-mechanism is also exploited during lymphocyte maturation to ensure a high level of Ig diversity. *NBS1* mutation leads therefore to the combination of radiosensitivity and immunodeficiency. On the other hand, nibrin clearly functions as a gatekeeper in signalling changes in genome status to the cell cycle, particularly to the intra-S-phase and G2/M checkpoints. This role in cell proliferation in the absence of exogenous damage might explain the growth retardation of NBS patients. The gatekeeper/caretaker combination is particularly precarious and leads to a high cancer risk for NBS patients and, in contrast to FA, for their heterozygous relatives also (Seemanova 1990; Steffen et al. 2004).

References

Akkari YM, Bateman RL, Reifsteck CA, D'Andrea AD, Olson SB, Grompe M (2001) The 4N cell cycle delay in Fanconi anemia reflects growth arrest in late S phase. Mol Genet Metab 74:403–412

Anderson DE, Trujillo KM, Sung P, Erickson HP (2001) Structure of the Rad50 Mre11 DNA repair complex from Saccharomyces cerevisiae by electron microscopy. J Biol Chem 276:37027–37033

Antoccia A, Stumm M, Saar K, Ricordy R, Maraschio P, Tanzarella C (1999) Impaired p53-mediated DNA damage response, cell-cycle disturbance and chromosome aberrations in Nijmegen breakage syndrome lymphoblastoid cell lines. Int J Radiat Biol 75:583–591

Arwert F, Rooimans MA, Westerveld A, Simons JW, Zdzienicka MZ (1991) The Chinese hamster cell mutant V-H4 is homologous to Fanconi anemia (complementation group A). Cytogenet Cell Genet 56:23–26

Bakkenist CJ, Kastan MB (2003) DNA damage activates ATM through intermolecular autophosphorylation and dimer dissociation. Nature 421:499–506

Bressan DA, Baxter BK, Petrini JH (1999) The Mre11-Rad50-Xrs2 protein complex facilitates homologous recombination-based double-strand break repair in Saccharomyces cerevisiae. Mol Cell Biol 19:7681–7687

Bridge WL, Vandenberg CJ, Franklin RJ, Hiom K (2005) The BRIP1 helicase functions independently of BRCA1 in the Fanconi anemia pathway for DNA crosslink repair. Nat Genet 37:953–957

Buscemi G et al. (2001) Chk2 activation dependence on Nbs1 after DNA damage. Mol Cell Biol 21:5214–5222

Carney JP et al. (1998) The hMre11/hRad50 protein complex and Nijmegen breakage syndrome: linkage of double-strand break repair to the cellular DNA damage response. Cell 93:477–486

Chen HT et al. (2000) Response to RAG-mediated VDJ cleavage by NBS1 and gamma-H2AX. Science 290:1962–1965

Chen PL, Chen CF, Chen Y, Xiao J, Sharp ZD, Lee WH (1998) The BRC repeats in BRCA2 are critical for RAD51 binding and resistance to methyl methanesulfonate treatment. Proc Natl Acad Sci USA 95:5287–5292

Clatworthy AE, Valencia-Burton MA, Haber JE, Oettinger MA (2005) The MRE11-RAD50-XRS2 complex, in addition to other non-homologous end-joining factors, is required for V(D)J joining in yeast. J Biol Chem 280:20247–20252

Cressman VL, Backlund DC, Avrutskaya AV, Leadon SA, Godfrey V, Koller BH (1999) Growth retardation, DNA repair defects, and lack of spermatogenesis in BRCA1-deficient mice. Mol Cell Biol 19:7061–7075

de Jager M, van Noort J, van Gent DC, Dekker C, Kanaar R, Wyman C (2001) Human Rad50/Mre11 is a flexible complex that can tether DNA ends. Mol Cell 8:1129–1135

Deans B, Griffin CS, Maconochie M, Thacker J (2000) Xrcc2 is required for genetic stability, embryonic neurogenesis and viability in mice. Embo J 19:6675–6685

Demuth I et al. (2004) An inducible null mutant murine model of Nijmegen breakage syndrome proves the essential function of NBS1 in chromosomal stability and cell viability. Hum Mol Genet 13:2385–2397

Deng CX, Wang RH (2003) Roles of BRCA1 in DNA damage repair: a link between development and cancer. Hum Mol Genet 12 Spec No 1:R113–123

Digweed M et al. (2002) Attenuation of the formation of DNA-repair foci containing RAD51 in Fanconi anaemia. Carcinogenesis 23:1121–1126

Donoho G et al. (2003) Deletion of Brca2 exon 27 causes hypersensitivity to DNA crosslinks, chromosomal instability, and reduced life span in mice. Genes Chromosomes Cancer 36:317–331

Dronkert ML, Kanaar R (2001) Repair of DNA interstrand cross-links. Mutat Res 486:217–247

Dumon-Jones V et al. (2003) Nbn heterozygosity renders mice susceptible to tumor formation and ionizing radiation-induced tumorigenesis. Cancer Res 63:7263–7269

Folias A et al. (2002) BRCA1 interacts directly with the Fanconi anemia protein FANCA. Hum Mol Genet 11:2591–2597

Frappart PO et al. (2005) An essential function for NBS1 in the prevention of ataxia and cerebellar defects. Nat Med 11:538–544

Freie BW et al. (2004) A role for the Fanconi anemia C protein in maintaining the DNA damage-induced G2 checkpoint. J Biol Chem 279:50986–50993

Garcia-Higuera I et al. (2001) Interaction of the Fanconi anemia proteins and BRCA1 in a common pathway. Mol Cell 7:249–262

Gatei M et al. (2000) ATM-dependent phosphorylation of nibrin in response to radiation exposure. Nat Genet 25:115–119

Gennery AR et al. (2004) The clinical and biological overlap between Nijmegen Breakage Syndrome and Fanconi anemia. Clin Immunol 113:214–219

Girard PM, Riballo E, Begg AC, Waugh A, Jeggo PA (2002) Nbs1 promotes ATM dependent phosphorylation events including those required for G1/S arrest. Oncogene 21:4191–4199

Godthelp BC, Artwert F, Joenje H, Zdzienicka MZ (2002) Impaired DNA damage-induced nuclear Rad51 foci formation uniquely characterizes Fanconi anemia group D1. Oncogene 21:5002–5005

Hopfner KP, Karcher A, Craig L, Woo TT, Carney JP, Tainer JA (2001) Structural biochemistry and interaction architecture of the DNA double-strand break repair Mre11 nuclease and Rad50-ATPase. Cell 105:473–485

Howlett NG et al. (2002) Biallelic inactivation of BRCA2 in Fanconi anemia. Science 297:606–609

Hussain S et al. (2004) Direct interaction of FANCD2 with BRCA2 in DNA damage response pathways. Hum Mol Genet 13:1241–1248

Hussain S, Witt E, Huber PA, Medhurst AL, Ashworth A, Mathew CG (2003) Direct interaction of the Fanconi anaemia protein FANCG with BRCA2/FANCD1. Hum Mol Genet 12:2503–2510

Ito A et al. (1999) Expression of full-length NBS1 protein restores normal radiation responses in cells from Nijmegen breakage syndrome patients. Biochem Biophys Res Commun 265:716–721

Ivanov EL, Sugawara N, Fishman-Lobell J, Haber JE (1996) Genetic requirements for the single-strand annealing pathway of double-strand break repair in Saccharomyces cerevisiae. Genetics 142:693–704

Joenje H et al. (1997) Evidence for at least eight Fanconi anemia genes. Am J Hum Genet 61:940–944

Johnson RD, Liu N, Jasin M (1999) Mammalian XRCC2 promotes the repair of DNA double-strand breaks by homologous recombination. Nature 401:397–399

Jongmans W, Vuillaume M, Chrzanowska K, Smeets D, Sperling K, Hall J (1997) Nijmegen breakage syndrome cells fail to induce the p53-mediated DNA damage response following exposure to ionizing radiation. Mol Cell Biol 17:5016–5022

Kitagawa R, Bakkenist CJ, McKinnon PJ, Kastan MB (2004) Phosphorylation of SMC1 is a critical downstream event in the ATM-NBS1-BRCA1 pathway. Genes Dev 18:1423–1438

Kracker S et al. (2005) Nibrin functions in Ig class-switch recombination. Proc Natl Acad Sci USA 102:1584–1589

Lahdesmaki A, Taylor AM, Chrzanowska KH, Pan-Hammarstrom Q (2004) Delineation of the role of the Mre11 complex in class switch recombination. J Biol Chem 279:16479–16487

Lankenau D-H, Gloor GB (1998) In vivo gap repair in Drosophila: a one-way street with many destinations. Bioessays 20:317–327

Larminat F, Germanier M, Papouli E, Defais M (2004) Impairment of homologous recombination control in a Fanconi anemia-like Chinese hamster cell mutant. Biol Cell 96:545–552

Lee JH, Paull TT (2004) Direct activation of the ATM protein kinase by the Mre11/Rad50/Nbs1 complex. Science 304:93–96

Lee JH, Paull TT (2005) ATM activation by DNA double-strand breaks through the Mre11-Rad50-Nbs1 complex. Science 308:551–554

Levitus M et al. (2004) Heterogeneity in Fanconi anemia: evidence for 2 new genetic subtypes. Blood 103:2498–2503

Levitus M et al. (2005) The DNA helicase BRIP1 is defective in Fanconi anemia complementation group. J Nat Genet 37:934–935

Levran O et al. (2005) The BRCA1-interacting helicase BRIP1 is deficient in Fanconi anemia. Nat Genet 37:931–933

Lewis LK, Karthikeyan G, Westmoreland JW, Resnick MA (2002) Differential suppression of DNA repair deficiencies of Yeast rad50, mre11 and xrs2 mutants by EXO1 and TLC1 (the RNA component of telomerase). Genetics 160:49–62

Lobachev K, Vitriol E, Stemple J, Resnick MA, Bloom K (2004) Chromosome fragmentation after induction of a double-strand break is an active process prevented by the RMX repair complex. Curr Biol 14:2107–2112

Ma JL, Kim EM, Haber JE, Lee SE (2003) Yeast Mre11 and Rad1 proteins define a Ku-independent mechanism to repair double-strand breaks lacking overlapping end sequences. Mol Cell Biol 23:8820–8828

Manolis KG, Nimmo ER, Hartsuiker E, Carr AM, Jeggo PA, Allshire RC (2001) Novel functional requirements for non-homologous DNA end joining in Schizosaccharomyces pombe. Embo J 20:210–221

Maser RS, Zinkel R, Petrini JH (2001) An alternative mode of translation permits production of a variant NBS1 protein from the common Nijmegen breakage syndrome allele. Nat Genet 27:417–421

Meetei AR et al. (2003) A novel ubiquitin ligase is deficient in Fanconi anemia. Nat Genet 35:165–170

Meetei AR et al. (2005) A human ortholog of archaeal DNA repair protein Hef is defective in Fanconi anemia complementation group M. Nat Genet 37:958–963

Moore JK, Haber JE (1996) Cell cycle and genetic requirements of two pathways of non-homologous end-joining repair of double-strand breaks in Saccharomyces cerevisiae. Mol Cell Biol 16:2164–2173

Moreau S, Ferguson JR, Symington LS (1999) The nuclease activity of Mre11 is required for meiosis but not for mating type switching, end joining, or telomere maintenance. Mol Cell Biol 19:556–566

Moynahan ME, Cui TY, Jasin M (2001a) Homology-directed dna repair, mitomycin-c resistance, and chromosome stability is restored with correction of a Brca1 mutation. Cancer Res 61:4842–4850

Moynahan ME, Pierce AJ, Jasin M (2001b) BRCA2 is required for homology-directed repair of chromosomal breaks. Mol Cell 7:263–272

Nakanishi K et al. (2002) Interaction of FANCD2 and NBS1 in the DNA damage response. Nat Cell Biol 4:913–920

Nakanishi K et al. (2005) Human Fanconi anemia monoubiquitination pathway promotes homologous DNA repair. Proc Natl Acad Sci USA 102:1110–1115

Nelms BE, Maser RS, MacKay JF, Lagally MG, Petrini JH (1998) In situ visualization of DNA double-strand break repair in human fibroblasts. Science 280:590–592

New HV et al. (2005) Nijmegen breakage syndrome diagnosed as Fanconi anaemia. Pediatr Blood Cancer 44:494–499

Niedzwiedz W, Mosedale G, Johnson M, Ong CY, Pace P, Patel KJ (2004) The Fanconi anaemia gene FANCC promotes homologous recombination and error-prone DNA repair. Mol Cell 15:607–620

Pan Q, Petit-Frere C, Lahdesmaki A, Gregorek H, Chrzanowska KH, Hammarstrom L (2002) Alternative end joining during switch recombination in patients with ataxia-telangiectasia. Eur J Immunol 32:1300–1308

Paull TT, Gellert M (1999) Nbs1 potentiates ATP-driven DNA unwinding and endonuclease cleavage by the Mre11/Rad50 complex. Genes Dev 13:1276–1288

Petersen S et al. (2001) AID is required to initiate Nbs1/gamma-H2AX focus formation and mutations at sites of class switching. Nature 414:660–665

Petrini JH, Donovan JW, Dimare C, Weaver DT (1994) Normal V(D)J coding junction formation in DNA ligase I deficiency syndromes. J Immunol 152:176–183

Pichierri P, Averbeck D, Rosselli F (2002) DNA cross-link-dependent RAD50/MRE11/NBS1 subnuclear assembly requires the Fanconi anemia C protein. Hum Mol Genet 11:2531–2546

Pichierri P, Rosselli F (2004) The DNA crosslink-induced S-phase checkpoint depends on ATR-CHK1 and ATR-NBS1-FANCD2 pathways. Embo J 23:1178–1187

Richardson C, Elliott B, Jasin M (1999) Chromosomal double-strand breaks introduced in mammalian cells by expression of I-Sce I endonuclease. Methods Mol Biol 113:453–463

Rothfuss A, Grompe M (2004) Repair kinetics of genomic interstrand DNA cross-links: evidence for DNA double-strand break-dependent activation of the Fanconi anemia/BRCA pathway. Mol Cell Biol 24:123–134

Saar K et al. (1997) The gene for the ataxia-telangiectasia variant, Nijmegen breakage syndrome, maps to a 1-cM interval on chromosome 8q21. Am J Hum Genet 60:605–610

Schroeder TM, German J (1974) Bloom's syndrome and Fanconi's anemia: demonstration of two distinctive patterns of chromosome disruption and rearrangement. Humangenetik 25:299–306

Seemanova E (1990) An increased risk for malignant neoplasms in heterozygotes for a syndrome of microcephaly, normal intelligence, growth retardation, remarkable facies, immunodeficiency and chromosomal instability. Mutat Res 238:321–324

Seyschab H et al. (1995) Comparative evaluation of diepoxybutane sensitivity and cell cycle blockage in the diagnosis of Fanconi anemia. Blood 85:2233–2237

Shima H, Suzuki M, Shinohara M (2005) Isolation and characterization of novel xrs2 mutations in Saccharomyces cerevisiae. Genetics 170:71–85

Smith S, Gupta A, Kolodner RD, Myung K (2005) Suppression of gross chromosomal rearrangements by the multiple functions of the Mre11-Rad50-Xrs2 complex in Saccharomyces cerevisiae. DNA Repair (Amst) 4:606–617

Stark JM, Pierce AJ, Oh J, Pastink A, Jasin M (2004) Genetic steps of mammalian homologous repair with distinct mutagenic consequences. Mol Cell Biol 24:9305–9316

Steffen J et al. (2004) Increased cancer risk of heterozygotes with NBS1 germline mutations in Poland. Int J Cancer 111:67–71

Stiff T, Reis C, Alderton GK, Woodbine L, O'Driscoll M, Jeggo PA (2005) Nbs1 is required for ATR-dependent phosphorylation events. Embo J 24:199–208

Strathdee CA, Gavish H, Shannon WR, Buchwald M (1992) Cloning of cDNAs for Fanconi's anaemia by functional complementation. Nature 358:434

Sullivan KE, Veksler E, Lederman H, Lees-Miller SP (1997) Cell cycle checkpoints and DNA repair in Nijmegen breakage syndrome. Clin Immunol Immunopathol 82:43–48

Taniguchi T, Garcia-Higuera I, Andreassen PR, Gregory RC, Grompe M, D'Andrea AD (2002) S-phase-specific interaction of the Fanconi anemia protein, FANCD2, with BRCA1 and RAD51. Blood 100:2414–2420

Tauchi H et al. (2002) Nbs1 is essential for DNA repair by homologous recombination in higher vertebrate cells. Nature 420:93–98

The Fanconi anaemia/breast cancer consortium (1996) Positional cloning of the Fanconi anaemia group A gene. Nat Genet 14:324–328

Timmers C et al. (2001) Positional cloning of a novel Fanconi anemia gene, FANCD2. Mol Cell 7:241–248

Tsuzuki T et al. (1996) Targeted disruption of the Rad51 gene leads to lethality in embryonic mice. Proc Natl Acad Sci USA 93:6236–6240

Tutt A et al. (2001) Mutation in Brca2 stimulates error-prone homology-directed repair of DNA double-strand breaks occurring between repeated sequences. Embo J 20:4704–4716

Uziel T, Lerenthal Y, Moyal L, Andegeko Y, Mittelman L, Shiloh Y (2003) Requirement of the MRN complex for ATM activation by DNA damage. Embo J 22:5612–5621

van Engelen BG, Hiel JA, Gabreels FJ, van den Heuvel LP, van Gent DC, Weemaes CM (2001) Decreased immunoglobulin class switching in Nijmegen Breakage syndrome due to the DNA repair defect. Hum Immunol 62:1324–1327

Varon R et al. (1998) Nibrin, a novel DNA double-strand break repair protein, is mutated in Nijmegen breakage syndrome. Cell 93:467–476

Wang X, Andreassen PR, D'Andrea AD (2004) Functional interaction of monoubiquitinated FANCD2 and BRCA2/FANCD1 in chromatin. Mol Cell Biol 24:5850–5862

Wang Y, Cortez D, Yazdi P, Neff N, Elledge SJ, Qin J (2000) BASC, a super complex of BRCA1-associated proteins involved in the recognition and repair of aberrant DNA structures. Genes Dev 14:927–939

Weksberg R, Buchwald M, Sargent P, Thompson MW, Siminovitch L (1979) Specific cellular defects in patients with Fanconi anemia. J Cell Physiol 101:311–323

Xu B, Kim S, Kastan MB (2001) Involvement of Brca1 in S-phase and G(2)-phase checkpoints after ionizing irradiation. Mol Cell Biol 21:3445–3450

Yabuki M, Fujii MM, Maizels N (2005) The MRE11-RAD50-NBS1 complex accelerates somatic hypermutation and gene conversion of immunoglobulin variable regions. Nat Immunol 6:730–736

Yamaguchi-Iwai Y et al. (1999) Mre11 is essential for the maintenance of chromosomal DNA in vertebrate cells. Embo J 18:6619–6629

Yamamoto K et al. (2003) Fanconi anemia FANCG protein in mitigating radiation- and enzyme-induced DNA double-strand breaks by homologous recombination in vertebrate cells. Mol Cell Biol 23:5421–5430

Yamazaki V, Wegner RD, Kirchgessner CU (1998) Characterization of cell cycle checkpoint responses after ionizing radiation in Nijmegen breakage syndrome cells. Cancer Res 58:2316–2322

Yang YG et al. (2005) The Fanconi anemia group A protein modulates homologous repair of DNA double-strand breaks in mammalian cells. Carcinogenesis 26:1731–1740

Yeo TC et al. (2000) V(D)J rearrangement in Nijmegen breakage syndrome. Mol Immunol 37:1131–1139

You Z, Chahwan C, Bailis J, Hunter T, Russell P (2005) ATM Activation and Its Recruitment to Damaged DNA Require Binding to the C Terminus of Nbs1. Mol Cell Biol 25:5363–5379

Yuan SS, Chang HL, Lee EY (2003) Ionizing radiation-induced Rad51 nuclear focus formation is cell cycle-regulated and defective in both ATM(-/-) and c-Abl(-/-) cells. Mutat Res 525:85–92

Zhu J, Petersen S, Tessarollo L, Nussenzweig A (2001) Targeted disruption of the Nijmegen breakage syndrome gene NBS1 leads to early embryonic lethality in mice. Curr Biol 11:105–109

Part E
Radiation and other Genotoxic Stress

Genome Dyn Stab (1)
D.-H. Lankenau: Genome Integrity
DOI 10.1007/7050_014/Published online: 10 February 2006

Induction of DNA Repair Genes in Mammalian Cells in Response to Genotoxic Stress

Markus Christmann · Gerhard Fritz · Bernd Kaina (✉)

Department of Toxicology, University of Mainz, Obere Zahlbacher Straße 67, 55131 Mainz, Germany
kaina@uni-mainz.de

Abstract Genotoxic agents provoke the activation of receptor-triggered pathways and DNA damage-related functions. Here we review data on immediate-early cellular responses and transcriptional activation of DNA repair genes following exposure of mammalian cells to genotoxic stress. Functional consequences of induction of DNA repair genes are also briefly discussed.

1 Introduction

Mammalian cells respond on exposure to ultraviolet (UV) light, ionizing radiation (IR) or chemical DNA-damaging agents with the modification and activation of different transcription factors eventually leading to up-regulation of genes controlling cell cycle arrest, DNA repair and survival or, if this fails, cell death. Depending on the time sequence of their appearance, these responses can be subgrouped into immediate-early, delayed and late responses. The immediate-early response includes activation of pre-existing signaling proteins and transcription factors (such as Jun kinase, p38 kinase, ATM, ATR, BRCA1 and p53) by post-translational mechanisms as well as transcriptional up-regulation of genes belonging to the fos/jun family (e.g. c-fos, c-jun). Thus, the immediate-early response upon genotoxic stress sets the cell in a state of being prepared to activate genes involved in defense and protection. The delayed genotoxic response is represented by the activation of repair functions resulting in the removal of DNA adducts or favouring the bypass of DNA lesions during replication. The late response of cells provoked by genotoxins is characterized by the induction of genotoxic effects (e.g. clastogenicity and sister-chromatid exchange formation), gene mutations, gene amplification and reproductive cell death. Presently, more than 120 different DNA repair genes, including genes encoding genuine DNA repair enzymes and proteins associated with DNA repair, have been identified (Christmann et al. 2003). Although most of them have been cloned and studied as to their function, data on regulation of repair genes are still scarce. Only a few DNA mammalian repair genes have been reported to be subject to transcriptional regulation upon genotoxic stress; these are reviewed here.

2
Activation of Transcription Factors by Genotoxic Stress (Early Response)

The early genotoxic stress response is characterized by the activation of a complex cascade of protein phosphorylations leading to rapid activation of transcription factors. The most important transcription factors, which become activated within a few minutes after exposure of mammalian cells to genotoxins, are AP-1 and p53. The main pathways of activation of these transcription factors by DNA-damaging insults is illustrated in Fig. 1.

p53 is a sequence-specific transcription factor (el-Deiry et al. 1992) that plays a pivotal role not only in DNA repair but also in the regulation of cell cycle progression (Janus et al. 1999). p53 becomes activated upon DNA replication arrest and DNA double-strand break (DSB) formation which are provoked by DNA-damaging compounds. Recognition of DNA double-strand breaks and replication blocking lesions is accomplished by ATM (ataxia telangiectasia mutated) and ATR (ataxia telangiectasia related) protein, respectively (Zou et al. 2003). This mechanism requires RAD17 and the 9-1-1 complex; the latter consisting of RAD9, HUS1 and RAD1 (Zou and Elledge 2003; Zou et al. 2003). ATM phosphorylates checkpoint kinase-2 (CHK2)

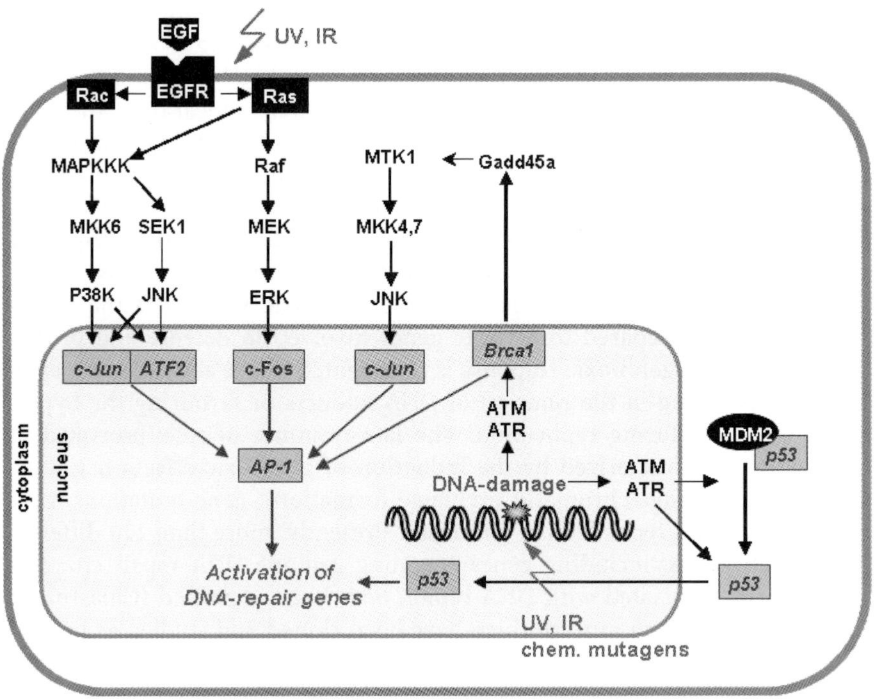

Fig. 1 Model of cellular responses to genotoxic stress. For explanation see text

upon the formation of DSBs (Matsuoka et al. 2000; Zhou et al. 2000), whereas ATR predominantly triggers CHK1 phosphorylation at Ser 345 after DNA replication blockade (Guo et al. 2000; Liu et al. 2000). Interestingly, ATM and CHK2 also become activated in early tumorigenesis. Whether this is due to DSBs is unclear. It may indicate, however, that DNA damage has occurred and that the response may establish an anti-cancer barrier (Bartkova et al. 2005). Both CHK2 and CHK1 are able to phosphorylate and thereby activate p53 (Shieh et al. 2000). Phosphorylation of p53 by checkpoint kinases masks the MDM2 binding site thus leading to a release of p53 from MDM2. Since MDM2 normally targets p53 for ubiquitin-dependent degradation, the dissociation of the MDM2-p53 complex results in stabilization of the p53 protein (Chehab et al. 2000; Chehab et al. 1999; Hirao et al. 2000; Unger et al. 1999). ATM and ATR may also directly phosphorylate p53, thereby increasing its transactivation activity (Banin et al. 1998; Canman et al. 1998). Phosphorylation also provokes down-regulation of MDM2 (Khosravi et al. 1999). As a consequence, the nuclear translocation of p53 and its DNA binding activity become enhanced and target genes become transcriptionally activated.

Stimulation of AP-1-regulated gene expression depends on two different mechanisms, namely increase in the expression of the proteins the AP-1 complex is composed of (e.g. c-JUN, c-FOS) and the post-translational activation of the complex by phosphorylation-based mechanisms. In mammalian cells chemical agents inducing bulky replication blocking lesions, radiation (UV light and ionizing radiation) as well as simple methylating agents such as MMS and MNNG act as activators of *c-fos* and *c-jun* gene expression (Dosch and Kaina 1996). Phosphorylation of the translated proteins is mediated either in a DNA damage-dependent or DNA damage-independent manner. It is believed that notably UV-C light causes activation of growth factor receptors, such as the EGF receptor (Knebel et al. 1996; Sachsenmaier et al. 1994). In this process, reactive oxygen species are thought to play a key role (Huang et al. 1996). Activated (i.e. tyrosine-phosphorylated) receptors stimulate the activity of GTP-binding proteins (i.e. RAS and the Ras-homologous GTPase RAC), which in turn trigger activation of a MAP kinase cascade finally leading to an enhanced activity of mitogen-activated protein kinases (MAPK) such as ERK1/2, JNK and p38K. These protein kinases phosphorylate and thereby stimulate the DNA binding and transactivating activity of numerous transcription factors, including the AP-1 family.

In addition to this immediate-early signaling evoked by genotoxic stress, AP-1 appears to be activated by DNA damage-related signal mechanisms, sharing initial steps with the activation of p53. As mentioned earlier, ATM and ATR become activated by DSBs and DNA replication blockade, respectively. ATM and ATR phosphorylate and activate the breast cancer associated protein 1 (BRCA1) (Tibbetts et al. 2000; Xu et al. 2002) which is implicated in the regulation of several cellular functions and chromatin remodeling as well as DNA repair (Starita and Parvin 2003). Furthermore, BRCA1 associates with RNA polymerase II within the RNA polymerase II holoenzyme complex (Anderson et al. 1998) thus having impact on transcription *in vitro* (Haile and Parvin 1999). BRCA1 was shown to regulate the expression of p21 (Somasundaram et al. 1997) and GADD45 (Harkin et al. 1999; Jin et al. 2000). Additionally, BRCA1 physically interacts with p53 and stimulates its transcriptional activity (Ouchi et al. 1998; Zhang et al. 1998). A functional link between ATM/ATR and JNK signaling is provided by GADD45a. GADD45a is transcriptionally activated in a BRCA1 dependent manner and increases the activity of MTK1/MEKK4 by binding to it (Mita et al. 2002; Takekawa and Saito 1998). MTK1/MEKK4 is the major upstream regulatory kinase of MKK4 and MKK7 which control the activity of stress-activated protein kinases/c-Jun-N-terminal kinases (SAPK/JNK) (Fig. 1). Interestingly, MKK7 activity can be inhibited by interaction with another member of the GADD family, namely GADD45b (Papa et al. 2004a,b).

Despite growing knowledge about early genotoxic response pathways, their inter-relationship with downstream-regulated genes is still fragmentary. Nevertheless, there is growing evidence for a role of AP-1 and p53 in the regulation of DNA repair genes, which are summarized below.

3 Transcriptional Activation of DNA Repair Genes (Delayed Response)

Up to now only a small number of DNA repair genes (to our knowledge 11 out of $\sim$ 130 DNA repair genes) have been described to be inducible in mammalian cells upon their exposure to DNA-damaging agents. These genes are summarized in Table 1 and concisely described below.

3.1 O^6-Methylguanine-DNA Methyltransferase (MGMT)

O^6-Alkylguanine is the major mutagenic and cytotoxic DNA lesion induced by alkylating agents. This lesion is repaired in a suicide reaction by O^6-methylguanine-DNA methyltransferase (alkyltransferase; MGMT). Induction (i.e. transcriptional activation) of MGMT occurs 12–24 h after treatment with UV-C, γ-rays or alkylating agents (Fritz et al. 1991; Grombacher and Kaina 1995). In HeLa S3 cells, MGMT expression level was increased three to five fold upon treatment with different activators of protein kinase C (PKC) such as phorbol-12-myristate-13-acetate (TPA) and 1,2-diacylglycerol (DAG) (Boldogh et al. 1998). The MGMT promoter harbours two AP-1 binding sites whose deletion attenuates MGMT promoter activation (Boldogh et al. 1998). The MGMT promoter can be transactivated by co-expression of c-FOS and c-JUN in F9 cells. Beside regulation by c-FOS and c-JUN, p53 was also shown to be required for MGMT gene induction upon ionizing radiation (IR). Thus,

Table 1 Genotoxin-induced DNA repair genes

Repair gene	Inducing agents	Transcription factor involved
ape	H_2O_2, γ-ray	AP-1
ddb2	UV-C, adriamycin, cisplatin	p53, BRCA1
ercc1	γ-ray, arsenic compounds, EGF	AP-1
fen1	UV-C, MMS	p53
lig1	UV-C	?
mgmt	γ-ray, UV-C, MMS, TPA glucocorticoids	p53, AP-1
mpg	UV-C, TPA	?
msh2	UV-B, TPA	p53, AP-1
ogg1	MMS, γ-ray	NF-YA
poli	UV-C	?
rev3	arsenic compounds, γ-ray MNNG	?
xrcc1	γ-ray, EGF	?
xpc	UV-C, γ-ray, MMS	p53, BRCA1

p53 knockout mice do not show MGMT induction upon whole body irradiation (Rafferty et al. 1996) and the MGMT promoter was found to be activated in p53 wild-type but not p53-deficient mouse fibroblasts (Grombacher et al. 1998). Up-regulation of MGMT by pretreatment of rat hepatoma cells with either IR or an alkylating agent protects these cells from the mutagenic effect of a subsequent treatment with O^6-methylguanine generating compounds such as N-methyl-N'-nitro-N-nitrosoguanidine (MNNG) (Fritz et al. 1991). Although the induction is not a specific response to alkylating agents, the protective effect provoked by MGMT is reminiscent to the adaptive response of *Escherichia coli* (Jeggo et al. 1977; Samson and Cairns 1977). Interestingly, the promoter of the MGMT gene contains glucocorticoid responsive elements (GREs), and the gene has been shown to be inducible by corticosteroids as well (Grombacher et al. 1996). Whether MGMT is inducible by genotoxic stress in human cells is still an unresolved question.

3.2
8-Oxo-guanine-DNA Glycosylase (OGG1)

8-Oxo-guanine-DNA glycosylase (OGG1) is the major glycosylase involved in removing the oxidative premutagenic lesion 8-oxo-guanine from DNA. OGG1 was shown to be inducible in HCT116 cells 24–48 h after exposure to 100–200 μM methyl methanesulfonate (MMS) on RNA and protein level (Lee et al. 2004). Induction by MMS is mediated via two inverted CCAAT motifs and the transcription factor NF-YA. In addition, OGG1 was found to be inducible 6 h after treatment with ionizing radiation (1 Gy) in the lung tissue of mice (Risom et al. 2003).

3.3 Apurinic/apyrimidinic Endonuclease (APE1, APEX, Ref-1)

The DNA repair enzyme apurinic/apyrimidinic endonuclease is involved in base excision repair (BER). It converts apurinic/apyrimidinic sites, which are created by different DNA glycosylases upon removal of modified purines and pyrimidines (for review see Christmann et al. 2003), into DNA single strand breaks. The expression level of *ape* mRNA is increased in rat liver cells (H4) upon exposure to γ-rays and in CHO cells and V79 fibroblasts upon treatment with agents generating oxidative species such as hydrogen peroxide and sodium hypochloride (Grosch et al. 1998; Ramana et al. 1998). Genotoxic stress-triggered induction of APE was observed on the level of the APE promoter, mRNA, protein and enzyme activity. Analysis of the human APE promoter revealed a CREB binding site to be responsible for the oxidative stress-induced transcriptional activation of the *ape* gene. This CRE element is recognized by c-JUN and ATF-2 (Grosch and Kaina 1999). APE induction protects cells from clastogenic effects provoked by oxidative compounds, showing that induction of APE is part of an adaptive response against oxidative stress. APE was found to be expressed at a higher level in the inflammatory tissue of the colon (colitis ulcerosa) which indicates that reactive oxygen species generated during inflammation may up-regulate APE (Hofseth et al. 2003).

3.4 Flap Endonuclease 1 (FEN1)

Flap endonuclease 1 (FEN1) is indispensable for long-patch BER (Klungland and Lindahl 1997) and stimulates strand displacement and repair synthesis by DNA polymerase β (Prasad et al. 2000). Furthermore, it plays a critical role in DNA replication because it is required for the removal of Okazaki fragments (Bae et al. 2001). FEN1 is induced in MEFs 6–12 h after exposure to UV-C light as shown on mRNA and protein level (Christmann et al. 2005). The FEN1 promoter is activated by co-transfection with p53 and harbours a potential p53 binding site as identified by EMSA. FEN1 induction is observed in wild-type but not in p53-deficient MEFs. Increased expression of FEN1 attenuates the UV-C-induced S-phase arrest. It, therefore, plays an important role in the recovery of cells from UV-C-induced replication blockade (Christmann et al. 2005). In addition to UV-C, FEN1 expression is also induced by MMS as shown in microarray-based analyses using L5178Y mouse lymphoma cells (Islaih et al. 2004). Interestingly, as shown in the same study, this induction was not observed in human lymphoblastoid TK6 cells, indicating that induction of FEN1 by MMS is presumably restricted to mouse cells.

3.5 X-ray Repair Cross-Complementing Protein-1 (XRCC1)

X-ray repair cross-complementing protein-1 (XRCC1) interacts with DNA ligase III and Polβ and is involved in short patch BER (Caldecott et al. 1996). In addition, it is one of the main components required for DNA single-strand break repair (Caldecott 2003). In human prostate cancer cells, XRCC1 is induced by ionizing radiation and the mitogen EGF (Yacoub et al. 2001, 2003). Induction of XRCC1 was blocked by EGFR and MEK inhibitors, indicating a role for the MAPK pathway in regulating XRCC1 gene expression.

3.6 DNA Ligase I (LIG1)

DNA ligase I (LIG1) is involved in the ligation step during long-patch BER. Furthermore, it catalyses the ligation step during NER. Induction of LIG1 was observed in human primary fibroblasts 24 h after exposure to UV-C light (20 J/m^2) (Montecucco et al. 1995).

3.7 DNA Damage Binding Protein 2 (DDB2, p48)

The DNA damage binding protein 2 (DDB2, p48) is part of the heterodimeric DNA damage recognition complex XPE that is part of the nucleotide excision repair (NER) machinery. DDB2 together with DDB1 recognizes UV-induced cyclobutan pyrimidine dimers (CPDs) but only marginally detects (6–4) photoproducts. The DDB2 mRNA expression increases 16–24 h after exposure to 10 J/m^2 UV-C and 4–8 h after exposure to ionizing radiation (2 Gy) in wild-type but not in p53-deficient human fibroblasts. In p53-deficient cells even basal DDB2 expression was not detectable, and transfection with p53 restored DDB2 expression (Hwang et al. 1999). While p53-dependent induction of DDB2 was shown in human cells, in mice DDB2 seems not to be inducible by UV-C light. This is in accordance with the fact that the mouse DDB2 promoter lacks p53 binding sites (Tan and Chu 2002). Induction of DDB2 can also be mediated by BRCA1, either directly in the absence of p53 (Hartman and Ford 2002) or by increasing the binding affinity of p53 to the DDB2 promoter (Takimoto et al. 2002). Beside UV-C light, DDB2 expression was also reported to be induced 16 h after exposure to 0.2 $\mu g/ml$ adriamycin and 10 μM cisplatin (Takimoto et al. 2002).

3.8 Xeroderma Pigmentosum C (XPC)

The xeroderma pigmentosum C (XPC) protein is part of a heterodimeric recognition complex that plays a crucial role in NER. XPC together with HR23B recognizes UV light-induced (6–4) photoproducts, but not (or to a significantly lesser extent) cyclobutane pyrimidine dimers (CPDs) (Hey et al. 2002; Kusumoto et al. 2001). The *xpc* gene is induced following UV-C treatment 6–24 h after exposure to 15 J/m^2 (Adimoolam and Ford 2002). The induction was enhanced by increasing the dose up to 40 J/m^2. It was not observed in p53-deficient cells. Similar to DDB2, XPC expression can be stimulated by BRCA1 (Hartman and Ford 2002). XPC expression is also inducible by ionizing radiation (4–24 h after treatment with 10 Gy) and by MMS (12 h after treatment with 50 μg/ml) (Amundson et al. 2002) as well as by treatment with benzo[a]pyrene diol epoxide (4 h after exposure to 1 μM) (Wang et al. 2003).

3.9 Excision Repair Cross-Complementing Protein 1 (ERCC1)

Excision repair cross-complementing protein 1 (ERCC1) interacts with XPF and performs within this complex the 5′-incision step during NER (Sijbers et al. 1996). ERCC1 was shown to be inducible in various tissues of mice. Its expression is enhanced by treatment of mice with arsenic compounds 3 h after exposure in the liver (Liu et al. 2001) and 6 h after treatment with IR (1 Gy) in the lung (Risom et al. 2003). In human prostate cancer cells (the lines DU145 and LNCaP have been used for the experiments), ERCC1 expression is induced by ionizing radiation and epidermal growth factor (EGF) (Yacoub et al. 2001, 2003). ERCC1 induction can be blocked by EGF receptor and MEK inhibitors, indicating that the MAPK pathway is involved in the regulation of ERCC1 induction. In addition, ERCC1 expression is up-regulated by activated c-Ha-RAS. ERCC1 promoter analysis revealed that the transcription factor AP1 is required for this response (Youn et al. 2004).

3.10 MutS Homolog Protein 2 (MSH2)

MutS homolog protein 2 (MSH2) is involved in mismatch repair (MMR), which is responsible for removal of base mismatches caused by spontaneous and induced base deamination, oxidation, methylation and replication errors. Together with MSH6 it forms the MutSα complex, which is responsible for the recognition of DNA mispairs. The cloned and transfected *msh2* promoter was reported to be activated 6 h after exposure of cells to 50 or 100 J/m^2 UV-B in a p53 and c-JUN dependent manner (Scherer et al. 2000). Different potential p53 binding sites were described to be present in the *msh2* promoter in Saos-2 cells (Scherer et al. 1996) and in A2780 ovarian cancer cells (Warnick et al. 2001). Although *msh2* appears to be inducible on the level of the cloned promoter, as demonstrated in reporter gene assays, evidence is still lacking that the endogenous *msh2* gene becomes up-regulated in response to genotoxic stress. The relevance of p53 for MSH2 regulation is supported by the fact that p53 overexpression correlates with increased MSH2 expression in non-small cell lung cancer (Xinarianos et al. 2002). Beside p53, AP-1 is also involved in the regulation of *msh2* gene expression. Thus, in myeloid leukemic U937 cells, the expression of MSH2 protein was increased upon treatment with 50 nM TPA which occurred via protein kinase C (PKC) activation (Humbert et al. 2002). Mutation of the AP-1 binding sites within the *msh2* promoter or expression of dominant negative c-JUN abrogated the TPA triggered induction of MSH2 (Humbert et al. 2003).

3.11 Polymerase Iota (POL*I*)

Polymerase iota (POL*I*) is a translesion polymerase involved in the bypass of UV light-induced DNA lesions as well as AP sites. POL*I* was reported to be induced on RNA and protein level 30 – 120 min after exposure to 100 J/m^2 UV-C light (Yang et al. 2004).

3.12 Polymerase Zeta (POL*Z*)

DNA Polymerase Zeta (POLZ) consists of two subunits (REV3 and REV7), which cooperate with REV1 during translesion synthesis. POL*Z* is involved in the translesion bypass of (6–4) photoproducts (Johnson et al. 2001) and benzo[a]pyrene diol epoxide-induced DNA lesions (Li et al. 2002). The expression of the human REV3 gene was increased in human cells upon treatment with MNNG (Yu et al. 2004; Zhu et al. 2003).

4 Implications for Cell Survival (Late Response)

Both p53 and AP-1 are physiologically relevant for cellular defence against DNA damage-induced cell death. Thus, p53-deficient and c-Fos-deficient cells are hypersensitive to the cytotoxic effect of UV-C and MMS (Haas and Kaina 1995; Lackinger et al. 2001; Lackinger and Kaina 2000; Smith et al. 2000) and respond with a markedly elevated level of UV light-induced chromosomal aberrations and apoptosis. In addition, p53-deficient cells are impaired in several DNA repair mechanisms, including BER and NER. This is attributed to both a direct role of p53 in the DNA repair process and a role of p53 as transcription factor.

p53-deficient cells are impaired in global genomic nucleotide excision repair (GGR) (Ford and Hanawalt 1997). This finding was explained by the fact that the nucleotide excision repair genes DDB2 (alias p48, XPE) and XPC are not transcriptionally activated by UV-C in p53-deficient cells (Adimoolam and Ford 2002; Hwang et al. 1999). Another p53 regulated factor relevant for NER is GADD45a. GADD45a binds to UV-damaged chromatin and increases the accessibility of repair factors to the site of DNA damage (Smith et al. 2000). Accordingly, GADD45a-deficient cells display reduced repair of UV-damaged DNA. In contrast to the role p53 plays in NER, its role in BER does not only depend on its function as a transcription factor. Thus, recombinant p53 stimulates BER, notably the Polβ-mediated repair synthesis (Offer et al. 2001), and a physical interaction between p53 and Polβ has been suggested, which stabilizes the interaction between Polβ and abasic DNA (Zhou et al. 2001). In addition, p53 is involved in the regulation of the expression of Polβ since Polβ expression is drastically reduced in p53 mutant and p53-null cells. This is likely due to the reduced stability of Polβ in the absence of p53 (Seo et al. 2002).

Regarding the role of AP-1 in DNA repair, only limited data are available to date. c-Jun-deficient cells were reported to show a reduced frequency of apoptosis after MMS treatment (Kolbus et al. 2000). c-Jun knockout MEFs grow very slowly unless they lose p53 (own unpublished data). Since the toxic effect of many genotoxins is strictly bound on DNA replication, c-Jun-deficient MEFs can hardly be compared with other cell types as to their genotoxic response. As opposed to c-Jun (p53wt)-deficient cells, c-Fos (p53wt)-deficient MEFs grow in a similar way to the wild-type. Both primary *c-fos* knockout MEFs and established fibroblasts were shown to be hypersensitive to UV light and chemical genotoxins except ionising radiation (Haas and Kaina 1995; Kaina et al. 1997, Lackinger and Kaina 2000; Lackinger et al. 2001). This has been explained by impaired regulation of S-phase recovery upon the induction of specific DNA lesions (Kaina et al. 1997). More recent data from this group revealed that c-Fos-deficient cells display a defect in NER that is due to reduced expression of XPF (Christmann and Kaina 2005, in preparation).

The role of p53 and AP-1 in the regulation of expression of DNA repair enzymes can be analyzed using the DNA microarray technology. In Fig. 2, the result of a typical microarray experiment using arrays containing 120 DNA repair genes and 30 cell cycle regulated genes is shown. The arrays were hybridized with labelled cDNA from wild-type, p53-deficient and c-Fos-deficient mouse embryonal fibroblasts (MEFs) that have been exposed to 20 J/m^2 UVC. The induction factors for selected DNA repair genes upon UVC exposure of MEFs are depicted. Apparently, exposure to UVC light leads to up-regulation of **cdk1a** (p21), **gadd45a** and **gadd45b** in all cell lines. This can be explained by an induction of these genes involving either p53 or BRCA1

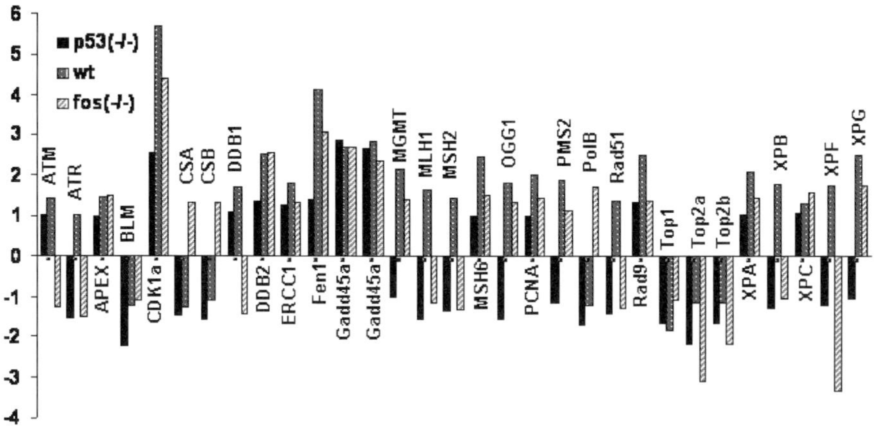

Fig. 2 Expression of DNA repair genes upon UV-C exposure of mouse fibroblasts wild-type for p53, and deficient for c-Fos and p53. Mouse fibroblast cell lines which are wild-type (wt), deficient for c-Fos (fos–/–) or deficient for p53 (p53–/–) were irradiated with UV-C light (20 J/m^2) or were left unirradiated (control). 6 h after exposure, total RNA was isolated and subjected to microarray analysis. The induction factors of selected DNA repair genes are presented as columns. Signal detection was performed using an Affymetrix 428 Array Scanner and quantitation was performed using the ImaGene 4.2 and GeneSight 3.2 software. As internal control, gapdh, β-actin and α-tubuline were used

(el-Deiry et al. 1993; Zhan 2005). For UVC-mediated induction of **cdkn1a**, p53 is obviously more important than BRCA1 as indicated by its reduced induction in p53-deficient cells. A p53-dependent induction was observed for **ddb2**, **fen1** and **xpg**, representing the molecular reason for attenuated NER and replication recovery in p53-deficient cells. A negative regulation of the genes **top1**, **top2a**, **top2b** is observed in all cell lines and the down-regulation of **xpf** is only observed in c-FOS-deficient cells.

The data presented are an example to demonstrate the potential usefulness of microarray technology for the identification of DNA repair-related cellular responses induced by specific DNA-damaging treatments. Change in regulation of DNA repair genes on transcriptional level is an important but not the only effect exerted by genotoxic agents. Post-translational modifications of repair enzymes and regulatory proteins should also be taken into account as contributing to the cellular genotoxic stress response. Alterations of the expression of DNA repair genes and post-translational regulation may be related to the development of cancer. Changes in DNA repair gene expression may also determine the effectiveness of cancer therapy upon the application of DNA-damaging treatments.

Acknowledgements Our work was supported by DFG Ka 724/13-1 and SFB 432/B7.

References

Adimoolam S, Ford JM (2002) p53 and DNA damage-inducible expression of the xeroderma pigmentosum group C gene. Proc Natl Acad Sci USA 99:12985–12990

Amundson SA, Patterson A, Do KT, Fornace AJ Jr (2002) A nucleotide excision repair master-switch: p53 regulated coordinate induction of global genomic repair genes. Cancer Biol Ther 1:145–149

Anderson SF, Schlegel BP, Nakajima T, Wolpin ES, Parvin JD (1998) BRCA1 protein is linked to the RNA polymerase II holoenzyme complex via RNA helicase A. Nat Genet 19:254–256

Bae SH, Bae KH, Kim JA, Seo YS (2001) RPA governs endonuclease switching during processing of Okazaki fragments in eukaryotes. Nature 412:456–461

Banin S et al. (1998) Enhanced phosphorylation of p53 by ATM in response to DNA damage. Science 281:1674–1677

Bartkova J et al. (2005) DNA damage response as a candidate anti-cancer barrier in early human tumorigenesis. Nature 434:864–870

Boldogh I et al. (1998) Regulation of expression of the DNA repair gene O6-methylguanine-DNA methyltransferase via protein kinase C-mediated signaling. Cancer Res 58:3950–3956

Caldecott KW (2003) XRCC1 and DNA strand break repair. DNA Repair (Amst) 2:955–969

Caldecott KW, Aoufouchi S, Johnson P, Shall S (1996) XRCC1 polypeptide interacts with DNA polymerase beta and possibly poly (ADP-ribose) polymerase, and DNA ligase III is a novel molecular nick-sensor invitro. Nucleic Acids Res 24:4387–4394

Canman CE et al. (1998) Activation of the ATM kinase by ionizing radiation and phosphorylation of p53. Science 281:1677–1679

Chehab NH, Malikzay A, Appel M, Halazonetis TD (2000) Chk2/hCds1 functions as a DNA damage checkpoint in G(1) by stabilizing p53. Genes Dev 14:278–288

Chehab NH, Malikzay A, Stavridi ES, Halazonetis TD (1999) Phosphorylation of Ser-20 mediates stabilization of human p53 in response to DNA damage. Proc Natl Acad Sci USA 96:13777–13782

Christmann M, Tomicic MT, Origer J, Kaina B (2005) Fen1 is induced p53 dependently and involved in the recovery from UV light-induced replication inhibition. Oncogene 24:8304:8313

Christmann M, Tomicic MT, Roos WP, Kaina B (2003) Mechanisms of human DNA repair: an update. Toxicology 193:3–34

Dosch J, Kaina B (1996) Induction of c-fos, c-jun, junB and junD mRNA and AP-1 by alkylating mutagens in cells deficient and proficient for the DNA repair protein O6-methylguanine-DNA methyltransferase (MGMT) and its relationship to cell death, mutation induction and chromosomal instability. Oncogene 13:1927–1935

el-Deiry WS, Kern SE, Pietenpol JA, Kinzler KW, Vogelstein B (1992) Definition of a consensus binding site for p53. Nat Genet 1:45–49

el-Deiry WS et al. (1993) WAF1, a potential mediator of p53 tumor suppression. Cell 75:817–825

Ford JM, Hanawalt PC (1997) Expression of wild-type p53 is required for efficient global genomic nucleotide excision repair in UV-irradiated human fibroblasts. J Biol Chem 272:28073–28080

Fritz G, Tano K, Mitra S, Kaina B (1991) Inducibility of the DNA repair gene encoding O6-methylguanine-DNA methyltransferase in mammalian cells by DNA-damaging treatments. Mol Cell Biol 11:4660–4668

Grombacher T, Eichhorn U, Kaina B (1998) p53 is involved in regulation of the DNA repair gene O6-methylguanine-DNA methyltransferase (MGMT) by DNA damaging agents. Oncogene 17:845–851

Grombacher T, Kaina B (1995) Constitutive expression and inducibility of O6-methylguanine-DNA methyltransferase and N-methylpurine-DNA glycosylase in rat liver cells exhibiting different status of differentiation. Biochim Biophys Acta 1270:63–72

Grombacher T, Mitra S, Kaina B (1996) Induction of the alkyltransferase (MGMT) gene by DNA damaging agents and the glucocorticoid dexamethasone and comparison with the response of base excision repair genes. Carcinogenesis 17:2329–2336

Grosch S, Fritz G, Kaina B (1998) Apurinic endonuclease (Ref-1) is induced in mammalian cells by oxidative stress and involved in clastogenic adaptation. Cancer Res 58:4410–4416

Grosch S, Kaina B (1999) Transcriptional activation of apurinic/apyrimidinic endonuclease (Ape, Ref-1) by oxidative stress requires CREB. Biochem Biophys Res Commun 261:859–863

Guo Z, Kumagai A, Wang SX, Dunphy WG (2000) Requirement for Atr in phosphorylation of Chk1 and cell cycle regulation in response to DNA replication blocks and UV-damaged DNA in Xenopus egg extracts. Genes Dev 14:2745–2756

Haas S, Kaina B (1995) c-Fos is involved in the cellular defence against the genotoxic effect of UV radiation. Carcinogenesis 16:985–991

Haile DT, Parvin JD (1999) Activation of transcription in vitro by the BRCA1 carboxyl-terminal domain. J Biol Chem 274:2113–2117

Harkin DP et al. (1999) Induction of GADD45 and JNK/SAPK-dependent apoptosis following inducible expression of BRCA1. Cell 97:575–586

Hartman AR, Ford JM (2002) BRCA1 induces DNA damage recognition factors and enhances nucleotide excision repair. Nat Genet 32:180–184

Hey T, Lipps G, Sugasawa K, Iwai S, Hanaoka F, Krauss G (2002) The XPC-HR23B complex displays high affinity and specificity for damaged DNA in a true-equilibrium fluorescence assay. Biochemistry 41:6583–6587

Hirao A et al. (2000) DNA damage-induced activation of p53 by the checkpoint kinase Chk2. Science 287:1824–1827

Hofseth LJ et al. (2003) The adaptive imbalance in base excision-repair enzymes generates microsatellite instability in chronic inflammation. J Clin Invest 112:1887–1894

Huang RP, Wu JX, Fan Y, Adamson ED (1996) UV activates growth factor receptors via reactive oxygen intermediates. J Cell Biol 133:211–220

Humbert O, Achour I, Lautier D, Laurent G, Salles B (2003) hMSH2 expression is driven by AP1-dependent regulation through phorbol-ester exposure. Nucleic Acids Res 31:5627–5634

Humbert O et al. (2002) Implication of protein kinase C in the regulation of DNA mismatch repair protein expression and function. J Biol Chem 277:18061–18068

Hwang BJ, Ford JM, Hanawalt PC, Chu G (1999) Expression of the p48 xeroderma pigmentosum gene is p53-dependent and is involved in global genomic repair. Proc Natl Acad Sci USA 96:424–428

Islaih M et al. (2004) Comparison of gene expression changes induced in mouse and human cells treated with direct-acting mutagens. Environ Mol Mutagen 44:401–419

Janus F, Albrechtsen N, Dornreiter I, Wiesmuller L, Grosse F, Deppert W (1999) The dual role model for p53 in maintaining genomic integrity. Cell Mol Life Sci 55:12–27

Jeggo P, Defais TM, Samson L, Schendel P (1977) An adaptive response of E. coli to low levels of alkylating agent: comparison with previously characterised DNA repair pathways. Mol Gen Genet 157:1–9

Jin S et al. (2000) BRCA1 activation of the GADD45 promoter. Oncogene 19:4050–4057
Johnson RE, Haracska L, Prakash S, Prakash L (2001) Role of DNA polymerase zeta in the bypass of a (6–4) TT photoproduct. Mol Cell Biol 21:3558–3563
Kaina B, Haas S, Kappes H (1997) A general role for c-Fos in cellular protection against DNA-damaging carcinogens and cytostatic drugs. Cancer Res 57:2721–2731
Khosravi R, Maya R, Gottlieb T, Oren M, Shiloh Y, Shkedy D (1999) Rapid ATM-dependent phosphorylation of MDM2 precedes p53 accumulation in response to DNA damage. Proc Natl Acad Sci USA 96:14973–14977
Klungland A, Lindahl T (1997) Second pathway for completion of human DNA base excision-repair: reconstitution with purified proteins and requirement for DNase IV (FEN1). Embo J 16:3341–3348
Knebel A, Rahmsdorf HJ, Ullrich A, Herrlich P (1996) Dephosphorylation of receptor tyrosine kinases as target of regulation by radiation, oxidants or alkylating agents. Embo J 15:5314–5325
Kolbus A, Herr I, Schreiber M, Debatin KM, Wagner EF, Angel P (2000) c-Jun-dependent CD95-L expression is a rate-limiting step in the induction of apoptosis by alkylating agents. Mol Cell Biol 20:575–582
Kusumoto R et al. (2001) Diversity of the damage recognition step in the global genomic nucleotide excision repair in vitro. Mutat Res 485:219–227
Lackinger D, Eichhorn U, Kaina B (2001) Effect of ultraviolet light, methyl methanesulfonate and ionizing radiation on the genotoxic response and apoptosis of mouse fibroblasts lacking c-Fos, p53 or both. Mutagenesis 16:233–241
Lackinger D, Kaina B (2000) Primary mouse fibroblasts deficient for c-Fos, p53 or for both proteins are hypersensitive to UV light and alkylating agent-induced chromosomal breakage and apoptosis. Mutat Res 457:113–123
Lee MR et al. (2004) Transcription factors NF-YA regulate the induction of human OGG1 following DNA-alkylating agent methylmethane sulfonate (MMS) treatment. J Biol Chem 279:9857–9866
Li Z, Zhang H, McManus TP, McCormick JJ, Lawrence CW, Maher VM (2002) hREV3 is essential for error-prone translesion synthesis past UV or benzo[a]pyrene diol epoxide-induced DNA lesions in human fibroblasts. Mutat Res 510:71–80
Liu J, Kadiiska MB, Liu Y, Lu T, Qu W, Waalkes MP (2001) Stress-related gene expression in mice treated with inorganic arsenicals. Toxicol Sci 61:314–320
Liu Q et al. (2000) Chk1 is an essential kinase that is regulated by Atr and required for the G(2)/MDNA damage checkpoint. Genes Dev 14:1448–1459
Matsuoka S, Rotman G, Ogawa A, Shiloh Y, Tamai K, Elledge SJ (2000) Ataxia telangiectasia-mutated phosphorylates Chk2 in vivo and in vitro. Proc Natl Acad Sci USA 97:10389–10394
Mita H, Tsutsui J, Takekawa M, Witten EA, Saito H (2002) Regulation of MTK1/MEKK4 kinase activity by its N-terminal autoinhibitory domain and GADD45 binding. Mol Cell Biol 22:4544–4555
Montecucco A, Savini E, Biamonti G, Stefanini M, Focher F, Ciarrocchi G (1995) Late induction of human DNA ligase I after UV-C irradiation. Nucleic Acids Res 23:962–966
Offer H et al. (2001) Structural and functional involvement of p53 in BER in vitro and in vivo. Oncogene 20:581–589
Ouchi T, Monteiro AN, August A, Aaronson SA, Hanafusa H (1998) BRCA1 regulates p53-dependent gene expression. Proc Natl Acad Sci USA 95:2302–2306
Papa S et al. (2004a) Gadd45 beta mediates the NF-kappa B suppression of JNK signalling by targeting MKK7/JNKK2. Nat Cell Biol 6:146–153

Papa S, Zazzeroni F, Pham CG, Bubici C, Franzoso G (2004b) Linking JNK signaling to NF-kappaB: a key to survival. J Cell Sci 117:5197–5208
Prasad R, Dianov GL, Bohr VA, Wilson SH (2000) FEN1 stimulation of DNA polymerase beta mediates an excision step in mammalian long patch base excision repair. J Biol Chem 275:4460–4466
Rafferty JA, Clarke AR, Sellappan D, Koref MS, Frayling IM, Margison GP (1996) Induction of murine O6-alkylguanine-DNA-alkyltransferase in response to ionising radiation is p53 gene dose dependent. Oncogene 12:693–697
Ramana CV, Boldogh I, Izumi T, Mitra S (1998) Activation of apurinic/apyrimidinic endonuclease in human cells by reactive oxygen species and its correlation with their adaptive response to genotoxicity of free radicals. Proc Natl Acad Sci USA 95:5061–5066
Risom L, Moller P, Vogel U, Kristjansen PE, Loft S (2003) X-ray-induced oxidative stress: DNA damage and gene expression of HO-1, ERCC1 and OGG1 in mouse lung. Free Radical Res 37:957–966
Sachsenmaier C, Radler-Pohl A, Zinck R, Nordheim A, Herrlich P, Rahmsdorf HJ (1994) Involvement of growth factor receptors in the mammalian UVC response. Cell 78:963–972
Samson L, Cairns J (1977) A new pathway for DNA repair in Escherichia coli. Nature 267:281–283
Scherer SJ et al. (2000) p53 and c-Jun functionally synergize in the regulation of the DNA repair gene hMSH2 in response to UV. J Biol Chem 275:37469–37473
Scherer SJ, Welter C, Zang KD, Dooley S (1996) Specific in vitro binding of p53 to the promoter region of the human mismatch repair gene hMSH2. Biochem Biophys Res Commun 221:722–728
Seo YR, Fishel ML, Amundson S, Kelley MR, Smith ML (2002) Implication of p53 in base excision DNA repair: in vivo evidence. Oncogene 21:731–737
Shieh SY, Ahn J, Tamai K, Taya Y, Prives C (2000) The human homologs of checkpoint kinases Chk1 and Cds1 (Chk2) phosphorylate p53 at multiple DNA damage-inducible sites. Genes Dev 14:289–300
Sijbers AM et al. (1996) Xeroderma pigmentosum group F caused by a defect in a structure-specific DNA repair endonuclease. Cell 86:811–822
Smith ML et al. (2000) p53-mediated DNA repair responses to UV radiation: studies of mouse cells lacking p53, p21, and/or gadd45 genes. Mol Cell Biol 20:3705–3714
Somasundaram K et al. (1997) Arrest of the cell cycle by the tumour-suppressor BRCA1 requires the CDK-inhibitor p21WAF1/CiP1. Nature 389:187–190
Starita LM, Parvin JD (2003) The multiple nuclear functions of BRCA1: transcription, ubiquitination and DNA repair. Curr Opin Cell Biol 15:345–350
Takekawa M, Saito H (1998) A family of stress-inducible GADD45-like proteins mediate activation of the stress-responsive MTK1/MEKK4 MAPKKK. Cell 95:521–530
Takimoto R, MacLachlan TK, Dicker DT, Niitsu Y, Mori T, el-Deiry WS (2002) BRCA1 transcriptionally regulates damaged DNA binding protein (DDB2) in the DNA repair response following UV irradiation. Cancer Biol Ther 1:177–186
Tan T, Chu G (2002) p53 Binds and activates the xeroderma pigmentosum DDB2 gene in humans but not mice. Mol Cell Biol 22:3247–3254
Tibbetts RS et al. (2000) Functional interactions between BRCA1 and the checkpoint kinase ATR during genotoxic stress. Genes Dev 14:2989–3002
Unger T et al. (1999) Critical role for Ser20 of human p53 in the negative regulation of p53 by Mdm2. Embo J 18:1805–1814

Wang A et al. (2003) Response of human mammary epithelial cells to DNA damage induced by BPDE: involvement of novel regulatory pathways. Carcinogenesis 24:225–234

Warnick CT, Dabbas B, Ford CD, Strait KA (2001) Identification of a p53 response element in the promoter region of the hMSH2 gene required for expression in A2780 ovarian cancer cells. J Biol Chem 276:27363–27370

Xinarianos G et al. (2002) p53 status correlates with the differential expression of the DNA mismatch repair protein MSH2 in non-small cell lung carcinoma. Int J Cancer 101:248–252

Xu B, O'Donnell AH, Kim ST, Kastan MB (2002) Phosphorylation of serine 1387 in Brca1 is specifically required for the Atm-mediated S-phase checkpoint after ionizing irradiation. Cancer Res 62:4588–4591

Yacoub A, McKinstry R, Hinman D, Chung T, Dent P, Hagan MP (2003) Epidermal growth factor and ionizing radiation up-regulate the DNA repair genes XRCC1 and ERCC1 in DU145 and LNCaP prostate carcinoma through MAPK signaling. Radiat Res 159:439–452

Yacoub A, Park JS, Qiao L, Dent P, Hagan MP (2001) MAPK dependence of DNA damage repair: ionizing radiation and the induction of expression of the DNA repair genes XRCC1 and ERCC1 in DU145 human prostate carcinoma cells in a MEK1/2 dependent fashion. Int J Radiat Biol 77:1067–1078

Yang J, Chen Z, Liu Y, Hickey RJ, Malkas LH (2004) Altered DNA polymerase iota expression in breast cancer cells leads to a reduction in DNA replication fidelity and a higher rate of mutagenesis. Cancer Res 64:5597–5607

Youn CK et al. (2004) Oncogenic H-Ras up-regulates expression of ERCC1 to protect cells from platinum-based anticancer agents. Cancer Res 64:4849–4857

Yu Y, Yang J, Zhu F, Xu F (2004) Response of REV3 promoter to N-methyl-*N*′-nitro-N-nitrosoguanidine. Mutat Res 550:49–58

Zhan Q (2005) Gadd45a, a p53- and BRCA1-regulated stress protein, in cellular response to DNA damage. Mutat Res 569:133–143

Zhang H et al. (1998) BRCA1 physically associates with p53 and stimulates its transcriptional activity. Oncogene 16:1713–1721

Zhou BB et al. (2000) Caffeine abolishes the mammalian G(2)/MDNA damage checkpoint by inhibiting ataxia-telangiectasia-mutated kinase activity. J Biol Chem 275:10342–10348

Zhou J, Ahn J, Wilson SH, Prives C (2001) A role for p53 in base excision repair. Embo J 20:914–923

Zhu F, Yang J, Xu F, Yu YN (2003) Cloning and bioinformatics of human REV3 gene promoter region and its response to carcinogen N-methyl-*N*′-nitro-N-nitrosoguanidine. Zhejiang Da Xue Xue Bao Yi Xue Ban 32:393–397

Zou L, Elledge SJ (2003) Sensing DNA damage through ATRIP recognition of RPA-ssDNA complexes. Science 300:1542–1548

Zou L, Liu D, Elledge SJ (2003) Replication protein A-mediated recruitment and activation of Rad17 complexes. Proc Natl Acad Sci USA 100:13827–13832

Genome Dyn Stab (1)
D.-H. Lankenau: Genome Integrity
DOI 10.1007/7050_011/Published online: 22 December 2005

Pathways of DNA Double-Strand Break Repair in Mammalian Cells after Ionizing Radiation

Andrea Krempler (✉) · Markus Löbrich

Fachrichtung Biophysik, Universität des Saarlandes, 66421 Homburg/Saar, Germany
andrea.krempler@uniklinik-saarland.de

Abstract Exposure to ionizing radiation induces a considerable amount of double-strand breaks in mammalian DNA. Because this type of lesion substantially affects chromosomal integrity, double-strand breaks are considered as one of the most harmful DNA lesions. In order to avoid serious negative consequences for the cell and the whole organism, these breaks need to be reconstituted by highly efficient and precise repair mechanisms. Here, we review our current understanding of how the two main double-strand break repair mechanisms, non-homologous end joining and homologous recombination, operate to rejoin breaks induced by ionizing radiation.

1 Repair of DNA Double-Strand Breaks

Genetic instability and cancer development are consequences of DNA damage as well as failed repair and/or surveillance mechanisms. In this context, DNA double-strand breaks (DSBs) constitute the most serious threat for the cell, as both strands of the DNA are affected and need to be rejoined in the correct way. DSBs not only arise through exogenous factors such as ionizing radiation (IR) or some chemicals, but are also produced during endogenous processes such as replication, V(D)J recombination, and meiosis. Normally, cells induce cell cycle arrest, apoptosis, or mitotic cell death if DSBs are left unrepaired, to prevent propagation of damaged genomic material. Consequently, loss of surveillance mechanisms leads to chromosomal aberrations and lays the basis for malignant transformation. In addition, incorrectly repaired DSBs result in chromosomal rearrangements (Olive 1998; Hoeijmakers 2001; van Gent et al. 2001).

1.1 Homologous Recombination

Two distinct mechanisms are largely responsible for coping with DSBs in higher eukaryotic cells: non-homologous end-joining (NHEJ) and homologous recombination (HR; Fig. 1). HR is a very precise process that uses the information of a homologous sequence to reconstitute a break site faithfully. Surprisingly, it is not the major repair pathway in higher eukaryotes (see be-

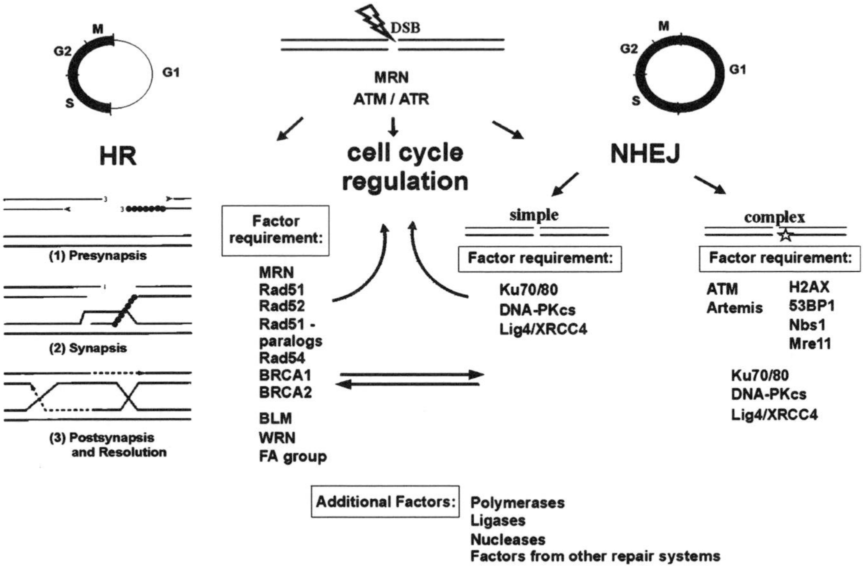

Fig. 1 Mechanisms of DSB repair. The induction of a DSB leads to the activation of the checkpoint kinases ATM or ATR. Depending on the DSB and the cell cycle phase, the signal is then transferred to the DSB repair pathways and to the cell cycle checkpoint proteins. The two major pathways involved in DSB repair are HR (exemplified here by gene conversion) and NHEJ. Each repair pathway is characterized by specific factors whose requirements depend on the nature of the break and may include the recruitment of additional proteins (e.g. from other repair pathways). Cross-talk to cell cycle regulatory mechanisms ensures the faithful restoration of the break-site and substitution of one repair pathway by the other is also highly possible

low). Depending on the nature of the DSB and the availabilty of homologous sequences, several types of HR have been described. Single-stand annealing is the simplest process and is used when two homologous (repetitive) sequences flank a DSB. After resection of the break site, the homologous regions are annealed, resulting in a single, restored element and the loss of the second repeat. Break induced replication (BIR) is initiated when only one site of a DSB is able to engage in HR, whereas during gene conversion both ends of a DSB participate in the invasion of homologous regions. Gene conversion is the most conservative and faithful mechanism of DSB repair (Paques and Haber 1999; Haber 2000). In the classical view, the processes during HR can be subdivided into three main steps (Fig. 1): during pre-synapsis (1), the DNA-ends are processed to yield recombination proficient single stranded overhangs which form a joint molecule with the DNA template during synapsis (2). Post-synapsis and resolution (3) involve the repair of the DNA strands and the separation of the recombined DNA molecules. The precise molecular mechanism of HR is best understood in *E. coli* (Eggleston and West 1996; Wyman et al. 2004). The search for identical factors in eukaryotes revealed a related but more complicated mechanism with a multitude of associated factors. Biochemical and biophysical studies on the individual components led to the current molecular model of DSB repair through HR in eukaryotes. The main factors involved in HR are proteins of the Rad52 epistasis group, including the homologs of *Saccharomyces cerevisiae* Rad51, Rad52, and Rad54, and the Rad51 paralogs XRCC2, XRCC3, Rad51B, Rad51C, and Rad51D (Symington 2002; Thompson and Schild 2002). The relative importance of the proteins is reflected by the viability and phenotypes of their knockouts. Total knockouts of Rad51 are not viable, probably due to its essential role in the mitotic and meiotic cell cycles of most eukaryotic organisms (Lim and Hasty 1996; Tsuzuki et al. 1996). In contrast, Rad52 knockouts in mammals do not cause embryonic death and show almost no increase in sensitivity to DNA-damaging agents (Rijkers et al. 1998; Yamaguchi-Iwai et al. 1998). Knockout mutants of the Rad51 paralogs which are thought to serve as mediators for Rad51 could be generated in the DT40 chicken cell line and isolated as Chinese hamster ovary (CHO) cell mutants (Jones et al. 1987; Liu et al. 1998; Takata et al. 2001). Moderate sensitivity to IR and strong sensitivity to crosslinking agents as well as chromosomal instability are universal features of these Rad52 group mutants. Interestingly, XRCC2, Rad51B and Rad51D knockouts are viable in the chicken cell system whereas homozygous mutant mice are embryonically lethal (Shu et al. 1999; Deans et al. 2000; Pittman and Schimenti 2000).

The initial step in eukaryotic homology directed repair is the recognition and binding of the break ends by a complex consisting of Mre11, Rad50, and Nbs1 (MRN complex). Mre11 and Rad50 form discrete nuclear foci after exposure to IR that are thought to represent sites of DSB repair (Maser et al. 1997). However, the precise function of MRN, especially during HR, is still under discussion. Before strand invasion can take place, the DNA ends need to be processed to yield 3′-single strand (ss) DNA overhangs. A nuclease activity must therefore be present at the break site. Mre11 has been shown to possess exo- and endonuclease activity, but a function in the generation of the 3′-ssDNA overhangs could never be demonstrated unambiguously (Haber 1998). The complex of Rad50 and Nbs1 also possesses activity for partial unwinding of duplex DNA, which is dependent on Rad50's ATPase activity (Paull and Gellert 1998, 1999; Trujillo et al. 1998). Recent studies point to a more general role of the MRN complex as a DSB sensor that may also bridge two DNA molecules such as sister chromatids or broken DNA ends (see below and contribution of H.-P. Hopfner in this issue). After processing, the stretch of ssDNA is covered with Rad51 recombinase, resulting in a nucleoprotein filament. Rad51 can be visualized as discrete nuclear foci in mitotic S-phase cells and after exposure to DNA damaging agents. Rad51 is homologous in structure and function to the bacterial RecA protein, the most important factor for HR in *E. coli* (Bianco et al. 1998). Replication protein A (RPA) with its high affinity for ssDNA is also involved in the formation of the nucleoprotein filament. A current model suggests that RPA initially protects the ssDNA and resolves secondary structures before it is displaced by Rad51 with the help of Rad52 and the breast cancer associated genes BRCA1 and BRCA2 (Song and Sung 2000; Krejci et al. 2003; Shivji and Venkitaraman 2004; Ting and Lee 2004). The Rad51 nucleoprotein filament, together with the mediator proteins from the Rad52 epistasis group, is then capable of interacting with a second DNA molecule (Baumann et al. 1996; Sung et al. 2003). The invasion of intact, homologous duplex DNA by ssDNA and the formation of a D-loop is the most likely model for the reactions following in vivo (McIlwraith et al. 2000). Rad54 may be required during synapsis to promote unwinding of the donor duplex (Solinger et al. 2001). The 3′-end is then extended by DNA synthesis. In theory, the D-loop is able to pair with the other side of the DSB, and in this way the 3′-end of the non-invading strand can also be extended by DNA synthesis, forming a double-Holliday-junction intermediate. However, alternative models such as synthesis-dependent strand annealing (SDSA) are also conceivable (Paques and Haber 1999; van den Bosch et al. 2002). The following events, branch migration and resolution of the crossed DNA strands, are currently not well understood in eukaryotic HR and the topic of extensive study (Hiom 2001; Constantinou et al. 2002). Proteins that may be involved in these processes are the Bloom-helicase (BLM) and the product of the WRN gene (Saintigny et al. 2002; Heyer et al. 2003; Liu and West 2004). Recently, Rad51C and XRCC3 were identified as constituents of the so-called resolvasome (Liu et al. 2004). Another group of proteins that is tightly associated with HR, at least after exposure to crosslinking agents, is the Fanconi anemia (FA) complimentation group (Symington 2002; Thompson and Schild 2002; Wyman et al. 2004). Strikingly, FANCD1 has been identified to represent BRCA2, the interaction partner of Rad51 (Howlett et al. 2002; Wang and D'Andrea 2004; Nakanishi et al. 2005).

1.2 NHEJ Core Components

NHEJ is the major DSB repair mechanism in multicellular eukaryotes and also effects rearrangements during V(D)J recombination, a process that is responsible for antibody and T-cell receptor diversity. NHEJ is a very fast process, but can produce errors at the rejoining site (Pfeiffer et al. 2004). It involves the DNA-dependent protein kinase (DNA-PK) holoenzyme, consisting of the DNA end-binding heterodimer Ku70/Ku80 and the catalytic subunit DNA-PKcs (Jeggo 1998; Jackson 2002), and the XRCC4/DNA Ligase IV (XRCC4/Lig4) complex (Fig. 1). Cell lines defective in any of these genes are generally highly IR sensitive and show diminished V(D)J recombination activity (Chang et al. 1993; Kurimasa et al. 1999; Riballo et al. 1999; Wachsberger et al. 1999). In addition, mutations in core components of the NHEJ pathway result in a marked DSB repair defect at early times post-IR (Kuhne et al. 2004; Riballo et al. 2004).

NHEJ repairs broken ends with little or no requirement for sequence homology because proteins and enzymes involved in NHEJ recognize DNA ends based on their structure rather than on their sequence. Major insights into the mechanism of NHEJ have been gained from in vitro assays with cell-free extracts or purified native or recombinant proteins (Labhart 1999a,b; Feldmann et al. 2000; Odersky et al. 2002; Ma et al. 2004; Pfeiffer et al. 2005; also see contribution of P. Pfeiffer in this issue). NHEJ in eukaryotes is thought to begin with the binding of the heterodimer Ku70/Ku80 to the DNA ends. The structural resolution of Ku70/Ku80 bound to DNA revealed that the heterodimer forms an asymmetric ring around the DNA helix (Walker et al. 2001). Ku then recruits DNA-PKcs, which exhibits serine-threonine kinase (Lees-Miller et al. 1990; Gottlieb and Jackson 1993) and perhaps also end-bridging activity (DeFazio et al. 2002). The DNA-PK holoenzyme, phosphorylates the Ku heterodimer and is necessary to activate the XRCC4/Lig4 complex which is responsible for the break resealing step to complete the joining (Jeggo 1998; Smith and Jackson 1999; Jackson 2002; Drouet et al. 2005). This sequence of events is sufficient to ligate compatible or blunt ends. However, many DSBs, especially those resulting from exogenous agents, are incompatible and may need processing prior to ligation. Therefore, additional factors are likely to be involved in the end joining process.

1.3 Additional NHEJ Factors

The preferential usage of one repair mechanism suggests that the process is reasonably accurate. The spectrum of DSB-ends after IR ranges from simple blunt ends to complicated, non-compatible multiple-damaged sites. Extensive sequence homology is not the main criterion for NHEJ rejoining. However, the joining of incompatible ends by NHEJ follows some rules to reconstitute the break-site as precise as possible (Thode et al. 1990; Nicolas et al. 1995; Odersky et al. 2002; Raghavan and Raman 2004; Budman and Chu 2005; see also the contribution of P. Pfeiffer in this issue). In addition, the ends of a DSB have to be held in close proximity. There is some discussion about a "bridging" component or "alignment" factor in NHEJ DSB repair (Pfeiffer and Vielmetter 1988; Thode et al. 1990). The Ku-heterodimer, DNA-PKcs, and γH2AX were described to possess synapsis activity (Cary et al. 1997; Pang et al. 1997; Yaneva et al. 1997; Sedelnikova et al. 2003; Weterings et al. 2003). Currently, the MRN complex is discussed as a sensor molecule for DSBs as well as an alignment factor for break ends (Bakkenist and Kastan 2003; Lisby and Rothstein 2004). Notably, the Rad50 component contains a hook-like Zn-coordinating structure that is able to interact with the same structure of a second molecule (Wiltzius et al. 2005; see also the contribution of K.-P. Hopfner in this issue). Therefore, by binding to each side of a DSB a Rad50 homodimeric interaction could establish a stable connection between the two ends of a break. There is also the possibility that the structural arrangement of chromatin into nucleosomes and fibers prevents broken ends from drifting apart (Lieber et al. 2003; see also the contribution of A. Friedl).

During the subsequent rejoining steps, a polymerase would be needed for any NHEJ events that require fill-in of gaps or extension of the 3′-end at 5′-overhangs (Lieber et al. 2004). In *S. cerevisiae*, POL4, a member of the Pol X family, is responsible for a substantial fraction of the fill-in synthesis during NHEJ (Wilson and Lieber 1999). Only recently, it has been demonstrated for the mammalian system that Pol X family members participate in the end joining reaction of incompatible ends in a Ku-dependent manner (Ma et al. 2004).

Another NHEJ associated protein is the nuclease Artemis. This factor has initially been identified as an essential protein in the V(D)J recombination process that generates antibody diversity (Moshous et al. 2001). V(D)J recombination is a cellular process that introduces DNA DSBs enzymatically und uses the NHEJ repair machinery to reseal the breaks (Alt et al. 1992; Gellert 2002). Consequently, NHEJ-deficient cells usually also display immune defects. In the course of V(D)J recombination, two types of DNA ends arise. The DNA ends of the excised spacer sequences between the coding elements represent simple blunt ends (signal ends) whereas the ends of coding sequences for the immunoglobulin or T cell receptors are protected by hairpin structures (coding ends). These hairpin loops need to be opened before the joining step, a process that is dependent on the nuclease Artemis. Therefore, Artemis-deficient immune cells are defective in rejoining V(D)J coding junctions but are proficient in signal joint formation. During V(D)J recombination the hairpin-opening activity of Artemis is activated through phosphorylation by DNA-PKcs (Ma et al. 2002). A failure in the V(D)J recombination process results in a condition termed SCID (severe combined immunodeficiency). Mutations in the DSB-inducing endonucleases RAG1 and RAG2 account for some of the SCID cases that are not sensitive to IR in humans (Schwarz et al. 1996; Corneo et al. 2000; Villa et al. 2001). A large proportion of SCID patients are also sensitive to IR, termed RS-SCID. Nearly all of these RS-SCID cases result from null mutations in the Artemis gene (Nicolas et al. 1998; Moshous et al. 2001; Ma et al. 2002; Ege et al. 2005). Recently, it has been shown that Artemis is also an integral part of a DSB repair mechanism, which explains the pronounced radiosensitivity of RS-SCID patients (see below, Riballo et al. 2004).

1.4 Deficiency in NHEJ Components

To date, no heritable human disease is known that involves the total failure of any of the core NHEJ components, demonstrating the absolute requirement of this pathway in human cell development. LIG4 syndrome, the only described human disorder involving core NHEJ components is based on a homozygous mutation in Lig4. The mutation renders the enzyme nearly completely inactive and the cells highly radiosensitive when exposed to IR (Badie et al. 1997; Riballo et al. 1999). However, the small residual activity allows repair of DSBs with very slow kinetics that seems to be sufficient to cope with the small numbers of lesion that arise during V(D)J recombination or to remove a substantial number of breaks when enough time for repair is provided (Riballo et al. 1999; Kuhne et al. 2004). The hypomorphic character of the mutation is also evident in cell survival experiments, where delayed plating (cell cycle arrested "plateau phase" cells are allowed to repair after irradiation before they are plated for survival) substantially increases cell survival of Lig4-defective cells. However, the slow repair kinetics is accompanied by a higher misrejoining activity (Kuhne et al. 2004). The only human cell line known to lack DNA-PKcs is MO59J, which was isolated from a malignant glioma. The M059J cell line is often used in combination with its DNA-PKcs-proficient sister cell line, M059K, that was isolated from the same tumor (Allalunis-Turner et al. 1995; Lees-Miller et al. 1995). Unfortunately, the two cell lines are not isogenic, as they differentially express genes in addition to DNA-PKcs, one example being ATM (Galloway and Allalunis-Turner 2000; Ai et al. 2003).

In contrast to the human situation, knockout mice for NHEJ core components could be developed. Ku70 and Ku80 knockouts in mice are viable. In addition to the immune defects, the hypersensitivity to IR, and a nearly complete deficiency in DSB repair, the animals are small in size and show a premature aging phenotype (Nussenzweig et al. 1996; Zhu et al. 1996). DNA-PKcs deficiency accounts for the classical SCID phenotype in mice, horses and dogs (Wiler et al. 1995; Taccioli et al. 1998; Meek et al. 2001; Perryman 2004). This is in clear contrast to humans, where DNA-PKcs mutations were never associated with human SCID. DSB repair studies in G1 phase DNA-PKcs$^{-/-}$ mouse cells using γH2AX foci analysis show a partial repair defect where about 50% of the breaks are repaired within 1 day. Therefore, DNA-PKcs seems to play a facilitating role in the rejoining of most DSBs (Riballo et al. 2004). The total knockouts of Lig4 and XRCC4 are not viable, but in a p53$^{-/-}$ background, Lig4-deficient mouse embryonic fibroblasts (MEFs) could be explanted for repair studies in vitro (Barnes et al. 1998; Frank et al. 1998). Consistent with the embryonic lethality, these cells show a dramatic repair defect for DSBs in G1-phase cells upon irradiation, demonstrating the requirement for Lig4 in the NHEJ pathway (Riballo et al. 2004). In vitro systems using purified NHEJ core components also absolutely rely on the XRCC4/Lig4 complex (Ma et al. 2004). However, a backup pathway for NHEJ with slow repair kinetics seems to exist that involves ligase 3 (Gottlich et al. 1998; Wang et al. 2003; Audebert et al. 2004; Wang H et al. 2005).

2 The ATM- and Artemis-Dependent Sub-Pathway of DSB Repair

2.1 ATM

ATM (ataxia telangiectasia mutated) protein is a member of the phosphatidylinositol 3-kinase-like family of serine-threonine protein kinases (PIKKs). Other members of this protein family include ATM- and Rad3-related (ATR) and DNA-PKcs. ATM deficiency results in the rare human heritable disorder ataxia telangiectasia (A-T). A-T is characterized by progressive cerebellar ataxia and telangiectasia. In addition, A-T patients display clinical radiosensitivity, immunodeficiency, and show an elevated risk for cancer induction (Taylor et al. 1996). ATM kinase activity increases two to threefold after IR (Canman et al. 1998). This increase in activity is accompanied by autophosphorylation of serine1981 in ATM and dissociation of ATM from an inactive dimeric (or higher order) form to an active monomeric form (Bakkenist and Kastan 2003). These findings place ATM as one of the earliest responders to DSBs. Although ATM has the ability to interact with DNA in vitro (Smith et al. 1999), recent findings suggest that the long-sought primary sensor of DSBs is not ATM but the trimeric MRN complex. Cells defective in Mre11 or Nbs1 have many features in common with A-T cells and autophosphorylation of ATM is defective in cells that are compromised for Nbs1 or Mre11 (Carson et al. 2003; Uziel et al. 2003). Moreover, the MRN complex directly recruits and activates ATM upon DSB induction (Lee and Paull 2004, 2005; Paull and Lee 2005). Kinase-active, monomeric ATM phosphorylates a multitude of substrates present at or recruited to the DSBs (Kurz and Lees-Miller 2004). ATM is the central component of the signal transduction pathway that leads to cell cycle checkpoint control and DNA repair (Lavin and Khanna 1999; Shiloh 2003). Consequently, A-T cells display cell cycle checkpoint defects following IR, resulting in an inability to arrest at the G1/S, S, and G2/M checkpoints. Although earlier studies already presumed a small repair defect in addition to the checkpoint defects in A-T cells, it could only recently be demonstrated that these cells leave DSBs unrepaired after exposure to IR. In contrast to cells deficient in core NHEJ proteins, the repair defect in A-T cells only becomes obvious after longer repair times and the repair of most DSBs is comparable with wild type cells (Foray et al. 1997; Kuhne et al. 2004). The small but significant fraction of breaks remain unrepaired for several days, suggesting that ATM is an essential component of a repair pathway for a subset of IR-induced DSBs (Fig. 1). As A-T cell lines display radiosensitivity that is even higher than the sensitivity seen in any of the core NHEJ-deficient cells, it has been suggested that most of the defects in A-T cells must result from checkpoint failures and not from the rather small repair defect. However, studies using premature chromosome condensation of

G1-phase cells demonstrated a level of unrepaired chromosome breaks that was sufficient to explain the radiosensitivity of A-T cells (Cornforth and Bedford 1985; Sasai et al. 1994; Jeggo et al. 1998). In addition, delayed plating survival experiments, in which checkpoint effects are minimized, usually do not result in an increase in cell survival of irradiated A-T cells as has been observed for repair proficient cell lines (Weichselbaum et al. 1978; Cox et al. 1981; Arlett and Priestley 1983).

2.2 Artemis

The involvement of ATM in a DSB repair pathway raised the question of which substrates of the kinase participate in this repair function. In addition to the immune defect described above, Artemis-deficient cells are also highly radiation sensitive, and the Artemis protein undergoes hyperphosphorylation upon irradiation with IR (Nicolas et al. 1998; Riballo et al. 2004; Zhang et al. 2004; Wang J et al. 2005). In contrast to V(D)J recombination, the phosphorylation of Artemis during IR-induced DSB repair is PIKK3-dependent, but not restricted to DNA-PKcs (Riballo et al. 2004; Zhang et al. 2004; Wang J et al. 2005). Interestingly, Artemis-deficient, non-dividing G1/G0-phase cells also display a small but significant repair defect upon exposure to IR (Riballo et al. 2004; Wang J et al. 2005). In fact A-T and Artemis-deficient cells harbor the same small DSB repair defect with the majority of breaks being repaired normally while a subset of about 10% of all breaks remains unrepaired. These findings placed ATM and Artemis in the same repair pathway (Foray et al. 1997; Kuhne et al. 2004; Riballo et al. 2004). This was verified by epistasis-like studies with small inhibitors specific for ATM. Treatment of wild type cells augments the number of residual breaks after IR to the level observed in A-T and Artemis deficient cells, whereas the drug shows no further increase in the repair defect in Artemis-deficient cells. Transfection of full-length Artemis into Artemis-deficient cells restores the repair activity whereas a nuclease-dead variant of the protein fails to restore the repair defect. In addition, the magnitude of the repair defect correlates with the complexity of the induced breaks because the application of agents that produce chemically complex DSBs (e.g. α-particles) results in a higher proportion of residual breaks than exposure to drugs that produce simple DSBs. Together, these results suggest that Artemis also functions as an end-processing nuclease during DSB repair. All studies on IR-induced DSB repair were performed in cell cycle arrested, non-dividing cells where NHEJ is the exclusive repair mechanism. This raised the possibility that the ATM and Artemis-dependent repair pathway represents a sub-pathway of NHEJ. Indeed, Lig4 and DNA-PKcs are required for this sub-pathway of DSB repair (Fig. 1; Riballo et al. 2004; Jeggo and Lobrich 2005; Lobrich and Jeggo 2005a).

2.3 Proteins Locating to γH2AX Foci

It was tempting to speculate that other ATM substrates participate in the ATM dependent repair pathway. One of the earliest steps in the cellular response to DSBs is the phosphorylation of serine139 of H2AX, a subclass of eukaryotic histone proteins that are part of the nucleoprotein structure (Rogakou et al. 1998). Phosphorylation of H2AX after IR exposure is preferentially carried out by ATM kinase but DNA-PKcs can substitute because A-T deficient cells show H2AX phosphorylation, except for a very short time window immediately after exposure (Kuhne et al. 2004; Stiff et al. 2004). Phosphorylated H2AX, termed γH2AX, can be observed over several megabases flanking the DSB (Rogakou et al. 1999). These phosphorylation events can be visualized as discrete nuclear foci with a specific fluorescent antibody. The number of γH2AX foci correlates with the number of expected DSBs after irradiation, suggesting that each focus may represent an individual break and that each DSB may form a focus (Rogakou et al. 1999; Modesti and Kanaar 2001; Redon et al. 2002). Comparative analysis of DSB induction and repair by means of pulsed field gel electrophoresis (PFGE) and γH2AX analysis substantiated the idea that γH2AX foci formation and disappearance can be used to measure the repair of individual DSBs even at very low doses (Rothkamm and Lobrich 2003; Kuhne et al. 2004; Riballo et al. 2004). Induction of γH2AX foci follows a linear relationship between the number of foci induced per cell and the IR dose delivered. MEFs lacking H2AX display radiosensitivity and genomic instability and also show a small DSB repair defect (Bassing et al. 2002; Celeste et al. 2002; Riballo et al. 2004). Quantification of the repair defect in $H2AX^{-/-}$ MEFs by PFGE analysis revealed that it resembles the defect observed in A-T and Artemis-deficient cells after IR (Riballo et al. 2004).

Additional proteins including 53BP1, the MRN complex, MDC1, and BRCA1 accumulate at the sites of H2AX phosphorylation (Paull et al. 2000; Bassing et al. 2002; Celeste et al. 2002). These proteins can also be visualized as repair foci, albeit some of them follow a different time-course of appearance/disappearance than γH2AX foci. Foci formation of these proteins is dependent on phosphorylated H2AX. Strikingly, two human disorders, Nijmegen breakage syndrome (NBS) and ataxia telangiectasia-like disorder (ATDL), which result from defects in the MRN complex proteins Nbs1 and Mre11, respectively, show an A-T like phenotype. Repair studies demonstrated that repair of IR-induced breaks in ATDL and NBS cells follows the repair kinetics observed in A-T cells, indicating that these proteins are also epistatic to the ATM-dependent DSB repair pathway. Similar results were obtained when MEFs deficient for 53BP1 were assayed. Therefore, proteins that form IR-induced foci function in this newly discovered repair pathway (Fig. 1; Riballo et al. 2004).

3 DSB Repair and Cell Cycle

Detailed studies on the mechanisms of the NHEJ repair pathway are possible in plateau phase cells because NHEJ is the only repair pathway activated. Nevertheless, repair studies in cycling cells are necessary to understand radiosensitivity, genomic instability, and cancer development. In particular, the contribution of each of the two main DSB repair pathways, HR and NHEJ, during the different phases of the cell cycle and the interplay between repair and checkpoint control are still not completely understood. In general, HR appears to be less important than NHEJ for the repair of IR-induced breaks in higher eukaryotes. This is in contrast to the situation in the yeast *S. cerevisiae* where HR constitutes the major repair mechanism for DSBs. As diploid organisms carry two copies of a chromosome, it is surprising that HR does not play a more general role for IR-induced DSBs. Rather, HR between homologous chromosomes seems to be down-regulated during G0, G1 and early S phase, the reason of which is still under discussion (Takata et al. 1998). It seems that the complexity of the genome with a large amount of repetitive elements (which are not present in yeast cells) makes this repair process too slow and dangerous because homologous crossover events between repetitive regions could result in a chromosomal translocation or loss of heterozygosity. HR only gains importance for DSB repair during phases of the cell cycle when homologous sequences such as sister chromatids are found in close proximity (late S, G2 and M). This is supported by the fact that the sister chromatid is used about 100 times more often for HR repair than a homologous chromosome (Richardson et al. 1998; Johnson and Jasin 2001).

DSB repair analysis in cycling human diploid fibroblasts is complicated by the fact that cells deficient in HR are hardly available. Therefore, other cell systems have been exploited to study DSB repair in cycling cells. Two of them proved to be extremely valuable: DT40 chicken B-lymphocytic cells (Takata et al. 1998) and CHO cells (Iliakis and Okayasu 1990; Rothkamm et al. 2003; Kruger et al. 2004). Cell survival experiments in these cell systems demonstrated that both, NHEJ and HR, contribute to the radioresistance of asynchronous cells. In double mutants, the inactivation of both repair pathways shows a dramatic increase in radiosensitivity when compared with the respective single mutants (Takata et al. 1998; Essers et al. 2000; Fukushima et al. 2001). Similar results were obtained in studies with single and double knockouts of Lig4 or Ku80 and Rad54 in the mouse (Couedel et al. 2004; Mills et al. 2004). Furthermore, conditional depletion of Rad51 in DT40 cells leads to an increase in chromosomal breaks before they undergo cell death, and chromosomal breaks also occur in cells deficient in genes of the Rad52 epistasis group (Liu et al. 1998; Sonoda et al. 1998, 2001; Takata et al. 1998, 2000, 2001; Griffin et al. 2000). In contrast to asynchronous cell populations, synchronisation in G1 phase reduces the sensitivity of HR-defective but not NHEJ-defective mutants (Fukushima et al. 2001). In line with the cell survival experiments, PFGE and γH2AX foci assays show that in G1-phase CHO cells only NHEJ-defective mutants display a pronounced repair defect, whereas a mutant in HR has nearly the same repair kinetics as wild type cells (Rothkamm et al. 2003). This strongly indicates that NHEJ is the predominant repair mechanism in G1-phase cells. HR obtains considerable importance in terms of cell survival when the cells are synchronized in late S/G2 prior to irradiation (Fig. 1; Fuller and Painter 1988; Stamato et al. 1988; Whitmore et al. 1989; Jeggo 1990; Takata et al. 1998; Essers et al. 2000; Rothkamm et al. 2003; Saleh-Gohari and Helleday 2004). Compared with G1/early S-phase cells, late S/G2-phase mammalian cells are more tolerant to IR. This increased IR tolerance depends on Rad54 (Takata et al. 1998; Sonoda et al. 2001) and XRCC2 (Cheong et al. 1994; Liu et al. 1998). Furthermore, the formation of Rad51 foci is only induced by IR in S/G2 phases but not in the G1 phase (Tarsounas et al. 2004). Moreover, the cell cycle regulating cyclin dependent kinases (cdks) seem to be involved in the activation of HR during specific phases of the cell cylce (Esashi et al. 2005). Because G1- and G2-phase cells are equally susceptible to radiation-induced breakage (Kruger et al. 2004), repair can be directly compared between these two phases by standard techniques. Repair of DSBs in wild type CHO cells irradiated in mid to late S/G2 is improved when compared to repair kinetics and fidelity in G1 phase cells. The DSB repair defect of NHEJ-deficient and HR-deficient cells is equal. This suggests that both NHEJ and HR contribute to radiation-induced DSB repair during late S/G2. Taken together, the relative importance of HR in repairing IR-induced DSBs increases from G1 through S and into late S/G2. Notably, replication-associated breaks seem to be repaired mainly by an HR-dependent repair mechanism (Paques and Haber 1999; Haber 2000; Arnaudeau et al. 2001; Rothkamm et al. 2003). This is supported by the presence of a number of Rad51 foci at S phase (Haaf et al. 1999; Tashiro et al. 2000; Saintigny et al. 2001), an increase in sister chromatide exchange (SCE) events in HR mutants (Sonoda et al. 1999, 2001), and an increase in γH2AX foci number in HR mutants after aphidicolin treatment (Rothkamm et al. 2003). Recent studies identified the ATR-Chk1 signaling cascade as an important stimulator of homology directed repair (Sorensen et al. 2005). It is well established that ATR is activated in response to genotoxic and replicational stress at stalled DNA replication forks and that Chk1 is a central regulator of cell cycle delays in S and G2 phases of the cell cycle (Liu et al. 2000; Takai et al. 2000; Bartek and Lukas 2003). There are also indications that cells can progress from G2 into mitosis and reach the G1 phase with a considerable amount of unrepaired DSBs. In conclusion, NHEJ is important for survival after IR in all phases of the cell cycle, while HR primarily contributes to radioresistance in the late S/G2 phase (Rothkamm et al. 2003; Kruger et al. 2004; Hinz et al. 2005).

4 Perspectives

From what has been learned about the contribution of the different DSB repair pathways during the cell cycle and the existence of the ATM- and Artemis-dependent repair pathway several interesting questions arise: First, it will be interesting to investigate, whether the ATM- and Artemis-dependent repair pathway plays a role in all phases of the cell cycle, especially when HR contributes to DSB repair. So far, most repair experiments with A-T and Artemis cell lines were conducted in stationary fibroblasts that represent G1/G0-phase cells where NHEJ is the only active repair mechanism. Therefore, the ATM- and Artemis-dependent repair pathway probably represents a subpathway of NHEJ. This conclusion is substantiated by the epistasis-like investigations in Lig4- and DNA-PKcs-defective cells (see above and Riballo et al. 2004). Preliminary studies in our laboratory with cycling primary fibroblasts reveal a DSB repair defect in ATM- and Artemis-deficient cell lines also during G2 phase. If this defect is coupled to the NHEJ-dependent pathway or whether ATM and Artemis also contribute to homology directed DSB repair has to be investigated in cell systems where mutants for HR are available.

A second question relates to the dual function of ATM as a checkpoint and a repair protein and the reasons why evolution has coupled these two processes. A-T and Artemis cells show equal radiosensitivity in cell survival assays of primary human fibroblasts. However, whereas A-T cells have lost a number of DNA damage checkpoints, Artemis cells are fully proficient in checkpoint induction and even show prolonged G2/M accumulation (Riballo et al. 2004, Wang J et al. 2005). Artemis is therefore not part of the ATM-dependent checkpoint signalling pathway and the similarity in radiosensitivity is not easy to understand. It has to be considered that a large fraction of cells in a primary fibroblast cell population is in G1 phase. This cell cycle phase is rather long and provides enough time for the rapid, ATM-independent fraction of DSB repair even without induction of a checkpoint. Consequently, at the time when most G1 phase A-T and Artemis cells cross the G1/S border, only the slow repairing DSBs will benefit from the checkpoint. However, Artemis cells are unable to repair these breaks and show, therefore, radiosensitivity similar to A-T cells. The situation may be different when cells are irradiated in short cell cycle phases like G2. Due to the defective ATM-dependent G2/M checkpoint, A-T cells with a substantial amount of breaks will appear in mitosis because the length of G2 phase per se does not provide enough time for the repair of the ATM-independent fraction of DSBs. In contrast, Artemis cells fully induce the G2/M damage checkpoint to repair the majority of DSBs except for those requiring Artemis function. These residual breaks might even prolong the checkpoint (Lobrich and Jeggo 2005b). First cytological data in our group indeed demonstrate an increase in chromosomal breaks in mitotic A-T cells after short repair times when compared to Artemis deficient cells. Generally, these results lead to the assumption that a long cell cycle phase provides sufficient time for the repair of the majority of breaks and the checkpoint function is only important for the small fraction of slowly repaired breaks. In short cell cycle phases. the checkpoint gains greater importance to allow repair of all breaks. Therefore, the coupling of checkpoint control with the slow component of DSB repair ensures completion of the repair and stabilizes genomic integrity.

Another question of major interest that is already addressed by some groups is how repair pathways for different types of DNA lesions work together to reconstitute the DNA sequence. As has been demonstrated for Artemis, the ends of a double-strand break may need specific processing. There is an extensive search for additional factors that participate in DSB repair and proteins from other pathways such as base excision repair have been identified to participate in the rejoining of DSBs (Boulton and Jackson 1996a,b; Wu et al. 1999; Delacote et al. 2002; Frank-Vaillant and Marcand 2002; Allen et al. 2003; Audebert et al. 2004; Goetz et al. 2005). If all these repair pathways work together, one system might even substitute the other under certain conditions. Hence, proper repair is not the result of a strict separation of pathways, but an interplay of different factors that contribute to the reconstitution of the genome.

References

Ai R, Sandoval A, Labhart P (2003) Differential gene expression in human glioma cells: correlation with presence or absence of DNA-dependent protein kinase. Gene Expr 11:35–45

Allalunis-Turner MJ, Zia PK, Barron GM, Mirzayans R, Day RS III (1995) Radiation-induced DNA damage and repair in cells of a radiosensitive human malignant glioma cell line. Radiat Res 144:288–293

Allen C, Halbrook J, Nickoloff JA (2003) Interactive competition between homologous recombination and non-homologous end joining. Mol Cancer Res 1:913–920

Alt FW, Oltz EM, Young F, Gorman J, Taccioli G, Chen J (1992) VDJ recombination. Immunol Today 13:306–314

Arlett CF, Priestley A (1983) Defective recovery from potentially lethal damage in some human fibroblast cell strains. Int J Radiat Biol Relat Stud Phys Chem Med 43:157–167

Arnaudeau C, Lundin C, Helleday T (2001) DNA double-strand breaks associated with replication forks are predominantly repaired by homologous recombination involving an exchange mechanism in mammalian cells. J Mol Biol 307:1235–1245

Audebert M, Salles B, Calsou P (2004) Involvement of poly(ADP-ribose) polymerase-1 and XRCC1/DNA ligase III in an alternative route for DNA double-strand breaks rejoining. J Biol Chem 279:55117–55126

Badie C, Goodhardt M, Waugh A, Doyen N, Foray N, Calsou P, Singleton B, Gell D, Salles B, Jeggo P, Arlett CF, Malaise EP (1997) A DNA double-strand break defective fibroblast cell line (180BR) derived from a radiosensitive patient represents a new mutant phenotype. Cancer Res 57:4600–4607

Bakkenist CJ, Kastan MB (2003) DNA damage activates ATM through intermolecular autophosphorylation and dimer dissociation. Nature 421:499–506
Barnes DE, Stamp G, Rosewell I, Denzel A, Lindahl T (1998) Targeted disruption of the gene encoding DNA ligase IV leads to lethality in embryonic mice. Curr Biol 8:1395–1398
Bartek J, Lukas J (2003) Chk1 and Chk2 kinases in checkpoint control and cancer. Cancer Cell 3:421–429
Bassing CH, Chua KF, Sekiguchi J, Suh H, Whitlow SR, Fleming JC, Monroe BC, Ciccone DN, Yan C, Vlasakova K, Livingston DM, Ferguson DO, Scully R, Alt FW (2002) Increased ionizing radiation sensitivity and genomic instability in the absence of histone H2AX. Proc Natl Acad Sci USA 99:8173–8178
Baumann P, Benson FE, West SC (1996) Human Rad51 protein promotes ATP-dependent homologous pairing and strand transfer reactions in vitro. Cell 87:757–766
Bianco PR, Tracy RB, Kowalczykowski SC (1998) DNA strand exchange proteins: a biochemical and physical comparison. Front Biosci 3:D570–D603
Boulton SJ, Jackson SP (1996a) Identification of a *Saccharomyces cerevisiae* Ku80 homologue: roles in DNA double-strand break rejoining and in telomeric maintenance. Nucleic Acids Res 24:4639–4648
Boulton SJ, Jackson SP (1996b) *Saccharomyces cerevisiae* Ku70 potentiates illegitimate DNA double-strand break repair and serves as a barrier to error-prone DNA repair pathways. EMBO J 15:5093–5103
Budman J, Chu G (2005) Processing of DNA for nonhomologous end-joining by cell-free extract. EMBO J 24:849–860
Canman CE, Lim DS, Cimprich KA, Taya Y, Tamai K, Sakaguchi K, Appella E, Kastan MB, Siliciano JD (1998) Activation of the ATM kinase by ionizing radiation and phosphorylation of p53. Science 281:1677–1679
Carson CT, Schwartz RA, Stracker TH, Lilley CE, Lee DV, Weitzman MD (2003) The Mre11 complex is required for ATM activation and the G2/M checkpoint. EMBO J 22:6610–6620
Cary RB, Peterson SR, Wang J, Bear DG, Bradbury EM, Chen DJ (1997) DNA looping by Ku and the DNA-dependent protein kinase. Proc Natl Acad Sci USA 94:4267–4272
Celeste A, Petersen S, Romanienko PJ, Fernandez-Capetillo O, Chen HT, Sedelnikova OA, Reina-San-Martin B, Coppola V, Meffre E, Difilippantonio MJ, Redon C, Pilch DR, Olaru A, Eckhaus M, Camerini-Otero RD, Tessarollo L, Livak F, Manova K, Bonner WM, Nussenzweig MC, Nussenzweig A (2002) Genomic instability in mice lacking histone H2AX. Science 296:922–927
Chang C, Biedermann KA, Mezzina M, Brown JM (1993) Characterization of the DNA double-strand break repair defect in scid mice. Cancer Res 53:1244–1248
Cheong N, Wang X, Wang Y, Iliakis G (1994) Loss of S-phase-dependent radioresistance in irs-1 cells exposed to X-rays. Mutat Res 314:77–85
Constantinou A, Chen XB, McGowan CH, West SC (2002) Holliday junction resolution in human cells: two junction endonucleases with distinct substrate specificities. EMBO J 21:5577–5585
Corneo B, Moshous D, Callebaut I, de Chasseval R, Fischer A, de Villartay JP (2000) Three-dimensional clustering of human RAG2 gene mutations in severe combined immune deficiency. J Biol Chem 275:12672–12675
Cornforth MN, Bedford JS (1985) On the nature of a defect in cells from individuals with ataxia-telangiectasia. Science 227:1589–1591
Couedel C, Mills KD, Barchi M, Shen L, Olshen A, Johnson RD, Nussenzweig A, Essers J, Kanaar R, Li GC, Alt FW, Jasin M (2004) Collaboration of homologous recombina-

tion and nonhomologous end-joining factors for the survival and integrity of mice and cells. Genes Dev 18:1293–1304

Cox R, Masson WK, Weichselbaum RR, Nove J, Little JB (1981) The repair of potentially lethal damage in X-irradiated cultures of normal and ataxia telangiectasia human fibroblasts. Int J Radiat Biol Relat Stud Phys Chem Med 39:357–365

Deans B, Griffin CS, Maconochie M, Thacker J (2000) Xrcc2 is required for genetic stability, embryonic neurogenesis and viability in mice. EMBO J 19:6675–6685

DeFazio LG, Stansel RM, Griffith JD, Chu G (2002) Synapsis of DNA ends by DNA-dependent protein kinase. EMBO J 21:3192–3200

Delacote F, Han M, Stamato TD, Jasin M, Lopez BS (2002) An xrcc4 defect or Wortmannin stimulates homologous recombination specifically induced by double-strand breaks in mammalian cells. Nucleic Acids Res 30:3454–3463

Drouet J, Delteil C, Lefrancois J, Concannon P, Salles B, Calsou P (2005) DNA-dependent protein kinase and XRCC4-DNA ligase IV mobilization in the cell in response to DNA double-strand breaks. J Biol Chem 280:7060–7069

Ege M, Ma Y, Manfras B, Kalwak K, Lu H, Lieber MR, Schwarz K, Pannicke U (2005) Omenn syndrome due to ARTEMIS mutations. Blood 105:4179–4186

Eggleston AK, West SC (1996) Exchanging partners: recombination in *E. coli*. Trends Genet 12:20–26

Esashi F, Christ N, Gannon J, Liu Y, Hunt T, Jasin M, West SC (2005) CDK-dependent phosphorylation of BRCA2 as a regulatory mechanism for recombinational repair. Nature 434:598–604

Essers J, van Steeg H, de Wit J, Swagemakers SM, Vermeij M, Hoeijmakers JH, Kanaar R (2000) Homologous and non-homologous recombination differentially affect DNA damage repair in mice. EMBO J 19:1703–1710

Feldmann E, Schmiemann V, Goedecke W, Reichenberger S, Pfeiffer P (2000) DNA double-strand break repair in cell-free extracts from Ku80-deficient cells: implications for Ku serving as an alignment factor in non-homologous DNA end joining. Nucleic Acids Res 28:2585–2596

Foray N, Priestley A, Alsbeih G, Badie C, Capulas EP, Arlett CF, Malaise EP (1997) Hypersensitivity of ataxia telangiectasia fibroblasts to ionizing radiation is associated with a repair deficiency of DNA double-strand breaks. Int J Radiat Biol 72:271–283

Frank KM, Sekiguchi JM, Seidl KJ, Swat W, Rathbun GA, Cheng HL, Davidson L, Kangaloo L, Alt FW (1998) Late embryonic lethality and impaired V(D)J recombination in mice lacking DNA ligase IV. Nature 396:173–177

Frank-Vaillant M, Marcand S (2002) Transient stability of DNA ends allows nonhomologous end joining to precede homologous recombination. Mol Cell 10:1189–1199

Fukushima T, Takata M, Morrison C, Araki R, Fujimori A, Abe M, Tatsumi K, Jasin M, Dhar PK, Sonoda E, Chiba T, Takeda S (2001) Genetic analysis of the DNA-dependent protein kinase reveals an inhibitory role of Ku in late S-G2 phase DNA double-strand break repair. J Biol Chem 276:44413–44418

Fuller LF, Painter RB (1988) A Chinese hamster ovary cell line hypersensitive to ionizing radiation and deficient in repair replication. Mutat Res 193:109–121

Galloway AM, Allalunis-Turner J (2000) cDNA expression array analysis of DNA repair genes in human glioma cells that lack or express DNA-PK. Radiat Res 154:609–615

Gellert M (2002) V(D)J recombination: RAG proteins, repair factors, and regulation. Annu Rev Biochem 71:101–132

Goetz JD, Motycka TA, Han M, Jasin M, Tomkinson AE (2005) Reduced repair of DNA double-strand breaks by homologous recombination in a DNA ligase I-deficient human cell line. DNA Repair (Amst) 4:649–654

Gottlich B, Reichenberger S, Feldmann E, Pfeiffer P (1998) Rejoining of DNA double-strand breaks in vitro by single-strand annealing. Eur J Biochem 258:387–395

Gottlieb TM, Jackson SP (1993) The DNA-dependent protein kinase: requirement for DNA ends and association with Ku antigen. Cell 72:131–142

Griffin CS, Simpson PJ, Wilson CR, Thacker J (2000) Mammalian recombination-repair genes XRCC2 and XRCC3 promote correct chromosome segregation. Nat Cell Biol 2:757–761

Haaf T, Raderschall E, Reddy G, Ward DC, Radding CM, Golub EI (1999) Sequestration of mammalian Rad51-recombination protein into micronuclei. J Cell Biol 144:11–20

Haber JE (1998) The many interfaces of Mre11. Cell 95:583–586

Haber JE (2000) Partners and pathways: repairing a double-strand break. Trends Genet 16:259–264

Heyer WD, Ehmsen KT, Solinger JA (2003) Holliday junctions in the eukaryotic nucleus: resolution in sight? Trends Biochem Sci 28:548–557

Hinz JM, Yamada NA, Salazar EP, Tebbs RS, Thompson LH (2005) Influence of double-strand-break repair pathways on radiosensitivity throughout the cell cycle in CHO cells. DNA Repair (Amst) 4:782–792

Hiom K (2001) Recombination: homologous recombination branches out. Curr Biol 11:R278-R280

Hoeijmakers JH (2001) Genome maintenance mechanisms for preventing cancer. Nature 411:366–374

Howlett NG, Taniguchi T, Olson S, Cox B, Waisfisz Q, Die-Smulders C, Persky N, Grompe M, Joenje H, Pals G, Ikeda H, Fox EA, D'Andrea AD (2002) Biallelic inactivation of BRCA2 in Fanconi anemia. Science 297:606–609

Iliakis GE, Okayasu R (1990) Radiosensitivity throughout the cell cycle and repair of potentially lethal damage and DNA double-strand breaks in an X-ray-sensitive CHO mutant. Int J Radiat Biol 57:1195–1211

Jackson SP (2002) Sensing and repairing DNA double-strand breaks. Carcinogenesis 23:687–696

Jeggo PA (1990) Studies on mammalian mutants defective in rejoining double-strand breaks in DNA. Mutat Res 239:1–16

Jeggo PA (1998) Mutant rodent cells defective in DNA double-strand break repair. In: Nickoloff JA, Hoekstra MF (eds) DNA damage and repair, vol 2, DNA repair in higher eukaryotes. Humana Press, Totowa, NJ, pp 317–334

Jeggo PA, Lobrich M (2005) Artemis links ATM to double-strand break rejoining. Cell Cycle 4:359–362

Jeggo PA, Carr AM, Lehmann AR (1998) Splitting the ATM: distinct repair and checkpoint defects in ataxia-telangiectasia. Trends Genet 14:312–316

Johnson RD, Jasin M (2001) Double-strand-break-induced homologous recombination in mammalian cells. Biochem Soc Trans 29:196–201

Jones NJ, Cox R, Thacker J (1987) Isolation and cross-sensitivity of X-ray-sensitive mutants of V79-4 hamster cells. Mutat Res 183:279–286

Krejci L, Chen L, Van Komen S, Sung P, Tomkinson A (2003) Mending the break: two DNA double-strand break repair machines in eukaryotes. Prog Nucleic Acid Res Mol Biol 74:159–201

Kruger I, Rothkamm K, Lobrich M (2004) Enhanced fidelity for rejoining radiation-induced DNA double-strand breaks in the G2 phase of Chinese hamster ovary cells. Nucleic Acids Res 32:2677–2684

Kuhne M, Riballo E, Rief N, Rothkamm K, Jeggo PA, Lobrich M (2004) A double-strand break repair defect in ATM-deficient cells contributes to radiosensitivity. Cancer Res 64:500–508

Kurimasa A, Kumano S, Boubnov NV, Story MD, Tung CS, Peterson SR, Chen DJ (1999) Requirement for the kinase activity of human DNA-dependent protein kinase catalytic subunit in DNA strand break rejoining. Mol Cell Biol 19:3877–3884

Kurz EU, Lees-Miller SP (2004) DNA damage-induced activation of ATM and ATM-dependent signaling pathways. DNA Repair (Amst) 3:889–900

Labhart P (1999a) Ku-dependent nonhomologous DNA end joining in *Xenopus* egg extracts. Mol Cell Biol 19:2585–2593

Labhart P (1999b) Nonhomologous DNA end joining in cell-free systems. Eur J Biochem 265:849–861

Lavin MF, Khanna KK (1999) ATM: the protein encoded by the gene mutated in the radiosensitive syndrome ataxia-telangiectasia. Int J Radiat Biol 75:1201–1214

Lee JH, Paull TT (2004) Direct activation of the ATM protein kinase by the Mre11/Rad50/Nbs1 complex. Science 304:93–96

Lee JH, Paull TT (2005) ATM activation by DNA double-strand breaks through the Mre11-Rad50-Nbs1 complex. Science 308:551–554

Lees-Miller SP, Chen YR, Anderson CW (1990) Human cells contain a DNA-activated protein kinase that phosphorylates simian virus 40 T antigen, mouse p53, and the human Ku autoantigen. Mol Cell Biol 10:6472–6481

Lees-Miller SP, Godbout R, Chan DW, Weinfeld M, Day RS, III, Barron GM, Allalunis-Turner J (1995) Absence of p350 subunit of DNA-activated protein kinase from a radiosensitive human cell line. Science 267:1183–1185

Lieber MR, Ma Y, Pannicke U, Schwarz K (2003) Mechanism and regulation of human non-homologous DNA end-joining. Nat Rev Mol Cell Biol 4:712–720

Lieber MR, Ma Y, Pannicke U, Schwarz K (2004) The mechanism of vertebrate nonhomologous DNA end joining and its role in V(D)J recombination. DNA Repair (Amst) 3:817–826

Lim DS, Hasty P (1996) A mutation in mouse rad51 results in an early embryonic lethal that is suppressed by a mutation in p53. Mol Cell Biol 16:7133–7143

Lisby M, Rothstein R (2004) DNA repair: keeping it together. Curr Biol 14:R994–R996

Liu N, Lamerdin JE, Tebbs RS, Schild D, Tucker JD, Shen MR, Brookman KW, Siciliano MJ, Walter CA, Fan W, Narayana LS, Zhou ZQ, Adamson AW, Sorensen KJ, Chen DJ, Jones NJ, Thompson LH (1998) XRCC2 and XRCC3, new human Rad51-family members, promote chromosome stability and protect against DNA cross-links and other damages. Mol Cell 1:783–793

Liu Q, Guntuku S, Cui XS, Matsuoka S, Cortez D, Tamai K, Luo G, Carattini-Rivera S, DeMayo F, Bradley A, Donehower LA, Elledge SJ (2000) Chk1 is an essential kinase that is regulated by Atr and required for the G(2)/M DNA damage checkpoint. Genes Dev 14:1448–1459

Liu Y, Masson JY, Shah R, O'Regan P, West SC (2004) RAD51C is required for Holliday junction processing in mammalian cells. Science 303:243–246

Liu Y, West SC (2004) Happy Hollidays: 40th anniversary of the Holliday junction. Nat Rev Mol Cell Biol 5:937–944

Lobrich M, Jeggo PA (2005a) Harmonising the response to DSBs: a new string in the ATM bow. DNA Repair (Amst) 4:749–759

Lobrich M, Jeggo PA (2005b) The two edges of the ATM sword: co-operation between repair and checkpoint functions. Radiother Oncol 76:112–118

Ma Y, Pannicke U, Schwarz K, Lieber MR (2002) Hairpin opening and overhang processing by an Artemis/DNA-dependent protein kinase complex in nonhomologous end joining and V(D)J recombination. Cell 108:781–794

Ma Y, Lu H, Tippin B, Goodman MF, Shimazaki N, Koiwai O, Hsieh CL, Schwarz K, Lieber MR (2004) A biochemically defined system for mammalian nonhomologous DNA end joining. Mol Cell 16:701–713

Maser RS, Monsen KJ, Nelms BE, Petrini JH (1997) hMre11 and hRad50 nuclear foci are induced during the normal cellular response to DNA double-strand breaks. Mol Cell Biol 17:6087–6096

McIlwraith MJ, Van Dyck E, Masson JY, Stasiak AZ, Stasiak A, West SC (2000) Reconstitution of the strand invasion step of double-strand break repair using human Rad51 Rad52 and RPA proteins. J Mol Biol 304:151–164

Meek K, Kienker L, Dallas C, Wang W, Dark MJ, Venta PJ, Huie ML, Hirschhorn R, Bell T (2001) SCID in Jack Russell terriers: a new animal model of DNA-PKcs deficiency. J Immunol 167:2142–2150

Mills KD, Ferguson DO, Essers J, Eckersdorff M, Kanaar R, Alt FW (2004) Rad54 and DNA Ligase IV cooperate to maintain mammalian chromatid stability. Genes Dev 18:1283–1292

Modesti M, Kanaar R (2001) DNA repair: spot(light)s on chromatin. Curr Biol 11:R229–R232

Moshous D, Callebaut I, de Chasseval R, Corneo B, Cavazzana-Calvo M, Le Deist F, Tezcan I, Sanal O, Bertrand Y, Philippe N, Fischer A, de Villartay JP (2001) Artemis, a novel DNA double-strand break repair/V(D)J recombination protein, is mutated in human severe combined immune deficiency. Cell 105:177–186

Nakanishi K, Yang YG, Pierce AJ, Taniguchi T, Digweed M, D'Andrea AD, Wang ZQ, Jasin M (2005) Human Fanconi anemia monoubiquitination pathway promotes homologous DNA repair. Proc Natl Acad Sci USA 102:1110–1115

Nicolas AL, Munz PL, Young CS (1995) A modified single-strand annealing model best explains the joining of DNA double-strand breaks mammalian cells and cell extracts. Nucleic Acids Res 23:1036–1043

Nicolas N, Moshous D, Cavazzana-Calvo M, Papadopoulo D, de Chasseval R, Le Deist F, Fischer A, de Villartay JP (1998) A human severe combined immunodeficiency (SCID) condition with increased sensitivity to ionizing radiations and impaired V(D)J rearrangements defines a new DNA recombination/repair deficiency. J Exp Med 188:627–634

Nussenzweig A, Chen C, da CS, V, Sanchez M, Sokol K, Nussenzweig MC, Li GC (1996) Requirement for Ku80 in growth and immunoglobulin V(D)J recombination. Nature 382:551–555

Odersky A, Panyutin IV, Panyutin IG, Schunck C, Feldmann E, Goedecke W, Neumann RD, Obe G, Pfeiffer P (2002) Repair of sequence-specific 125I-induced double-strand breaks by nonhomologous DNA end joining in mammalian cell-free extracts. J Biol Chem 277:11756–11764

Olive PL (1998) The role of DNA single- and double-strand breaks in cell killing by ionizing radiation. Radiat Res 150:S42–S51

Pang D, Yoo S, Dynan WS, Jung M, Dritschilo A (1997) Ku proteins join DNA fragments as shown by atomic force microscopy. Cancer Res 57:1412–1415

Paques F, Haber JE (1999) Multiple pathways of recombination induced by double-strand breaks in *Saccharomyces cerevisiae*. Microbiol Mol Biol Rev 63:349–404

Paull TT, Gellert M (1998) The 3′ to 5′ exonuclease activity of Mre 11 facilitates repair of DNA double-strand breaks. Mol Cell 1:969–979

Paull TT, Gellert M (1999) Nbs1 potentiates ATP-driven DNA unwinding and endonuclease cleavage by the Mre11/Rad50 complex. Genes Dev 13:1276–1288

Paull TT, Lee JH (2005) The Mre11/Rad50/Nbs1 complex and its role as a DNA double-strand break sensor for ATM. Cell Cycle 4:737–740

Paull TT, Rogakou EP, Yamazaki V, Kirchgessner CU, Gellert M, Bonner WM (2000) A critical role for histone H2AX in recruitment of repair factors to nuclear foci after DNA damage. Curr Biol 10:886–895

Perryman LE (2004) Molecular pathology of severe combined immunodeficiency in mice, horses, and dogs. Vet Pathol 41:95–100

Pfeiffer P, Feldmann E, Odersky A, Kuhfittig-Kulle S, Goedecke W (2005) Analysis of DNA double-strand break repair by nonhomologous end joining in cell-free extracts from mammalian cells. Methods Mol Biol 291:351–371

Pfeiffer P, Goedecke W, Kuhfittig-Kulle S, Obe G (2004) Pathways of DNA double-strand break repair and their impact on the prevention and formation of chromosomal aberrations. Cytogenet Genome Res 104:7–13

Pfeiffer P, Vielmetter W (1988) Joining of nonhomologous DNA double-strand breaks in vitro. Nucleic Acids Res 16:907–924

Pittman DL, Schimenti JC (2000) Midgestation lethality in mice deficient for the RecA-related gene, Rad51d/Rad51l3. Genesis 26:167–173

Raghavan SC, Raman MJ (2004) Nonhomologous end joining of complementary and noncomplementary DNA termini in mouse testicular extracts. DNA Repair (Amst) 3:1297–1310

Redon C, Pilch D, Rogakou E, Sedelnikova O, Newrock K, Bonner W (2002) Histone H2A variants H2AX and H2AZ. Curr Opin Genet Dev 12:162–169

Riballo E, Critchlow SE, Teo SH, Doherty AJ, Priestley A, Broughton B, Kysela B, Beamish H, Plowman N, Arlett CF, Lehmann AR, Jackson SP, Jeggo PA (1999) Identification of a defect in DNA ligase IV in a radiosensitive leukaemia patient. Curr Biol 9:699–702

Riballo E, Kuhne M, Rief N, Doherty A, Smith GC, Recio MJ, Reis C, Dahm K, Fricke A, Krempler A, Parker AR, Jackson SP, Gennery A, Jeggo PA, Lobrich M (2004) A pathway of double-strand break rejoining dependent upon ATM, Artemis, and proteins locating to gamma-H2AX foci. Mol Cell 16:715–724

Richardson C, Moynahan ME, Jasin M (1998) Double-strand break repair by interchromosomal recombination: suppression of chromosomal translocations. Genes Dev 12:3831–3842

Rijkers T, Van Den OJ, Morolli B, Rolink AG, Baarends WM, Van Sloun PP, Lohman PH, Pastink A (1998) Targeted inactivation of mouse RAD52 reduces homologous recombination but not resistance to ionizing radiation. Mol Cell Biol 18:6423–6429

Rogakou EP, Pilch DR, Orr AH, Ivanova VS, Bonner WM (1998) DNA double-stranded breaks induce histone H2AX phosphorylation on serine 139. J Biol Chem 273:5858–5868

Rogakou EP, Boon C, Redon C, Bonner WM (1999) Megabase chromatin domains involved in DNA double-strand breaks in vivo. J Cell Biol 146:905–916

Rothkamm K, Lobrich M (2003) Evidence for a lack of DNA double-strand break repair in human cells exposed to very low X-ray doses. Proc Natl Acad Sci USA 100:5057–5062

Rothkamm K, Kruger I, Thompson LH, Lobrich M (2003) Pathways of DNA double-strand break repair during the mammalian cell cycle. Mol Cell Biol 23:5706–5715

Saintigny Y, Delacote F, Vares G, Petitot F, Lambert S, Averbeck D, Lopez BS (2001) Characterization of homologous recombination induced by replication inhibition in mammalian cells. EMBO J 20:3861–3870

Saintigny Y, Makienko K, Swanson C, Emond MJ, Monnat RJ Jr (2002) Homologous recombination resolution defect in Werner syndrome. Mol Cell Biol 22:6971–6978

Saleh-Gohari N, Helleday T (2004) Conservative homologous recombination preferentially repairs DNA double-strand breaks in the S phase of the cell cycle in human cells. Nucleic Acids Res 32:3683–3688

Sasai K, Evans JW, Kovacs MS, Brown JM (1994) Prediction of human cell radiosensitivity: comparison of clonogenic assay with chromosome aberrations scored using premature chromosome condensation with fluorescence in situ hybridization. Int J Radiat Oncol Biol Phys 30:1127–1132

Schwarz K, Gauss GH, Ludwig L, Pannicke U, Li Z, Lindner D, Friedrich W, Seger RA, Hansen-Hagge TE, Desiderio S, Lieber MR, Bartram CR (1996) RAG mutations in human B cell-negative SCID. Science 274:97–99

Sedelnikova OA, Pilch DR, Redon C, Bonner WM (2003) Histone H2AX in DNA damage and repair. Cancer Biol Ther 2:233–235

Shiloh Y (2003) ATM and related protein kinases: safeguarding genome integrity. Nat Rev Cancer 3:155–168

Shivji MK, Venkitaraman AR (2004) DNA recombination, chromosomal stability and carcinogenesis: insights into the role of BRCA2. DNA Repair (Amst) 3:835–843

Shu Z, Smith S, Wang L, Rice MC, Kmiec EB (1999) Disruption of muREC2/RAD51L1 in mice results in early embryonic lethality which can be partially rescued in a p53(–/–) background. Mol Cell Biol 19:8686–8693

Smith GC, Cary RB, Lakin ND, Hann BC, Teo SH, Chen DJ, Jackson SP (1999) Purification and DNA binding properties of the ataxia-telangiectasia gene product ATM. Proc Natl Acad Sci USA 96:11134–11139

Smith GC, Jackson SP (1999) The DNA-dependent protein kinase. Genes Dev 13:916–934

Solinger JA, Lutz G, Sugiyama T, Kowalczykowski SC, Heyer WD (2001) Rad54 protein stimulates heteroduplex DNA formation in the synaptic phase of DNA strand exchange via specific interactions with the presynaptic Rad51 nucleoprotein filament. J Mol Biol 307:1207–1221

Song B, Sung P (2000) Functional interactions among yeast Rad51 recombinase, Rad52 mediator, and replication protein A in DNA strand exchange. J Biol Chem 275:15895–15904

Sonoda E, Sasaki MS, Buerstedde JM, Bezzubova O, Shinohara A, Ogawa H, Takata M, Yamaguchi-Iwai Y, Takeda S (1998) Rad51-deficient vertebrate cells accumulate chromosomal breaks prior to cell death. EMBO J 17:598–608

Sonoda E, Sasaki MS, Morrison C, Yamaguchi-Iwai Y, Takata M, Takeda S (1999) Sister chromatid exchanges are mediated by homologous recombination in vertebrate cells. Mol Cell Biol 19:5166–5169

Sonoda E, Takata M, Yamashita YM, Morrison C, Takeda S (2001) Homologous DNA recombination in vertebrate cells. Proc Natl Acad Sci USA 98:8388–8394

Sorensen CS, Hansen LT, Dziegielewski J, Syljuasen RG, Lundin C, Bartek J, Helleday T (2005) The cell-cycle checkpoint kinase Chk1 is required for mammalian homologous recombination repair. Nat Cell Biol 7:195–201

Stamato TD, Dipatri A, Giaccia A (1988) Cell-cycle-dependent repair of potentially lethal damage in the XR-1 gamma-ray-sensitive Chinese hamster ovary cell. Radiat Res 115:325–333

Stiff T, O'Driscoll M, Rief N, Iwabuchi K, Lobrich M, Jeggo PA (2004) ATM and DNA-PK function redundantly to phosphorylate H2AX after exposure to ionizing radiation. Cancer Res 64:2390–2396

Sung P, Krejci L, Van Komen S, Sehorn MG (2003) Rad51 recombinase and recombination mediators. J Biol Chem 278:42729–42732

Symington LS (2002) Role of RAD52 epistasis group genes in homologous recombination and double-strand break repair. Microbiol Mol Biol Rev 66:630–70, table

Taccioli GE, Amatucci AG, Beamish HJ, Gell D, Xiang XH, Torres Arzayus MI, Priestley A, Jackson SP, Marshak RA, Jeggo PA, Herrera VL (1998) Targeted disruption of the catalytic subunit of the DNA-PK gene in mice confers severe combined immunodeficiency and radiosensitivity. Immunity 9:355–366

Takai H, Tominaga K, Motoyama N, Minamishima YA, Nagahama H, Tsukiyama T, Ikeda K, Nakayama K, Nakanishi M, Nakayama K (2000) Aberrant cell cycle checkpoint function and early embryonic death in Chk1(–/–) mice. Genes Dev 14:1439–1447

Takata M, Sasaki MS, Sonoda E, Morrison C, Hashimoto M, Utsumi H, Yamaguchi-Iwai Y, Shinohara A, Takeda S (1998) Homologous recombination and non-homologous end-joining pathways of DNA double-strand break repair have overlapping roles in the maintenance of chromosomal integrity in vertebrate cells. EMBO J 17:5497–5508

Takata M, Sasaki MS, Sonoda E, Fukushima T, Morrison C, Albala JS, Swagemakers SM, Kanaar R, Thompson LH, Takeda S (2000) The Rad51 paralog Rad51B promotes homologous recombinational repair. Mol Cell Biol 20:6476–6482

Takata M, Sasaki MS, Tachiiri S, Fukushima T, Sonoda E, Schild D, Thompson LH, Takeda S (2001) Chromosome instability and defective recombinational repair in knockout mutants of the five Rad51 paralogs. Mol Cell Biol 21:2858–2866

Tarsounas M, Davies AA, West SC (2004) RAD51 localization and activation following DNA damage. Philos Trans R Soc Lond B Biol Sci 359:87–93

Tashiro S, Walter J, Shinohara A, Kamada N, Cremer T (2000) Rad51 accumulation at sites of DNA damage and in postreplicative chromatin. J Cell Biol 150:283–291

Taylor AM, Metcalfe JA, Thick J, Mak YF (1996) Leukemia and lymphoma in ataxia telangiectasia. Blood 87:423–438

Thode S, Schafer A, Pfeiffer P, Vielmetter W (1990) A novel pathway of DNA end-to-end joining. Cell 60:921–928

Thompson LH, Schild D (2002) Recombinational DNA repair and human disease. Mutat Res 509:49–78

Ting NS, Lee WH (2004) The DNA double-strand break response pathway: becoming more BRCAish than ever. DNA Repair (Amst) 3:935–944

Trujillo KM, Yuan SS, Lee EY, Sung P (1998) Nuclease activities in a complex of human recombination and DNA repair factors Rad50, Mre11, and p95. J Biol Chem 273:21447–21450

Tsuzuki T, Fujii Y, Sakumi K, Tominaga Y, Nakao K, Sekiguchi M, Matsushiro A, Yoshimura Y, MoritaT (1996) Targeted disruption of the Rad51 gene leads to lethality in embryonic mice. Proc Natl Acad Sci USA 93:6236–6240

Uziel T, Lerenthal Y, Moyal L, Andegeko Y, Mittelman L, Shiloh Y (2003) Requirement of the MRN complex for ATM activation by DNA damage. EMBO J 22:5612–5621

van den Bosch M, Lohman PH, Pastink A (2002) DNA double-strand break repair by homologous recombination. Biol Chem 383:873–892

van Gent DC, Hoeijmakers JH, Kanaar R (2001) Chromosomal stability and the DNA double-stranded break connection. Nat Rev Genet 2:196–206

Villa A, Sobacchi C, Vezzoni P (2001) Recombination activating gene and its defects. Curr Opin Allergy Clin Immunol 1:491–495

Wachsberger PR, Li WH, Guo M, Chen D, Cheong N, Ling CC, Li G, Iliakis G (1999) Rejoining of DNA double-strand breaks in Ku80-deficient mouse fibroblasts. Radiat Res 151:398–407

Walker JR, Corpina RA, Goldberg J (2001) Structure of the Ku heterodimer bound to DNA and its implications for double-strand break repair. Nature 412:607–614

Wang H, Perrault AR, Takeda Y, Qin W, Wang H, Iliakis G (2003) Biochemical evidence for Ku-independent backup pathways of NHEJ. Nucleic Acids Res 31:5377–5388

Wang H, Rosidi B, Perrault R, Wang M, Zhang L, Windhofer F, Iliakis G (2005) DNA ligase III as a candidate component of backup pathways of nonhomologous end joining. Cancer Res 65:4020–4030

Wang J, Pluth JM, Cooper PK, Cowan MJ, Chen DJ, Yannone SM (2005) Artemis deficiency confers a DNA double-strand break repair defect and Artemis phosphorylation status is altered by DNA damage and cell cycle progression. DNA Repair (Amst) 4:556–570

Wang X, D'Andrea AD (2004) The interplay of Fanconi anemia proteins in the DNA damage response. DNA Repair (Amst) 3:1063–1069

Weichselbaum RR, Nove J, Little JB (1978) Deficient recovery from potentially lethal radiation damage in ataxia telengiectasia and xeroderma pigmentosum. Nature 271:261–262

Weterings E, Verkaik NS, Bruggenwirth HT, Hoeijmakers JH, van Gent DC (2003) The role of DNA dependent protein kinase in synapsis of DNA ends. Nucleic Acids Res 31:7238–7246

Whitmore GF, Varghese AJ, Gulyas S (1989) Cell cycle responses of two X-ray sensitive mutants defective in DNA repair. Int J Radiat Biol 56:657–665

Wiler R, Leber R, Moore BB, VanDyk LF, Perryman LE, Meek K (1995) Equine severe combined immunodeficiency: a defect in V(D)J recombination and DNA-dependent protein kinase activity. Proc Natl Acad Sci USA 92:11485–11489

Wilson TE, Lieber MR (1999) Efficient processing of DNA ends during yeast nonhomologous end joining. Evidence for a DNA polymerase beta (Pol4)-dependent pathway. J Biol Chem 274:23599–23609

Wiltzius JJ, Hohl M, Fleming JC, Petrini JH (2005) The Rad50 hook domain is a critical determinant of Mre11 complex functions. Nat Struct Mol Biol 12:403–407

Wu X, Wilson TE, Lieber MR (1999) A role for FEN-1 in nonhomologous DNA end joining: the order of strand annealing and nucleolytic processing events. Proc Natl Acad Sci USA 96:1303–1308

Wyman C, Ristic D, Kanaar R (2004) Homologous recombination-mediated double-strand break repair. DNA Repair (Amst) 3:827–833

Yamaguchi-Iwai Y, Sonoda E, Buerstedde JM, Bezzubova O, Morrison C, Takata M, Shinohara A, Takeda S (1998) Homologous recombination, but not DNA repair, is reduced in vertebrate cells deficient in RAD52. Mol Cell Biol 18:6430–6435

Yaneva M, Kowalewski T, Lieber MR (1997) Interaction of DNA-dependent protein kinase with DNA and with Ku: biochemical and atomic-force microscopy studies. EMBO J 16:5098–5112

Zhang X, Succi J, Feng Z, Prithivirajsingh S, Story MD, Legerski RJ (2004) Artemis is a phosphorylation target of ATM and ATR and is involved in the G2/M DNA damage checkpoint response. Mol Cell Biol 24:9207–9220

Zhu C, Bogue MA, Lim DS, Hasty P, Roth DB (1996) Ku86-deficient mice exhibit severe combined immunodeficiency and defective processing of V(D)J recombination intermediates. Cell 86:379–389

Genome Dyn Stab (1)
D.-H. Lankenau: Genome Integrity
DOI 10.1007/7050_010/Published online: 2 December 2005

Genomic Instability: Signaling Pathways Orchestrating the Responses to Ionizing Radiation and Cisplatin

Elza T. Sakamoto-Hojo[1,2,5] (✉) · Stephano S. Mello[2] · Carmen L. Bassi[2] · Igor M. Merchi[2] · Patricia O. Carminati[2] · Ana L. Fachin[2] · Eduardo A. Donadi[3] · Geraldo A.S. Passos[2,4]

[1]Departamento de Biologia, Faculdade de Filosofia Ciências e Letras de Ribeirão Preto, Ribeirão Preto, SP, Brazil
etshojo@usp.br

[2]Departamento de Genética, Ribeirão Preto, SP, Brazil
etshojo@usp.br

[3]Departamento de Clínica Médica, Faculdade de Medicina de Ribeirão Preto, Ribeirão Preto, SP, Brazil

[4]Faculdade de Odontologia de Ribeirão Preto, Universidade São Paulo (USP), Ribeirão Preto, SP, Brazil

[5]Departamento Biologia, Faculdade de Filosofia, Ciências e Letras de Ribeirão Preto, Universidade São Paulo (USP), Av. Bandeirantes 3900, 14040-901 Ribeirão Preto, SP, Brazil
etshojo@usp.br

Abstract Eukaryotic cells have developed several types of cellular defense mechanisms in the face of injuries caused by a variety of genotoxic agents. Each type of DNA lesion can be recognized and processed by specialized repair pathways. Although the signaling responses to ionizing radiation (IR) and the cross-linking drug cisplatin have been extensively studied in different cell types, the mechanisms are still unclear. Following IR, the ATM-CHK2 pathway is preferentially activated, while the ATR-CHK1 seems to be the alternative pathway in response to cisplatin treatment, similarly to UV-light exposure. While NER (nucleotide excision repair) and HR (homologous recombination) seem to constitute the main repair processes for cisplatin-induced DNA-adducts, HR and NHEJ (non-homologous end-joining) are important mechanisms for radiation-induced DSB repair; in addition, IR-induced oxidative damage can be repaired by NER and BER pathways. IR and cisplatin induce cell cycle arrest in all phases, while the induction of apoptosis mainly depends on the cell type and treatment conditions. It is still unclear how these processes act in concert, in spite of a considerable body of information in the literature that has emerged in the last decade. The data regarding gene expression analyzed at transcription level associated to cellular endpoints may indicate at least some interconnection among DNA repair, cell-cycle checkpoints and apoptosis. All these studies have been carried out in normal cells, with the aim of clarifying the mechanisms leading to genomic instability and carcinogenesis, in contrast to those performed in tumor cells, which are relevant for understanding drug or IR responses towards improvement in therapeutic strategies.

1 Genomic Instability in Cells Submitted to Genotoxic Stress

Organism growth and development involves the correct maintenance of genome stability and cell reproduction, requiring an accurate replication of the genome, as well as equal division of chromosomes between daughter cells. The fidelity of the eukaryotic genome is maintained by coordinated actions of cellular pathways, including DNA repair, chromatin remodeling, apoptosis, and cell cycle checkpoints (Li and Zou 2005). It is not surprising that the occurrence of errors in the mitotic machinery may originate genetic instability and consequently, the development of cancer (Pihan and Doxsey 1999; Rasnick and Duesberg 1999; Smith et al. 2003). Thus, eukaryotic cells have developed many types of cellular defense mechanisms, in order to overcome the adverse conditions imposed by the exposure to genotoxic agents (originated from exogenous and endogenous sources), which continuously affect the organisms, causing several types of DNA damage.

DNA lesions can be classified according to the chemical nature of base modifications. Major categories of these lesions include dimers, base modifications (oxidative damage, aberrant methylation, depurination, etc.), bulky adducts, intrastrand and interstrand cross-links, single- and double-strand breaks (DSBs) (Friedberg et al. 1995). DSBs constitute the lesions of major significance for irradiated cells (Chaubey et al. 2001; Natarajan et al. 1993; Suzuki et al. 2003), but in addition, oxidative DNA damage generated by reactive oxygen species (ROS) can also be produced (Bohr and Dianov 1999; Slupphaug et al. 2003). The other types of lesions can be induced by a variety of chemical and physical agents.

Each type of DNA lesion can be recognized and processed by specialized repair pathways. Direct DNA repair can occur by photoreversal of UV-induced pyrimidine dimers by DNA photolyase and the removal of the O^6-methyl group from O^6-methyl guanine in DNA by methylguanine DNA methyltransferase (MGMT) enzyme; while photolyase is not present in many species, including humans, MGMT has nearly universal distribution in living organisms (Sancar et al. 2004). In eukaryotes, there are five main DNA-repair processes: nucleotide excision repair (NER), base excision repair (BER), mismatch repair (MMR), non-homologous end-joining (NHEJ) and homologous recombination (HR), according to the kind of DNA damage (Sancar et al. 2004; Sengupta and Harris 2005). While most base modifications are cleared by NER, BER and MMR (Cline and Hanawalt 2003), DSBs are primarily repaired by HR and NHEJ processes (Rothkamm et al. 2003). A lack of DNA repair, or failures in some of the steps, can lead to cell death, mutations, deletions and chromosomal rearrangements, contributing to cellular transformation and tumor development.

In recent decades, genomic instability has been studied by several endpoints, including analyses of chromosomal rearrangements and aberrations, gene amplification, aneuploidy, micronucleus formation, microsatellite instability and gene mutations. There has been considerable body of evidence concerning the importance of genomic instability in human cancer and its temporal occurrence in the carcinogenic process (Harms Ringdahl 1998; Huang et al. 2003; Pfeiffer 1998; Smith et al. 2003).

Persistence of genetic alterations, such as chromosomal rearrangements, may represent a risk to cancer development, as in case of humans accidentally exposed to ^{137}cesium (Camparoto et al. 2003; Natarajan et al. 1998; Sakamoto-Hojo et al. 1999), and victims of the Chernobyl accident (Johnson et al. 1999; Scarpato et al. 1997), since carcinogenesis induced by ionizing radiation (IR) was reported to be induced in A-bomb survivors in Japan, several years after radiation exposure (Preston et al. 1994; Thompson et al. 1994).

Recently, improvements in anticancer therapies have implied the exposure of patients to high doses of ionizing radiation and chemotherapeutic drugs, whose mutagenic effects have been well established, but second malignant neoplasias have been reported with increasing frequency in cancer patients (Bhatia et al. 2002; Koishi et al. 1998). This fact has stimulating in vivo studies on genomic instability in cancer survivors, years after therapy (Brassesco et al. 2004), as well as in vitro experiments with antitumoral drugs in different cell types, so as to clarify the mechanisms underlying the cellular responses to these agents.

The recent advances in genomics and the development of microarray techniques have led to high expectations regarding the elucidation of mechanisms implicated in genomic instability and carcinogenesis, as well as for the characterization of cellular responses to genotoxic agents in different cell types. In tumors, these approaches are relevant for improvements in therapeutic regimens, but they also require studies in normal cells, as well as the application of several assays which should be performed at cellular and molecular levels. On the basis of literature data regarding genomic instability caused by physical and chemical agents, the cellular responses include several processes occurring in a complex and intricate genetic cascade. In this chapter, we attempt to highlight some aspects towards the understanding of signaling responses to ionizing radiation and the alkylating drug cisplatin, with emphasis on the increasing evidence indicating a functional connection between cell-cycle checkpoints, repair pathways and cell survival, or alternatively cell death (apoptosis).

2
Activation of Cell-Cycle Checkpoints in Response to DNA Damage

Cell-cycle delay (or arrest) is an important response to induced DNA damage, and alterations in this mechanism may result in the accumulation of genetic alterations and/or deregulated cell growth, leading to genomic instability and neoplastic development (Murakami et al. 1995; Tlsty et al. 1995). Many gene products participating in the damage and replication checkpoints in eukaryotes were identified in budding and fission yeast systems (Hartwell et al. 1994), and these studies established the overall framework of the checkpoints. The original concept of the cell cycle checkpoint held that these mechanisms are in place to ensure the orderly progression in the presence of DNA damage or replication interference. The elimination of checkpoints may result in cell death, infidelity in the distribution of chromosomes or other organelles, or increased susceptibility to environmental perturbations such as DNA damaging agents (Hartwell and Weinert 1989).

Thus, genomic instability arising from compromised checkpoint mechanism has been linked to cancer development (Hartwell and Kastan 1994). In mammalian cells, cell-cycle arrest in the G1, S, and G2 phases occurs after exposure to genotoxic agents, and appears to protect cells from permanent genetic damage and transformation. The cell-cycle arrest after DNA damage promotes survival by allowing time for cells to repair DNA lesions. Following DNA damage induction, a signal transduction pathway is activated, involving the participation of proteins, which have been divided into sensors, transducers and effectors (Iliakis et al. 2003). These proteins act in a sequence of steps including damage recognition (signal initiation by damage sensors), signal mediation and processing (mediators and signaling kinases), and targeting of effector processes (Li and Zou 2005), which signal towards cell-cycle arrest, DNA repair, and/or apoptosis, as schematically represented in Fig. 1. It should be noted that DNA repair pathways are even functional in the absence of damage-induced cell-cycle arrest, and alternatively, apoptosis can occur independently of the cell-cycle arrest machinery (Sancar et al. 2004).

There are functional checkpoints for the G1/S and G2/M transitions, as well as for the S-phase progression. The signal transducers activate TP53 and inactivate cyclin-dependent kinases to inhibit cell cycle progression. Different types of genotoxic stress differentially activate cell-cycle checkpoints. For example, ionizing radiation triggers both G1/S, intra-S and G2/M checkpoints, within hours after exposure (Bartek et al. 2004; Han et al. 1995; Kuerbitz et al. 1992), whereas the methylating agent temozolomide cause a G2/M arrest after the second S-phase following drug treatment (Hirose et al. 2001). The TP53 tumor suppressor plays an important role in DNA damage induced G1/S checkpoint. The specific cell-cycle phase where the blockage occurs depends largely on the *TP53* status. Wild-type cells for *TP53* predominantly suffer an arrest in G1 phase. In contrast, the mutant-*TP53* cells suffer an accumulation in both S and G2 phases (Eastman 2004). DNA damage checkpoints require the recognition of DNA damage to initiate subsequent events. It has been demonstrated that the arrest in the cell cycle is mediated by activation of TP53 by ATM or ATR (Canman et al. 1998; Turenne et al. 2001).

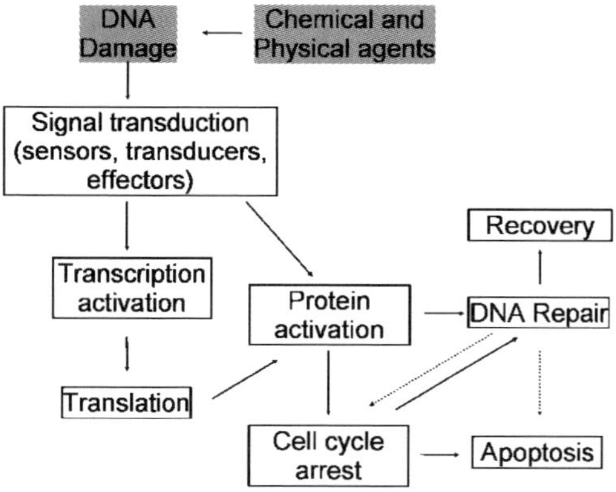

Fig. 1 Cellular responses triggered by the induction of DNA damage in mammalian cells. The *dotted lines* represent alternative pathways, which can be activated in case of extensive damage and/or failure in repair mechanisms

In mammalian cells, ATM and ATR play critical roles in early signal recognition and transduction through all three cell-cycle checkpoints: G1/S, intra-S-phase and G2/M, and while ATM protein kinase activity is stimulated in vivo by agents that induce DSBs, ATR is activated by UV-light and agents that stall DNA replication (Abraham 2001; Sancar et al. 2004; Yang et al. 2003).

ATM protein is a central component to the signaling response to DSBs induced by IR, being essential for activating divergent molecular responses that involve transcriptional regulation, cell cycle arrest and modulation of DNA repair (Iliakis et al. 2003; Li et al. 2001). Following IR, ATM is activated and phosphorylates many proteins, including TP53, H2AX, CHK2, RPA and BRCA1 (Iliakis et al. 2003; Riballo et al. 2004; Shiloh 2003), leading to the initiation and maintenance of G1/S arrest (Iliakis et al. 2003; Sancar et al. 2004). Riballo et al. (2004) showed that H2AX phosphorylation by DNA-PKcs and ATM in the vicinity of the DSB results in the accumulation of additional proteins, including 53BP1 and the MRN complex (MRE11/RAD50/NBS1), at the break site. MRN was recently shown to stimulate the kinase activity of ATM in vitro towards its substrates TP53, CHK2 and histone H2AX, and NBS1 was shown to directly contact ATM (Lee and Paull 2005). Furthermore, MRN is considered as a central player in most aspects of the cellular response to DSBs (Assenmacher and Hopfner 2004).

On the other hand, as already mentioned, ATR is mainly activated in response to UV light, or UV-mimetic agents (Abraham 2001) and replication fork blockers (Nyberg et al. 2002; Zhou and Elledge 2000). The signal is sensed by ATR, RAD17-RFC and 9-1-1 complex, following the phosphorylation of CHK1, which phosphorylates CDC25, leading to G1 arrest. Whether the initial arrest is caused by the ATM-CHK2-CDC25A or the ATR-CHK1-CDC25A pathways, this rapid response is followed by the TP53 mediated maintenance of G1/S arrest

(Li and Zou 2005; Sancar et al. 2004). TP53 activates many targets, including CDKN1A (p21), which inhibits the S-phase promoting CDK2/cyclin E complex (Lukas et al. 2001; Sherr and Roberts 1995).

Recently, Falck et al. (2005) identified related and conserved carboxy-terminal motifs in human NBS1, ATRIP and KU80 proteins that are required for their interaction with ATM, ATR and DNA-PKcs, respectively; these motifs are essential not only for efficient recruitment of ATM, ATR and DNA-PKcs to sites of damage, but are also critical for ATM-, ATR- and DNA-PKcs-mediated signaling events that trigger cell cycle checkpoints and DNA repair.

During DNA replication, an intra-S-phase checkpoint is activated when DNA lesions are induced or remain unrepaired after the G1/S checkpoint, leading to a blockage in the replication mechanism (Sancar et al. 2004). For DSBs, ATM, the MRE11-RAD50-NBS1 complex (Gatei et al. 2000), and also BRCA1 (Quinn et al. 2003; Xu et al. 2001) act as sensors in the intra-S checkpoint activation. In addition, in vitro studies showed that DNA-PK also plays a role as a molecular sensor of DNA damage, which enhances the signal transduction via phosphorylation of many downstream targets (Collis et al. 2005). It was demonstrated that DNA-PK plays a redundant, overlapping role with ATM in the retention of several proteins (MDC1, 53BP1, MRN complex, BRCA1) at the damage site by contributing to H2AX phosphorylation (Stiff et al. 2004). Other activities have been attributed to DNA-PK (for more details, see Collis et al. 2005; Lees Miller and Meek 2003).

Differently from the IR-induced S-phase checkpoint, which has been extensively investigated by many authors, little is known about the molecular mechanisms involved in the S-phase checkpoint elicited in mammalian cells by DNA cross-links agents. There is evidence that these agents can activate ATR, and subsequently the CHK1 protein in human cells (Feijoo et al. 2001; Zhao and Piwnica-Worms 2001). Recently, it was demonstrated that RAD9 is required for the activation of S-phase checkpoint and the maintenance of chromosomal stability, and there is evidence that ATR, RAD17 and RAD9 participate in the control of DNA replication and monitoring the presence of DNA damage and errors induced by replication blockers during the S-phase (Dang et al. 2005).

The control of mitotic entry is performed by the G2/M checkpoint, which can occur by a TP53-dependent pathway by up-regulating *CDKN1A*, *GADD45* and *SFN* (14-3-3σ) and by directly repressing cyclin B transcription (Innocente et al. 1999; Peng et al. 1997; Taylor and Stark 2001; Wang et al. 1999). Additionally, the TP53-independent pathway can also occur, via ATR-CHK1-CDC25C (O'Connell and Cimprich 2005; Sanchez et al. 1997), in parallel with the ATM-CHK2-CDC25C pathway (Matsuoka et al. 2000), for UV-induced lesions, or IR-induced DSBs, respectively.

The connection between G2/M checkpoint and DNA repair is still unclear. Recently, the ATM/ATR signaling pathways controlling the cell cycle progression, DNA replication and repair have been reported, focusing on the integration between the checkpoint activation and DNA repair mechanisms (for review, see McGowan and Russell 2004).

Similarly, the signaling pathways connected to the apoptosis process are not well understood. TP53 appears to mediate cell death through a variety of mechanisms that include either down-regulation of anti-apoptotic genes (*BCL-2* and *MAP4*), or up-regulation of pro-apoptotic genes, such as *BAX*, *IGF-BP3*, *DR5*, *FAS*, *APAF1* and various other apoptosome components (Slee et al. 2004; Zabkiewicz and Clarke 2004). The identification of proteins that modulate apoptosis has generated insights into the molecular events relevant for anticancer therapy, and a balance between the expression of pro-apoptotic and anti-apoptotic functionally antagonistic proteins is thought to determine whether cells will survive or undergo apoptosis (Shu et al. 1998). It is becoming clear that the apoptotic response reflects the cumulative action of numerous TP53-induced signals (Vousden and Lu 2002).

Many cancer cells lack a functional TP53 protein and, as a result, do not suffer G1-phase blockage; however, these cells seem to maintain their ability to arrest in G2. A complete understanding of TP53-independent pathways that are involved in G2 checkpoint control is interesting, since TP53-deficient cancer cells are more vulnerable to G2 checkpoint abrogation than cells containing a functional TP53 (Fan et al. 1995; Wang et al. 1996).

In addition to the role of TP53 in facilitating DNA damage repair by promoting cell-cycle arrest, there is evidence that it may also participate in the regulation of DNA repair process (Pestell et al. 2000; Wrighton et al. 2004). While BER is markedly stimulated by TP53 (Offer et al. 2002), NER also involves protein products of TP53-regulated genes, such as *DDB2, DDB1*, *XPC* and *GADD45* (Cheo et al. 1997; Hwang et al. 1999; Smith et al. 2000). According to a recent review, TP53 might function as a "molecular node" that resides at the intersection of upstream signaling cascades and downstream DNA-repair and -recombination pathways, by activating or repressing a large number of targets (Sengupta and Harris 2005).

Activation of cell-cycle checkpoints following DNA damage plays an integral role in the preservation of genomic stability by allowing cells to limit the propagation of deleterious mutations, but the understanding on how these processes act in concert with other stress responses, such as DNA repair and apoptosis, is still a hard task, in spite of a considerable body of information in the literature that has emerged in the last decade.

3
Signaling Pathways in Response to Ionizing Radiation

Cell exposure to IR elicits a complex network of cellular responses, including the activation of membrane receptors, stimulation of protein kinases, transcriptional activation for several classes of genes, and altered cell cycle regulation. For many years, the cell cycle has been linked to IR-induced cellular responses; however, only in the past decade has the concept of a coordinated DNA damage response integrating the processes underlying damage recognition, cell cycle checkpoints and DNA repair begun to be elucidated (Schmidt-Ullrich et al. 2000; Wilson 2004).

The recent studies conducted on the profiles of gene expression by using the microarray method have demonstrated modulation of genes controlling the cell cycle, DNA repair processes, apoptosis, metabolism, and genes playing roles in the maintenance of genome integrity following IR-induced DNA damage in mammalian cells (Amundson and Fornace 2001; Amundson et al. 2003; Cardoso et al. 2002; Chaudhry et al. 2003; Heinloth et al. 2003; Kitahara et al. 2002; Mayer et al. 2002; Momota et al. 2003; Sakamoto-Hojo et al. 2003; Yin et al. 2003). Due to the different experimental conditions and several cell types investigated, an apparent variety of results has been obtained for the sets of modulated genes. In a recent review, Snyder and Morgan (2004) reported that very few genes have been found to be consistently up-regulated by IR, and that set includes *GADD45A*, *CDKN1A* and genes associated with the NER pathway.

GADD45A protein has been reported to interact with multiple important cellular proteins, including CDC2 protein kinase, proliferating cell nuclear antigen (PCNA), and CDKN1A (p21Waf1/Cip1), indicating its importance in the control of cell cycle checkpoint, DNA repair, and signaling transduction, required for the maintenance of genomic stability (Zhan 2005).

Recently, Rieger and Chu (2004) demonstrated that 59% of the significantly modulated genes presented unexpected functions that were not previously associated with IR or UV responses, such as those related to signal transduction, RNA binding and editing, protein synthesis and degradation, energy metabolism, metabolism of macromolecular precursors, and cell structure and adhesion. The authors used cell lines derived from 15 different individuals and a rigorous statistical method, which provided an accurate portrait of the transcriptional response to DNA damage. However, in this study, the cell lines were immortalized by Epstein-Barr virus, which confers some degree of genomic instability due to the transformation process per se. They found that, while *ATM* and *ATR* showed only minor transcriptional responses, DNA repair genes (such as *XPC*, *DDB2*, *PCNA* and *GADD45A*) were significantly induced; these genes are involved in global genomic repair, a pathway for nucleotide excision repair of non-transcribed DNA. In addition, as reported by the authors, genes involved in HR, except *RAD51C*, and NHEJ were not affected by IR.

NHEJ and HR are the major pathways of DSB repair in mammalian cells (Lees-Miller and Meek 2003; Pastwa and Blasiak 2003; Valerie and Povirk 2003). Although NHEJ repairs the majority of DSBs in G0 and G1 phases of the cell cycle (Haber 2000; Valerie and Povirk 2003), it is important in all phases, while HR is particularly important at late S/G2 (Lees-Miller and Meek 2003; Rothkamm et al. 2003; Valerie and Povirk 2003). We studied the occurrence of TRBV8.1-DB2.1 V(D)J recombination in murine fetal thymus organ culture (FTOC), and the transcriptional profiles were analyzed by cDNA microarrays, so as to evaluate the modulation of genes implicated in TRBV8.1-BD2.1 rearrangements in irradiated-thymus organ culture mimicking days 15–17 of development. Clustering analysis showed similar expression profiles for genes implicated in both processes, V(D)J recombination and DNA DSB repair, such as *XRCC4, RAG-2, DCLRE1C* (Artemis) and *PRKDC* (DNA-PKcs) (Cardoso et al. 2005). These proteins are included within the components of the NHEJ core protein (Jeggo and O'Neill 2002; Jeggo et al. 1999; Karran 2000; Ma et al. 2002), thus reinforcing the role of NHEJ pathway in irradiated cells, and the overlap between the two processes (Chaudhuri and Alt 2004; Jeggo and Concannon 2001; Schlissel et al. 1993).

While NHEJ and HR are important processes to repair IR-induced DSBs, there is evidence that NER and BER pathways are the main processes involved in the repair of IR-induced oxidized bases (Kow et al. 1990; Lin and Sancar 1989; Yacoub et al. 2001). NER was initially identified as a repair system for DNA damage that induces gross distortions in the DNA molecule, such as UV-induced pyrimidine dimers and large chemical adducts (Sancar and Sancar 1988). In DU145 human prostate cancer cells, it was observed a MAPK-dependent increase in the protein levels of *ERCC1* and *XRCC1* in irradiated cells, indicating the participation of NER proteins in response to irradiation (Yacoub et al. 2001).

Transcriptional modulation in the expression of genes involved in NER pathway following irradiation has been recently reported, at least for few genes: *DDB2* and *XPC* (Amundson et al. 2000), *DDB2* and *GADD45A* (Jen and Cheung 2003), *DDB2, XPC, GADD45* and *PCNA* (Tusher et al. 2001), suggesting that NER proteins may somehow participate in the response to γ-rays. However, there is evidence that the induction in the expression of NER genes is not necessarily associated to the requirement of the formal NER mechanism, but to the repair of radiation-induced cross-links, such as DNA-protein cross-links, as suggested by Murray and Rosenberg (1996). Although the amount of IR-induced oxidative DNA damage (produced by the indirect effect of γ-rays) might be relevant, the possibility exists that the mechanism involved in the repair of these lesions may not involve the NER pathway; this is an unsolved question which still needs to be investigated.

Some special features of IR, such as radiation quality and half-life, dose and dose rate (DR), may cause different biological responses (Dainiak 2002; Hall 2000). By analyzing the induction of micronuclei and inhibition of proliferation, as indicators of biological response, it has been demonstrated that DR may influence the cell response in human osteosarcoma cells (Magae et al. 2003). For X- or γ-rays, DR is one of the main factors determining the biological consequences of a given absorbed dose, and this is related to DNA repair processes. For Chinese hamster cells (CHL-F line), a high DR has induced an increase in cell death, compared with low DR (Hall 2000). Studies on gene expression carried out by Amundson et al. (2003) showed that the reduction of DR over 3 orders of magnitude (0.28; 2.4; 29.0 cGy/min) caused some protection against the induction of apoptosis. The same study indicated a group of up-regulated genes in a dose rate-dependent fashion and the majority of these genes are involved in apoptosis regulation. However, the literature is still scarce regarding DR effects in different cell types, especially in terms of gene expression profiles.

Profiles of gene expression were studied in our laboratory, by irradiating confluent normal human fibroblasts (most of them at G0-phase) with 4 Gy of γ-rays, delivered at two different DR (0.5 and 2.0 Gy/min) in cells sub-cultured soon after irradiation, collected at 6 h later, most probably undergoing the G1-phase (Table 1 and Figs. 2A and B, Merchi et al., unpublished data). Experiments carried out at the same conditions indicated that the cell survival was 83.8% for low DR (0.5 Gy/min) and 55.5% for high DR (2.0 Gy/min). The results on gene expression indicated the modulation of DNA repair genes, but only for those belonging to the NER process, such as *ERCC1*, *ERCC3* (*XPB*) and *ERCC4* (*XPF*). These genes play a role in the NER mechanism that occurs in association to transcription (Transcription Coupled Repair: TCR). For fibroblasts irradiated at high DR, *ERCC3* and *ERCC4* genes were down-regulated in comparison with the results obtained for irradiation at low DR. In addition, the inhibition of genes involved in transcription (*FYB* and *CREB1*) and DNA replication (*RFC5*) was also observed at high DR. These results indicated the predominance of the induction of NER genes, while modulation of NHEJ genes was not detected in primary fibroblasts irradiated at low DR; this aspect still needs further investigation. The transcriptional activation of DNA repair genes was then predominantly demonstrated in irradiated cells at low DR.

Biological effects of low radiation doses to human population are poorly investigated, although IR has been extensively studied in different test systems. In a previous study, we reported alterations in the profiles of gene expression for human lymphocytes from hospital workers occupationally exposed to low-doses of IR for many years. Several classes of genes, such as cell cycle control (*TRRAP*, *TP53*), DNA repair (*LIG4*), stress response (*DNAJA1*), and signal transduction (*MAPK10* and *MAPK8IP1*), were modulated at transcription level (Sakamoto-Hojo et al. 2003).

The profiles of gene expression for low radiation doses were also analyzed in human lymphocytes from healthy donors, irradiated in vitro with 10, 25 and 50 cGy of γ-rays (time collection: 48 h post-irradiation) by using the cDNA microarray method in samples collected from six different healthy donors. Among the modulated genes for different doses, some of them can be associated with damage response to IR, such as *LIG4* (DNA repair), *TGFB2* (growth factor), *PTPN1* and *CCND2* (cell cycle regulation), *GSTT1* (detoxification), while genes related to other functions, such as *USP53* (ubiquitin protease) and *CALM1* (transducer of calcium signaling), were also found to

Table 1 Genes presenting modulation in the expression profiles in cells submitted to genotoxic stress

Gene name	Symbol	Chromosome location	Function
Radiation-exposed murine fetal thymus organ culture, 4 Gy (Cardoso et al. 2005)			
X-ray repair complementing defective repair in Chinese hamster cells 4	XRCC4	13C3	DNA repair
Recombination activating gene 2	RAG-2	2	V(D)J recombination
DNA cross-link repair 1C (PSO2 homolog, *S. cerevisiae*)	DCLREIC (Artemis)	2A1	V(D)J recombination/ DNA repair
Protein kinase, DNA-activated, catalytic polypeptide	PRKDC (DNA-PKcs)	16	Apoptosis/ DNA repair
Radiation-exposed primary fibroblasts (4 Gy) at dose rates of 0.5 and 2.0 Gy/min, harvesting time of 6 h (Merchi et al., unpublished data)			
0.5 Gy/min			
Excision repair cross-complementing rodent repair deficiency, complementation group 1	ERCC1	19q13.2-q13.3	DNA repair
Excision repair cross-complementing rodent repair deficiency, complementation group 3	ERCC3 (XPB)	2q21	DNA repair
Excision repair cross-complementing rodent repair deficiency, complementation group 4	ERCC4 (XPF)	16p13.3-p13.11	DNA repair
2 Gy/min			
FYN binding protein (FYB-120/130)	FYB	5p13.1	Signal transduction
CAMP responsive element binding protein 1	CREB1	2q34	Transcription factor
Replication factor C (activator 1) 5, 36.5 kDa	RFC5	12q24.2-q24.3	Replication factor

induced genes
repressed genes

Table 1 (continued)

Gene name	Symbol	Chromosome location	Function
Radiation-exposed primary fibroblasts, irradiated versus unirradiated cells (2 Gy; cells harvested at 1, 6, 9 and 24 h; Sakamoto-Hojo et al. 2003)			
TGFB inducible early growth response (TIEG)	TIEG	8q22.2	Oxidative damage-apoptosis
Ubiquinol-cytochorome c reductase complex (7.2 Kda)	HSPC051	22cen-q12.3	Oxidative damage-apoptosis
v-jun sarcoma virus 17 oncogene homolog (avian)	JUN	1p32-p31	Oxidative damage-apoptosis
CASP2 and RIPK1 domain containing adaptor with death domain	CRADD	12q21.33	Apoptosis
Caspase 10, apoptosis-related cysteine protease	CASP10	2q33-q34	Apoptosis
Tumor necrosis factor receptor sperfamily, member 6	TNFRSF6	10q24.1	Apoptosis/ anti-apoptosis
Protein kinase, DNA-activated, catalytic polypeptide	PRKDC (DNA-PKcs)	8q11	Apoptosis/ repair
DUSP1	DUSP1	5q.34	Growth arrest
E2F trancription factor 4, p107/p130-bibding	E2F4	16q21-q22	Cell cycle regulation
RAP1A, member of RAS oncogenes family	RAP1A	1p13.3	Cell cycle regulation/ anti Ras
APEX nuclease (multifunctional DNA repair enzyme)1	APEX1	14q11.2-q12	Multifunctional repair enzyme
Radiation-exposed primary versus transformed fibroblasts (2 Gy; cells harvested at 1, 6, 9 and 24 h; Mello et al., unpublished data)			
DNA mismatch repair protein Mlh1	MLH1	3p21.3	DNA repair

induced genes
repressed genes

Table 1 (continued)

Gene name	Symbol	Chromosome location	Function
Excision repair cross-complementing rodent repair deficiency, complementation group 5 (xeroderma pigmentosum, complementation group G (Cockayne syndrome))	ERCC5	13q22	DNA repair
RecQ protein-like (DNA helicase Q1-like)	RECQL	12p12	DNA repair
ADP-ribosyl-transferase-like 1	PARP4	13q11	DNA repair
Ubiquitin-conjugating enzyme E2B (RAD6 homolog)	UBE2B	5q23-q31	DNA repair
Ligase III, DNA, ATP-dependent	LIG3	17q11.2-q12	DNA repair
O-6-methylguanine-DNA methyltransferase	MGMT	10q26	DNA repair
X-ray repair complementing defective repair in Chinese hamster cells 1	XRCC1	19q13.2	DNA repair
Protein kinase, DNA-activated, catalytic polypeptide	DNA-PKcs	8q11	DNA repair
Radiation-exposed occupationally hospital workers (Sakamoto-Hojo et al., 2003)			
Transformation/transcription domain-associated protein	TRRAP	7q21.2-q22.1	Cell cycle regulation
Tumor protein p53 (Li-Fraumeni syndrome)	TP53	17p13.1	Cell cycle regulation
Ligase IV, DNA, ATP-dependent	LIG4	13q33-q34	Double strand repair
DnaJ (Hsp40) homolog, subfamily A, member 1	DNAJA1	9p13-p12	Protein import into mitochondria
Mitogen-activated protein kinase 10	MAPK10	4q22.1-q23	Signal transduction
Mitogen-activated protein kinase 8 interacting protein 1	MAPK8IP1	11p12-p11-2	Signal transduction
Radiation-exposed lymphocytes (10, 25 and 50 cGy); 48 h post-irradiation (Fachin et al., unpublished data)			
Calmodulin 1 (phosphorylase kinase, delta)	CALM1	14q24-q31	Signal transduction

induced genes
repressed genes

Table 1 (continued)

Gene name	Symbol	Chromosome location	Function
Glutathione S-transferase theta 1	GSTT1	22q11.23	Transferase
Cyclin D2	CCND2	12p13	Cell cycle control
Ligase IV, DNA, ATP-dependent	LIG4	13q33-q34	DNA repair
Protein tyrosine phosphatase, non-receptor type 1	PTPN1	20q13.1-q13.2	Signal transduction
Transforming growth factor, beta 2	TGFB2	1q41	Growth factor
Ubiquitin specific protease 53	USP53	4q26	Protein metabolism
Radiation-exposed (1 Gy) glioma U343-MGa, collection time: 6 h (Bassi et al., unpublished data)			
Heat shock 27kDa protein 1	HSPB1	7q11.23	Stress resistance/ actin organization
Breast cancer 1 (BRCA1)	BRCA1	17q21	Oxidative damage repair
H2A histone family, member X	H2AX	11q23.2-q23.3	Basal histone
Fanconi anemia, complementation group G	FANCG	9p13	DNA repair
Cisplatin-treated (25 μM) AT primary fibroblasts, collection time: 16 h (Mello et al., unpublished data)			
Three prime repair exonuclease 1	TREX1	3p21.3-p21.2	DNA repair
Excision repair cross-complementing rodent repair deficiency, complementation group 6	ERCC6	10q11.23	DNA repair
Flap structure-specific endonuclease 1	FEN1	11q12	DNA repair
Breast cancer 1 (BRCA1)	BRCA1	17q21	Oxidative damage repair
Tumor protein p53 (Li-Fraumeni syndrome)	TP53	17p13.1	Cell cycle regulation

induced genes
repressed genes

be modulated (Table 1, Fachin et al., unpublished data). Similarly to the results reported by Rieger and Chu (2004), some differentially expressed genes are involved in processes indirectly related to IR response.

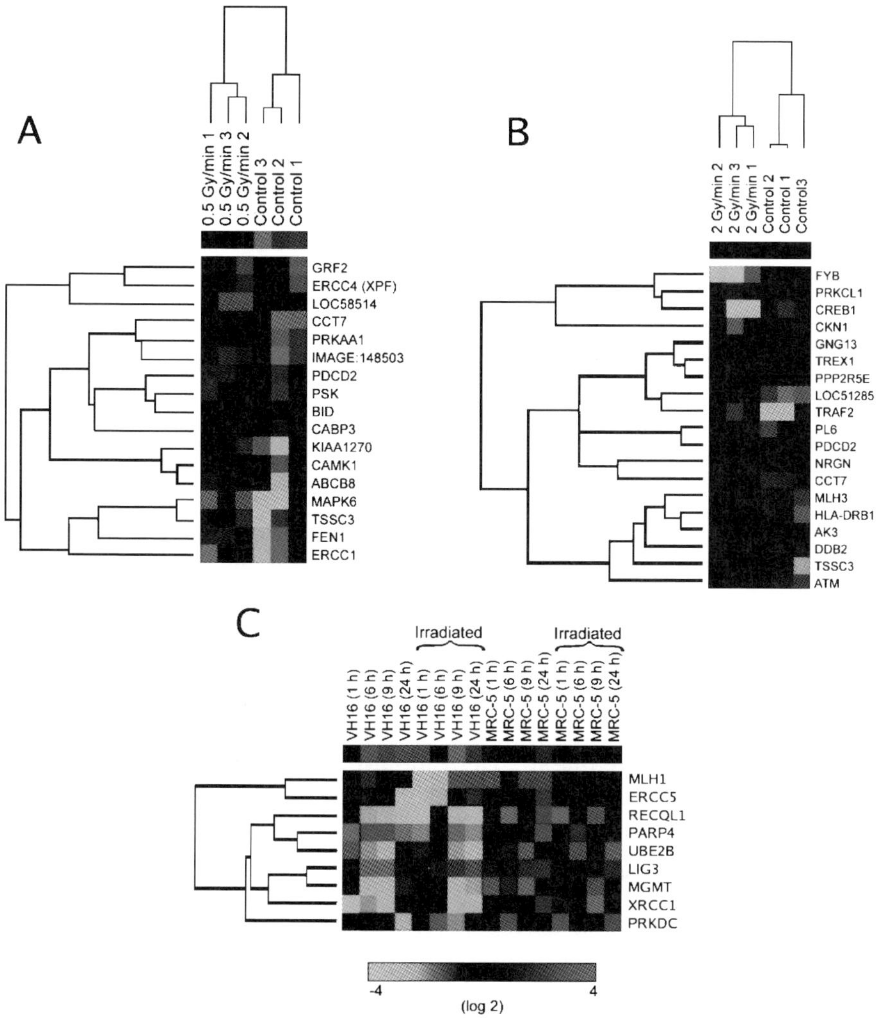

Fig. 2 Hierarchical clustering of the gene expression data obtained for primary (normal) and SV40-transformed cells irradiated with γ-rays. FHN primary cells were irradiated with 4 Gy at different dose rates: **A** 0.5 Gy/min and **B** 2 Gy/min, and compared with control mock-irradiated cultures. VH16 primary fibroblasts were also compared with SV40-transformed cells (MRC-5) following irradiation with 2 Gy at a single dose-rate, and compared with controls (**C**). In the dose-rate experiments (**A** and **B**) cells were harvested at 6 h after irradiation; SV40-transformed and VH16 primary normal cells were harvested at 1, 6, 9 and 24 h following exposure. The normalized expression value for each gene is depicted according to the scale; while the *red colour* indicates expression levels greater than the median, *green* indicates levels lower than the median (– 4 to 4 in log base 2 units), and *grey* indicates excluded values. In panels **A** and **B** the *column headers* (0.5 Gy/min, 2 Gy/min and Control) indicate the irradiation condition, and the *numbers* (1, 2 and 3) specify different experiments. In panel **C** the *column headers* (VH16 and MRC-5) indicate the two cells lines and the harvesting time (1, 6, 9 and 24 h)

Cellular responses following IR exposure varies according to the cell type, whether cells are normal, transformed, preneoplastic, or derived from tumors, since the level of genomic instability determined by single-, or multiple mutational events, may substantially interfere in the signaling pathways underlying the responses to the induction of DNA damage. For example, we have found that primary normal fibroblasts exposed to 2 Gy of γ-rays showed modulation in the expression of several genes (Table 1), whose functions are mainly related to oxidative damage (*TIEG*, *HSPC051* and *JUN*), apoptosis (*CRADD*, *CASP10* and *TNFRSF6*), cell cycle control (*RAP1A*, *DUSP1* and *E2F4*) and DNA repair (*PRKDC* and *APEX1*), among others (Sakamoto-Hojo et al. 2003). Interestingly, some genes with different functions, such as *PRKDC* and apoptosis-related genes, *CASP10*, *CRADD* and *TNFRSF6*, were grouped in the same cluster, and displayed an induction at 24 h post-irradiation. This similarity in the transcriptional profiles suggests the inter-connection among repair and apoptosis pathways in response to DNA damage.

The presence of a functional *TP53* gene seems to be an important aspect to be taken into account. According to Mori et al. (2004), most of the genes which were strongly up-regulated in CD4+ T lymphocytes (irradiated with 1 Gy of X-rays, and collected at 8 h post-irradiation) were predominantly targets for TP53. In contrast, Jurkat leukemic cells, which lack a functional *TP53* gene, presented a partially different pattern, indicating the modulation of genes belonging to the Rho and cytokine signaling pathways.

Experiments carried out with SV40-transformed cells (which present alterations in TP53 function due to inactivation by the large-T antigen produced by SV40-virus) showed that the transformation status may be important in terms of transcriptional response to IR (Mello et al. unpublished data), compared with primary normal cells, since the expression profiles varies depending on the transformation status in irradiated and mock-irradiated cells. Several genes were found up-regulated in transformed cells, and some of them are represented in Fig. 2C and listed in Table 1. Many of these genes are associated with DNA repair mechanisms, such as *PRKDC* (*DNA-PKcs*), *MHL1*, *XRCC1* and *LIG3* (Ligase 3).

Experiments on cellular responses to IR were also carried out in glioma U343-MGa cells (wild-type for *TP53*) by irradiating cell cultures with 1 Gy of γ-rays. A significant increase in apoptotic cells was not observed at different time collection between 6 and 72 h, while the survival fraction determined by clonogenic assay was 58% for cells irradiated with 1 Gy. The analysis of gene expression profiles by cDNA microarrays was also carried out in irradiated (1 Gy) U343 glioma cells harvested after 6 h. The results indicated the induction of stress response/DNA repair related genes, such as *HSPB1*, *BRCA1*, *H2AX*, and *FANCG* (Table 1, Bassi et al. unpublished data), which could play a role in the maintenance of cell survival in glioma cells. Parallel experiments on cell cycle kinetics showed an arrest at G2-phase observed at 6 h after irradiation, and this blockage was completely released after 24 h.

Induction of the *HSPB1* gene may be associated with apoptosis inhibition in glioma cells, since it prevents the correct formation of apoptosome complex, and interferes with the release of cytochrome c from mitochondria (Concannon et al. 2001; Paul et al. 2002). Regarding *BRCA1* and *H2AX* genes, their functions are related to the activation of G2 checkpoint (Momota et al. 2003; Xu et al. 2001; Yarden et al. 2002) and HR repair (Paull et al. 2000). FANCG, which was also highly expressed, seems to be required for efficient HR repair (Yamamoto et al. 2003). These results may suggest that HR was probably activated in glioma cells. Although only few genes were found to be up-regulated, these results indicate the inter-connection among the different stress responses at transcriptional levels (including DNA repair, cell-cycle checkpoint and cell death) when they are analyzed in confront with the results obtained in terms of cell-cycle arrest and a lack of apoptosis induction, which may strengthen the integration between processes.

Cell ability to repair radiation-induced DSBs usually correlates with sensitivity to IR-induced damage (Gangopadhyay et al. 2002). It has been reported that radioresistant glioblastoma cell lines showed up-regulated DNA repair genes involved in NHEJ and HR following irradiation, as demonstrated in experiments with DNA microarrays (Otomo et al. 2004). There is evidence that the activation of HR contributes to radioresistance (Lehnert and Chow 1997). The authors demonstrated that low-doses of IR (5–100 cGy) induced an increase in the frequency of double-strand recombinational events in radioresistant, but not in radiosensitive human cell lines, suggesting that the extent of HR activation could be related with the resistance to genotoxic agents. These results are compatible with the information in the literature about the influence of dose level and dose rate in the IR-activation of DNA repair mechanisms (Hall 2000), which seems to be facilitated at low dose, or low dose rate, as we already mentioned before for normal fibroblasts.

Recently, it has been reported that the activation of HR, which is facilitated by G2-arrest, showed a more pronounced role towards the protection of radiation-induced cell death compared to NHEJ, and this may contribute to determine the degree of radiosensitivity (Wang et al. 2004, 2005). The sensitivity to IR can also be a consequence of *TP53* status. Although the mechanisms of cell resistance are not well elucidated, the information about the activation of repair processes in tumors, and the influence of *TP53* mutations, or other mutational events, is relevant for cancer therapy.

Therefore, the signaling responses to IR-induced DNA damage depend predominantly on the extent and types of DNA lesions, cell types, phase of the cell cycle when cells are damaged, as well as the dose rate and other aspects of experimental protocols. Recently, several studies reported the transcriptional modulation of genes, or set of genes, which were not previously reported as associated with IR-response. It seems that these genes are indirectly participating in the complex network of the genetic cascade. This also indicates the existence of a large number of pathways that are still unknown. Thus, the studies on the molecular mechanisms underlying the cellular responses to IR are in progress aiming to investigate a variety of factors influencing the stress responses, but a substantial amount of information has emerged in the last decade, towards the understanding of the intricate network of signal transduction in response to DNA damage and the coordination among the different pathways.

4 Signaling Pathways in Response to Cisplatin

A broad spectrum of anticancer drugs is commonly used in chemotherapy. The mechanisms by which these drugs interact with the genetic material producing DNA lesions determine the cellular responses in terms of cytotoxicity, activation of cell-cycle checkpoints, DNA repair processes and apoptosis induction. In addition, many other factors can modify the response to drugs, and may interfere in the efficacy of the chemotherapeutic agent; among them, the genetic background of each cell type seems to be an important feature. Hence, comparison between different cell types, and normal versus tumor cells, may shed some light to clarify the molecular pathways underlying the cellular response to drug-treatment and, possibly, the mechanisms of resistance.

Cisplatin (cis-diammine-dichloro-platinum) is an alkylating chemotherapeuthic drug widely used in the therapy of patients with several types of tumor (ovary, testis, bladder, head and neck, lung, cervix and endometrium) (Borch 1987). It is also used in adjuvant chemotherapy for patients with glioblastoma (Poppenborg et al. 1999). This drug interacts with the DNA molecule forming inter- and intra-strand adducts, leading to topological alterations in the double helix (Jordan and Carmo-Fonseca 2000). Similar to other DNA damaging agents, cisplatin-induced DNA adducts seem to activate multiple signal transduction pathways, culminating in several cellular responses, including cell-cycle arrest, DNA repair and apoptosis.

It has been reported that the activation of cell-cycle checkpoints by TP53, leading to either G1/S, S or G2/M arrest, would provide to cells an opportunity to repair the DNA damage, thus reducing their sensitivity to chemotherapeutic agents (Datta et al. 2004; Qin and Ng 2002). In normal human fibroblasts (NHFs), cisplatin and UV irradiation clearly induced a G2/M arrest that is linked to TP53 protein induction, but its inactivation can lead to a marked arrest, possibly due to the abrogation of a G1/S block and/or to the inhibition of G2 exit, suggesting that the G2/M checkpoint is unrelated to TP53 induction following DNA damage (Ceraline et al. 1998). However, according to Datta et al. (2004), the abrogation of *TP53* sensitizes glioma cells treated with cisplatin, leading to cell death by apoptosis, while *TP53* wild-type cell lines are capable of arresting at G2/M. The authors suggested that chemotherapy-induced apoptosis by the abrogation of *TP53* can be a viable therapeutic strategy in some cases. Therefore, in addition to the cell type, the deficiency of this protein may elicit different responses in cells treated with cisplatin.

An important question regarding cellular responses to DNA damage is how the extent of lesions is recognized by cells as excessive, thus triggering the shift from repair to apoptotic cell death (Bernstein et al. 2002). The general perception is that the extent of DNA damage and/or repair capacity may be the major factor in such decision, and this seems to be dependent on drug concentration, or radiation dose (Offer et al. 2002).

Studies carried out in our laboratory with the U343 MG-a glioma cell line (wild-type for *TP53*) showed a different response to cisplatin and temozolomide (which induces DNA methyl-adducts) treatments. Cisplatin (12.5–300 μM) induced 20–80% reduction in cell survival following treatment during 24 h (cytotoxicity test), while the survival fraction was reduced to less than 1% after 5 days (Carminati et al. unpublished data). In contrast, U343 glioma cells showed resistance to temozolomide treatment, since survival rates remained at about 30–40%, independently of the concentration (up to 300 μM) (Bahia et al. unpublished data). In addition, U343 MG-a cells presented a significant increase in apoptosis frequency (24%) induced by 50 μM cisplatin, while temozolomide treatment did not induce apoptotic cell death, even for high drug concentrations. Apoptotic induction was also reported by Choi et al. (2004) in A172 human glioma cells treated with cisplatin at high drug concentration (300 μM).

Therefore, these results indicated that the same cell line, which is proficient for *TP53*, displayed distinct responses to different drugs, reinforcing the importance of the kind of DNA lesions and drug-DNA interactions in activating stress response mechanisms.

Following the formation of cisplatin DNA adducts, cells are unable to continue normal metabolic processes, such as replication (Heiger-Bernays et al. 1990; Oswald et al. 1990) and transcription (Cullinane et al. 1999; Tornaletti et al. 2003), until the recognition and removal of these lesions by the DNA repair machinery. Alternatively, some cells proficient in translesion synthesis are able to tolerate such damage (Mamenta et al. 1994; Vaisman et al. 1999). The mechanisms by which cisplatin adducts are repaired are still unclear. The most accepted model of cross-link repair is essentially based on HR and NER processes. Intrastrand cross-links are mainly a substrate for the human NER system, since they induce DNA distortions that are recognized by the human excinuclease, resulting in excision of these lesions (Huang et al. 1994; Zamble et al. 1996). Interstrand cross-links also cause strand distortions, which are preferentially repaired by HR mechanism, with the participation of some NER proteins (Cole et al. 1976; De Silva et al. 2000; Mu et al. 2000; Thompson 1996). According to the authors, RAD52, RAD54, XRCC2 and XRCC3 (HR proteins) are mainly involved in the repair process, as well as XPF and ERCC1

(NER proteins). However, recent studies have demonstrated that other proteins, such as CHK1 (Sorensen et al. 2005), CHK2 (Zhang et al. 2004) and mitotic CDKs (Aylon et al. 2004; Esashi et al. 2005; Ira et al. 2004), may participate in the activation and formation of nuclear foci involving HR-proteins after exposure to damaging agents.

ATM and ATR can be activated in response to DNA damage, acting as damage sensors for different types of lesions (Iliakis et al. 2003; McGowan and Russell 2004). The ATM protein (370kDa) is a member of the phosphatidylinositol 3-kinase (PI3K) family, and it is a key protein in the cellular response to genotoxic stress, mitotic and meiotic stability and cell cycle progression (Shiloh and Kastan 2001). It is well known that AT cells are deficient in activating the cell-cycle checkpoints in γ-irradiated cells (Chun and Gatti 2004; Lim et al. 2000; Painter and Young 1980). Recently, it has been reported that the ATM protein may participate in the activation of DNA repair mechanisms through the interaction with ARTEMIS, NBS1, MRE11, H2AX and 53BP1, which are proteins required for DSB repair (Jeggo and Lobrich 2005). Studies in DT40 cells demonstrated that ATM deficiency impairs HR-mediated DSB repair and that this protein may link the cell-cycle checkpoints to HR activation (Morrison et al. 2000; Yuan et al. 2003). In general, the deficiency of AT cells in cell-cycle checkpoints has been studied in irradiated cells, but their responses to drugs are still unclear.

Zhao and Piwnica-Worms (2001) demonstrated that ATR is a direct upstream activator of CHK1, a protein necessary for the intra-S-phase checkpoint in cisplatin-treated human cells (Feijoo et al. 2001). In ATM deficient cells, the activation of TP53 seems to occur via an ATM-independent pathway (Siddik 2003; Siddik et al. 1998), which is probably controlled by ATR and also by the FANC protein complex (Rosselli et al. 2003; Siddik 2003).

Studies are in progress in order to verify the influence of the *ATM* mutation in AT primary cells treated with the cross-linking agent cisplatin, in comparison with normal and XPA fibroblast cell line, which is hyper-sensitive to cisplatin. AT showed a higher survival frequency after drug treatment, when compared to XPA cells, and survival rates were above those values presented by normal cells. The experiments performed for apoptosis induction demonstrated that AT and normal cells did not undergo apoptosis until 48 h after treatment, while 53.6% of XPA cells underwent apoptosis at similar conditions. In addition, the analysis of gene expression profiles by the cDNA microarray method indicated the modulation of several classes of genes, including DNA repair/stress response genes, following cisplatin treatment (25 μM) in cells collected after 16 h (Mello et al., unpublished data). Both *BRCA1* and *TP53* genes were found down-regulated and *TREX1*, *FEN1* and *ERCC6* were up-regulated in AT cells, relative to the normal primary cells (Table 1, Mello et al., unpublished data).

It is known that BRCA1 and TP53 products play an important role in the checkpoint activation and also in DNA repair processes. While TP53 activates proteins belonging to different DNA repair pathways (Sengupta and Harris 2005), BRCA1 acts as an adapter, being necessary for the recruitment of other proteins that take part in HR system (Rosselli et al. 2003). In addition to the down-regulation of *BRCA1* and *TP53*, the lack of a functional ATM protein constitutes a serious impairment for the activation of their products. However, it has been suggested that ATR is responsible for the phosphorylation of BRCA1 and TP53 in response to DNA cross-links (Pichierri and Rosselli 2004; Rosselli et al. 2003), allowing the activation of these proteins even in AT cells. The consequences of down-regulation at the transcriptional level presented by *TP53* and *BRCA1* genes observed in cisplatin-treated AT cells was not confirmed at protein level, but may possibly be related to the lack of apoptosis induction observed in AT cells, since both genes are directly involved in this process (Slee et al. 2004; Thangaraju et al. 2000), or can also be related to the deficiency in DNA repair and cell-cycle control.

FEN1 is an endonuclease essential for correct DNA replication, being responsible for the excision of the 5′ flap generated when the DNA polymerase finds the 5′ prime of an Okasaki fragment (Gordenin et al. 1997), and the up-regulation of *FEN1* in AT cells can be probably related to the activation of BER (Huggins et al. 2002) and/or HR (Spiro and McMurray 2003) pathways. TREX1 protein interacts with ATR and cooperates in the activation of checkpoints, sensing the DNA damage by the recognition of the complex single-strand DNA-RPA, and participates in DNA repair (Zou and Elledge 2003). Finally, *ERCC6*, another induced gene in cisplatin-treated AT cells, encodes the CSB protein, which is important for the transcription-coupled repair (TCR) mechanism, and probably, this gene is one of the responsible for the RNA polymerase II uncoupling from the DNA, allowing access to the TCR machinery (Hanawalt 2000).

Alterations in gene expression patterns associated to the results obtained at cellular level possibly suggest that ATM deficient cells may present an alternative ATM-independent pathway for the processing of cisplatin-induced cross-links, which may confer an apparent resistant phenotype that was not expected for AT cells. Probably, this alternative pathway occurs via ATR, and the modulated genes might explain why AT cells respond differently of ATM-proficient cells in the processing and repair of DNA lesions induced by cisplatin. However, evidence towards the understanding about the integration between the processes of damage response is still incipient.

5 Concluding Remarks

Signaling responses to IR-induced DNA damage depend predominantly on the extent and types of DNA lesions, as well as the cell type and inherent genetic background. In addition, other factors, such as the cell cycle phase in which cells are damaged, the dose level and several aspects of experimental protocols, are also important to be taken into account. These differences make difficult the establishment of comparisons among the results presented by different authors, eventually generating some conflicting results and interpretations.

Although a considerable amount of information has emerged in the last decade towards the understanding of the intricate network of signal transduction in response to physical and chemical genotoxic agents, there is a number of molecular pathways underlying the cellular responses that is still unknown. Finally, data regarding gene expression analyzed at transcriptional level associated to those obtained at cellular level may indicate at least some inter-connection among cell cycle arrest, DNA repair pathways and apoptosis signaling. However, its demonstration requires a more detailed investigation for sets of genes, in particular, and the application of methods at protein level. While the studies in normal cells may clarify the mechanisms leading to genomic instability and carcinogenesis, in contrast, in tumor cells, they are relevant for an understanding about drug or IR responses, which can be useful in terms of improvement towards the efficacy of therapies, and the evaluation of sensitivity or resistance to a certain therapy regimen.

Acknowledgements The authors would like to thank all members and collaborators of the laboratories involved in the work on gene expression, in particular, Dr Cristina M. Junta, Dr Paula Sandrin-Garcia and Dr Renato S. Cardoso. The research Grant from FAPESP, CNPq, CAPES and FAEPA (HC-FMRP, USP) is also acknowledged. The authors also apologize to those investigators whose articles were not cited due to space constraints.

References

Abraham RT (2001) Cell cycle checkpoint signaling through the ATM and ATR kinases. Genes Dev 15:2177–2196

Amundson SA et al. (2000) Identification of potential mRNA biomarkers in peripheral blood lymphocytes for human exposure to ionizing radiation. Radiat Res 154:342–346

Amundson SA, Fornace AJ Jr (2001) Gene expression profiles for monitoring radiation exposure. Radiat Prot Dosimetry 97:11–16

Amundson SA et al. (2003) Differential responses of stress genes to low dose-rate gamma irradiation. Mol Cancer Res 1:445–452

Assenmacher N, Hopfner KP (2004) MRE11/RAD50/NBS1: complex activities. Chromosoma 113:157–166

Aylon Y, Liefshitz B, Kupiec M (2004) The CDK regulates repair of double-strand breaks by homologous recombination during the cell cycle. Embo J 23:4868–4875

Bartek J, Lukas C, Lukas J (2004) Checking on DNA damage in S phase. Nat Rev Mol Cell Biol 5:792–804

Bernstein C, Bernstein H, Payne CM, Garewal H (2002) DNA repair/pro-apoptotic dual-role proteins in five major DNA repair pathways: fail-safe protection against carcinogenesis. Mutat Res 511:145–178

Bhatia S, Sather HN, Pabustan OB, Trigg ME, Gaynon PS, Robison LL (2002) Low incidence of second neoplasms among children diagnosed with acute lymphoblastic leukemia after 1983. Blood 99:4257–4264

Bohr VA, Dianov GL (1999) Oxidative DNA damage processing in nuclear and mitochondrial DNA. Biochimie 81:155–160

Borch RF (1987) The platinum antitumor drugs. In: Powis G, Prough RA (eds) Metabolism and action of anticancer drugs. Taylor & Francis, London p 163–193

Brassesco MS, Camparoto ML, Tone LG, Sakamoto-Hojo ET (2004) Analysis of ETV6/RUNX1 fusions for evaluating the late effects of cancer therapy in ALL (acute lymphoblastic leukemia) cured patients. Cytogenet Genome Res 104:346–351

Camparoto ML, Ramalho AT, Natarajan AT, Curado MP, Sakamoto-Hojo ET (2003) Translocation analysis by the FISH-painting method for retrospective dose reconstruction in individuals exposed to ionizing radiation 10 years after exposure. Mutat Res 530:1–7

Canman CE et al. (1998) Activation of the ATM kinase by ionizing radiation and phosphorylation of p53. Science 281:1677–1679

Cardoso RS, Espanhol AR, Passos GA, Sakamoto-Hojo ET (2002) Differential gene expression in gamma-irradiated BALB/3T3 fibroblasts under the influence of 3-aminobenzamide, an inhibitior of parp enzyme. Mutat Res 508:33–40

Cardoso RS et al. (2005) Hybridization signatures of gamma-irradiated murine fetal thymus organ culture (FTOC) reveal modulation of genes associated with T-cell receptor V(D)J recombination and DNA repair. Mol Immunol (in press) Available online 13 April 2005 at http://www.sciencedirect.com/science/journal/01615890

Ceraline J et al. (1998) Inactivation of p53 in normal human cells increases G2/M arrest and sensitivity to DNA-damaging agents. Int J Cancer 75:432–438

Chaubey RC, Bhilwade HN, Rajagopalan R, Bannur SV (2001) Gamma ray induced DNA damage in human and mouse leucocytes measured by SCGE-Pro: a software developed for automated image analysis and data processing for Comet assay. Mutat Res 490:187–197

Chaudhry MA, Chodosh LA, McKenna WG, Muschel RJ (2003) Gene expression profile of human cells irradiated in G1 and G2 phases of cell cycle. Cancer Lett 195:221–233

Chaudhuri J, Alt FW (2004) Class-switch recombination: interplay of transcription, DNA deamination and DNA repair. Nat Rev Immunol 4:541–552

Cheo DL et al. (1997) Characterization of defective nucleotide excision repair in XPC mutant mice. Mutat Res 374:1–9

Choi BK, Choi CH, Oh HL, Kim YK (2004) Role of ERK activation in cisplatin-induced apoptosis in A172 human cells. Neurotoxicology 25:915–924

Chun HH, Gatti RA (2004) Ataxia-telangiectasia, an evolving phenotype. DNA Repair (Amst) 3:1187–1196

Cline SD, Hanawalt PC (2003) Who's on first in the cellular response to DNA damage? Nat Rev Mol Cell Biol 4:361–372

Cole RS, Levitan D, Sinden RR (1976) Removal of psoralen interstrand cross-links from DNA of *Escherichia coli*: mechanism and genetic control. J Mol Biol 103:39–59

Collis SJ, DeWeese TL, Jeggo PA, Parker AR (2005) The life and death of DNA-PK. Oncogene 24:949–961

Concannon CG, Orrenius S, Samali A (2001) Hsp27 inhibits cytochrome c-mediated caspase activation by sequestering both pro-caspase-3 and cytochrome c. Gene Expr 9:195–201

Cullinane C, Mazur SJ, Essigmann JM, Phillips DR, Bohr VA (1999) Inhibition of RNA polymerase II transcription in human cell extracts by cisplatin DNA damage. Biochemistry 38:6204–6212

Dainiak N (2002) Hematologic consequences of exposure to ionizing radiation. Exp Hematol 30:513–528

Dang T, Bao S, Wang XF (2005) Human Rad9 is required for the activation of S-phase checkpoint and the maintenance of chromosomal stability. Genes Cells 10:287–295

Datta K, Shah P, Srivastava T, Mathur SG, Chattopadhyay P, Sinha S (2004) Sensitizing glioma cells to cisplatin by abrogating the p53 response antisense oligonucleotides. Cancer Gene Ther 11:525–531

De Silva IU, McHugh PJ, Clingen PH, Hartley JA (2000) Defining the roles of nucleotide excision repair and recombination in the repair of DNA interstrand cross-links in mammalian cells. Mol Cell Biol 20:7980–7990

Eastman A (2004) Cell cycle checkpoints and their impact on anticancer therapeutic strategies. J Cell Biochem 91:223–231

Esashi F et al. (2005) CDK-dependent phosphorylation of BRCA2 as a regulatory mechanism for recombinational repair. Nature 434:598–604

Falck J, Coates J, Jackson SP (2005) Conserved modes of recruitment of ATM, ATR and DNA-PKcs to sites of DNA damage. Nature 434:605–611

Fan S et al. (1995) Disruption of p53 function sensitizes breast cancer MCF-7 cells to cisplatin and pentoxifylline. Cancer Res 55:1649–1654

Feijoo C et al. (2001) Activation of mammalian Chk1 during DNA replication arrest: a role for Chk1 in the intra-S phase checkpoint monitoring replication origin firing. J Cell Biol 154:913–923

Friedberg EC, Walker GC, Siede W (1995) DNA repair and mutagenesis. American Society of Microbiology Press, Washington D.C., p 698

Gangopadhyay S, Jalali F, Reda D, Peacock J, Bristow RG, Benchimol S (2002) Expression of different mutant p53 transgenes in neuroblastoma cells leads to different cellular responses to genotoxic agents. Exp Cell Res 275:122–131

Gatei M et al. (2000) ATM dependent phosphorylation of nibrin in response to radiation exposure. Nat Genet 25:115–119

Gordenin DA, Kunkel TA, Resnick MA (1997) Repeat expansion – all in a flap? Nat Genet 16:116–118

Haber JE (2000) Partners and pathways repairing a double-strand break. Trends Genet 16:259–264

Hall EJ (2000) Radiobiology for the radiologist, 5th edn. Lippincott Williams & Wilkins, Philadelphia, PA

Han Z et al. (1995) Evidence for a G2 checkpoint in p53-independent apoptosis induction by X-irradiation. Mol Cell Biol 15:5849–5857

Hanawalt PC (2000) DNA repair. The bases for Cockayne syndrome. Nature 405:415–416

Harms Ringdahl M (1998) Some aspects on radiation induced transmissible genomic instability. Mutat Res 404:27–33

Hartwell L, Weinert T, Kadyk L, Garvik B (1994) Cell cycle checkpoints, genomic integrity, and cancer. Cold Spring Harb Symp Quant Biol 59:259–263

Hartwell LH, Kastan MB (1994) Cell cycle control and cancer. Science 266:1821–1828

Hartwell LH, Weinert TA (1989) Checkpoints: controls that ensure the order of cell cycle events. Science 246:629–634

Heiger-Bernays WJ, Essigmann JM, Lippard SJ (1990) Effect of the antitumor drug cis-diamminedichloroplatinum(II) and related platinum complexes on eukaryotic DNA replication. Biochemistry 29:8461–8466
Heinloth AN et al. (2003) Identification of distinct and common gene expression changes after oxidative stress and gamma and ultraviolet radiation. Mol Carcinog 37:65–82
Hirose Y, Berger MS, Pieper RO (2001) p53 effects both the duration of G2/M arrest and the fate of temozolomide-treated human glioblastoma cells. Cancer Res 61:1957–1963
Huang JC, Zamble DB, Reardon JT, Lippard SJ, Sancar A (1994) HMG domain proteins specifically inhibit the repair of the major DNA adduct of the anticancer drug cisplatin by human excision nuclease. Proc Natl Acad Sci USA 91:10394–10398
Huang L, Snyder AR, Morgan WF (2003) Radiation-induced genomic instability and its implications for radiation carcinogenesis. Oncogene 22:5848–5854
Huggins CF, Chafin DR, Aoyagi S, Henricksen LA, Bambara RA, Hayes JJ (2002) Flap endonuclease 1 efficiently cleaves base excision repair and DNA replication intermediates assembled into nucleosomes. Mol Cell 10:1201–1211
Hwang BJ, Ford JM, Hanawalt PC, Chu G (1999) Expression of the p48 xeroderma pigmentosum gene is p53-dependent and involved in global genomic repair. Proc Natl Acad Sci USA 96:424–428
Iliakis G, Wang Y, Guan J, Wang H (2003) DNA damage checkpoint control in cells exposed to ionizing radiation. Oncogene 22:5834–5847
Innocente SA, Abrahamson JL, Cogswell JP, Lee JM (1999) p53 regulates a G2 checkpoint through cyclin B1. Proc Natl Acad Sci USA 96:2147–2152
Ira G et al. (2004) DNA end resection, homologous recombination and DNA damage checkpoint activation require CDK1. Nature 431:1011–1017
Jeggo PA, Concannon P (2001) Immune diversity and genomic stability: opposite goals but similar paths. J Photochem Photobiol B 65:88–96
Jeggo PA, Lobrich M (2005) Artemis links ATM to double strand break rejoining. Cell Cycle 4:359–362
Jeggo P, O'Neill P (2002) The Greek Goddess, Artemis, reveals the secrets of her cleavage. DNA Repair (Amst) 1:771–777
Jeggo P, Singleton B, Beamish H, Priestley A (1999) Double strand break rejoining by the Ku-dependent mechanism of non-homologous end-joining. C R Acad Sci III 322:109–112
Jen KY, Cheung VG (2003) Transcriptional response of lymphoblastoid cells to ionizing radiation. Genome Res 13:2092–2100
Johnson KL, Nath J, Pluth JM, Tucker JD (1999) The distribution of chromosome damage, non-reciprocal translocations and clonal aberrations in lymphocytes from Chernobyl clean-up workers. Mutat Res 439:77–85
Jordan P, Carmo-Fonseca M (2000) Molecular mechanisms involved in cisplatin cytotoxicity. Cell Mol Life Sci 57:1229–1235
Karran P (2000) DNA double strand break repair in mammalian cells. Curr Opin Genet Dev 10:144–150
Kitahara O, Katagiri T, Tsunoda T, Harima Y, Nakamura Y (2002) Classification of sensitivity or resistance of cervical cancers to ionizing radiation according to expression profiles of 62 genes selected by cDNA microarray analysis. Neoplasia 4:295–303
Koishi S et al. (1998) Biomarkers in long survivors of pediatric acute lymphoblastic leukemia patients: late effects of cancer chemotherapy. Mutat Res 422:213–222
Kow YW, Wallace SS, Van Houten B (1990) UvrABC nuclease complex repairs thymine glycol, an oxidative DNA base damage. Mutat Res 235:147–156

Kuerbitz SJ, Plunkett BS, Walsh WV, Kastan MB (1992) Wild-type p53 is a cell cycle checkpoint determinant following irradiation. Proc Natl Acad Sci USA 89:7491–7495
Lee JH, Paull TT (2005) ATM activation by DNA double-strand breaks through the Mre11-Rad50-Nbs1 complex. Science 308:551–554
Lees-Miller SP, Meek K (2003) Repair of DNA double strand breaks by non-homologous end joining. Biochimie 85:1161–1173
Lehnert S, Chow TY (1997) Low doses of ionizing radiation induce nuclear activity in human tumour cell lines which catalyzes homologous double-strand recombination. Radiat Environ Biophys 36:67–70
Li L, Zou L (2005) Sensing, signaling, and responding to DNA damage: organization of the checkpoint pathways in mammalian cells. J Cell Biochem 94:298–306
Li L, Story M, Legerski RJ (2001) Cellular responses to ionizing radiation damage. Int J Radiat Oncol Biol Phys 49:1157–1162
Lim DS et al. (2000) ATM phosphorylates p95/nbs1 in an S-phase checkpoint pathway. Nature 404:613–617
Lin JJ, Sancar A (1989) A new mechanism for repairing oxidative damage to DNA: (A)BC excinuclease removes AP sites and thymine glycols from DNA. Biochemistry 28:7979–7984
Lukas C et al. (2001) DNA damage-activated kinase Chk2 is independent of proliferation or differentiation yet correlates with tissue biology. Cancer Res 61:4990–4993
Ma J, Murphy M, O'Dwyer PJ, Berman E, Reed K, Gallo JM (2002) Biochemical changes associated with a multidrug-resistant phenotype of a human glioma cell line with temozolomide-acquired resistance. Biochem Pharmacol 63:1219–1228
Magae J, Hoshi Y, Furukawa C, Kawakami Y, Ogata H (2003) Quantitative analysis of biological responses to ionizing radiation, including dose, irradiation time, and dose rate. Radiat Res 160:543–548
Mamenta EL, Poma EE, Kaufmann WK, Delmastro DA, Grady HL, Chaney SG (1994) Enhanced replicative bypass of platinum-DNA adducts in cisplatin-resistant human ovarian carcinoma cell lines. Cancer Res 54:3500–3505
Matsuoka S, Rotman G, Ogawa A, Shiloh Y, Tamai K, Elledge SJ (2000) Ataxia telangiectasia-mutated phosphorylates Chk2 in vivo and in vitro. Proc Natl Acad Sci USA 97:10389–10394
Mayer C et al. (2002) DNA repair capacity after gamma-irradiation and expression profiles of DNA repair genes in resting and proliferating human peripheral blood lymphocytes. DNA Repair (Amst) 1:237–250
McGowan CH, Russell P (2004) The DNA damage response: sensing and signaling. Curr Opin Cell Biol 16:629–633
Momota H et al. (2003) Histone H2AX sensitizes glioma cells to genotoxic stimuli by recruiting DNA double-strand break repair proteins. Int J Oncol 23:311–315
Mori M, Benotmane MA, Vanhove D, van Hummelen P, Hooghe-Peters EL, Desaintes C (2004) Effect of ionizing radiation on gene expression in CD4+ T lymphocytes and in Jurkat cells: unraveling novel pathways in radiation response. Cell Mol Life Sci 61:1955–1964
Morrison C, Sonoda E, Takao N, Shinohara A, Yamamoto K, Takeda S (2000) The controlling role of ATM in homologous recombinational repair of DNA damage. Embo J 19:463–471
Mu D et al. (2000) DNA interstrand cross links induce futile repair synthesis in mammalian cell extracts. Mol Cell Biol 20:2446–2454

Murakami MS, Strobel MC, Vande Woude GF (1995) Cell cycle regulation, oncogenes, and antineoplastic drugs. In: Mendelson J, Howley PM, Israel MA, Liotta LA (eds) The molecular basis of cancer. Saunders, Philadelphia, PA, p 574

Murray D, Rosenberg E (1996) The importance of the ERCC1/ERCC4[XPF] complex for hypoxic-cell radioresistance does not appear to derive from its participation in the nucleotide excision repair pathway. Mutat Res 364:217–226

Natarajan AT, Darroudi F, Jha AN, Meijers M, Zdzienicka MZ (1993) Ionizing radiation induced DNA lesions which lead to chromosomal aberrations. Mutat Res 299:297–303

Natarajan AT et al. (1998) 137Cesium-induced chromosome aberrations analyzed by fluorescence in situ hybridization: eight years follow up of the Goiania radiation accident victims. Mutat Res 400:299–312

Nyberg KA, Michelson RJ, Putnam CW, Weinert TA (2002) Toward maintaining the genome: DNA damage and replication checkpoints. Annu Rev Genet 36:617–656

O'Connell MJ, Cimprich KA (2005) G2 damage checkpoints: what is the turn-on? J Cell Sci 118:1–6

Offer H et al. (2002) The onset of p53-dependent DNA repair or apoptosis is determined by the level of accumulated damaged DNA. Carcinogenesis 23:1025–1032

Oswald CB, Chaney SG, Hall IH (1990) Inhibition of DNA synthesis in P388 lymphocytic leukemia cells of BDF1 mice by cis-diamminedichloroplatinum(II) and its derivatives. J Pharm Sci 79:875–880

Otomo T, Hishii M, Arai H, Sato K, Sasai K (2004) Microarray analysis of temporal gene responses to ionizing radiation in two glioblastoma cell lines: up-regulation of DNA repair genes. J Radiat Res (Tokyo) 45:53–60

Painter RB, Young BR (1980) Radiosensitivity in ataxia-telangiectasia: a new explanation. Proc Natl Acad Sci USA 77:7315–7317

Pastwa E, Blasiak J (2003) Non-homologous DNA end joining. Acta Biochim Pol 50:891–908

Paul C, Manero F, Gonin S, Kretz-Remy C, Virot S, Arrigo AP (2002) Hsp27 as a negative regulator of cytochrome C release. Mol Cell Biol 22:816–834

Paull TT, Rogakou EP, Yamazaki V, Kirchgessner CU, Gellert M, Bonner WM (2000) A critical role for histone H2AX in recruitment of repair factors nuclear foci after DNA damage. Curr Biol 10:886–895

Peng CY, Graves PR, Thoma RS, Wu Z, Shaw AS, Piwnica-Worms H (1997) Mitotic and G2 checkpoint control: regulation of 14-3-3 protein phosphorylation of Cdc25C on serine-216. Science 277:1501–1505

Pestell KE, Hobbs SM, Titley JC, Kelland LR, Walton MI (2000) Effect of p53 status on sensitivity to platinum complexes in a human ovarian cancer cell line. Mol Pharmacol 57:503–511

Pfeiffer P (1998) The mutagenic potential of DNA double-strand break repair. Toxicol Lett 96–97:119–129

Pichierri P, Rosselli F (2004) Fanconi anemia proteins and the s phase checkpoint. Cell Cycle 3:698–700

Pihan GA, Doxsey SJ (1999) The mitotic machinery as a source of genetic instability in cancer. Semin Cancer Biol 9:289–302

Poppenborg H, Knupfer MM, Preiss R, Wolff JE, Galla HJ (1999) Cisplatin (CDDP)-induced radiation resistance is not associated with resistance in 86HG39 and A172 malignant glioma cells. Eur J Cancer 35:1150–1154

Preston DL et al. (1994) Cancer incidence in atomic bomb survivors. Part III. Leukemia, lymphoma and multiple myeloma, 1950–1987. Radiat Res 137:S68–S97

Qin LF, Ng IO (2002) Induction of apoptosis by cisplatin and its effect on cell cycle-related proteins and cell cycle changes in hepatoma cells. Cancer Lett 175:27–38

Quinn JE et al. (2003) BRCA1 functions as a differential modulator of chemotherapy-induced apoptosis. Cancer Res 63:6221–6228
Rasnick D, Duesberg PH (1999) How aneuploidy affects metabolic control and causes cancer. Biochem J 340(Pt3):621–630
Riballo E et al. (2004) A pathway of double-strand break rejoining dependent upon ATM, Artemis, and proteins locating to gamma-H2AX foci. Mol Cell 16:715–724
Rieger KE, Chu G (2004) Portrait of transcriptional responses to ultraviolet and ionizing radiation in human cells. Nucleic Acids Res 32:4786–4803
Rosselli F, Briot D, Pichierri P (2003) The Fanconi anemia pathway and the DNA interstrand cross-links repair. Biochimie 85:1175–1184
Rothkamm K, Kruger I, Thompson LH, Lobrich M (2003) Pathways of DNA double-strand break repair during the mammalian cell cycle. Mol Cell Biol 23:5706–5715
Sakamoto-Hojo ET, Natarajan AT, Curado MP (1999) Chromosome translocations in lymphocytes from individuals exposed to Cs-137 75 years after the accident in Goiania (Brazil). Radiat Prot Dosim 86:25–32
Sakamoto-Hojo ET et al. (2003) Gene expression profiles in human cells submitted to genotoxic stress. Mutat Res 544:403–413
Sancar A, Sancar GB (1988) DNA repair enzymes. Annu Rev Biochem 57:29–67
Sancar A, Lindsey-Boltz LA, Unsal-Kacmaz K, Linn S (2004) Molecular mechanisms of mammalian DNA repair and the DNA damage checkpoints. Annu Rev Biochem 73:39–85
Sanchez Y et al. (1997) Conservation of the Chk1 checkpoint pathway in mammals: linkage of DNA damage to Cdk regulation through Cdc25. Science 277:1497–1501
Scarpato R, Lori A, Panasiuk G, Barale R (1997) FISH analysis of translocations in lymphocytes of children exposed to the Chernobyl fallout: preferential involvement of chromosome 10. Cytogenet Cell Genet 79:153–156
Schlissel M, Constantinescu A, Morrow T, Baxter M, Peng A (1993) Double-strand signal sequence breaks in V(D)J recombination are blunt, 5′-phosphorylated, RAG-dependent, and cell cycle regulated. Genes Dev 7:2520–2532
Schmidt-Ullrich RK, Dent P, Grant S, Mikkelsen RB, Valerie K (2000) Signal transduction and cellular radiation responses. Radiat Res 153:245–257
Sengupta S, Harris CC (2005) p53: traffic cop at the crossroads of DNA repair and recombination. Nat Rev Mol Cell Biol 6:44–55
Sherr CJ, Roberts JM (1995) Inhibitors of mammalian G1 cyclin-dependent kinases. Genes Dev 9:1149–1163
Shiloh Y (2003) ATM and related protein kinases: safeguarding genome integrity. Nat Rev Cancer 3:155–168
Shiloh Y, Kastan MB (2001) ATM: genome stability, neuronal development, and cancer cross paths. Adv Cancer Res 83:209–254
Shu HK, Kim MM, Chen P, Furman F, Julin CM, Israel MA (1998) The intrinsic radioresistance of glioblastoma-derived cell lines is associated with a failure of p53 to induce p21(BAX) expression. Proc Natl Acad Sci USA 95:14453–14458
Siddik ZH (2003) Cisplatin: mode of cytotoxic action and molecular basis of resistance. Oncogene 22:7265–7279
Siddik Z, Mims B, Lozano G, Thai G (1998) Independent pathways of p53 induction by cisplatin and X-rays in a cisplatin-resistant ovarian tumor cell line. Cancer Res 58:698–703
Slee EA, O'Connor DJ, Lu X (2004) To die or not to die: how does p53 decide? Oncogene 23:2809–2818
Slupphaug G, Kavli B, Krokan HE (2003) The interacting pathways for prevention and repair of oxidative DNA damage. Mutat Res 531:231–251

Smith LE, Nagar S, Kim GJ, Morgan WF (2003) Radiation-induced genomic instability: radiation quality and dose response. Health Phys 85:23–29

Smith ML et al. (2000) p53-mediated DNA repair responses to UV radiation: studies of mouse cells lacking p53, p21, and/or gadd45 genes. Mol Cell Biol 20:3705–3714

Snyder AR, Morgan WF (2004) Gene expression profiling after irradiation: clues to understanding acute and persistent responses? Cancer Metastasis Rev 23:259–268

Sorensen CS et al. (2005) The cell-cycle checkpoint kinase Chk1 is required for mammalian homologous recombination repair. Nat Cell Biol 7:195–201

Spiro C, McMurray CT (2003) Nuclease-deficient FEN-1 blocks Rad51/BRCA1-mediated repair and causes trinucleotide repeat instability. Mol Cell Biol 23:6063–6074

Stiff T, O'Driscoll M, Rief N, Iwabuchi K, Lobrich M, Jeggo PA (2004) ATM and DNA-PK function redundantly to phosphorylate H2AX after exposure to ionizing radiation. Cancer Res 64:2390–2396

Suzuki K, Ojima M, Kodama S, Watanabe M (2003) Radiation-induced DNA damage and delayed induced genomic instability. Oncogene 22:6988–6993

Taylor WR, Stark GR (2001) Regulation of the G2/M transition by p53. Oncogene 20:1803–1815

Thangaraju M, Kaufmann SH, Couch FJ (2000) BRCA1 facilitates stress-induced apoptosis in breast and ovarian cancer cell lines. J Biol Chem 275:33487–33496

Thompson DE et al. (1994) Cancer incidence in atomic bomb survivors. Part II: Solid tumors, 1958–1987. Radiat Res 137:S17–S67

Thompson LH (1996) Evidence that mammalian cells possess homologous recombinational repair pathways. Mutat Res 363:77–88

Tlsty TD et al. (1995) Genomic instability and cancer. Mutat Res 337:1–7

Tornaletti S, Patrick SM, Turchi JJ, Hanawalt PC (2003) Behavior of T7 RNA polymerase and mammalian RNA polymerase II at site-specific cisplatin adducts in the template DNA. J Biol Chem 278:35791–35797

Turenne GA, Paul P, Laflair L, Price BD (2001) Activation of p53 transcriptional activity requires ATM's kinase domain and multiple N-terminal serine residues of p53. Oncogene 20:5100–5110

Tusher VG, Tibshirani R, Chu G (2001) Significance analysis of microarrays applied to the ionizing radiation response. Proc Natl Acad Sci USA 98:5116–5121

Vaisman A et al. (1999) Effect of DNA polymerases and high mobility group protein 1 on the carrier ligand specificity for translesion synthesis past platinum-DNA adducts. Biochemistry 38:11026–11039

Valerie K, Povirk LF (2003) Regulation and mechanisms of mammalian double-strand break repair. Oncogene 22:5792–5812

Vousden KH, Lu X (2002) Live or let die: the cell's response to p53. Nat Rev Cancer 2:594–604

Wang H, Powell SN, Iliakis G, Wang Y (2004) ATR affecting cell radiosensitivity is dependent on homologous recombination repair but independent of nonhomologous end joining. Cancer Res 64:7139–7143

Wang H, Hu B, Liu R, Wang Y (2005) CHK1 affecting cell radiosensitivity is independent of non-homologous end joining. Cell Cycle 4:300–303

Wang Q, Fan S, Eastman A, Worland PJ, Sausville EA, O'Connor PM (1996) UCN-01: a potent abrogator of G2 checkpoint function in cancer cells with disrupted p53. J Natl Cancer Inst 88:956–965

Wang XW et al. (1999) GADD45 induction of a G2/M cell cycle checkpoint. Proc Natl Acad Sci USA 96:3706–3711

Wilson GD (2004) Radiation and the cell cycle, revisited. Cancer Metastasis Rev 23:209–225

Wrighton KH, Prele CM, Sunters A, Yeudall WA (2004) Aberrant p53 alters DNA damage checkpoints in response to cisplatin: downregulation of CDK expression and activity. Int J Cancer 112:760–770

Xu B, Kim S, Kastan MB (2001) Involvement of Brca1 in S-phase and G(2)-phase checkpoints after ionizing irradiation. Mol Cell Biol 21:3445–3450

Yacoub A, Park JS, Qiao L, Dent P, Hagan MP (2001) MAPK dependence of DNA damage repair: ionizing radiation and the induction of expression of the DNA repair genes XRCC1 and ERCC1 in DU145 human prostate carcinoma cells in a MEK1/2 dependent fashion. Int J Radiat Biol 77:1067–1078

Yamamoto K et al. (2003) Fanconi anemia FANCG protein in mitigating radiation- and enzyme-induced DNA double-strand breaks by homologous recombination in vertebrate cells. Mol Cell Biol 23:5421–5430

Yang J, Yu Y, Hamrick HE, Duerksen-Hughes PJ (2003) ATM, ATR and DNA-PK: initiators of the cellular genotoxic stress responses. Carcinogenesis 24:1571–1580

Yarden RI, Pardo-Reoyo S, Sgagias M, Cowan KH, Brody LC (2002) BRCA1 regulates the G2/M checkpoint by activating Chk1 kinase upon DNA damage. Nat Genet 30:285–289

Yin E, Nelson DO, Coleman MA, Peterson LE, Wyrobek AJ (2003) Gene expression changes in mouse brain after exposure to low-dose ionizing radiation. Int J Radiat Biol 79:759–775

Yuan SS, Chang HL, Lee EY (2003) Ionizing radiation-induced Rad51 nuclear focus formation is cell cycle-regulated and defective in both ATM(– / –) and c-Abl(– / –) cells. Mutat Res 525:85–92

Zabkiewicz J, Clarke AR (2004) DNA damage-induced apoptosis: insights from the mouse. Biochim Biophys Acta 1705:17–25

Zamble DB, Mu D, Reardon JT, Sancar A, Lippard SJ (1996) Repair of cisplatin DNA adducts by the mammalian excision nuclease. Biochemistry 35:10004–10013

Zhan Q (2005) Gadd45a, a p53- and BRCA1-regulated stress protein, in cellular response to DNA damage. Mutat Res 569:133–143

Zhang J et al. (2004) Chk2 phosphorylation of BRCA1 regulates DNA double-strand break repair. Mol Cell Biol 24:708–718

Zhao H, Piwnica-Worms H (2001) ATR-mediated checkpoint pathways regulate phosphorylation and activation of human Chk1. Mol Cell Biol 21:4129–4139

Zhou BB, Elledge SJ (2000) The DNA damage response: putting checkpoints in perspective. Nature 408:433–439

Zou L, Elledge SJ (2003) Sensing DNA damage through ATRIP recognition of RPA-ssDNA complexes. Science 300:1542–1548

Genome Dyn Stab (1)
D.-H. Lankenau: Genome Integrity
DOI 10.1007/7050_015/Published online: 13 May 2006

Ion Irradiation as a Tool to Reveal the Spatiotemporal Dynamics of DNA Damage Response Processes

Gisela Taucher-Scholz (✉) · Burkhard Jakob

Biophysics, Gesellschaft für Schwerionenforschung,
Planckstraße 1, 64291 Darmstadt, Germany
G.Taucher-Scholz@gsi.de

Electronic Supplementary Material
The electronic reference of this chapter is http://dx.doi.org/10.1007/7050_015. The online version of this chapter includes electronic supplementary material. This material is available to authorised users and can be accessed by means of the ESM button beneath the abstract. To cite or link to this chapter you can use the above reference.

Abstract In response to genotoxic exposure, mammalian cells have developed efficient DNA repair and cell-cycle checkpoint mechanisms that allow coping with both endogenous and exogenous sources of DNA damage. Imaging approaches to visualize the dynamics of repair-related proteins at locally restricted regions of DNA damage have substantially contributed to the understanding of the biochemical processes involved in these damage response pathways. Taking advantage of their extremely localized dose deposition, charged particle beams have recently emerged as a tool to generate discrete sites of subnuclear damage. The application of this focal type of irradiation to study spatiotemporal aspects of DNA damage recognition and repair processes is presented in the context of recent advancements in this field.

1 Introduction

DNA damage and in particular double-strand breaks (DSBs) pose a major threat to the preservation of genome integrity (Shiloh 2003). DSBs are potentially detrimental lesions, as they disrupt the genetic information of both DNA strands. While they are the principal deleterious lesions induced by environmental genotoxic agents such as ionizing radiation (IR) or radiomimetic drugs, DSBs may also arise endogenously as byproducts of normal intracellular metabolism (Jackson 2002; Vilenchik and Knudson 2003).

Failure to repair DSBs can lead to cell death via the generation of lethal chromosomal aberrations (van Gent et al. 2001; Thompson and Schild 2002) or the induction of apoptosis (Norbury and Zhivotovsky 2004). On the other hand, inadequate repair of DSBs in surviving cells may result in mutations or genomic rearrangements ultimately leading to genomic instability and cancer (Khanna and Jackson 2001; van Gent et al. 2001; Kastan and Bartek 2004).

Efficient DNA damage response mechanisms have therefore evolved that enable a cell either to eliminate or cope with the damage (Hoeijmakers 2001; Christmann et al. 2003; Sancar et al. 2004). The cellular reactions involved in these responses include: (1) the detection of DNA damage; (2) the activation of DNA damage checkpoints to coordinate cell cycle progression with damage removal; (3) the repair of the DNA damage to restore the genetic information; and/or (4) eventually a permanent withdrawal of heavily damaged cells from proliferation to reduce the likelihood of malignant transformation. Depending on the cell type and the extent of persisting damage, the mechanisms preventing cell cycle progression may lead to the induction of permanent arrest, premature differentiation, cellular senescence or apoptosis (Bernstein et al. 2002; Norbury and Zhivotovsky 2004; Park and Gerson 2005; von Zglinicki et al. 2005).

The orderly course of events in response to DNA damage, regulated by the interplay of genomic surveillance or checkpoint pathways and repair mechanisms, is of primordial importance to ensure the accurate transmission of genetic information from one cell generation to the next (Dasika et al. 1999; Zhou and Elledge 2000; Rouse and Jackson 2002). Great progress in the understanding of genetic and biochemical aspects of these pathways has recently been achieved and extensively discussed in a number of reviews (Valerie and Povirk 2003; Bartek et al. 2004; Kastan and Bartek 2004; Sancar et al. 2004; Stracker et al. 2004; Stucki and Jackson 2004; Ward and Chen 2004; Lavin et al. 2005; Löbrich and Jeggo 2005; Sengupta and Harris 2005). A large number of proteins are responsible for these processes, acting as sensors, mediators, signal transducers or effectors in damage repair (Iliakis et al. 2003; Koundrioukoff et al. 2004), as schematically outlined in Fig. 1.

Despite comprehensive studies on the cellular response to irradiation, the sequence of early events involving the recognition of DNA damage and the initial activation of repair and checkpoint pathways are presently not fully understood. The focal organization of these processes at a microscopic level has recently become apparent by the visualization of distinct ionizing radiation-induced foci (IRIF) at the sites of DNA damage (Maser et al. 1997; Mirzoeva and Petrini 2001). One of the most important findings aiding these studies has been the observation that exposure to ionizing radiation causes rapid phosphorylation of the minor nucleosomal histone variant H2AX at sites of DSBs, generating γ-H2AX (Rogakou et al. 1998). Using antibody staining and fluorescence microscopy, discrete visible foci containing thousands of γ-H2AX molecules in a megabase chromatin region around the break can be detected in cell nuclei, thus providing a means to recognize the sites of DSBs (Rogakou et al. 1999; Modesti and Kanaar 2001; Pilch et al. 2003; Takahashi and Ohnishi 2005).

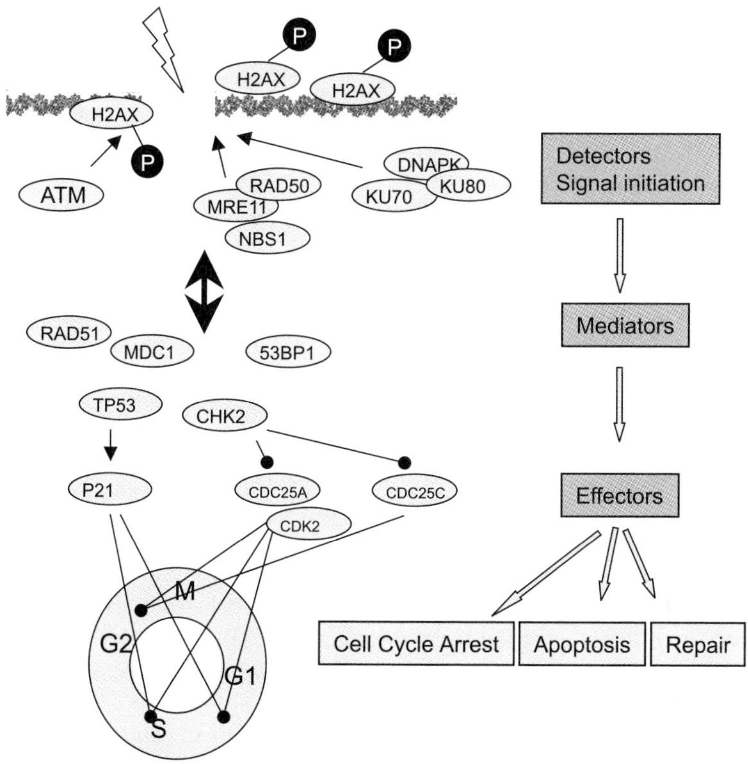

Fig. 1 Proteins important for the cellular response to DNA double-strand breaks. The very simplified schematic drawing shows some relevant features of the DNA damage response induced by ionizing radiation, focusing on proteins mentioned in the text in relation to their focal assembly at DNA DSBs: ATM (Bakkenist and Kastan 2003), Mre11/Rad50/Nbs1 (Stracker et al. 2004), MDC1 (Stucki and Jackson 2004), 53BP1 (Ward and Chen 2004), p21 (cyclin dependent kinase inhibitor CDKN1A, also $p21^{waf/cip}$, Jakob et al. 2000), Rad51 (Haaf et al. 1995). Phosphorylation of the histone variant H2AX occurs specifically at DSBs (Rogakou et al. 1998). Additional signalling and effector proteins are included for better understanding (for more details refer to reviews in the text)

Important insights into how and when the DNA damage response proteins interact with the damaged sites have been obtained by monitoring the hierarchy of protein accumulation at IRIF, e.g. in colocalization studies using various repair-related proteins and γ-H2AX as a marker for DSBs (Fernandez-Capetillo et al. 2004; Lisby and Rothstein 2004). Furthermore, the analysis of focus formation after the induction of DNA damage in genetic null backgrounds has allowed description of the interdependencies of repair and checkpoint proteins during their recruitment to DSBs (Fernandez-Capetillo et al. 2002; Celeste et al. 2003; Ward et al. 2003). Combined with biochemical analyses, these studies have helped to identify a series of nuclear proteins relocalizing and interacting at DNA lesions upon DNA damage. This includes proteins classified as potential sensors of DNA damage, like the Mre11/Rad50/Nbs1 complex (Petrini and Stracker 2003), others with a postulated function as mediators of protein recruitment (e.g. 53BP1 (Ward and Chen 2004) and MDC1 (Stucki and Jackson 2004)) and also proteins with a role in the transduction of DNA damage signals to activate and carry out damage repair (Paull et al. 2000; Lukas et al. 2003; Fernandez-Capetillo et al. 2004).

However, one of the problems encountered in this type of experimental system is that DSBs and other lesions randomly induced by conventional ionizing radiation, like gamma or X-rays, are homogeneously distributed through the nucleus, and "lesion-containing" and "lesion-free" regions cannot be defined. In addition, depending on the experimental system (i.e. proteins and antibodies) the labelled protein of interest concentrated at the IRIF may be hard to distinguish from protein molecules retained by internal structures (e.g. nuclear bodies) or, at high levels of expression, from the basal level of nuclear protein. These factors and the non-recurrent observation inherent to fixed samples are hindering a follow up of the dynamical behavior of IR-induced foci with time. Moreover, occasional foci of repair proteins, also observed in cells not exposed to genotoxic agents, may interfere with the analysis (Vilenchik and Knudson 2003; Sedelnikova et al. 2004; McManus and Hendzel 2005).

A useful way to overcome some of these limitations in this type of studies is the production of discernable regions of DNA damage, which can be studied by microscopic analysis of the repair-related proteins recruited to these sites even above a certain background signal and in a further step, in the combination of localized irradiation and live cell observation. Various techniques are currently available to generate DSBs in (defined) spatially restricted nuclear volumes.

Laser microirradiation was described a few decades ago as a means of producing local cellular damage in the context of studies on spatial chromosomal organization (Berns et al. 1969; Cremer et al. 1982; Cremer and Cremer 1986). In the repair research field, laser technology in combination with the procedure of Limoli and Ward (1993) using Hoechst-stained DNA sensitized with BrdU was first applied for introducing defined areas of subnuclear damage including DSBs by Rogakou et al. (1999). The refinement and general availability of this technique have now led to its application in many laboratories as summarized in Table 1 and briefly outlined below.

The irradiation of subnuclear areas has also been applied in the field of nucleotide excision repair (NER) studies, based on conventional UV-lamps and microirradiation using polycarbonate pore-filters, a technique first introduced by Moné et al. (2001).

Pioneering work regarding partial volume irradiation with ionizing radiation as a source to induce DNA damage was reported by Nelms et al.

Table 1 Microirradiation assays for the visualization of spatiotemporal repair dynamics in mammalian cells. For particle irradiation, the characteristic physical parameters of the ions used are indicated

Technique				Refs.
X-rays				
Ultrasoft X-rays/gold microgrid				Nelms et al. 1998
UV-irradiation				
Local UV-damage using micropore filters				Moné et al. 2001, 2004
				Katsumi et al. 2001
				Volker et al. 2001
				Green and Almouzni 2003
				Okano et al. 2003
				Rademakers et al. 2003
				Rubbi and Milner 2003a,b
				Thorel et al. 2004
				Dunand-Sauthier et al. 2005
				Essers et al. 2005
Laser Microirradiation				
UVA (or two photon) laser irradiation of sensitized cells (halogenated thymidine analogs and Hoechst dye)				Rogakou et al. 1999
				Tashiro et al. 2000
				Walter et al. 2003a
				Lukas et al. 2003, 2004
				Celeste et al. 2003
				Fernandez-Capetillo et al. 2004
				Bradshaw et al. 2005
				Mortusewicz et al. 2005
				Pryde et al. 2005
				Bekker-Jensen et al. 2005
				Difilippantonio et al. 2005
UVA laser irradiation in non-sensitized cells				Lan et al. 2004
Laser irradiation with visible light (Nd:YAG laser) in non-sensitized cells				Kim et al. 2002, 2005
				Mikhailov et al. 2002
α-particle irradiation (^{241}Am; < 1.37 MeV/u)				Aten et al. 2004
Heavy-ion irradiation				
Ion Species	Energy (MeV/u)	LET* (keV/μm)	Dose (Gy)	
C to U	3.1–9.6	170–14 350	0.8–87	Jakob et al. 2000, 2002, 2003, 2005
Fe	95	1000	3	Goto et al. 2002
C; Ar; U	4.8–9.1	320–14 350	10–46	Taucher-Scholz et al. 2003
C; Ar	200–400	16–100	10	
N		125	0.5–2	Karlsson and Stenerlöw 2004
O	6.25	380		Hauptner et al. 2004
Si; Fe	581–600	54–176	0.5–1.7	Desai et al. 2005

* Based on the LET (linear energy transfer) the number of DSBs induced along the ion track can be roughly estimated according to Jakob et al. (2003); e.g. a LET of 170 keV/μm yields about 2.2 DSBs/μm assuming 35 DSBs/Gy after photon irradiation and a nuclear volume of about 400 μm^3 for G1 cells. The values given above can be correspondingly scaled.

(1998). Using synchrotron-generated ultrasoft X-rays and irradiation through a grid to compare irradiated areas with non-irradiated ones, the authors were the first to show the recruitment of the Mre11/Rad50/Nbs1 complex to DSB-containing nuclear subvolumes in human fibroblasts. Although more advanced and precise techniques have become available for focal irradiation, this study stimulated the subsequent research on the redistribution of repair proteins. The recent development of an ultrasoft X-ray microprobe further emphasizes the potential use of ultrasoft X-rays aiding this purpose (Folkard et al. 2001; Schettino et al. 2002)[1].

Recently, we have shown that the irradiation of cell nuclei with energetic heavy ions induces extremely localized DNA damage at the sites of particle traversal, as visualized by the recruitment of repair-related proteins to ion-induced foci (Jakob et al. 2000; 2002). As specified in the following, this provides a new tool for the production of locally confined subnuclear lesions using ionizing radiation.

The technical developments in relation to the induction of DNA damage in restricted areas of the nucleus have further been complemented by the use of GFP-tagged checkpoint and repair proteins to perform these studies in living cells. With real-time monitoring, great progress in studies on the protein dynamics at defined sites of damage can be achieved (Lukas et al. 2004; 2005).

In the following chapters we will present an overview of microirradiation assays that have recently been developed with the aim of studying the spatiotemporal dynamics of damage response proteins in mammalian cells. Special emphasis will be given to the application of ion irradiation as a tool serving this purpose. Several recent reports comprising current progress on this rapidly expanding topic are given in Table 1.

2
Local Irradiation Approaches to Reveal Spatiotemporal DNA Damage Responses

2.1
Microirradiation Using Laser Technology

In the early 1990s Limoli and Ward introduced a new method for the generation of DNA DSBs by UVA-irradiation of cell nuclei sensitized by incorporated halogenated thymidine analogs and/or intercalating bisbenzimide dyes (e.g. Hoechst 33 258) (Limoli and Ward 1993). On the basis of these sensitization mechanisms and the development and availability of UVA laser systems, microirradiation of subnuclear compartments and the generation of localized

[1] Ultrasoft X-rays: X-rays in the energy region below 5 keV, interacting in tissue to produce low-energy electrons with a very short range and high biological efficiency.

damage including DSBs were achieved by several groups (Table 1). In addition UVA (Lan et al. 2004) and Nd – YAG lasers (Kim et al. 2002; 2005; Mikhailov et al. 2002) have been used for that purpose without additional sensitization (Table 1). However, despite the similarities of laser-evoked effects to known DSB-induced checkpoint and repair factor activation (Kim et al. 2005), the cellular responses after such extensive damage might differ in timing and duration from those after more physiological doses of IR. The various laser-based microirradiation techniques used and the biological goals addressed were a constituent of a recent review by Lukas et al. (2005).

2.2
Localized Radiation-induced Damage Produced by Charged Particles

A primary feature characterizing charged particle radiation on a microscopic scale is the different spatial distribution of energy deposition compared to conventional ionizing radiation like photons. Accelerated ions interacting with matter produce tracks with extremely localized energy deposition and high local ionization density along the trajectory of the primary ion (Krämer and Kraft 1994; Nikjoo et al. 1998). The locally deposited dose decreases steeply (proportional to $1/r^2$) with increasing radial distance from the track center, whereby the limit of the radial dose distribution is determined by the specific energy of the primary particle and can accordingly be defined (Kiefer and Straaten 1986).

In view of these considerations, when charged particles hit a biological target like the cell nucleus, extremely high doses are locally deposited at the sites of the traversing tracks, while areas with minor or no dose deposition can occur a short radial distance away from the ion hits. This leads to an inhomogeneous spatial pattern of dose distribution, in contrast to photons where energy deposition events are randomly distributed throughout volumes of cellular dimensions. To allow for a more intuitive understanding of the particular effects of charged particles, the microscopic distributions of energy deposition are compared for photons and ions over an area typical of a mammalian cell nucleus (Fig. 2a). Noteworthy, for the low energy ions used the dimensions of the ionization tracks show maximum radial extensions typically around 1 μm, with the bulk of energy being deposited within a radial spread well below this value (Scholz 2003). Consequently, charged particles are potentially capable of intersecting very small fractions of the nuclear volume, in the range of biologically relevant structures of the nuclear chromatin environment.

Track structure and the associated distinct pattern of dose deposition are known to be critical determinants of the biological action of charged particle radiation (reviewed in Kraft 1987; Goodhead 1994; Scholz 2003). Over the last few years a large body of evidence has emerged indicating that the extremely localized energy deposition of accelerated ions leads to particularly

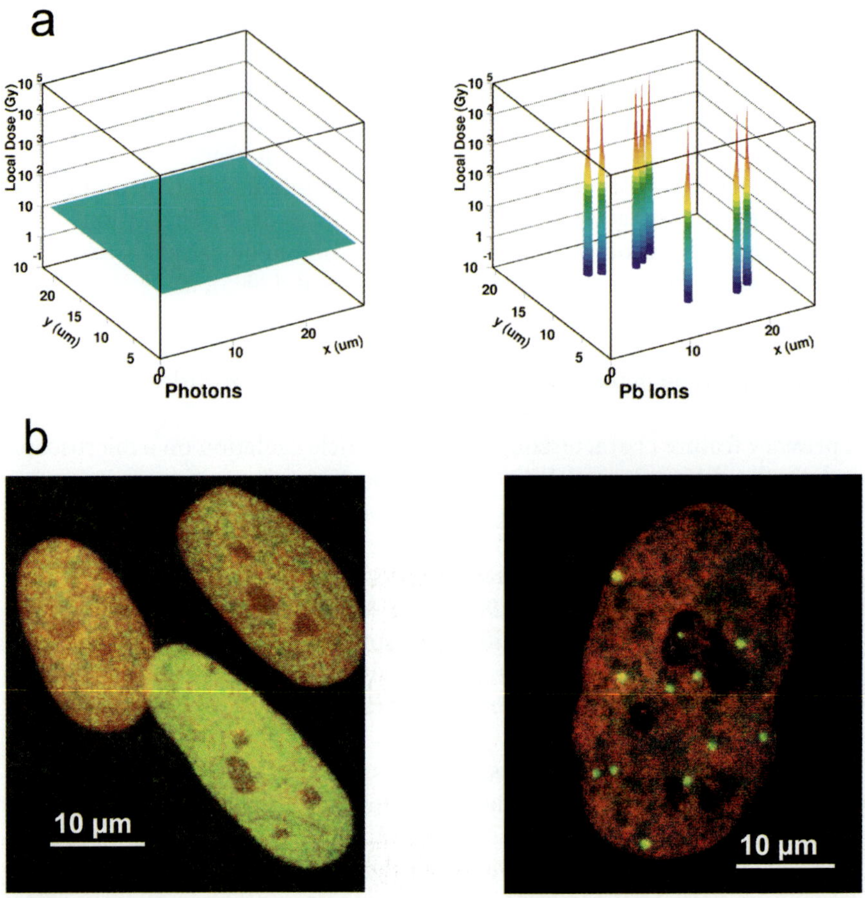

Fig. 2 Calculated and in situ observed radiation-response following irradiation of a cell nucleus with charged particles. **a** Simulation of the microscopic dose deposition after exposure of an area corresponding to the typical size of mammalian cell nuclei to photons (*left*) and low energy lead ions (2×10^6 p/cm^2, *right*). Note the logarithmic scale of the *z*-axis. **b** Biological visualization of the radiation-induced response of the CDKN1A (p21) protein in normal human fibroblast nuclei by means of immunocytochemical staining (for methods see Jakob et al. 2000). The images of immunostained CDKN1A (p21) (*green*) reflect the microscopic dose distribution of the corresponding radiation type: *Left*: After exposure to photons (10 Gy X-rays), at 90 min postirradiation a global increase in the nuclear CDKN1A signal intensity is observed. *Right*: Irradiation with 2×10^6 p/cm^2 uranium ions leads to a fast relocalization of CDKN1A to the sites of ion traversals, as observed already 2 min after irradiation (*green foci*). *Red*: DNA counterstain

complex and severe types of DNA damage (Löbrich et al. 1998; Stenerlöw et al. 2000). The production of so-called "clustered damage", consisting of multiple lesions like damaged bases, single- and double-strand breaks in close proximity (Ward 1985; Goodhead and Nikjoo 1989), has been theoretically predicted and also supported by experimental results (Jenner et al. 2001; Prise et al. 2001). However, only recently the spatial distribution of ion-induced biological damage could directly be visualized in subcellular dimensions (Jakob et al. 2000). Using appropriate immunofluorescence staining techniques based on antibodies against DNA damage response proteins, the damaged sites appear as discrete fluorescent foci that could be clearly assigned to the positions of particle traversals (Scholz et al. 2001). The pattern of fluorescent foci induced in the cell nucleus directly reflects the microscopic distribution of dose (Taucher-Scholz et al. 2003), as visualized in Fig. 2b for photons (left) and heavy ions (right). In contrast to the grainy but homogeneously distributed pattern of foci obtained after X-irradiation, discrete spots of damage can be observed in a human fibroblast nucleus after ion traversal, as predicted based on the expectation values of energy deposition depicted in Fig. 2a. A direct assignment of induced foci to individual ion hits is achieved by retrospective etching of tracks (Scholz et al. 2001), as shown below (Sect. 2.3). The correlation of induced foci and charged particle traversal was further confirmed by Karlsson and Stenerlöw (2004) using the DSB marker γ-H2AX for immunostaining. In addition, based on morphological criteria, Goto et al. (2002) have assigned foci induced by energetic ions to either simple or clustered type of damage. However, caution should prevail as the appearance depends on the protein studied (van Veelen et al. 2005). Nevertheless, the inherent ability of charged particles to produce discernable regions of damage within cell nuclei is confirmed throughout all of these studies.

Recently, a more precise targeting of charged particle radiation has become possible due to the application of ion microbeams (Brenner and Hall 2002). Microprobe facilities offer the advantage of delivering counted ions to individual cells within a population and, depending on the configuration, to defined regions within cells.

At the facilities of the Gray Cancer Institute (Prise et al. 2003) and Columbia University (Hall and Hei 2003) collimated proton and helium particle beams have been used for the defined irradiation of single cells or cellular compartments, mainly with the aim of studying the so-called bystander effect, i.e. the response of the neighboring non-irradiated cells (Morgan 2003). A charged particle microbeam delivering accelerated protons was used to locally irradiate mammalian cells to show the focal activation of poly(ADP-ribose) polymerase concomitant with the recruitment of repair proteins to sites of DNA damage (Tartier et al. 2003).

Microprobes capable of delivering heavier charged particles to individual biological cells have recently been described in various facilities (Hauptner et al. 2004; Funayama et al. 2005; Heiß et al. 2006). At the SNAKE microprobe in Munich, Hauptner et al. (2004) used oxygen ions to generate a broad geometrical pattern of irradiated spots that were visualized by Rad51 immunostaining. The focused heavy ion microbeam produced at the GSI facility was used for accurate targeting of mammalian cell nuclei with a defined

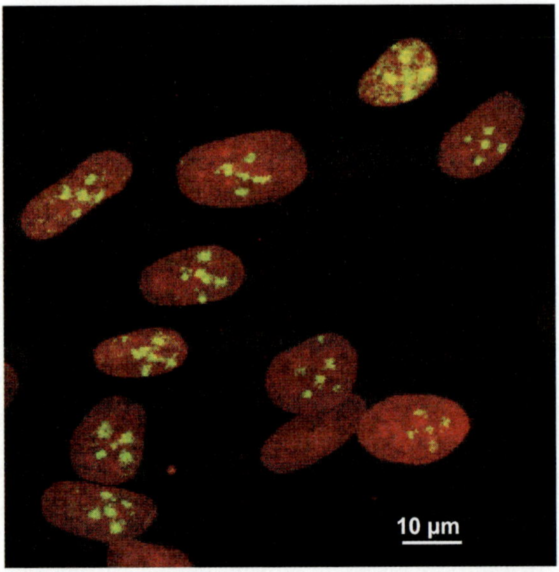

Fig. 3 Aimed irradiation with heavy ions at the GSI heavy-ion microprobe. Human fibroblast nuclei are shown after defined irradiation with 4.8 MeV/u carbon ions in a cross formation (5 hits/spot) with 3 μm distance between the positions. After irradiation, fixation (30 min postirradiation) and immunostaining for γ-H2AX (*green*), the targeted cells show a clearly identifiable cross pattern of ion-induced foci. *Red*: DNA counterstain

pattern and number of ions (Heiß et al. 2006). Exemplarily, the irradiation of fibroblast nuclei with a cross formation of ion spots is depicted in Fig. 3. This type of irradiation producing defined subnuclear patterns of damage emphasizes the remarkable potential of charged particles as a tool to approach spatio-temporal aspects of the DNA damage response.

2.3 Spatio-temporal Protein Dynamics at Sites of Ion-induced DNA Damage

Despite the increasing knowledge of biochemical mechanisms and proteins involved in the processing of DNA lesions, our understanding of these events regarding succession in space and time is rather limited (Lukas et al. 2005). Especially in the case of charged particle irradiation, where we have to deal with restricted regions of damage comprising multiple nearby DNA lesions of various types, the order and potential sterical interference of the participating proteins, possibly belonging to different repair pathways, is quite unclear. A basic question in the context of the repair of clustered lesions is whether different types of DNA damage are recognized and processed independently or, if this is not the case, by which means the cell regulates the priority of the damage response actions.

A way to address the hierarchical and spatio-temporal organization of DNA lesion processing is studying early protein aggregation at sites of ion traversals. For this purpose we developed a remotely controlled microscopic system located directly at the beamline, capable of imaging early responses during irradiation with heavy ions in living cells (Jakob et al. 2005). The use of live cell microscopy in combination with fusion constructs of the protein of interest with derivatives of the green fluorescent protein, circumvents limitations related to immunostaining techniques as a rather slow and static approach. In different systems the application of this technique has led to a plethora of new insights regarding the dynamics of cellular protein responses.

To determine the kinetics of protein accumulation in so called IRIF after ion impacts, we used for example cells expressing the protein chimera of enhanced green fluorescent protein (EGFP) and aprataxin. The recruitment of this repair-related protein to ion-induced DNA lesions has recently been described (Gueven et al. 2004). Representative time-lapse images of an EGFP-aprataxin-expressing A549 cell are shown for selected time points from a sequence obtained with the new beamline microscope before and after irradiation with low energy uranium ions (Fig. 4a). To cope with the fast evolution of the focal signal, a frame rate of 0.26 frames per second was chosen and the irradiation time was constricted to about 3 s. The results were similar to those described after nickel ion irradiation of EGFP-aprataxin-expressing HeLa cells (Jakob et al. 2005) showing a fast recruitment of aprataxin to sites of ion traversals. From the normalized ratio of the focal signal intensity to the global nuclear intensity, the dynamical development of the recruitment process can be quantified by fitting mathematical saturation curves to the data points (Lukas et al. 2004; Bekker-Jensen et al. 2005). The data points for the kinetics of aprataxin recruitment are exemplarily shown in Fig. 4d, using the time series depicted in Fig. 4a. For the GFP-aprataxin recruitment, a one parameter saturation function yielded a time constant of about 30 s after irradiation with 4.2 MeV/u nickel ions[2] (Jakob et al. 2005).

Notably, once the protein aggregations at the ion tracks have formed, they remained stable in position during the course of the observation time of several minutes (Fig. 4a). This observation is not supportive of major changes in chromatin structure after the ion impact, a topic discussed below in more detail.

In order to confirm that the fluorescent protein aggregations corresponded to sites of ion traversals, the cells were fixed after the observation period of 5 min and immunostained for the DSB-marker γ-H2AX (Fig. 4b). In addition, the polycarbonate foil used as the base of the irradiation chamber served also

[2] 4.2 MeV/u nickel ions: Accelerated, positively charged nickel ions with a specific kinetic energy of 4.2 MeV per atomic mass unit. For more details see Table 1.

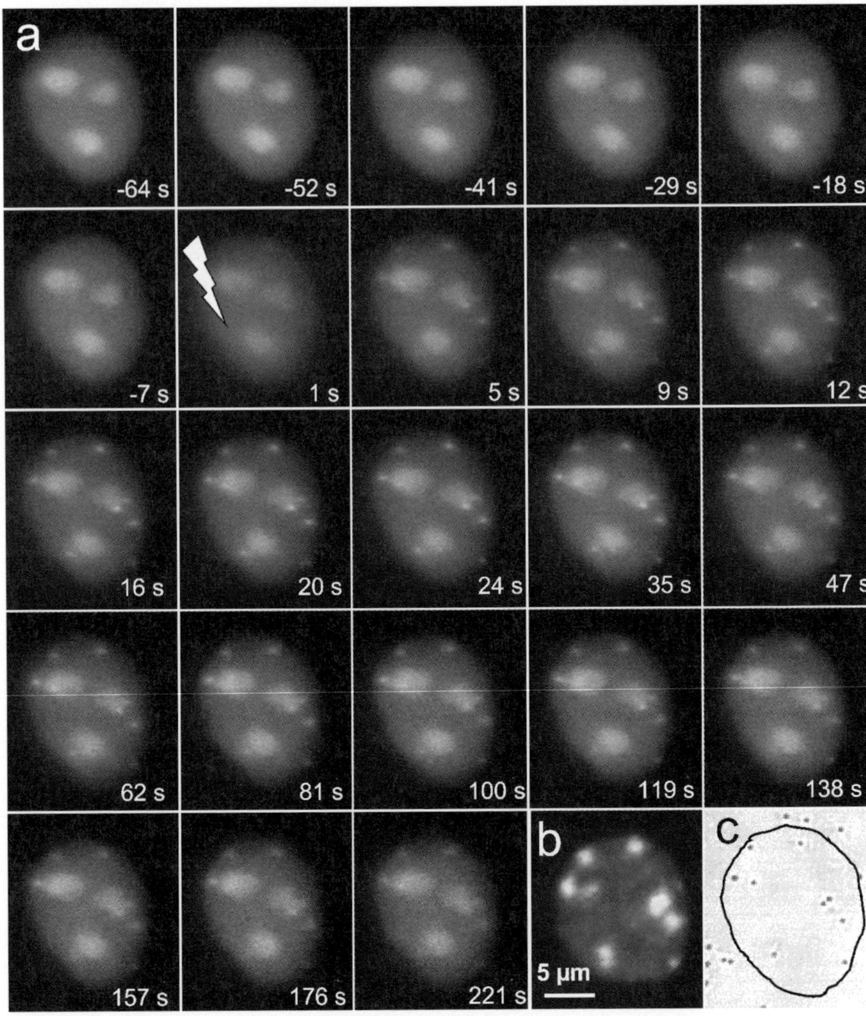

Fig. 4 Measurement of the kinetics of protein recruitment after ion irradiation at the beamline microscope. **a** Representative images of a time-lapse experiment showing an EGFP-aprataxin expressing A549 cell nucleus during irradiation with low energy uranium ions (cells were kindly provided by N. Gueven, Queensland Institute of Medical Research, Australia). The image acquisition rate was 0.26 Hz. Time intervals are indicated relative to the start of irradiation at time point 0 s. The *arrow* indicates the first image taken during the irradiation period of 3 s. The accumulation of EGFP-aprataxin can be detected in less than 5 s after irradiation. **b** Same nucleus as in a after fixation and immunostaining for γ-H2AX showing the presence of DSBs at sites of ion traversal. Cells were fixed 7 min after irradiation. **c** The polycarbonate foil used as the cellular substrate served as a nuclear track detector for the retrospective determination of ion traversals (*dots*) (see Jakob et al. 2005). Each traversal can be attributed to both an aprataxin and a γ-H2AX focus of a and b, respectively. *Thick line*: outline of the nucleus of a

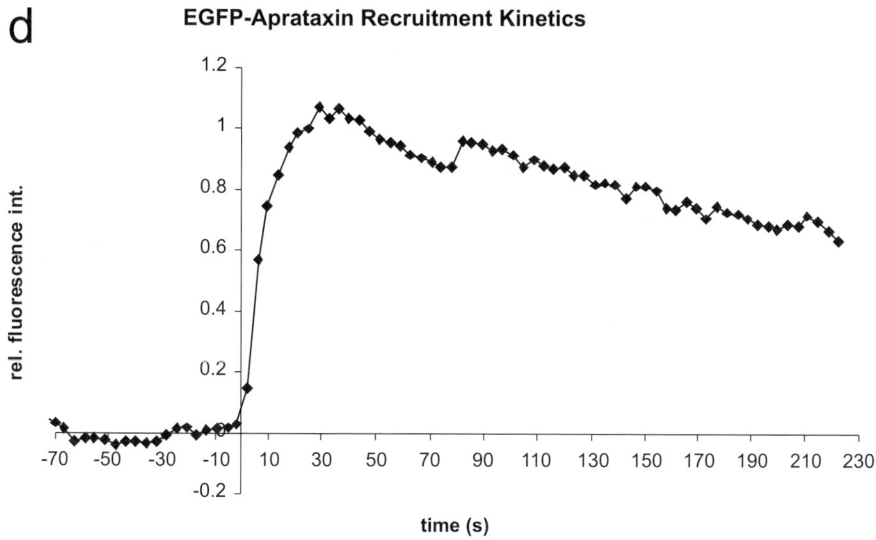

Fig. 4 d Kinetics of EGFP-aprataxin recruitment determined from the images of the A549 nucleus in Fig. 4a. The normalized fluorescence intensity at the sites of ion traversals is plotted as a function of time to derive the kinetic parameters

as a nuclear track detector by etching the latent tracks with ethanolic NaOH to show the positions of traversing ions at the location of the nucleus (Fig. 4c). Matching of the various patterns indeed demonstrates that GFP-protein accumulations as well as γ-H2AX co-localize with the ion tracks. Moreover, by comparing the inter-track distances on the polycarbonate detector to the observed inter-foci distances of the protein aggregates, possible displacements of irradiation induced foci with time would become apparent, but were not observed during the first minutes after the ion insults.

2.4 Introduction of Linear Tracks of Damage Along the Ion Trajectory in the Cell Nucleus

In addition to the discrete spots of localized damage produced by incident ions hitting the nucleus perpendicular to the plane of the cells, visualization of the biological response along the particle trajectory has been made possible by changing the irradiation geometry. In contrast to common irradiation setups, the incoming beam direction is applied at a very small angle with respect to the cell layer, leading to the production of linear tracks of damage across the cell nuclei (Jakob et al. 2003). These tracks can be visualized by immunostaining of repair-related proteins (e.g. PCNA, Mre11; γ-H2AX) revealing streaks of fluorescent protein clusters but displaying grainy substructures,

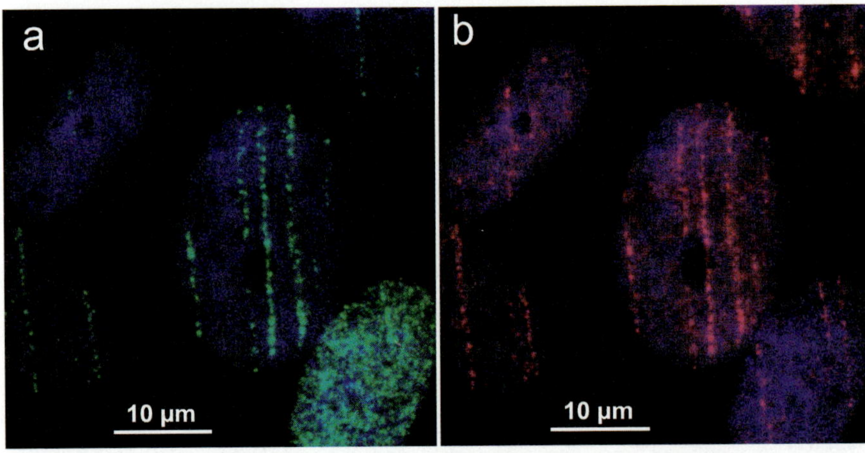

Fig. 5 Visualization of repair protein accumulation along ion trajectories. Human fibroblasts were exposed to 250 MeV/nucleon uranium ions, with the incoming beam applied parallel to the cell monolayer (for details see Jakob et al. 2003). Immunocytochemical staining of PCNA (*green*) and Mre11 (*red*) in normal human fibroblasts 30 min after irradiation reveals streaks of protein aggregates stretching across the nucleus. S-phase cells are indicated by intense green PCNA staining over the whole nucleus (*lower right*). *Blue*: DNA counterstain

with protein aggregates and gaps alternating along the ion trajectories, as illustrated in Fig. 5.

The analysis of the spatial distribution of repair protein clusters along charged particle trajectories in fibroblast cells has pointed out some notable features. On the one hand, some variability of the substructure of the induced streaks is observed even for multiple tracks of a particular ion within the same cell. However, strikingly similar patterns of protein clustering are obtained when varying ion species of varying linear energy transfer (LET) are used for irradiation. The LET of the charged particle represents the rate of energy deposition in the longitudinal direction of the traversed material and characterizes the particle in terms of the ionization density along the track. For the purpose of simplification, the particle's LET can be considered proportional to the expected average density of lesions produced, e.g. DSBs induced per unit track length (Krämer and Kraft 1994). Accordingly, the choice of adequate charged particle beams allows selective modulation of the density of lesions induced along the track. Comparing different ion species covering a wide range of LETs, it seems remarkable that charged particle tracks differing over 40fold in the density of induced DSBs show quite comparable distributions of alternating protein clusters and gaps along the particle trajectories (Jakob et al. 2003). Varying patterns of linear tracks were also detected by Aten and coworkers (Aten et al. 2004) using alpha-particle irradiation; but the overall patterns of clustering observed were again similar to those obtained using heavier ions. In a study aimed at testing the influence of DNA damage density using high energy ion beams, Desai et al. (2005) reported a distinction between two ion species differing 3fold in LET, based on the qualitative analysis of γ-H2AX immunosignals induced along the particle tracks at different times after irradiation.

Moreover, similar patterns of protein aggregation are also observed when proteins expected to concentrate at different repair sites due to their role in different repair pathways, are used for immunostaining (Jakob et al. 2003). Specifically, PCNA (required for excision repair) and γ-H2AX (a well known DSB marker) show a clear colocalization of fluorescent signals along the linear tracks, albeit with the generally more spread out appearance of γ-H2AX foci (Jakob et al. 2003). Although the yields of ionizing radiation-induced lesions associated to each of these repair pathways are very different, this is not reflected in the density of protein foci detected along the particle tracks, giving further indication that the protein aggregates are not directly correlated to the individual lesions induced. Rather, such a behavior not reproducing the stochastics in the distribution of induced damage neither for increasing LET nor for different lesions, illustrates a non-random formation of repair clusters possibly determined by the underlying chromatin organization. Hence the results are supportive of a link between damage response processes and structural chromatin effects (Gontijo et al. 2003; Koundrioukoff et al. 2004).

Interestingly, clear changes in track morphology with time are observed when cells containing linear tracks of damage are incubated for repair. One effect is the disappearance of repair foci with varying timing for proteins belonging to different pathways (Jakob et al. 2002). In addition, changes in size and distribution of foci, suggesting a migration and merging of damaged chromatin domains, have been reported during lesion processing using α-particle irradiated HeLa cells (Aten et al. 2004) or human fibroblasts exposed to high energy heavy ions (Desai et al. 2005). A possible influence of LET and cell type on these effects still needs to be explored. On the other hand, the variability of foci and the heterogeneous morphology of individual patterns of protein clustering along linear tracks as supported by all studies are complicating the interpretation. In conclusion, the complexity of the results emphasizes the need of real time monitoring to address this topic.

3
Role of Structural Chromatin Changes in the DNA Damage Response

Growing evidence indicates that the dynamic organization and the accessibility of chromatin are closely related to many nuclear processes including DSB repair (Gontijo et al. 2003). Indeed, different types of post-translational modifications of histone proteins and the action of chromatin remodelling complexes have been shown to play important roles in facilitating DNA repair (reviewed in Gontijo et al. 2003; Peterson and Cote 2004).

The variable but non-random distribution of protein clusters detected along charged particle trajectories (see above Sect. 2.4) also suggests a connection between the spatial organization of DNA damage response factors and the underlying chromatin topology. The irradiation-induced foci visualized by indirect immunofluorescence, most likely containing thousands of molecules of repair factors, are not reflecting the individual induced DSBs, but may represent the gathering of multiple DNA lesions, as also suggested by others (Lisby and Rothstein 2004; Aten et al. 2004). The intriguing question is whether this apparent clustering of lesions within protein foci is a consequence of local chromatin heterogeneity or rather mediated by radiation-induced changes in chromatin structure.

Chromosomal domains normally become stably positioned at early G1 and undergo only locally confined movements during the remaining interphase, without being redistributed to other nuclear sites (Zink et al. 1998; Dimitrova and Gilbert 1999; Sadoni et al. 1999; Chubb et al. 2002; Walter et al. 2003b). This applies also to the S phase, during which the DNA remains relatively immobile, while the replication machinery assembles at changing nuclear sites (Leonhardt et al. 2000; Sporbert et al. 2002; Sadoni et al. 2004). A still open question after DNA injury is specifically to which extent visible chromatin dynamics participates in the recognition, repair and signalling processes involved in the repair of DNA lesions. In this context, different types of lesions and different levels of complexity, as induced by different types of radiation, are likely to have distinct chromatin modulation requirements for the detection and processing of DNA damage. In the case of NER, bending of DNA by bulky lesions has been suggested to be the initial signal for the binding of repair protein complexes (Wood 1999; Sugasawa et al. 2001). Furthermore, at least on the nucleosomal level, changes of chromatin structure by unwinding have been described in order to increase the accessibility of lesions (Gontijo et al. 2003). Interestingly, damage-induced changes of chromatin structure itself have recently been proposed as a primary mechanism in damage sensing and triggering of signal transduction cascades (Bakkenist and Kastan 2003). On the other hand, global changes of chromatin compaction have also been postulated to occur after local damage (Rubbi and Milner 2003a).

Besides the impact of chromatin organization on the accessibility to repair factors, the stability of DNA positioning after the introduction of damage is an important issue, especially crucial concerning the repair of DSBs in damaged chromatin. As non-homologous end joining is the predominant pathway of DSB repair in somatic mammalian cells particularly in the G1 phase (Rothkamm et al. 2003), positional changes of DNA might lead to the joining of non-equivalent ends in cells harboring more than one DSB. Such misrejoining would lead to chromosomal rearrangements and translocations.

On the other hand, repair by homologous recombination is likely to require chromatin alterations to facilitate the homology search or the strand invasion and branch migration reactions (Paques and Haber 1999). A certain chromatin flexibility could provide the basis for homology search if the sister chromatid is not in the immediate vicinity. In fact, the interaction with distant regions of template DNA may require considerable movement, and the accessibility of the template DNA may influence the homology search process, as suggested by the Synthesis Dependent Strand Annealing model for recombinational repair (Lankenau and Gloor 1998; Paques and Haber 1999). Links between chromatin remodelling and homologous recombination are indeed suggested by biochemical and genetic studies in yeast (Peterson and Cote 2004) and *Drosophila* (Lankenau et al. 2000).

In connection with the introduction of linear tracks of damage in mammalian cells nuclei, in situ support for the mobility of damaged chromosome domains arises from changes in track morphology with time. In contrast to the stable positioning of DNA observed in undamaged cells, recent studies monitoring γ-H2AX foci along particle trajectories suggest substantial movement of the DSB-containing chromatin domains in the course of lesion processing, starting within minutes after irradiation (Aten et al. 2004; Desai et al. 2005). The clustering and merging of protein foci detected in a subpopulation of cells on a time scale of hours would be in line with the moving together of damaged chromatin-ends, providing a basis for the generation of complex chromosomal translocations (Anderson et al. 2002; Aten et al. 2004).

However, the studies discussed above were performed with fixed cells, rendering the interpretation of dynamic processes often difficult due to the huge variability of the staining patterns appearing. Only live cell studies are capable of giving a definite answer to the question how the components involved (DNA and repair/signalling factors) move in space and time. At a time scale early after irradiation, these studies can be performed with the aid of the beamline microscope described previously (Jakob et al. 2005; see above). Therefore, taking advantage of the production of locally confined DNA damage by charged particle irradiation, the question whether a radiation-induced chromatin distortion occurred directly after ion impact was addressed in living cells. Such chromatin remodelling could contribute to the recognition of DNA lesions and the assembly of multi-protein complexes indicated by the formation of IRIF.

As it is difficult to measure and analyze motion in uniformly labelled structures like intercalating chromatin dyes, radiation-induced changes were investigated in the punctuated pattern generated in nuclei of human fibroblasts by scratch replication labelling (Zink et al. 1998; Schermelleh et al. 2001). The uptake and incorporation of fluorescently labelled thymidine analogs during the S-phase leads to fluorescent signals in the chromatin depending on the progression in the S-phase. These dots can rather easily be tracked in space and time, a feature also exploited in fluorescent speckle microscopy (Danuser and Waterman-Storer 2003). Using this technique in combination with beamline microscopy, no radiation-induced motion exceeding the generally observed constrained Brownian motion type was apparent during the first minutes after the ion insults (Fig. 6). The figure displays representative images of a time series of Alexa488-dUTP labelled human fibroblast nuclei during the irradiation with nickel ions. The insert shows the color coded transition (green to red) of the replication foci during 10 minutes of observation, indicating none (yellow) or only minor displacement. A more detailed analysis of the movement of selected foci during irradiation with low energy carbon ions is described in Jakob

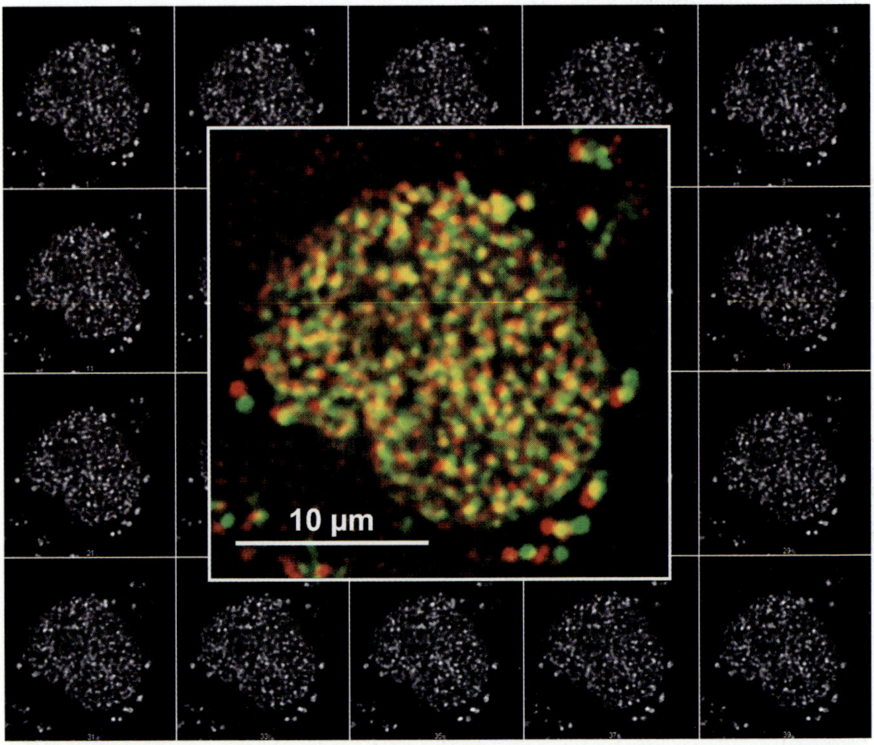

Fig. 6 Assessment of radiation-induced chromatin movement. Fluorescence-labelled nucleotides incorporated in nascent DNA during the S-phase by scratch replication techniques yield a pattern of bright fluorescent speckles which can be imaged and traced over time. *Background*: Representative images of a time-lapse experiment showing Alexa488-dUTP labelled chromatin in a human AG1522 fibroblast during irradiation with low energy nickel ions. Images were taken every 30 s going from *upper left* to *lower right*. Despite a general billowing movement of the speckles, no major displacements can be observed within the first 10 min after irradiation. *Overlay*: *Color*-coded displacement of the speckles (*green* before irradiation; *red* 8 min after irradiation). Only small displacements (< 1 μm) take place in the nucleus, independent of irradiation

et al. 2005. Furthermore, up to now, all these experiments gave neither indication of a local traversal-induced chromatin condensation being responsible for the formation of foci, nor supported a global chromatin relaxation serving as a cellular signal as proposed by Rubbi and Milner (2003a) after locally confined UV-irradiation. However, local changes in chromatin structure restricted to the sites of induced damage may still occur. Exploring the cellular response to DSBs in the context of chromatin is clearly a future challenge.

4
Concluding Remarks and Future Directions

The reorganization of checkpoint and repair proteins into focal assemblies at sites of DNA damage has provided the basis for new experimental means helping to elucidate the molecular events involved in the DNA damage response. Specifically in combination with techniques allowing production of locally confined subcellular damage, immunofluorescent analyses offer the opportunity to study spatiotemporal relationships between proteins participating in repair and damage signalling.

The development of DSB-generating lasers over the last years has successfully introduced this technology in the field of repair as a readily available tool for focal irradiation (Lukas et al. 2005). Most recently, charged particle irradiation has emerged as a powerful approach to generate locally restricted IR-induced lesions in subnuclear dimensions (Jakob et al. 2003; Hauptner et al. 2004; Aten et al. 2004). Besides the better known nature of the type of damage produced, the defined intrinsic physical properties of ion beams offer the advantage of modulating the density of lesions along the track through the variation of the charged particle species and LET.

The application of the focal irradiation techniques to studies on the recruitment of DNA repair factors to sites of damage has provided new insights into the cellular response to DNA damage. It is important to characterize further the dynamics of individual components as well as their spatial architecture and molecular function within repair foci. In particular, attempts must be emphasized to monitor these processes in real-time within living cells.

Finally, the use of charged particles to produce locally confined regions of DNA damage can provide an important contribution to the study of small-scale movements of damaged chromatin domains. This topic is highly relevant to the elucidation of key events like the initial activation of the DNA damage response and the accessibility of repair factors to the damaged sites.

References

Anderson RM, Stevens DL, Goodhead DT (2002) M-FISH analysis shows that complex chromosome aberrations induced by α-particle tracks are cumulative products of localized rearrangements. Proc Natl Acad Sci USA 99:12167–12172

Aten JA, Stap J, Krawczyk PM, Van Oven CH, Hoebe RA, Essers J, Kanaar R (2004) Dynamics of DNA double-strand breaks revealed by clustering of damaged chromosome domains. Science 303:92–95

Bakkenist CJ, Kastan MB (2003) DNA damage activates ATM through intermolecular autophosphorylation and dimer dissociation. Nature 421:499–506

Bartek J, Lukas C, Lukas J (2004) Checking on DNA damage in S phase. Nat Rev Mol Cell Biol 5:792–804

Bekker-Jensen S, Lukas C, Melander F, Bartek J, Lukas J (2005) Dynamic assembly and sustained retention of 53BP1 at the sites of DNA damage are controlled by Mdc1/NFBD1. J Cell Biol 170:201–211

Berns MW, Olson RS, Rounds DE (1969) In vitro production of chromosomal lesions with an argon laser microbeam. Nature 221:74–75

Bernstein C, Bernstein H, Payne CM, Garewal H (2002) DNA repair/pro-apoptotic dual-role proteins in five major DNA repair pathways: fail-safe protection against carcinogenesis. Mutat Res 511:145–178

Bradshaw PS, Stavropoulos DJ, Meyn MS (2005) Human telomeric protein TRF2 associates with genomic double-strand breaks as an early response to DNA damage. Nat Genet 37:193–197

Brenner DJ, Hall EJ (2002) Microbeams: a potent mix of physics and biology. Summary of the 5th International Workshop on Microbeam Probes of Cellular Radiation Response. Radiat Prot Dosimetry 99:283–286

Celeste A, Fernandez-Capetillo O, Kruhlak MJ, Pilch DR, Staudt DW, Lee A, Bonner RF, Bonner WM, Nussenzweig A (2003) Histone H2AX phosphorylation is dispensable for the initial recognition of DNA breaks. Nat Cell Biol 5:675–679

Christmann M, Tomicic MT, Roos WP, Kaina B (2003) Mechanisms of human DNA repair: an update. Toxicology 193:3–34

Chubb JR, Boyle S, Perry P, Bickmore WA (2002) Chromatin motion is constrained by association with nuclear compartments in human cells. Curr Biol 12:439–445

Cremer C, Cremer T (1986) Induction of chromosome shattering by ultraviolet light and caffeine: the influence of different distributions of photolesions. Mutat Res 163:33–40

Cremer T, Cremer C, Schneider T, Baumann H, Hens L, Kirsch-Volders M (1982) Analysis of chromosome positions in the interphase nucleus of Chinese hamster cells by laser-UV-microirradiation experiments. Hum Genet 62:201–209

Danuser G, Waterman-Storer CM (2003) Quantitative fluorescent speckle microscopy: where it came from and where it is going. J Microsc 211:191–207

Dasika GK, Lin SC, Zhao S, Sung P, Tomkinson A, Lee EY (1999) DNA damage-induced cell cycle checkpoints and DNA strand break repair in development and tumorigenesis. Oncogene 18:7883–7899

Desai N, Davis E, O'Neill P, Durante M, Cucinotta FA, Wu H (2005) Immunofluorescence detection of clustered γ-H2AX foci induced by HZE-particle radiation. Radiat Res 164:518–522

Difilippantonio S, Celeste A, Fernandez-Capetillo O, Chen HT, Reina SM, Van Laethem F, Yang YP, Petukhova GV, Eckhaus M, Feigenbaum L, Manova K, Kruhlak M, Camerini-Otero RD, Sharan S, Nussenzweig M, Nussenzweig A (2005) Role of Nbs1 in the activation of the Atm kinase revealed in humanized mouse models. Nat Cell Biol 7:675–685

Dimitrova DS, Gilbert DM (1999) The spatial position and replication timing of chromosomal domains are both established in early G1 phase. Mol Cell 4:983–993

Dunand-Sauthier I, Hohl M, Thorel F, Jaquier-Gubler P, Clarkson SG, Scharer OD (2005) The spacer region of XPG mediates recruitment to nucleotide excision repair complexes and determines substrate specificity. J Biol Chem 280:7030–7037

Essers J, Theil AF, Baldeyron C, van Cappellen WA, Houtsmuller AB, Kanaar R, Vermeulen W (2005) Nuclear dynamics of PCNA in DNA replication and repair. Mol Cell Biol 25:9350–9359

Fernandez-Capetillo O, Chen HT, Celeste A, Ward I, Romanienko PJ, Morales JC, Naka K, Xia Z, Camerini-Otero RD, Motoyama N, Carpenter PB, Bonner WM, Chen J, Nussenzweig A (2002) DNA damage-induced G2-M checkpoint activation by histone H2AX and 53BP1. Nat Cell Biol 4:993–997

Fernandez-Capetillo O, Lee A, Nussenzweig M, Nussenzweig A (2004) H2AX: the histone guardian of the genome. DNA Repair (Amst) 3:959–967

Folkard M, Schettino G, Vojnovic B, Gilchrist S, Michette AG, Pfauntsch SJ, Prise KM, Michael BD (2001) A Focused Ultrasoft X-Ray Microbeam for Targeting Cells Individually with Submicrometer Accuracy. Radiat Res 156:796–804

Funayama T, Wada S, Kobayashi Y, Watanabe H (2005) Irradiation of mammalian cultured cells with a collimated heavy-ion microbeam. Radiat Res 163:241–246

Gontijo AM, Green CM, Almouzni G (2003) Repairing DNA damage in chromatin. Biochimie 85:1133–1147

Goodhead DT, Nikjoo H (1989) Track structure analysis of ultrasoft X-rays compared to high- and low-LET radiations. Int J Radiat Biol 55:513–529

Goodhead DT (1994) Initial events in the cellular effects of ionizing radiations: clustered damage in DNA. Int J Radiat Biol 65:7–17

Goto S, Watanabe M, Yatagai F (2002) Delayed cell cycle progression in human lymphoblastoid cells after exposure to high-LET radiation correlates with extremely localized DNA damage. Radiat Res 158:678–686

Green CM, Almouzni G (2003) Local action of the chromatin assembly factor CAF-1 at sites of nucleotide excision repair in vivo. EMBO J 22:5163–5174

Gueven N, Becherel OJ, Kijas AW, Chen P, Howe O, Rudolph JH, Gatti R, Date H, Onodera O, Taucher-Scholz G, Lavin MF (2004) Aprataxin, a novel protein that protects against genotoxic stress. Hum Mol Genet 13:1081–1093

Haaf T, Golub EI, Reddy G, Radding CM, Ward DC (1995) Nuclear foci of mammalian Rad51 recombination protein in somatic cells after DNA damage and its localization in synaptonemal complexes. Proc Natl Acad Sci USA 92:2298–2302

Hall EJ, Hei TK (2003) Genomic instability and bystander effects induced by high-LET radiation. Oncogene 22:7034–7042

Hauptner A, Dietzel S, Drexler GA, Reichart P, Krucken R, Cremer T, Friedl AA, Dollinger G (2004) Microirradiation of cells with energetic heavy ions. Radiat Environ Biophys 42:237–245

Heiß M, Fischer BE, Jakob B, Fournier C, Becker G, Taucher-Scholz G (2006) Aimed irradiation of mammalian cells using a heavy ion microprobe. Radiat Res 165:231–239

Hoeijmakers JH (2001) DNA repair mechanisms. Maturitas 38:17–22

Iliakis G, Wang Y, Guan J, Wang H (2003) DNA damage checkpoint control in cells exposed to ionizing radiation. Oncogene 22:5834–5847

Jackson SP (2002) Sensing and repairing DNA double-strand breaks. Carcinogenesis 23:687–696

Jakob B, Scholz M, Taucher-Scholz G (2000) Immediate localized CDKN1A (p21) radiation response after damage produced by heavy-ion tracks. Radiat Res 154:398–405

Jakob B, Scholz M, Taucher-Scholz G (2002) Characterization of CDKN1A (p21) binding to sites of heavy-ion-induced damage: colocalization with proteins involved in DNA repair. Int J Radiat Biol 78:75–88

Jakob B, Scholz M, Taucher-Scholz G (2003) Biological imaging of heavy charged-particle tracks. Radiat Res 159:676–684

Jakob B, Rudolph JH, Gueven N, Lavin MF, Taucher-Scholz G (2005) Live cell imaging of heavy-ion-induced radiation responses by beamline microscopy. Radiat Res 163:681–690

Jenner TJ, Fulford J, O'Neill P (2001) Contribution of base lesions to radiation-induced clustered DNA damage: implication for models of radiation response. Radiat Res 156:590–593

Karlsson KH, Stenerlöw B (2004) Focus formation of DNA repair proteins in normal and repair-deficient cells irradiated with high-LET ions. Radiat Res 161:517–527

Kastan MB, Bartek J (2004) Cell-cycle checkpoints and cancer. Nature 432:316–323

Katsumi S, Kobayashi N, Imoto K, Nakagawa A, Yamashina Y, Muramatsu T, Shirai T, Miyagawa S, Sugiura S, Hanaoka F, Matsunaga T, Nikaido O, Mori T (2001) In situ visualization of ultraviolet-light-induced DNA damage repair in locally irradiated human fibroblasts. J Invest Dermatol 117:1156–1161

Khanna KK, Jackson SP (2001) DNA double-strand breaks: signaling, repair and the cancer connection. Nat Genet 27:247–254

Kraft G (1987) Radiobiological effects of very heavy ions: inactivation, induction of chromosome aberrations and strand breaks. Nucl Sci Appl 3:1–28

Kiefer J, Straaten H (1986) A model of ion track structure based on classical collision dynamics. Phys Med Biol 31:1201–1209

Kim JS, Krasieva TB, LaMorte V, Taylor AM, Yokomori K (2002) Specific recruitment of human cohesin to laser-induced DNA damage. J Biol Chem 277:45149–45153

Kim JS, Krasieva TB, Kurumizaka H, Chen DJ, Taylor AM, Yokomori K (2005) Independent and sequential recruitment of NHEJ and HR factors to DNA damage sites in mammalian cells. J Cell Biol 170:341–347

Koundrioukoff S, Polo S, Almouzni G (2004) Interplay between chromatin and cell cycle checkpoints in the context of ATR/ATM-dependent checkpoints. DNA Repair (Amst) 3:969–978

Krämer M, Kraft G (1994) Calculations of heavy-ion track structure. Radiat Environ Biophys 33:91–109

Lan L, Nakajima S, Oohata Y, Takao M, Okano S, Masutani M, Wilson SH, Yasui A (2004) In situ analysis of repair processes for oxidative DNA damage in mammalian cells. Proc Natl Acad Sci USA 101:13738–13743

Lankenau DH, Gloor GB (1998) In vivo gap repair in *Drosophila*: a one-way street with many destinations. BioEssays 20:317–327

Lankenau DH, Peluso MV, Lankenau S (2000) The Su(Hw) chromatin insulator protein alters double-strand break repair frequencies in the *Drosophila* germ line. Chromosoma 109:148–160

Lavin MF, Birrell G, Chen P, Kozlov S, Scott S, Gueven N (2005) ATM signaling and genomic stability in response to DNA damage. Mutat Res 569:123–132

Leonhardt H, Rahn HP, Weinzierl P, Sporbert A, Cremer T, Zink D, Cardoso MC (2000) Dynamics of DNA replication factories in living cells. J Cell Biol 149:271–280

Limoli CL, Ward JF (1993) A new method for introducing double-strand breaks into cellular DNA. Radiat Res 134:160–169

Lisby M, Rothstein R (2004) DNA repair: keeping it together. Curr Biol 14: R994–R996

Löbrich M, Cooper PK, Rydberg B (1998) Joining of correct and incorrect DNA ends at double-strand breaks produced by high-linear energy transfer radiation in human fibroblasts. Radiat Res 150:619–626
Löbrich M, Jeggo PA (2005) Harmonising the response to DSBs: a new string in the ATM bow. DNA Repair (Amst) 4:749–759
Lukas C, Falck J, Bartkova J, Bartek J, Lukas J (2003) Distinct spatiotemporal dynamics of mammalian checkpoint regulators induced by DNA damage. Nat Cell Biol 5:255–260
Lukas C, Melander F, Stucki M, Falck J, Bekker-Jensen S, Goldberg M, Lerenthal Y, Jackson SP, Bartek J, Lukas J (2004) Mdc1 couples DNA double-strand break recognition by Nbs1 with its H2AX-dependent chromatin retention. EMBO J 23:2674–2683
Lukas C, Bartek J, Lukas J (2005) Imaging of protein movement induced by chromosomal breakage: tiny local lesions pose great global challenges. Chromosoma 114:146–154
Maser RS, Monsen KJ, Nelms BE, Petrini JH (1997) hMre11 and hRad50 nuclear foci are induced during the normal cellular response to DNA double-strand breaks. Mol Cell Biol 17:6087–6096
McManus KJ, Hendzel MJ (2005) ATM-dependent DNA Damage-independent Mitotic Phosphorylation of H2AX in Normally Growing Mammalian Cells. Mol Biol Cell 16:5013–5025
Mikhailov A, Cole RW, Rieder CL (2002) DNA damage during mitosis in human cells delays the metaphase/anaphase transition via the spindle-assembly checkpoint. Curr Biol 12:1797–1806
Mirzoeva OK, Petrini JH (2001) DNA damage-dependent nuclear dynamics of the Mre11 complex. Mol Cell Biol 21:281–288
Modesti M, Kanaar R (2001) DNA repair: spot(light)s on chromatin. Curr Biol 11:R229–R232
Mone MJ, Volker M, Nikaido O, Mullenders LH, van Zeeland AA, Verschure PJ, Manders EM, van Driel R (2001) Local UV-induced DNA damage in cell nuclei results in local transcription inhibition. EMBO Rep 2:1013–1017
Mone MJ, Bernas T, Dinant C, Goedvree FA, Manders EM, Volker M, Houtsmuller AB, Hoeijmakers JH, Vermeulen W, van Driel R (2004) In vivo dynamics of chromatin-associated complex formation in mammalian nucleotide excision repair. Proc Natl Acad Sci USA 101:15933–15937
Morgan WF (2003) Non-targeted and delayed effects of exposure to ionizing radiation: I Radiation-induced genomic instability and bystander effects in vitro. Radiat Res 159:567–580
Mortusewicz O, Schermelleh L, Walter J, Cardoso MC, Leonhardt H (2005) Recruitment of DNA methyltransferase I to DNA repair sites. Proc Natl Acad Sci USA 102:8905–8909
Nelms BE, Maser RS, MacKay JF, Lagally MG, Petrini JH (1998) In situ visualization of DNA double-strand break repair in human fibroblasts. Science 280:590–592
Nikjoo H, Uehara S, Wilson WE, Hoshi M, Goodhead DT (1998) Track structure in radiation biology: theory and applications. Int J Radiat Biol 73:355–364
Norbury CJ, Zhivotovsky B (2004) DNA damage-induced apoptosis. Oncogene 23:2797–2808
Okano S, Lan L, Caldecott KW, Mori T, Yasui A (2003) Spatial and temporal cellular responses to single-strand breaks in human cells. Mol Cell Biol 23:3974–3981
Park Y, Gerson SL (2005) DNA repair defects in stem cell function and aging. Annu Rev Med 56:495–508
Paull TT, Rogakou EP, Yamazaki V, Kirchgessner CU, Gellert M, Bonner WM (2000) A critical role for histone H2AX in recruitment of repair factors to nuclear foci after DNA damage. Curr Biol 10:886–895

Paques F, Haber JE (1999) Multiple pathways of recombination induced by double-strand breaks in *Saccharomyces cerevisiae*. Microbiol Mol Biol Rev 63:349–404

Peterson CL, Cote J (2004) Cellular machineries for chromosomal DNA repair. Genes Dev 18:602–616

Petrini JH, Stracker TH (2003) The cellular response to DNA double-strand breaks: defining the sensors and mediators. Trends Cell Biol 13:458–462

Pilch DR, Sedelnikova OA, Redon C, Celeste A, Nussenzweig A, Bonner WM (2003) Characteristics of gamma-H2AX foci at DNA double-strand breaks sites. Biochem Cell Biol 81:123–129

Prise KM, Pinto M, Newman HC, Michael BD (2001) A review of studies of ionizing radiation-induced double-strand break clustering. Radiat Res 156:572–576

Prise KM, Folkard M, Michael BD (2003) A review of the bystander effect and its implications for low-dose exposure. Radiat Prot Dosimetry 104:347–355

Pryde F, Khalili S, Robertson K, Selfridge J, Ritchie AM, Melton DW, Jullien D, Adachi Y (2005) 53BP1 exchanges slowly at the sites of DNA damage and appears to require RNA for its association with chromatin. J Cell Sci 118:2043–2055

Rademakers S, Volker M, Hoogstraten D, Nigg AL, Mone MJ, van Zeeland AA, Hoeijmakers JH, Houtsmuller AB, Vermeulen W (2003) Xeroderma pigmentosum group A protein loads as a separate factor onto DNA lesions. Mol Cell Biol 23:5755–5767

Rogakou EP, Pilch DR, Orr AH, Ivanova VS, Bonner WM (1998) DNA double-stranded breaks induce histone H2AX phosphorylation on serine 139. J Biol Chem 273:5858–5868

Rogakou EP, Boon C, Redon C, Bonner WM (1999) Megabase chromatin domains involved in DNA double-strand breaks in vivo. J Cell Biol 146:905–916

Rothkamm K, Kruger I, Thompson LH, Löbrich M (2003) Pathways of DNA double-strand break repair during the mammalian cell cycle. Mol Cell Biol 23:5706–5715

Rouse J, Jackson SP (2002) Interfaces between the detection, signaling, and repair of DNA damage. Science 297:547–551

Rubbi CP, Milner J (2003a) p53 is a chromatin accessibility factor for nucleotide excision repair of DNA damage. EMBO J 22:975–986

Rubbi CP, Milner J (2003b) Disruption of the nucleolus mediates stabilization of p53 in response to DNA damage and other stresses. EMBO J 22:6068–6077

Sadoni N, Langer S, Fauth C, Bernardi G, Cremer T, Turner BM, Zink D (1999) Nuclear organization of mammalian genomes Polar chromosome territories build up functionally distinct higher order compartments. J Cell Biol 146:1211–1226

Sadoni N, Cardoso MC, Stelzer EH, Leonhardt H, Zink D (2004) Stable chromosomal units determine the spatial and temporal organization of DNA replication. J Cell Sci 117:5353–5365

Sancar A, Lindsey-Boltz LA, Unsal-Kacmaz K, Linn S (2004) Molecular mechanisms of mammalian DNA repair and the DNA damage checkpoints. Annu Rev Biochem 73:39–85

Schermelleh L, Solovei I, Zink D, Cremer T (2001) Two-color fluorescence labeling of early and mid-to-late replicating chromatin in living cells. Chrom Res 9:77–80

Schettino G, Folkard M, Prise KM, Vojnovic B, Michael BD (2002) Upgrading of the Gray Laboratory soft X ray microprobe and V79 survival measurements following irradiation of one or all cells with a CK X ray beam of different size. Radiat Prot Dosimetry 99:287–288

Scholz M, Jakob B, Taucher-Scholz G (2001) Direct evidence for the spatial correlation between individual particle traversals and localized CDKN1A (p21) response induced by high-LET radiation. Radiat Res 156:558–563

Scholz M (2003) Effects of ion radiation on cells and tissues in Adv Polym Sci 162:95–155, Springer, Heidelberg

Sedelnikova OA, Horikawa I, Zimonjic DB, Popescu NC, Bonner WM, Barrett JC (2004) Senescing human cells and ageing mice accumulate DNA lesions with unrepairable double-strand breaks. Nat Cell Biol 6:168–170

Sengupta S, Harris CC (2005) p53: traffic cop at the crossroads of DNA repair and recombination. Nat Rev Mol Cell Biol 6:44–55

Shiloh Y (2003) ATM and related protein kinases: safeguarding genome integrity. Nat Rev Cancer 3:155–168

Sporbert A, Gahl A, Ankerhold R, Leonhardt H, Cardoso MC (2002) DNA polymerase clamp shows little turnover at established replication sites but sequential de novo assembly at adjacent origin clusters. Mol Cell 10:1355–1365

Stenerlöw B, Hoglund E, Carlsson J, Blomquist E (2000) Rejoining of DNA fragments produced by radiations of different linear energy transfer. Int J Radiat Biol 76:549–557

Stracker TH, Theunissen JW, Morales M, Petrini JH (2004) The Mre11 complex and the metabolism of chromosome breaks: the importance of communicating and holding things together. DNA Repair (Amst) 3:845–854

Stucki M, Jackson SP (2004) MDC1/NFBD1: a key regulator of the DNA damage response in higher eukaryotes. DNA Repair (Amst) 3:953–957

Sugasawa K, Okamoto T, Shimizu Y, Masutani C, Iwai S, Hanaoka F (2001) A multistep damage recognition mechanism for global genomic nucleotide excision repair. Genes Dev 15:507–521

Takahashi A, Ohnishi T (2005) Does gammaH2AX foci formation depend on the presence of DNA double strand breaks? Cancer Lett 229:171–179

Tartier L, Spenlehauer C, Newman HC, Folkard M, Prise KM, Michael BD, Menissier-De Murcia J, De Murcia G (2003) Local DNA damage by proton microbeam irradiation induces poly(ADP-ribose) synthesis in mammalian cells. Mutagenesis 18:411–416

Tashiro S, Walter J, Shinohara A, Kamada N, Cremer T (2000) Rad51 accumulation at sites of DNA damage and in postreplicative chromatin. J Cell Biol 150:283–291

Taucher-Scholz G, Jakob B, Becker G, Scholz M (2003) Microscopic visualization of a biological response to charged particle traversal. Nucl Instrum Methods Phys Res Sec B 209:270–276

Thompson LH, Schild D (2002) Recombinational DNA repair and human disease. Mutat Res 509:49–78

Thorel F, Constantinou A, Dunand-Sauthier I, Nouspikel T, Lalle P, Raams A, Jaspers NG, Vermeulen W, Shivji MK, Wood RD, Clarkson SG (2004) Definition of a short region of XPG necessary for TFIIH interaction and stable recruitment to sites of UV damage. Mol Cell Biol 24:10670–10680

Valerie K, Povirk LF (2003) Regulation and mechanisms of mammalian double-strand break repair. Oncogene 22:5792–5812

van Gent DC, Hoeijmakers JH, Kanaar R (2001) Chromosomal stability and the DNA double-stranded break connection. Nat Rev Genet 2:196–206

van Veelen LR, Cervelli T, van de Rakt MW, Theil AF, Essers J, Kanaar R (2005) Analysis of ionizing radiation-induced foci of DNA damage repair proteins. Mutat Res 574:22–33

Vilenchik MM, Knudson AG (2003) Endogenous DNA double-strand breaks: production, fidelity of repair, and induction of cancer. Proc Natl Acad Sci USA 100:12871–12876

Volker M, Mone MJ, Karmakar P, van Hoffen A, Schul W, Vermeulen W, Hoeijmakers JH, van Driel R, van Zeeland AA, Mullenders LH (2001) Sequential assembly of the nucleotide excision repair factors in vivo. Mol Cell 8:213–224

von Zglinicki T, Saretzki G, Ladhoff J, d'Adda dF, Jackson SP (2005) Human cell senescence as a DNA damage response. Mech Ageing Dev 126:111–117
Walter J, Cremer T, Miyagawa K, Tashiro S (2003a) A new system for laser-UVA-microirradiation of living cells. J Microsc 209:71–75
Walter J, Schermelleh L, Cremer M, Tashiro S, Cremer T (2003b) Chromosome order in HeLa cells changes during mitosis and early G1, but is stably maintained during subsequent interphase stages. J Cell Biol 160:685–697
Ward IM, Minn K, van Deursen J, Chen J (2003) p53 binding protein 53BP1 is required for DNA damage responses and tumor suppression in mice. Mol Cell Biol 23:2556–2563
Ward I, Chen J (2004) Early events in the DNA damage response. Curr Top Dev Biol 63:1–35
Ward JF (1985) Biochemistry of DNA lesions. Radiat Res Suppl 8:103–111
Wood RD (1999) DNA damage recognition during nucleotide excision repair in mammalian cells. Biochimie 81:39–44
Zhou BB, Elledge SJ (2000) The DNA damage response: putting checkpoints in perspective. Nature 408:433–439
Zink D, Cremer T, Saffrich R, Fischer R, Trendelenburg MF, Ansorge W, Stelzer EH (1998) Structure and dynamics of human interphase chromosome territories in vivo. Hum Genet 102:241–251

Subject Index

Page number followed by "*t*" indicates table.
Page number followed by "*f*" indicates figure.
"pp" indicates several following pages.

B

C

E

F

I

J

K

O

P

Q

R

S

V

W

X

Y